SUSTAINABLE AGRICULTURE IN TEMPERATE ZONES

SUSTAINABLE AGRICULTURE IN TEMPERATE ZONES

Edited by

Charles A. Francis
University of Nebraska
Lincoln, Nebraska

Cornelia Butler Flora
Virginia Polytechnic Institute and State University
Blacksburg, Virginia

Larry D. King
North Carolina State University
Raleigh, North Carolina

A WILEY-INTERSCIENCE PUBLICATION
JOHN WILEY & SONS, INC.
New York • Chichester • Brisbane • Toronto • Singapore

Library of Congress Cataloging in Publication Data:

Sustainable agriculture in temperate zones / edited by Charles A. Francis, Cornelia Butler Flora, Larry D. King.
p. cm.
"A Wiley-Interscience publication."
Includes bibliographical references.
ISBN 0-471-62227-3
1. Sustainable agriculture. I. Francis, Charles A. II. Flora, Cornelia Butler, 1943– . III. King, Larry D.
S494.5.S86S87 1990
630′.912–dc20 89-39543
CIP

Printed in the United States of America

10 9 8 7 6 5 4 3 2

Printed and bound by Quinn - Woodbine, Inc.

CONTRIBUTORS

REBECCA W. ANDREWS
Rodale Research Center
Kutztown, Pennsylvania

GEORGE W. BIRD
Department of Entomology
Michigan State University
East Lansing, Michigan

JOHN W. DORAN
Department of Agronomy, ARS/USDA
University of Nebraska
Lincoln, Nebraska

FRANK DRUMMOND
Department of Entomology
University of Maine
Orono, Maine

THOMAS EDENS
Department of Resource Development
Michigan State University
East Lansing, Michigan

DERRICK EXNER
Department of Agronomy
Iowa State University
Ames, Iowa

CORNELIA BUTLER FLORA
Department of Sociology
Virginia Polytechnic Institute and State University
Blacksburg, Virginia

CHARLES A. FRANCIS
Department of Agronomy
University of Nebraska
Lincoln, Nebraska

ELEANOR GRODEN
Department of Entomology
University of Maine
Orono, Maine

WES JACKSON
The Land Institute
Salina, Kansas

RHONDA R. JANKE
Rodale Research Center
Kutztown, Pennsylvania

LARRY D. KING
Soil Science Department
North Carolina State University
Raleigh, North Carolina

MATT LIEBMAN
Department of Plant and Soil Sciences
University of Maine
Orono, Maine

WILLIAM LOCKERETZ
School of Nutrition
Tufts University
Medford, Massachusetts

J. PATRICK MADDEN
Glendale, California

BILL MURPHY
Department of Plant and Soil Science
University of Vermont
Burlington, Vermont

STEVEN E. PETERS
Rodale Research Center
Kutztown, Pennsylvania

James F. Power
Department of Agronomy, ARS/USDA
University of Nebraska
Lincoln, Nebraska

Warren W. Sahs
Agricultural Research Division
University of Nebraska
Lincoln, Nebraska

Richard Thompson
Boone, Iowa

Sharon Thompson
Boone, Iowa

Matthew R. Werner
Agroecology Program
University of California
Santa Cruz, California

Garth Youngberg
Institute for Alternative Agriculture
Greenbelt, Maryland

PREFACE

Sustainable agricultural systems for production of food, feed grains, fiber, and other human needs are essential for our long-term survival and well being. With all good intentions, we have met most of the food challenges of a growing global population to date through applications of science and investment of fossil fuel energy. New technology has provided the key for much of this modern miracle in agriculture. Yet some of the consequences of the application of this singular industrial approach or paradigm were not anticipated.

We are learning now that inappropriate applications of this technology in a highly specialized, industrial agriculture can have unintended and unfortunate consequences: nitrate in groundwater, pesticides in waterways and on food, soil erosion from sloping lands. We are beginning to appreciate the very critical nature of a finite supply of non-renewable fossil fuel energy sources and some essential nutrients that are crucial for plant growth. There are certain aspects of the structure of modern agriculture and the prevalent reward system that make it difficult to sustain this, one of our most important industries. Equally important, the migration of many people out of farming has brought into question the sustainability of rural communities and an entire way of life that many consider important to our future.

Sustainable Agriculture in Temperate Zones reviews the current thinking about the philosophy of sustaining agriculture into the future. With chapters written by acknowledged technical specialists in each area, we have brought together the research data and reference material from the literature to support the hypothesis that agriculture can be made productive, environmentally sound, and resource efficient. The focus is on temperate zones, although the U.S. literature is reviewed more carefully than that of other temperate countries. The principles that are explored here have application in any climatic or geographical region.

This is a book about alternatives, about options in the use of technology, about the future. To some degree, there is a reliance on the concepts of stewardship, of biodiversity, of reliance on mixed farming systems. Yet we do not advocate a return to the systems or methods of the past. We do focus on a future-oriented approach that incorporates the most modern advances in biotechnology, in engineering, in systems studies, and in other relevant areas of science. There is a clear emphasis on use of information whenever possible to substitute for non-renewable production inputs, in an attempt to search out the best management practices that are appropriate to each management situation. There is an intentional focus on environmental soundness of practices, on regeneration of the productive capacity of the soil resources, on long-term profitability of agriculture within an ecologically acceptable management system. There is focus on values of farm families and of society.

Can we afford a sustainable agriculture? The editors and contributing authors argue that *we cannot possibly afford any type of agriculture that is not sustainable into the indefinite future*. Within this framework, we present a definition in Chapter One that guides our thinking about future systems and how they may be evaluated:

> Sustainable agriculture is a philosophy based on human goals and on understanding the long-term impact of our activities on the environment and on other species. Use of this philosophy guides our application of prior experience and the latest scientific advances to create integrated, resource-conserving, equitable farming systems. These systems reduce environmental degradation, maintain agricultural productivity, promote economic viability in both the short and long term, and maintain stable rural communities and quality of life.

The editors and contributing authors express sincere thanks to the technical editors and production staff of John Wiley and Sons for their encouragement and timely recommendations, as well as an efficient production of the book. John Fancher and the staff of Publication Services in Champaign, Illinois, provided excellent typesetting of the book. Specific acknowledgments appear following some of the individual chapters.

CHARLES A. FRANCIS

Lincoln, Nebraska

CORNELIA BUTLER FLORA

Blacksburg, Virginia

LARRY D. KING

Raleigh, North Carolina

CONTENTS

SUSTAINABLE AGRICULTURE IN TEMPERATE ZONES

1 SUSTAINABLE AGRICULTURE—AN OVERVIEW

CHARLES A. FRANCIS

Department of Agronomy,
University of Nebraska,
Lincoln, Nebraska

GARTH YOUNGBERG

Institute for Alternative Agriculture,
Greenbelt, Maryland

RELEVANCE OF ALTERNATIVE AGRICULTURAL APPROACHES

Conventional, chemical-based, and energy-intensive agricultural production systems and technologies are being scrutinized to an unprecedented degree in the United States and in other parts of the temperate zone. Throughout most of the 1980s, this accelerating process of review and reevaluation has included important elements of the agricultural establishment: landgrant universities, ARS/USDA, commercial industry, farm press, and others. The 1970s emphasis on fencerow-to-fencerow, high-input production, particularly of such major cash grain crops as wheat and corn, has given way to a new sense of restraint and uncertainty among agriculturists that would have been impossible to imagine just a decade ago. Increasingly, agronomists and policy analysts are beginning to search for an appropriate blend of technologies and policies that can lead to a more resource-efficient, environmentally sound, and profitable agriculture.

This process of reexamination and exploration is occurring despite the widely acknowledged benefits of large-scale, highly specialized, chemical- and capital-intensive farming systems that have gradually come to dominate temperate zone agricultural production practices over the past 50 years. Clearly, cash grain monocultures dependent on fossil fuel–based synthetic fertilizers and pesticides and large confinement animal feeding operations have fostered a highly productive and labor-efficient agriculture.

Recognizing the obvious production benefits of conventional farming approaches, one must ask: What accounts for the growing skepticism about these methods and models among farmers, scientists, policy makers, consumers, and analysts? Why has alternative agriculture moved so quickly into the public agricultural discussion and policy agendas?

The rapidly accelerating interest in alternative farming systems over the past several years is mainly a result of agriculture's altered physical, economic, structural, and policy environment. These changes have given rise to a broad range of deepening concerns about the sustainability of current practices. Many of these concerns were summarized in the 1980 USDA *Report and Recommendations on Organic Farming:*

- Increased cost and uncertain availability of energy and farm chemicals.
- Increased resistance of weeds and insects to herbicides and insecticides.
- Decline in soil productivity from erosion and accompanying loss of organic matter and plant nutrients.
- Pollution of surface waters with agricultural chemicals and sediment.
- Destruction of wildlife, bees, and beneficial insects by pesticides.
- Hazards to human and animal health from pesticides and feed additives.
- Detrimental effects of agricultural chemicals on food quality.
- Depletion of finite reserves of concentrated plant nutrients (for example, phosphate rock).
- Decrease in number of farms, particularly family farms, and disappearance of localized and direct marketing systems.

One year after this publication, another major departmental report (USDA, 1981) emphasized such historical and ongoing structural trends in U.S. agriculture as (1) increase in average farm size, (2) concentration of farm sales and assets, (3) growth in capital intensity, and (4) increased mechanization and specialization of farm enterprise. These trends have continued.

Developments in agriculture over the past several years reflect a growing awareness of the social and environmental costs that are associated with large-scale farming enterprises but are borne by the public. These include growing disquiet over agriculturally related environmental degradation—particularly ground-water pollution, continuing economic pressure on U.S. farmers, uncertain farm export markets, growing pressure to reduce federal commodity price supports, and heightened consumer concerns over food safety and quality.

Agricultural economists have long recognized that soil conservation has social or public value (Bunce, 1942). But the costs of soil conservation are not fully reflected in the market price that farmers receive for their produce. This external cost results from market failure to accurately reflect all costs of production. Agricultural critics identify other social costs (health, environmental quality, stable rural communities) resulting from concentrated and specialized production systems that are borne by the public. Recognition of these external costs drives the search for sustainable production systems that either internalize or minimize externalities.

Moreover, a number of more recent studies, many bearing the official sanction of mainstream agricultural organizations, journal articles, and reports in the conventional farm press have provided additional, convincing evidence of the adverse effects of chemical-intensive farming systems on the environment and the economic well-being of agriculture. Much of this research is summarized in the chapters that follow. Other relevant, recent books are summarized in the proceedings of the North Central Regional Conference, *Sustainable Agriculture in the Midwest* (Francis and King, 1988), and in the International Conference on Sustainable Agricultural Systems (Edwards et al., 1989).

Many such analyses have helped to dispel the more commonly held misunderstandings and negative perceptions of the nature and potentials of modern alternative systems. The negative symbolism of low-input agriculture has long been cited as a major barrier to its wider-scale adoption (Youngberg, 1978). Now, however, it is the combination of a more dispassionate and realistic understanding of the role and potential of alternative systems in modern production agriculture, along with diminished confidence in the sustainability of conventional approaches, that accounts for the newly appreciated relevance of alternative systems (National Academy of Sciences, 1989).

WHAT IS SUSTAINABLE AGRICULTURE?

Much of the current debate about the nature and potential of sustainable agriculture centers around definitions. Some persons consider sustainable agriculture to be a philosophy, while others say it provides guidelines for choosing practices. Still others view it as a management strategy. Some opponents and supporters of the concept consider sustainable agriculture to be another name for organic farming. At the other end of the spectrum, some have defined this concept as maximum economic yield. Perhaps the term *sustainable agriculture* is too desirable. After all, no one would advocate that we develop a "nonsustainable" agriculture!

There is additional confusion about *sustainable agriculture* because of the varied groups and interests who are currently promoting different philosophies to guide agriculture in this period of change. There are scientists who prefer to define sustainability purely in farm production terms, while others insist on including an economic dimension. Some people broaden the scope of the definition to include concern about the local and distant environment, now and in the future. Others express concern about the current reliance on, and future prospects for, nonrenewable resource investment in agriculture. A time dimension is also critical. Sustainable for how long? Under what assumptions about availability and cost of fossil fuels? Who pays the total environmental costs of today's farming practices, and to what extent are we passing costs on to future generations?

Still other questions involve the social, community health, and equity issues related to agriculture. Will current conventional farming, increased size of operations, and fewer farms and farm families allow us to sustain both farming and farm communities? What changes will be required to meet the health concerns of farmers and consumers? What about food quality and food security in an agricultural industry whose ownership is concentrated in a few hands? These issues are addressed by Flora in Chapter 12 and by Lockeretz in Chapter 15. The evolution of philosophies and terms to describe them is well reviewed in the recent conference paper by Harwood (1989), who traces the roots of current activities to organic farming (Rodale, 1945; Rodale, 1973), sustainable agriculture (Jackson, 1980), and regenerative systems (Rodale, 1983).

In an editorial in *Science* magazine, Wittwer (1978) called for a research agenda that would develop new technologies to contribute to an economically, socially, and ecologically sound agriculture. Gips (1984) expanded the concept into a broad working definition: "A sustainable agriculture is ecologically sound, economically viable, socially just, and humane." The process of developing a sustainable agriculture was outlined by Rodale (1983), who later insisted on a system of food production that would improve, rather than continue to degrade or even maintain, the status quo of soils. He called this "regenerative agriculture" (Rodale, 1985).

Francis and Hildebrand (1988) summarized several definitions. University of Nebraska extension specialists have been using a practical definition that describes the production components or decisions that a farmer makes each season:

> A sustainable agricultural system is the result of a management strategy which helps the producer to choose hybrids and varieties, soil fertility packages including rotations, pest management approaches, tillage methods, and crop sequences to reduce costs of purchased inputs, minimize the impact of the system on the immediate and the off-farm environment, and provide a sustained level of production and profit from farming. (Francis et al., 1987; University of Nebraska, 1987).

Although this definition includes specific practices, as well as economic and environmental dimensions, it fails to address in an explicit manner the concerns many feel about the finite nature of fossil fuel resources and the dimensions of long-term sustainability of both production systems and people in their communities.

USAID (1988) developed a concept paper, "The Transition to Sustainable Agriculture: An Agenda for A.I.D.," in which were incorporated both the needs of people and the objective of regenerating production potential. Sustainability was defined as the ability of an agricultural system to meet evolving human needs without destroying and, if possible, by improving the natural resource base on which it depends. A concurrent report from BIFAD (1988) expands on this theme. Bird (1988) described sustainable agriculture in greater detail as an optimization process that "uses quality of life mandates, atmospheric resources, organic resources, sedimentary resources, external inputs, and material residues for the long-term sustainable production of appropriate amounts of reasonably priced food, feed, and fiber, while maintaining a long-term high quality environment." He stated that this process was not maximization of either crop yield or short-term profit, nor was it an external input minimization process. According to Bird, sustainable agriculture is both knowledge- and management-intensive, and it must "provide long-term added value to the biological, environmental and human capital on which agriculture is based."

Edwards (1988) said that sustainable agriculture involves "integrated systems of agricultural production less dependent on high inputs of energy and synthetic chemicals and more management-intensive than conventional monocultural systems. These systems maintain, or only slightly decrease, productivity, maintain or increase net income for the farmer, are ecologically desirable, and protect the environment" (Francis and King, 1988). In the International Symposium on Sustainable Agriculture, Harwood (1989) described, "an agriculture that can evolve indefinitely toward greater human utility, greater efficiency of resource use, and a balance with the environment that is favorable both to humans and to most other species."

The following definition was developed at the annual meeting of the American Society of Agronomy (1989): "A sustainable agriculture is one that, over the long term, enhances environmental quality and the resource base on which agriculture depends; provides for basic human food and fiber needs; is economically viable; and enhances the quality of life for farmers and society as a whole." In a special issue of *Futures* magazine from Michigan State University, Erb (1989) quotes the chairperson of the crop and soil science department, Dr. Eldor Paul: "Our concept of sustainable agriculture includes a system of agricultural production that is resource conserving, environmentally safe, and economically viable. At the same time, it must recognize human values, provide high-quality food, and support the family farm and rural communities as part of a healthy larger system."

In contrast, Hoeft and Nafziger (1988) proposed that sustainable agriculture is best defined or achieved by striving for maximum economic yield (MEY). This is the yield that gives highest net return for each set of soil and climate conditions, and thus the lowest unit cost of production. In this same direction, the Potash and Phosphate Institute (1989) described the low-point sustainable agriculture activity funded by the USDA as a program designed to "reduce inputs and to convert conventional farmers to organic farmers." The writers described four management changes suggested for the conversion process:

1. Diversification of cash grain farms to include livestock.
2. Rotation of row crops with legumes to provide nitrogen.
3. Increased use of animal manure to replace fertilizer.
4. Use of biological controls, cultivation, and labor as substitutes for purchased pesticides.

As an alternative, the Institute proposed use of optimum levels of inputs that define the best management practices (BMPs) for each location.

A compromise definition worked out by the North Central Regional Committee on Sustainable Agriculture (NCR-157, personal communication) in early 1989 included both efficiency of resource use and production, as well as environmental and social dimensions:

> Sustainable agriculture is a system that utilizes an understanding of natural processes along with the latest scientific advances to create integrated, resource-conserving farming systems. These systems will reduce environmental degradation, are economically viable, maintain a stable rural community, and provide a productive agriculture in both the short and the long term.

In the International Conference on Sustainable Agricultural Systems, Harwood (1989) described the complexity of introducing such a value-laden approach in the international development community, given the broad and emerging political and social agenda that confronts agriculture around the world. He suggested that the application of these concepts would depend on each country's policies and political agendas and how they influence development. There is no question about the diversity of interpretations in countries that are endowed with greatly varied natural and production resources and with different degrees of severity in current food supply.

In the same conference, Brady (1989) reviewed the emergence of sustainable agriculture within the international development community as "a topic whose time has come." Among the present concerns of scientists and policy people involved in agriculture, the dangers of excessive chemical fertilizer and pesticide use, problems of soil and water conservation, and general concerns about the environment were listed. Brady stated that officials of the U.S. Agency for International Development had determined that three elements were critical to agricultural sustainability:

- Income generation, especially among the poor.
- Expanded food availability and consumption, through both increased production and improved marketing.
- Conservation and enhancement of natural resources.

International dimensions of the soil erosion challenge were summarized by Brown and Wolf (1984). Concerns of soil scientists in the USSR have largely been ignored, while about a half million hectares of cropland are lost to wind erosion each year in that country. Up to two-thirds of the total cropland has suffered from various types of erosion. River siltation and dust storms are major problems in both India and China. Soil loss is truly an international problem in the temperate zone, and it is not a problem that is easily solved.

Definitions of terms and their use in different countries and languages add to the dilemma. Debate will continue, as described by Lockeretz (Chapter 15), until people decide on common terms of reference, including a time frame and a set of assumptions about resources. He explores in detail the reasons why there is difficulty sorting out philosophy from strategy, process from product, and specific cultural practices from systems approaches. Further, the value placed on environmental consequences of agriculture and long-term considerations of nonrenewable resource use must be determined before agreement can be reached. The debate is healthy and useful to agriculture and to society as a whole if it increases awareness of these broad concerns about food production and our environment. The debate raises the profile of sustainable agriculture on the national research and development agenda.

Some of these challenges can be partially met through appropriate research and extension programs, yet all of them are related to policy decisions in national governments. What is needed is a sincere commitment by policy makers to address the long-term issues surrounding sustainability of agriculture and the food supply. Diversity in languages, definitions, and stated objectives of development groups further complicates our understanding of sustainable agriculture.

Chapters in this book explore the current thinking and state of technology related to long-term viability of agriculture. Although there is some variation in the definitions used by authors, our working concept is as follows:

Sustainable agriculture is a philosophy based on human goals and on understanding the long-term impact of our activities on the environment and on other species. Use of this philosophy guides our application of prior experience and the latest scientific advances to create integrated, resource-conserving, equitable farming systems. These systems reduce environmental degradation, maintain agricultural productivity, promote economic viability in both the short and long term, and maintain stable rural communities and quality of life.

COMMON MISCONCEPTIONS ABOUT SUSTAINABLE AGRICULTURE

Some of the confusion about sustainable agriculture is being resolved by the continuing discussion over its character and potential. Much of the dialogue is being pursued in regional and national meetings and in the mainstream organizations in agriculture, including land grant universities. These groups are joining forces with nonpublic research and public information organizations that have long been interested in the resource implications and environmental impacts of current chemical- and fertilizer-intensive agricultural practices.

One important aspect of these discussions is education to correct a number of myths, stereotypes, or misconceptions about sustainable agriculture that have confused our understanding of alternative approaches. Some of these have been generated because of the association of reduced input approaches with specific organizations or individuals who have high profile and outspoken advocacy of a particular type of technology in agriculture. Such association is nonscientific, to say the least, and detrimental to communication and understanding of the issues. Here are some myths generated over the past several years and the facts that dispel such misunderstandings (Francis and King, 1988).

"Sustainable approaches are only for small farmers." Contrary to this perception, we note that wheat farmer Fred Kirschenmann (3,000 acres, North Dakota), livestock/crop producers Richard and Sharon Thompson (300 acres, Iowa; see Chapter 9), and cash crop farmer Ron Ellermeier (700 acres, Nebraska) are among many active and successful present-day operators who use either no purchased chemicals and fertilizers or very reduced levels of these inputs. These low-input alternatives are increasingly accepted by farmers across the full spectrum of farm sizes and enterprises.

"Reducing inputs means going 'cold turkey.'" Actually, there is no requirement that a farmer needs to immediately eliminate all purchased chemicals or fertilizers from the operation in order to meet most definitions of an evolving sustainable agriculture. It is biologically and economically rational to examine carefully the cost and benefit of each input in the total production system to determine where costs can be reduced. It is also important to weigh the long-term impacts of each input, both on the immediate production field and the broader surrounding areas.

"To go 'low-input' means to convert the entire farm." In fact, there is little rationality in making a complete change in a successful and profitable farming operation unless it is clear that such a change will increase profitability or sustainability of the farm. Just as with any change in technology—new hybrid or variety, modified tillage system, reduced fertilizer package, field scouting and threshold approach to insect or weed management—it is wise to try that practice in one field or even in a small area. Most managers would follow the same strategy in conventional farming.

"Substantial input use reduction is the same as going 'organic.'" While the goal of organic certification (which naturally requires the total elimination of synthetically compounded fertilizers and pesticides) is an increasingly attractive farming and marketing alternative for many growers, most current definitions of sustainable agriculture allow for some use of synthetic inputs. The organic foods industry views pure organic production as one important model within a sustainable agriculture system.

"Sustainable farmers must use older, open-pollinated varieties and not hybrids." This is obviously not true. Any producer who ignores the potentials of modern hybrids and the latest crop varieties is missing one of the most cost-effective ways to include new technology in a cropping system. Disease and insect resistance, drought tolerance, marginal resistance to some adverse soil problems, and a range in crop maturity are all benefits from current hybrids and varieties that contribute to reduced costs of production and greater sustainability of the entire system. There is value in preserving the genetic variability of older varieties, and this is being done by the USDA at the National Seed Storage Laboratory in Fort Collins, the Seed Savers Exchange in Northeast Iowa, and other public and private groups working on specific crops.

"Yields are reduced when chemical and fertilizer inputs are reduced." Any change in production practices may reduce crop yields if this modification is not understood, tested, and applied correctly in each operation. Some farmers report no change in yields, some a reduction, and others an increase in yields when systems are modified to involve fewer chemicals and more intensive management.

"Low-input approaches increase risk in farming." Actually, if a chemical or fertilizer input is removed or reduced in a system in which that input is solving a yield-limiting constraint, and if there is no tested alternative to solve that same problem, there may be increased risk. If any input is removed from the system that is not contributing to productivity (for example, a corn rootworm insecticide on first-year corn) or that has a low probability of yield response (for example, starter fertilizer in many soil conditions), then profit can be increased. If there is potential to substitute an internal, low-cost resource for the previous input, it is likely that a farmer can reduce the risk of lower returns and increase the likelihood of sustained profitability in the system.

"Current cash-grain crops and systems make most efficient use of inputs." Although most crops are well adapted to the areas where they currently

find wide acceptance, the majority of these crops were introduced from other areas. Of the major U.S. crops, corn is from Central America, soybeans from China, wheat from the Middle East, and grain sorghum from East Africa. Alternative crops and crop/livestock systems using both introduced and native species can make equally efficient, or even better, use of crop growth resources and thus provide a more diversified product output.

"Farmers change systems for philosophical and religious reasons." In truth, although some producers change their production practices for these reasons, a survey in *The New Farm* revealed that their readers reduced chemical inputs in farming primarily for economic and health/safety reasons. Most farmers modify practices and reduce inputs for a combination of reasons.

"Low-input farming means low management and low levels of production." Nothing could be further from the truth with today's alternative approaches to farming. As we develop a better technical understanding of the complex biological, soil, and climatic interactions that influence crop and animal production, we increase the potential for high levels of sustained production. These systems could be labeled "science-intensive," "information-intensive," or "management intensive" because they depend on research and the creative application of conventional wisdom about farming and the results of laboratory and field investigation.

"Total agricultural production would be drastically reduced by widespread application of low-input practices." In fact, much recent research indicates that this broad generalization grossly exaggerates and simplifies the likely consequences of such a shift. Many changes have taken place in the decade since CAST (1980) concluded that wide adoption of organic practices would reduce production and increase soil erosion by forcing the cultivation of more marginal lands. The most recent CAST report (1988) on this general subject, for example, presents a much more concerned and moderate evaluation of energy use and alternatives to promote the long-term viability of agriculture. Obviously, we now have a much better understanding of the potential of alternative systems. In the past decade, much research has focused on how to reduce inputs without adversely affecting yields. Equally important is the emergence of a more moderate philosophy about the use of fossil fuel inputs in all sectors of agriculture. For example, the chemical industry is seeking to develop new products and practices that reduce the negative effects of applied materials: lower rates, more specific targeting, biologically derived products, and integrated pest control methods. Finally, there is increased recognition that the incremental nature of these shifts ought to allow sufficient time for any necessary changes in overall input/output relationships.

Details on the development of practices that reduce chemicals and fertilizers and make agriculture more profitable, as well as more sustainable in the biological sense, are given in the chapters that follow. Experts in each field provide evidence of the changes that have occurred primarily over the past decade in the area of alternative, low-input, or sustainable agriculture.

HISTORICAL DEVELOPMENTS RELEVANT TO THE CONCEPT OF SUSTAINABLE AGRICULTURE

In thinking about the current and future needs and characteristics of a truly sustainable agriculture, it may be instructive to review briefly some of these issues in a historical perspective. How sustainable has agriculture been over the centuries, and how has this picture changed? Has agriculture ever been truly sustainable? What has been the impact of technology, and how have we arrived at the current dilemma? There is little debate about the relationship between current needs and the rapidly expanding human population. But this is a relatively recent phenomenon. Throughout much of human history, there was adequate land to produce the needed food.

In a review of the history of crop yields, Evans (1980) described productivity in preagricultural harvesting, in times of early plant domestication, in subsistence agriculture, and in current high-input systems. Much of the information on early systems is based on inference from archeological evidence, on calculations from scanty historical reports, or on speculation. Yet these estimates provide some insight into the historical development of agriculture and the changes that may have occurred in its sustainability over time.

Preagricultural harvests of plants supplemented the hunting for meat by prehistoric tribes. Evans (1980) speculates that plants were at least as important as animals in the diet. One recent observation on efficiency in these systems was made with the !Kung bushmen in Botswana, where vegetable foods make up about two-thirds of the diet. It is hypothesized that their current hunting and gathering systems are similar to those that have been used for thousands of years. Women gather about 15 calories of food for each calorie of energy invested (Lee, 1968), a situation typical of the efficiency of many methods of food gathering observed in other parts of the world. Harlan (1965) hand-harvested a mixed stand of wild relatives of barley, wheat, and oats in Turkey at the rate of about one kilogram of clean grain per hour. Yield was thus measured per unit of human energy invested or per hour of labor, rather than per hectare. Such measures make sense if we assume that human energy was scarcer than land during these times. It is likely that these systems were sustainable because human populations were small and the available territory for each tribe was large. It is equally likely that the people suffered during drought years, since they had little or no control over the production environment where their preferred plants grew. People who often migrated probably preserved and carried with them limited amounts of food.

With the beginning of more organized agriculture about 10,000 years ago, people began to consciously select the most desirable plants that they found in the fields or growing near their garbage dumps where plant refuse had been thrown. Plucknett and Smith (1986) describe how these systems may have developed, with increasing intensity of intentional planting and cultivation near dwellings and higher productivity where fertility was favorable for crop growth. (Development of crop varieties in these systems is discussed in Chapter 2, "Breeding Hybrids and Varieties for Sustainable Systems"). People harvested what they needed in these early organized systems, and perhaps produced some excess for barter or sale. This was subsistence agriculture evolving into a myriad of forms to suit local climate and soil conditions. Production systems were relatively independent of any machinery or other inputs from outside the family and the immediate ecosystem (Evans, 1976).

Yields often were reported as seed harvested per seed sown, but with some additional clues, authors have been able to derive yields per unit land area. Cereal yields of one-half to three-quarters ton per hectare probably were commonplace in the Middle Ages, although some were higher. Irrigated wheat yields of about two tons per hectare were recorded in Mesopotamia in 2400 B.C., but these apparently dwindled to under one ton per hectare by 1700 B.C. (Jacobsen and Adams, 1958). The higher yields were not sustained due to a probable increase in salinity of irrigation water or accumulation of salts in the soil.

Some wheat yields in Galilee during Roman times may have reached more than 3.5 tons per hectare (Feliks, 1963). These levels of productivity apparently were sustained as long as fertility from animal and green manure was maintained and the irrigation water was of adequate quantity and quality, yet historical records of droughts and insect attacks suggest that farmers through the ages have suffered from many climatic and biotic stresses. Reliability of food supply has thus been relatively precarious over the centuries.

Little is known about improvements in agriculture during the post-Roman years in Europe. Large civilizations based on agriculture thrived in Asia, Africa, and the Western Hemisphere. The rise and decline of these cultures probably depended in part on their stewardship of the land resource and the balance of the population with agricultural productivity. History recalls the famine in Ireland as the failure of the predominant potato crop, an example of concentration on a single species and the inability of people to sustain a food supply when confronted with a major biological catastrophe. We have since learned to control such plant disease epiphytotics.

Evans (1980) concludes that much of our improvement in productivity has resulted from better control of the cropping environment and selection

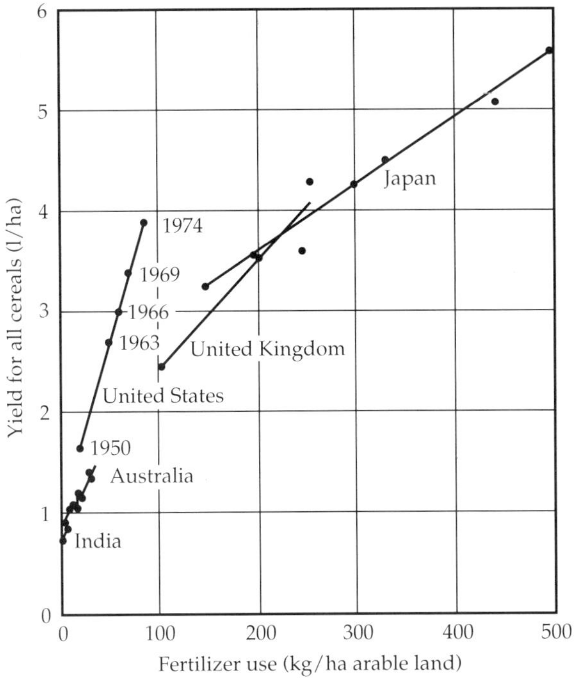

Figure 1.1 The average grain yield for all cereals increases with the amount of fertilizer used. For each country data are given for 1950, 1963, 1966, 1969, and 1974 (Fig. 2 from Evans, 1980).

of crops for adaptation to those growing conditions, not from a great increase in yield potential per se. Improvements in productivity over the past century have come from better soil fertility, improved weed and other pest control, irrigation and dryland water management in some areas, and hybrids or varieties of crops that could respond to these more stable and controlled growing conditions. Fehr (1984) discusses the genetic improvements of major crops as they contribute to productivity.

Yield per hectare of all cereals as a function of fertilizer use, for example, is shown in Figure 1.1 (from Evans, 1980). The response of cereals to increased use of fertilizer, certainly confounded with other management factors, is shown as a straight line increase in Japan, the United Kingdom, the United States, and Australia/India. Evans states that the use of phosphorus and potassium fertilizers has doubled every 15 years, while the application of nitrogen to crops has doubled every 8 years. Fertilizer use per acre in the United States has declined slightly since 1980, due to government set aside programs, higher fertilizer cost, and better management by farmers. It has been estimated that in the mid-1960s, the amount of synthetic nitrogen fertilizer applied to corn in the United States was about equal to the amount of nitrogen harvested; by the late 1980s, we were applying about 50 percent more nitrogen than was being harvested

TABLE 1.1 Phases in Agricultural Evolution and Their Characteristics.

	Institutional Development	Public Participation	Technologies
1. Hunting/gathering subsistence	Minimal	Entire population	Biological only
2. Participatory, partly commercial	Mostly local, concerned with land use; some markets starting to develop	Dominant portion, increasing with population growth	Mostly biological
3. Participatory, commercial	Generally equal distribution of national, regional, and local levels	Dominant portion but decreasing	Well-developed small-scale machinery and other industrial inputs
4. Industrial	Global	Small number actually producing	Industrial inputs substituting for biological
5. Post-industrial	Blend of local, regional, and global institutions based on balance of resource use efficiency and local employment needs	Stable	Biological controls predominant; systems interactions are intricate and quantified

Source: Harwood, 1989.

with the crop (G. A. Peterson, personal communication). The use of insecticides, fungicides, and especially herbicides likewise has increased at a rapid rate. Although productivity per unit land area in agriculture has continued to increase, the efficiency of use of applied inputs is decreasing.

This brings our discussion to the present date. In most of the developed world, we operate a highly structured agriculture that is predominantly dependent on production inputs derived from fossil fuels. We have learned to "dominate" the production environment, leveling out some of the vagaries of climate and reducing or eliminating pest populations with chemicals. This domination is possible only as long as we have an abundant and low-cost supply of fossil fuels. It is becoming increasingly evident, however, that the sustainability of this type of system is even more tenuous from an ecological, an environmental, and perhaps a social point of view (see Flora, Chapter 12). As described in the introduction to this chapter, we are learning more about the unintended and undesirable effects of input-intensive production systems.

The current quest for a more stable or sustainable food supply has been conducted by most scientists and policy makers within the prevailing fossil fuel input-intensive paradigm. We have assumed that there will always be another technological solution and an unlimited supply of fossil fuel energy to produce and market that solution. An increasing number of people—urban and rural, in both industrialized and developing nations—are becoming aware of environmental quality, the finite supply of nonrenewable resources, health and safety in agriculture, and food quality. Some even question the absolute reliance on an industrial paradigm for our continued successful production of food, given some of the problems and consequences listed above.

As the following chapters unfold, it would be useful to the reader to consider the agricultural evolutionary stages suggested by Harwood (1989). His list includes

- Subsistence agriculture (hunting and gathering).
- Participatory, partly commercial (beginning of a market economy).
- Participatory, commercial (smallholder production, evolution of trade).
- Industrial (major substitution of capital for labor, global markets).
- Postindustrial (yet to be explicitly defined).

The five stages are represented in Table 1.1. The world's agriculture largely operates in the industrial stage or mode, while Harwood suggests that an evolving resource base, social values, and market structure are moving us toward "something both qualitatively and quantitatively different." Harwood lists these changes as follows:

- Stabilization of the number of people employed in agricultural production.
- Growth of food services relative to the production sector.
- Shift from increasing specialization to some degree of enterprise integration.
- Shift toward biological/genetic control of biological processes rather than chemical control.
- Increase in prevalence of social and environmental considerations in public investment in agriculture.

Harwood (1989) implies that these changes will guide our decisions in agriculture in the future. Most changes are not comfortable for the participants, since there is much adjustment, retraining, and reorganization of

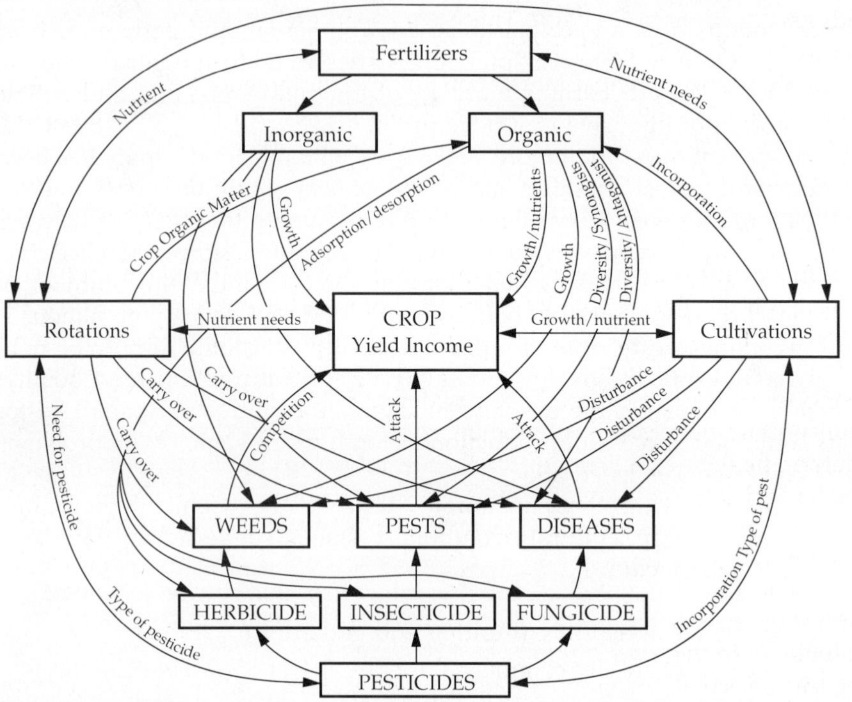

Figure 1.2 Interactions between inputs in a cropping system (from Fig. 5, Edwards et al., 1989).

institutions that must take place during this shift. Yet we must keep in mind that these are evolutionary changes that occur over time, at different rates in different places. The following chapters outline some of the underlying theories and specific changes in agricultural practices that will make such a paradigm shift possible.

SPECIFIC ISSUES AND TECHNOLOGIES RELEVANT TO THE AGRICULTURE OF THE FUTURE

Sustainability of agricultural production is dependent on a large number of biological, climatic, economic, and social factors. There is increasing awareness of how these factors interrelate. In contrast to most agricultural research, which has concentrated on specific components of technology and often organized along department and discipline lines, the efforts to improve agricultural systems will require more emphasis on interdisciplinary teams (Francis and King, 1988). There is growing appreciation of the numerous biological interactions that are active in crop and animal production and how certain "integration efficiencies" can be used to advantage (Harwood, 1984). These are especially relevant when crop and animal enterprises can be integrated on the farm.

Some of the principal interactions of specific practices with crop yields and income, as well as among practices, are shown in Figure 1.2 (from Edwards et al., 1989). Prominent among these interactions are the crop nutrient influences on pest management, effects of tillage and crop rotations on pests, and the many complexities involved with integrated pest management. These interactions are both spatial and temporal, with an additional degree of complexity introduced with the study of long-term effects of alternative combinations of crops and practices (Francis et al., 1986). Although much of the focus in Farming Systems Research and Extension and in practical research on cropping practices has begun to concentrate on integrated systems (Francis and Hildebrand, 1988), it is still important to study individual factors in the production system and determine their effects on productivity and net profit. The farmer still must make decisions on individual components of technology—hybrids, fertility sources and timing, approaches to pest management—even though these may now be evaluated in a broader systems context.

The chapters that follow explore individual component technologies, although these are most often considered through their effects on the sustainability of the total cropping or crop/animal system. For example, many of the objectives of crop breeding programs over the past half century have involved selection and testing for biological and climatic stress tolerance or resistance. How classical plant breeding methods, as well as new advances in biotechnology, can contribute to more sustainable production systems are described by Charles Francis in Chapter 2. Although crop improvement through breeding is a long-term investment, the results can be very cost-effective for types of resistance or increases in yield potential that are found to be highly heritable and easily manipulated by the plant breeder.

The broad concepts of insect and plant pathogen management are explored in Chapter 3 by George Bird, Tom Edens, Frank Drummond, and Eleanor Groden of Michigan State University and the University of Maine. The authors describe pests as a part of the total crop ecosystem, and how pests function as part of the energy and food chain in sustainable agricultural systems. The history of pest control is described through the synthetic pesticide era, the integrated pest management era, and finally the era of sustainable agriculture.

Sustainable weed management practices are defined and integrated in Chapter 4 by Matt Liebman and Rhonda Janke of the University of Maine and the Rodale Research Center. They discuss the potentials of crop rotations and intercropping systems, as well as the suppression of weeds by crops and by allelopathic cover crop species. Cultural practices such as crop density, spatial organization, cultivation, and fertility can also influence weed populations. Use of livestock in the farming operation and the integration of several of the above methods are seen as potential ways to reduce pesticide use and build toward a more sustainable and profitable system.

Much of the current concern about water quality centers on excess nitrate in both surface and ground water. In discussing sustainable soil fertility practices, Larry King from North Carolina State University describes in Chapter 5 the important cycles that nutrients pass through as well as the value of alternative sources of fertility. Both losses of nutrients and a range of sources must be considered in the total nutrient cycling within a field. It is important to consider how a system can be designed to make maximum use of all available sources and minimize biologically and economically undesirable outflow of nutrients from the system.

After several decades of evolving toward monoculture cereals in the United States and elsewhere, farmers are beginning to use more crop rotations and legumes in their cropping patterns. Legume rotations and short-term cover crops are now becoming an integral part of modern production systems, as described by James Power of ARS/USDA at the University of Nebraska (Chapter 6). He presents both national and field-level data on the prominence of legumes and their potential for fixing nitrogen for succeeding crops in a sequence. Different soil properties and climatic factors strongly influence nitrogen fixation, and information presented here complements the soil nutrient cycling described in the preceding chapter.

Another important aspect of soil fertility is the biological activity of a specific site and how it can be influenced by crop management. John Doran of ARS/USDA at the University of Nebraska and Matt Werner of the Agroecology Program, University of California, Santa Cruz discuss this complex field in Chapter 7. Soil microflora and fauna play a vital role in the cycling of nutrients, and their relative importance increases as we seek more biologically based and sustainable approaches to managing soil fertility. Populations of soil organisms are highly dynamic and strongly influenced by applied nutrients and other management options. These complex interactions are only beginning to be understood well enough to make their management possible.

Some of the most ecologically stable farming systems in the temperate zone are well-designed and managed pastures. Bill Murphy of the University of Vermont describes in Chapter 8 the dynamics of component species and nutrients under alternative management schemes. He provides details of how pasture systems can be organized and the importance of paddock size and rotational grazing on both the long-term biology of the pastures and on the economics of recommended systems.

Practical applications of the principles described in regard to fertility, pest management, use of rotations and legumes, and integration of animals into the system are discussed in Chapter 9. Richard and Sharon

Thompson, who farm near Boone, Iowa, and colleague Derrick Exner from Iowa State University give details on the operation of the Thompson farm. Their cattle and hog enterprises are integrated fully with the production of feed grains and pastures, and the farm has been essentially independent of chemical inputs for over two decades. The economics of production and distribution of labor are important details included in the chapter. The process of converting from a conventional to a sustainable farming activity shows how this can be done on a practical working farm in the heart of the U.S. cornbelt.

The experimental basis for this conversion process is summarized in Chapter 10 by Rebecca Andrews, Rhonda Janke, and Steven Peters of the Rodale Research Center and Warren Sahs of the University of Nebraska. Important details of long-term experiments describe the dynamics of nitrogen, potassium, and phosphorus in the systems, as well as the practical application of alternative pest management techniques. Two field conversion experiments, in Pennsylvania and Nebraska, are used as case studies.

Production systems cannot be sustained in the long term if they are not profitable in the year-to-year cash flow that must be managed by the farmer. Patrick Madden, formerly of Pennsylvania State University and CSRS/USDA, describes in Chapter 11 some strategies for making sustainable systems more profitable. From results of experiments and farmer surveys, he explores the factors of risk and net profits during a transition phase from conventional to low-input management. He further explores the potential impact of wide-spread adoption of these methods, as well as the current federal programs designed to stimulate their use.

Farm families are highly dependent on rural communities for much of their physical and social sustenance. Cornelia Butler Flora explores in Chapter 12 the implications of today's highly specialized and capitalized farms and agriculture for the future of these communities. She looks at patterns of employment by families and investments in agriculture, and how these are affecting the structure of the rural U.S. scene. The factors that appear to influence the viability of rural communities are listed and described, along with the influences that a more sustainable agricultural system would provide.

In Chapter 13, Dr. Flora discusses national policy issues and how they influence agricultural sustainability. Farmers in the temperate zone have become increasingly dependent on a confusing set of government controls and subsidies, and most producers feel that they cannot survive without this intervention. There are also nonagricultural policies that influence the decisions and success of farmers and ranchers. Finally, the current agricultural policy legislation is discussed, along with desirable modifications that would influence the future sustainability of agriculture.

Future dimensions of cropping and farming systems are discussed in Chapters 14 and 16. The long-term potential of an agriculture based on perennial crops and ecological principles is described by Wes Jackson, who draws from the model followed over the past decade at The Land Institute in Kansas. This work has passed sequentially from an inventory of available species to study of specific combinations of selected species, to detailed agronomic and physiological characterization of mixtures and their resource use, and to broader scale planting of larger plots that can begin to provide data on the biological and economic potential of these systems. The research and training described in this chapter provide a model for other types of studies on both cropping systems and ecological systems. Further, The Land Institute serves as a successful example of how critical research can be implemented outside of public institutions. The Institute allows people of vision and commitment to become involved in the total research effort.

Reasons for confusion of terms in discussions about sustainable agriculture are presented in Chapter 15 by William Lockeretz of Tufts University. He explores the development and evolution of terms as well as the distinction between sustainable agriculture as a philosophy and as a management strategy. Following the broad range of interpretations used by other authors, Lockeretz's chapter helps us sort out the unresolved issues and focus on what is important for the future.

Perspectives for the future of sustainable agriculture are explored in Chapter 16 by Charles Francis. The integration of these concepts and potential new technologies is critical for successful application of the principles. A study of measures by which this philosophy can be incorporated into our mainstream research and extension agenda is important for the immediate success of programs in universities in the temperate zone. For the long term, it is important to integrate this philosophy into teaching programs at both the graduate and the undergraduate levels. It is critically important that the next generations of teachers, researchers, and extension specialists take a broad look at the science and practice of agriculture, in terms of its reliance on fossil fuel inputs as well as its impact on the environment. We need an agriculture that can become a sustainable source of food, feed, fiber, and other human needs, without drastically affecting the total environment. The chapters that follow give details on this theme.

Acknowledgments

The authors appreciate the reviews and suggestions of Drs. Paul Marsh (Washington D.C.), Joan B. Youngquist, Jim Power and Robert Aiken (University of Nebraska), Thomas Barker and Margaret Smith (Cornell University), and Cornelia Flora (Kansas State University). JoAnn Collins's efforts in finalizing the manuscript are greatly appreciated.

REFERENCES

American Society of Agronomy. (1989). "Decisions reached on sustainable agriculture," *Agron. News* January: p. 15.

BIFAD. (1988). "Environment and Natural Resources." Board for Intl. Food and Agr. Devel., Occasional Paper No. 12, Agency for Intl. Devel., Washington, DC, February. 41 pp.

Bird, G. W. (1988). "Sustainable agriculture: current state and future trajectory. Congressional Testimony on Sustainable Agriculture." April 18, Washington, DC.

Brady, N. C. 1989. "Making agriculture a sustainable industry." In *Sustainable Agricultural Systems* (Eds. C. A. Edwards, R. Lal, P. Madden, R. H. Miller, and G. House). Soil & Water Conservation Society, Ankeny, Iowa (in press).

Brown, L. R., and E. C. Wolf. (1984). "Soil Erosion: Quiet Crisis in the World Economy." Worldwatch Paper No. 60, Worldwatch Institute, Washington, DC, September. 50 pp.

Bunce, A. C. (1942). *The Economics of Soil Conservation*. Iowa State Coll. Press, Ames. 227 pp.

CAST. (1980). *Organic and conventional farming compared*, Council for Agricultural Science and Technology, Report No. 24, Ames, Iowa, October. 32 pp.

CAST. (1988). *Long-term Viability of U.S. Agriculture*, Council for Agricultural Science and Technology, Report No. 114, Ames, Iowa, June.

Edwards, C. A. (1988). "The concept of integrated systems in lower input/sustainable agriculture." In *Sustainable Agriculture in the Midwest* (Eds. C. A. Francis and J. W. King). Proc. North Central Regional Conf., Agr. Res. Division and Coop. Ext. Service, Univ. Nebraska, Lincoln.

Edwards, C. A. (1989). "The integration of components of sustainable agriculture." In *Sustainable Agricultural Systems* (Eds. C. A. Edwards, R. Lal, P. Madden, R. H. Miller, and G. House). Soil & Water Conservation Society, Ankeny, Iowa (in press).

Edwards, C. A., R. Lal, P. Madden, R. H. Miller, and G. House, editors. (1989). *Sustainable Agricultural Systems*. Soil & Water Conservation Society, Ankeny, Iowa (in press).

Erb, C. (1989). "Farming for tomorrow." Agr. Exper. Stn., Michigan State Univ. *Futures* **7(1)**:4–7.

Evans, L. A. (1976). "Physiological adaptation to performance as crop plants." *Phil. Trans. Roy. Soc. London* B.275:71–83.

Evans, L. A. (1980). "The natural history of crop yield," *Amer. Sci.* **68**:388–97.

Fehr, W. (1984). *Genetic Contributions to Yield Gains of Five Major Crop Plants*. Crop Sci. Soc. Amer., Spec. Pub. No. 7, Madison, Wisconsin.

Feliks, J. (1963). *Agriculture in Palestine in the Period of the Mishna and Talmud*. Magnes Press, Jerusalem.

Francis, C. A., R. R. Harwood, and J. F. Parr. (1986). "The potential for regenerative agriculture in the developing world," *Amer. J. Alternative Agr.* **1**:65–74.

Francis, C. A., and P. E. Hildebrand. (1988). "Farming systems research and extension (FSR/E) in support of sustainable agriculture." In *Contributions of FSR/E towards Sustainable Agricultural Systems* (Eds. D. E. Voth and T. Westing) pp.

391–393. Farming Systems Research & Extension Symposium, Univ. Arkansas, Fayetteville. October 9–12.

Francis, C. A., and J. W. King, editors. (1988). *Sustainable Agriculture in the Midwest.* Proc. North Central Regional Conf., Agr. Res. Division and Coop. Ext. Service, Univ. Nebraska, Lincoln. 102 pp.

Francis, C. A., D. Sander, and A. Martin. (1987). "Search for a sustainable agriculture," *Crops & Soils* August–September:12–14.

Gips, T. (1984). "What is sustainable agriculture?" *Manna* July/August.

Harlan, J. R. (1965). "A wild wheat harvest in Turkey," *Archaeology* **20**:197–201.

Harwood, R. R. (1984). "The integration efficiencies of cropping systems." In *Sustainable Agriculture and Integrated Farming Systems* (Eds. T. C. Edens, C. Fridgen, and S. L. Battenfield). Michigan State Univ. Press, East Lansing.

Harwood, R. R. (1989). "History of sustainable agriculture: U.S. and international perspective." In *Sustainable Agricultural Systems* (Eds. C. A. Edwards, R. Lal, P. Madden, R. H. Miller, and G. House). Soil & Water Conservation Society, Ankeny, Iowa. (in press).

Hoeft, R. G., and E. D. Nafziger. (1988). "Sustainable Agriculture." In *Proc. 1988 Illinois Fertilizer Conf.* (Ed. R. G. Hoeft). pp. 7–11. Dept. Agronomy, Univ. Illinois, Urbana.

Jackson, W. (1980). *New Roots for Agriculture.* Friends of the Earth, San Francisco.

Jacobsen, T., and R. M. Adams. (1958). "Salt and silt in ancient Mesopotamian agriculture," *Science* **128**:1251–58.

Lee, R. B. (1968) "What hunters do for a living, or how to make out on scarce resources." In *Man the Hunter* (Eds. R. B. Lee and I. de Vore). pp. 30–48. Aldine Publ. Co., Hawthorne, NY.

National Academy of Sciences. (1989). *Alternative Agriculure.* National Research Council, National Academy Press, Washington DC. 448 pp.

Plucknett, D. L., and N. J. H. Smith. (1986). "Historical perspectives on Multiple cropping." In *Multiple Cropping Systems* (Ed. C. A. Francis). pp. 20–39. Macmillan Publ. Co., New York.

Potash & Phosphate Institute. (1989). "The challenge of ground water concerns with production agriculture—is LISA the answer?" *Fertilegrams,* Atlanta, Georgia. February.

Rodale, J. I. (1945). *Pay Dirt.* Rodale Press, Emmaus, Pennsylvania. 242 pp.

Rodale, R. (1973). "The basics of organic farming," *Crops & Soils* **26(3)**:5–7, 30.

Rodale, R. (1983). "Breaking new ground: the search for sustainable agriculture," *The Futurist* **17(1)**:15–20.

Rodale, R. (1985). "Internal resources and external inputs—the two sources of all production needs." In *Workshop on Regenerative Farming Systems,* USAID, Washington, DC, and Rodale Institute, Emmaus, Pennsylvania.

University of Nebraska. (1987). *Sustainable Agriculture . . . Wise and Profitable Use of Our Resources in Nebraska,* Dept. Agronomy, Coop. Extension Service, Lincoln, Nebraska. 202 pp.

USAID. (1988). "The Transition to Sustainable Agriculture: An Agenda for A. I. D." Paper prepared for Committee for Agr. Sust. for Devel. Countries., Washington, DC.

USDA. (1980). *Report and recommendations on organic farming,* U.S. Govt. Printing office, Agriculture, July. 94 pp.

USDA. (1981). *A Time to Choose: Summary Report on the Structure of Agriculture.* U.S. Govt. Printing Office, Washington, DC.

Witwer, S. H. (1978). "The next generation of agricultural research," *Science* **199(4327)**:375.

Youngberg, G. (1978). "Alternative agriculturists: ideology, politics, and prospects." In *The New Politics of Food* (Eds. D. F. Hadwiger and W. P. Browne), pp. 227–246. Lexington Books, D.C. Heath & Co., Lexington, Massachusetts.

2 BREEDING HYBRIDS AND VARIETIES FOR SUSTAINABLE SYSTEMS

CHARLES A. FRANCIS
Department of Agronomy,
University of Nebraska,
Lincoln, Nebraska

Historical development of crop varieties
Potential need for unique varieties and hybrids
Choice of breeding methods and materials
Choice of testing environments
Yield stability and system sustainability
Examples from breeding for multiple cropping
Future contributions of plant breeding

HISTORICAL DEVELOPMENT OF CROP VARIETIES

Crop cultivars have been selected for improved productivity, wider adaptation, and sustainable production since before the time of recorded history. With the advent of agriculture some 10,000 years ago, there were no doubt already established a number of desirable cultivars which small groups had been collecting, planting, and tending as they gradually settled into a series of more sedentary farming patterns (Plucknett and Smith, 1986).

These cultivars may have first appeared in the garbage mounds near campsites, and were later planted and cared for as their desirability and close availability to the camp were recognized. In the nutrient-rich earth near camp, there could have been substantial expression of yield potential by individual plants. Selection of seed from the best individuals to plant back again was probably done by the first plant breeders, the women of the clan. They undoubtedly selected plants with seed that was easy to process or that displayed disease, insect, or environmental stress tolerance. These plants may have shown an ability to produce large seeds or tubers in a short period of time, or under a range of stress conditions—in summary, those that had some genetic capacity to sustain production.

Although the appearance of multiple species in these garden plots and the process of early selection are purely speculative, there is little doubt that cultivars were chosen for their potential to produce seed and forage and to ease the burden of collecting food from distant sources. Variation in rainfall and other climatic factors aided in the selection of cultivars with adequate yield potential under the range of locally prevailing conditions.

There is evidence of mixtures of species in these campsite gardens, as summarized by Plucknett and Smith (1986). Mixtures of wild diploid emmer wheat (*Triticum dicoccoides*), wild barley (*Hordeum spontaneum*), wild einkorn (*Triticum boeoticum*), peas (*Pisum* spp.), and lentils (*Lens culinaris*) were found in mountains of the Middle East (Wright, 1976). In a limestone cavern in northwestern Thailand, archeologists have found remains of fruits, pulses, root crops, and other species from about 9,000 years ago (Gorman, 1969), though mixed plant remains do not prove that the crops were grown together. Mixed plots provided variety in the diet, and genetic diversity may have been recognized as similar to the natural ecosystem in which people hunted and foraged, and they could have used this principle to increase the security of production of food and other products.

In some isolated regions, subsistence farming systems that persist today may give us clues to what early cultivation looked like (Plucknett and Smith, 1986). The Kayapó of the Brazilian Amazon basin have semidomesticated more than 50 species of trees and shrubs in the forest from which they collect food and other materials (Posey, 1983). They also create resource islands in open grassland that contain more than 100 species transplanted from the forest (Kerr and Posey, 1984). Although the evidence of sustainability in these systems is somewhat speculative, recent reviews of multiple cropping systems suggest that such diversity may have a number of benefits such as improving soil fertility and protecting crops from insects and pathogens (Francis, 1986a).

These could be called the first sustainable farming systems, and the crops selected in them the first cultivars for a sustainable agriculture. Increasing human populations and scarcity of land for extensive cultivation and mixed hunting/gathering/cultivating subsistence activities have led to more intensive systems. The rest of the chapter deals with breeding crops for current and future farming systems in their wide array of complexity and applications. The principles of species and crop diversity, the evolution of cropping patterns, and the relative sustainability of early extensive systems should not be forgotten. Key references for this chapter are the recent books by Christiansen and Lewis (1982) and Blum (1988), and the review by Francis (1986b). This chapter illustrates how plant breeding can make a significant contribution to sustainable systems that cause minimal long-term negative impact on the local and global ecosystem. This is achieved by specific adaptation and efficient use of resources.

POTENTIAL NEED FOR UNIQUE VARIETIES AND HYBRIDS

There is a vital need to increase food production to meet the expanding human demand, while we work to find biologically effective and socially acceptable ways to limit human population. In the past, much of the increase in food production came from moving onto new lands. Sustained food supply came from moving onto fertile soils and into areas that were never cultivated or were planted only after a long fallow period. This is becoming increasingly difficult and expensive. The major increases in food production in the future will have to come from improved productivity of existing lands (Christiansen and Lewis, 1982).

In modern agriculture, we have enhanced productivity primarily by changing the cropping environment to fit our crop cultivars. Added fertilizers, pesticides, and irrigation water have removed or alleviated some of the most critical stress conditions for crops. This strategy is becoming increasingly expensive, both in cost of inputs and in environmental impact of these products, particularly when inappropriately applied to production fields. In the mid-1960s in the United States, for example, we applied approximately the same amount of nitrogen to maize as we harvested with the crop; today we apply *about 50% more* than is harvested with the crop (G. A. Peterson, personal communication). One consequence of this gross inefficiency is accumulation of nitrate in the ground water, and another is unnecessary economic cost to U.S. maize producers. In contrast to complete domination of the production environment, we are now making a greater effort to adapt crops to existing soil and climatic conditions with the goal of improving resource use efficiency and sustainability. What genetic materials are needed to effect this change?

Whether specific or different hybrids or varieties are needed for sustainable systems depends on how we define *sustainable* and under what assumed conditions and objectives. If continuous irrigated maize production with high levels of fertilizer and pesticide inputs in the Central Valley of California is considered sustainable, for example, then hybrids are needed that produce maximum yields under these high-input conditions. If nitrate or pesticide problems in ground water are linked to agricultural practices, then some new combination of hybrid and cropping system is needed that will produce acceptable yields with no net detrimental effects on the environment. Consistent with the definitions of sustainable agriculture presented in Chapter 1, hybrids and varieties are needed that are productive under a range of environmental conditions, generally including a number of different stress situations. As with any breeding program, success depends on understanding the cropping conditions where products of the program will be grown. The breeding process includes choosing parents, making appropriate crosses and selections, and testing promising genotypes under those conditions.

Implicit in any crop improvement effort is knowledge of the realistic time perspective for successful breeding programs to produce new cultivars. Breeding crops is a long-term process. Plant breeding may require up to ten generations to produce a new variety, including crossing, advancing generations, testing, and increasing seed for release. For production and release of new hybrids, it is necessary to make a large number of crosses between parents and to test their progeny under the required cropping conditions. Promising new varieties of a self-pollinated crop must be increased for testing. The inbred or parent lines of potentially promising hybrids need to be increased and the new hybrid seed produced in quantities for wide testing and eventual distribution.

In the case of either varieties or hybrids, it is essential to predict what types of systems and limiting constraints will prevail ten years from now, instead of relying on what is seen in the field today. There are other short-term agronomic solutions that may be more cost-effective to solve today's immediate productivity constraints, while breeding solutions are reserved for those problems of a more long-term nature.

Genetic Potential: Genotype × Climatic Stress Interactions

There is increasing evidence of the potential for building genetic tolerance into crop varieties and hybrids for some major stress conditions (Blum, 1985, 1988). This includes genetic variation for response to drought stress, heat stress, cold temperatures and freezing, and other extremes in climate found in principal cropping regions. Tolerance to one or more of these conditions can help improve the sustainability of productivity under adverse growing conditions.

In general, genetic differences that are simply inherited (that is, controlled by one or a few major genes) are easily incorporated into new varieties or hybrids. This is because such traits are highly heritable, thus the probability of making successful selections under a specified set of environmental conditions is high. The high heritability and small numbers of genes controlling a trait imply strong genetic control and little environmental effect on the expressed phenotype. On the other hand, the more complex the inheritance or the more difficult it is to create the needed environments for testing, the slower is the progress from breeding for tolerance or resistance to stress. Examples of each situation follow.

Genetic Variation in Drought Tolerance. Drought incidence, duration, and intensity are highly variable and rarely predictable (Quisenberry, 1982). There are drastically different genetic tolerances to drought among plant species, including domesticated crops, as seen by their centers of origin and regions of current use. One obvious solution to reducing effects of periodic drought is to change to a more tolerant crop species. For example, sorghum is more tolerant of drought than maize, and pearl millet is more tolerant than sorghum. An example from East Africa illustrates crop choice to avoid drought. Near Dodoma in the central region of Tanzania, pearl millet is the predominant cereal, while in the highlands farther east the main crop is maize. Sorghum is found in a transition zone between the other cereals, and there is often an intercrop of two appropriate species along this gradient. Another example occurred from 1955 to 1975, when grain sorghum virtually replaced maize in Southeast Nebraska due to its drought tolerance and yield stability.

Often it is more desirable to modify an existing crop than to convince farmers and buyers to change crops. Genetic differences in drought resistance have been reported in many crops including barley, wheat, grain sorghum, millet, soybeans, rice, cowpeas, alfalfa, and sunflower (Quisenberry, 1982; Blum, 1985). Blum (1988) further reviews the physiological basis of this resistance in his recent well-referenced book. One of the simplest methods of measuring differences in drought tolerance is to calculate a yield ratio between drought stress and nonstress conditions (Blum, 1973; Mederski and Jeffers, 1973). Blum (1985) maintains that greater efficiency of selection for drought tolerance can be achieved by use of specific physiological measurements instead of yield alone. In tropical maize populations, Bolaños and Edmeades (1988) found that shortest anthesis-silking interval was associated with greatest drought tolerance under three stress conditions. In the ICRISAT pearl millet breeding program in India, enough variability still exists in parents of elite hybrids to allow selection for drought tolerance (ICRISAT, 1985). There appears to be sufficient genetic variation in drought tolerance in most crops to warrant a plant breeding approach to at least partially relieve this production constraint (Quisenberry, 1982).

Genetic Variation in Mineral Stress Tolerance. Crops generally are grown where soils are fertile, but increasing population pressure has pushed production into less favorable areas. Devine (1982) outlines three approaches used to grow crops in these suboptimal conditions: (1) altering the soil environment by adding fertilizers or lime, (2) selecting for desirable agronomic traits within wild species that are adapted to the stress conditions, or (3) genetically adapting current crop species to the adverse soil environment. The first option has been most expedient where soil amendments were available and cost-effective, while the second has been used for centuries by farmers. The third option is receiving most attention by plant breeders who intend to develop hybrids and varieties as components of a sustainable agricultural system. Existing crop species are desirable because there are known cultivars with high yield potential, accepted agronomic practices, and ready markets for the product. None of these conditions normally exist when a new species is introduced. Relative costs and efficiencies of these alternative approaches may vary under different economic circumstances.

Mineral elements may be deficient for a given crop, or may be present in excessive levels that cause toxicity to plants. Clark (1982) has reviewed the major soils of the world and mapped the mineral deficiencies that commonly occur. The identification of genetic variation for response to problem soils as well as progress in breeding crops with resistance to those soils have been reviewed by Blum (1985, 1988). There has been some success in identifying cultivars of several species tolerant to low levels of iron, manganese, and zinc. Some studies found relatively simple inheritance for resistance to minor element deficiencies, and this may help make a breeding effort successful. Likewise there has been some success in breeding for mineral element toxicity. Blum implies that genetic control of tolerance to major nutrient deficiencies is more complicated because of the pervasive involvement of major nutrients in the plants metabolic processes.

There is recent evidence, however, of genetic differences in efficiency of nitrogen use (Castleberry et al., 1984; Muruli and Paulsen, 1981). Maize lines have shown differential responses to changes in soil fertility, for example (Hoffer, 1926). Muruli and Paulsen (1981) found that lines selected for high nitrogen use efficiency produced higher yields than lines selected for high yield under high nitrogen conditions. Two important systems under genetic control were described by Moll et al. (1982): ability of maize to take up nitrogen from the soil, and ability to produce dry matter once nitrogen is in the plant. Some maize lines were found to be efficient for one process, and some for the other (Moll et al., 1983). Working with different mixtures of flint and dent maize germplasm from the United States and Argentina, Brun and Dudley (1989) found higher predicted genetic gains and higher actual yields at two nitrogen levels when selections were made in a higher nitrogen level environment. From these studies, it appears that nitrogen use efficiency is under genetic control (Castleberry et al., 1984), but more needs to be learned about complex basic mechanisms and how to make selections before it will have practical significance.

Zweifel et al. (1987) found significant sorghum hybrid by N level interactions for grain yield and N use efficiency measures. It appears that genotypic differences in partitioning of *N* between grain and stover are greater than the several indices of N efficiency, for example, grain or total dry matter per unit N uptake (Maranville et al., 1980).

Genetic differences in phosphorus uptake, accumulation, and distribution in grain sorghum were observed by Clark et al. (1978). There were significant differences among cultivars in their distribution of phosphorus between roots and tops, and between upper and lower leaves. These were manifested as deficiency symptoms in some genotypes. Studies of grain sorghum inbred parents and their hybrids showed greater differences among male than among female parent lines tested (Furlani et al., 1978). Male lines had greater relative productivity than female lines under low phosphorus conditions, and this capacity was transferred to their hybrids indicating some importance of dominant genes, epistatic effects, or unique sequences of linked genes. The authors conclude that improvement of sorghum lines and hybrids for production at low soil phosphorus levels is possible in a breeding program.

From the preceding reviews we can conclude that genetic variation exists for crop response to both mineral element deficiencies and toxicities. Devine (1982) lists four requisites for success in a breeding program: (1) the character is heritable, (2) techniques are available for efficient screening of materials, (3) desirable genetic variation exists, either within the range of cultivated varieties or in closely related wild populations, and (4) the estimated potential for improvement is sufficient to make the project worthwhile. We could amplify the last point by urging the comparison of cost-effectiveness of breeding versus other approaches to solve the crop nutrition problem. Progress in improving the acid soil tolerance of sorghum, maize, and triticale illustrates the potentials of the genetic approach.

Genetic Variation in Reaction to Other Climatic Stress Conditions. In addition to the above conditions, high and low temperatures, excessive water, salt accumulation, available light, and even air pollutants can cause problems in plant growth and development. Ways to overcome these stress conditions through plant breeding have been reviewed in the books by Christiansen and Lewis (1982) and by Blum (1988). Some of these conditions may recur with frequency or become factors in cropping decisions as planting dates are modified for other reasons, and others are more chronic problems for which there may be genetic solutions. In either case, sustained productivity may depend on the plant breeder's ability to find new cultivars with tolerance to the stress conditions.

Tolerance to chilling or freezing temperatures can broaden the geographic adaptation of a crop species or allow seeding outside the normal planting season. In either case, genetic tolerance in a crop cultivar can provide additional options in planning a crop sequence and total cropping system as well as help the crop avoid excessive damage from unseasonal cold temperatures.

Difficulties with screening crops for low temperatures center on the complexity of creating a repeatable set of conditions in the field. Plant breeders and physiologists have used multiple locations and different planting dates to attempt to expose plants to cold stress conditions at different stages of plant development. Marshall (1982) described the difficulties of breeding for cold tolerance and the progress to date in some species.

Genetic sources for tolerance include wild relatives of the cultivated varieties in oats (Suneson and Marshall, 1967) and in barley (Grafius, 1981). Triticale, a synthetic cereal made by hybridizing wheat (*Triticum* sp.) with rye (*Secale cereale*) has shown substantial variation for cold tolerance and may have promise in some regions (Larter, 1973). There is genetic variation in wild *Solanum* species that could be used to improve the cold tolerance of cultivated potato varieties (Estrada, 1978). In general, heritability of cold tolerance has been low due to the difficulties in repeating the testing environments. Some significant genetic progress has been made in cereals (Blum, 1988; Marshall, 1982).

Tolerance to high temperature has received less attention than low temperature tolerance from plant breeders (Marshall, 1982; Blum, 1985). High temperature and drought stress often are confounded, which reduces the potential rate of genetic progress through selection. Avoidance of heat stress can be promoted agronomically through choice of planting date or by using shorter season cultivars. Another genetic approach is screening and selecting tolerant cultivars. Marshall (1982) lists references to heritable differences in sorghum (*Sorghum bicolor*), maize (*Zea mays*), soybeans (*Glycine max*), and oats (*Avena sativa*). He further observes that only a small part of the known collections of these crops have ever been screened for heat tolerance.

Differences in plant tolerance to excessive water are primarily among species, with few studies reported on intraspecific genetic variation in plant reaction (Krizek, 1982). The principal effect of waterlogging is to deprive the root system of oxygen, and viable screening techniques are needed that approximate the natural conditions to which plants will be exposed in the field.

Tolerance to salinity has received substantial attention from plant breeders because of the extent of irrigated land that now has accumulated salts and also because of the need in some areas to irrigate with brackish or saline water (Blum, 1985). The tendency in these agricultural areas is toward increased salinity over time, as irrigation water continues to evaporate, leaving the salts behind. Thus breeding can offer part of the solution, but it must be coupled with other management efforts to achieve a sustained productivity under these adverse conditions. Some of the pioneering work on irrigation with salt water has taken place in California on barley (Norlyn, 1980) and tomato (Rush and Epstein, 1981).

Air pollutants are known to injure plant leaves, alter growth and potential yield, and even reduce quality of plant products (Reinert et al., 1982). The major pollutants of concern are ozone, sulfur dioxide, nitrogen oxides, and hydrogen fluoride, which result from combustion of fossil fuels and different manufacturing processes, including fertilizer production. The authors list a number of species in which genetic variation for tolerance to pollutants has been evaluated under ambient or controlled conditions.

Although there are a few cases of single gene control of tolerance, most research indicates polygenic inheritance of this trait.

Genetic Potential: Genotype × Biotic Stress Interaction

Crop productivity is greatly influenced by insects, plant pathogens, nematodes, and weeds in the cropping environment. Much of the crop resistance or tolerance to these biotic factors in the environment has been developed through natural selection. In fact, the plant breeder often searches for sources of resistance at the center of origin of each crop species where maximum genetic variability exists in the crop and where pathogens and insects have been interacting with the crop for centuries. There has been significant progress in breeding for tolerance or resistance to many insects (Jenkins, 1982), pathogens (Bell, 1982), and nematodes (Sasser, 1982). These three chapters in the book by Christiansen and Lewis (1982) can serve as prime references to the factors that are important in setting up a screening or testing program. They also discuss the primary biotic tolerance mechanisms found in crop plants. (See also Chapter 3 by Bird et al. in this volume). There is less known about weed interactions with crops, although studies of allelopathy have become more common in recent years (see Liebman and Janke, Chapter 4).

Important to success in breeding for tolerance to insects, pathogens, and nematodes are genetic variation in reaction to the pest, heritability of tolerance or resistance, and testing methods that allow the breeder to look at large numbers of parents and progeny from crosses. Often it is difficult to repeat the same testing conditions. An accurate evaluation of the potential problems from each pest species, including frequency and severity of occurrence, also is crucial. Progress in a breeding program is inversely proportional to the number of traits for which the plant breeder is selecting. Therefore, setting priorities on which pest problems are most important is a critical step in the planning process, as is comparing these problems to others that limit productivity and stability of yields.

Weeds present a different type of challenge. Crops that germinate and grow quickly can produce a canopy that shades the soil surface and inhibits germination and growth of some weeds. This is one aspect of breeding for "resistance", or a competitive edge for the crop or crops in a mixture. Another dimension is the allelochemical interaction between crops and weed species, commonly called allelopathy (Altieri and Liebman, 1986; Liebman and Janke, Chapter 4). Farmers in Central America commonly plant maize and beans (*Phaseolus vulgaris*) intercrops, and squash is added to the mixture primarily as a method of controlling weeds in the desired crops (Gliessman, 1983). Potential for weed control in temperate zone soybeans using an associated planting of rye is now being studied. There may be many untapped potentials for biological weed control using allelopathic effects of one species on others.

Genetic Potential: Genotype × Cropping System Interaction

Another type of stress on crop species is the competition encountered by one species when grown in association or intercropped with one or more other crops. The total dry matter and grain production actually involves a series of competitive interactions for growth resources. In an intercropped maize/bean system, for example, this includes intraplant competition (for example, apical dominance), intraspecies competition (among dry bean plants at high density in the same row), and interspecies competition (between maize and dry bean plants). Experimental techniques to separate these various types of competition are complex and not well developed, especially for agroforestry and other long-term systems. Most commonly we measure the integrated final total of the products of this process by harvesting and measuring final yield of component crops and, at times, the individual plants to study the specific effects of competition.

To date the breeding of crop varieties specifically adapted to this type of system has been a highly empirical process (Francis et al., 1976; Francis, 1981; Smith and Francis, 1986). Although there are a number of ways to quantify the genotype by cropping system interaction, perhaps the most frequently employed method is analysis of variance. When there is a significant variety by environment interaction for a specific crop, the interpretation is that potential exists for selecting varieties with specific adaptation to each system. Most commonly this has been seen in the lower story crops (e.g., cowpea, dry bean, mung bean, sweet potato), especially when there is a significant portion of the growth and development cycle overlapping or coinciding with an associated crop. Bean/maize inter-cropping is an example in Latin America and Central Africa. In contrast, crops that are relay or double cropped and those that form the taller story (e.g., maize, sorghum, tall millets) are less apt to demonstrate a significant interaction. This is a good guideline for deciding whether a breeding program to select cultivars specifically adapted to intercropping is likely to be successful, although it is much more rational to gather information from the literature or from germplasm types on hand to confirm this preliminary conclusion. One complication is that researchers often evaluate results of each component species, while farmers evaluate yields and economics of the entire system. A research area that is virtually untapped is agroforestry, where little is known about crop species by system interactions.

Many of the land races or farmer varieties of crops planted in complex multispecies systems used by low-resource farmers were selected by those same farmers for adaptation to the prevalent systems. There have been so few studies conducted by trained plant breeders looking at yield potential under intercropping that it is difficult to generalize about success. Nevertheless, there is a need to make genetic selections of some crops commonly grown in association with other crops, and there should be great potential for improving the genetic package for greater sustainability of these systems. Further, many of the same traits that are useful to add stability to yields in monoculture (e.g., drought tolerance, insect and pathogen resistance) will be equally important for intercropping. The differences between the two general types of systems may be reflected in the relative priorities put on specific traits in a selection program.

Specific examples that illustrate significant interactions between genotype and cultural practices include maize hybrids tested under different tillage systems. With the introduction of zero and reduced tillage approaches in the Midwest, there is concern that different hybrids may be needed for optimum performance under these new conditions (Mock, 1982; Newhouse, 1985). In one of the first such comparisons with eight hybrids in Iowa, Mock and Erbach (1977) found no significant hybrid by tillage interactions for grain yield or other traits measured. Similar results were obtained by Mason (1983) in Missouri in a study of 18 hybrids and in Iowa in studies of 14 hybrids (Hallauer and Colvin, 1985) and 60 hybrids (Newhouse and Crosbie, 1987).

In contrast, a study in western Nebraska of 169 topcrosses over two years showed a highly significant genotype by tillage interaction (Brakke et al., 1983). This was confirmed by results of a test of unselected lines in Iowa (Newhouse and Crosbie, 1987), although the genotype by location interaction was still of higher magnitude than genotype by system. The conclusions by these authors are that significant interactions do exist, but that other factors are more important in setting priorities in a breeding program. Similar results were found in a test of 19 commercial grain sorghum hybrids tested in two years, two tillage systems, and two contrasting planting dates in northeast Nebraska (Francis et al., 1986). Although some progress could be made in specific adaptation to a reduced tillage (low-input) system, more progress in yield potential probably could be made by selecting for traits such as drought tolerance, insect and pathogen resistance, or other factors reducing yields (Newhouse, 1985). Preliminary results at Cornell University showed significant genotype by system interactions for germination and stand of 12 maize hybrids under conventional, ridge, and no-till planting (T. Barker and M. Smith, personal communication).

This is but one example of a rather substantial change in cultural practices; it happens to be one for which genotype by cultural practice interaction has been studied in a number of locations. We need to carefully assess whether it may be profitable to initiate a breeding program for new cultural conditions or new environments before making the long-term and costly decision to invest in that program. It is possible, and even likely, that current hybrids or varieties can be used effectively in a broad range of situations beyond where they were selected and tested. On the other hand, breeding crops for specific local environmental conditions provide the basis for seed company business in each area. For example, hybrid maize is developed for local photoperiods, insect and disease problems, and prevalent rainfall/temperature patterns.

CHOICE OF BREEDING METHODS AND MATERIALS

Plant breeding methodologies have been developed over the past several decades to deal efficiently with broad variability; large numbers of parents, crosses, and progeny; and selection for a range of useful characteristics. Techniques are summarized in recent texts such as those of Fehr (1987) and Jensen (1988), and specific methods of selection for stress tolerance are reviewed by Christiansen and Lewis (1982) and Blum (1988).

As with any breeding program, the first step in adapting cultivars for sustainable systems is carefully defining objectives and describing the systems and conditions under which the hybrids or varieties will be produced. Choice of parents depends on the traits needed for adaptation to these conditions, while correlated traits will either complicate or simplify the selection process. Whether the same or different alleles are functioning to provide yield potential under stress or nonstress conditions is important, but not easily determined. Some methods based on natural selection have been proposed and used, but these generally are long-term and less efficient than new techniques based on rapid screening and molecular genetics. Choice of breeding methods is a critical step in the crop improvement process.

Plant breeding programs are futuristic by nature. To initiate a crossing and selection program today, directed only at current production-limiting constraints, will reduce the chances of developing new cultivars adapted to production environments of ten years from now. We have to use our best predictions and projections to anticipate what the production environment and prevailing systems will be at least 20 years into the future in order to have much hope of a lasting impact with new genetic materials (Francis, 1980). This is not an easy task. It is useful to recall that plant breeders have made little progress in the last century in increasing total biomass production of most crops (Evans, 1976; Cox et al., 1988). Partitioning to yield organs, however, has been increased. Thus variation in partitioning of dry matter, especially during the latter part of the development cycle (grain fill), is essential to a breeding program that is concerned with yields in stress environments (Saeed et al., 1986).

Newhouse (1985) illustrated the gradual and successful adaptation of maize hybrids to fall plowing as compared to no-till planting by comparing sets of four hybrids released during each decade from the 1930s to the 1970s. Hybrids released during the last two decades showed yield reduction of 10% to 15% under no-till planting, while those released in the 1930s showed virtually no differences. He correctly points out that the tillage change to fall plowing is confounded with other components of a high-input, intensive system including high levels of nitrogen. Newhouse concludes: "It is possible that corn genotypes which were selected for response to highly fertile conventional tillage environments cannot respond to fertile no-till environments in the same fashion." The great majority of farmers now practice some form of reduced or no-till planting; in the past decade alone, use of reduced tillage has spread from about half the maize acres to well over three-quarters of the acres in the Midwest. However, there are still a number of breeding program test plots on conventionally tilled and planted fields.

This is one example of how we should anticipate from current trends what farmers will be doing one decade in the future (when a new hybrid or variety from today's crossing block will be commercially available) and adjust our cultural practices accordingly. Defining the future production environments, especially in terms of availability and cost of resources, is more difficult than merely predicting tillage system. Today's final testing of hybrids before release should use best available current technology, however, since these hybrids will be available in two to three years.

One dimension that will become increasingly important is adaptation to variable environments, as discussed by Quisenberry (1982). Working with small grains, Hurd (1971) showed that varieties with the greatest root development under nonstress conditions did not have the best root systems among a group of lines tested in moisture-stressed environments. Likewise, Briggle and Vogel (1968) found that high-yielding dwarf wheats with broad adaptation in the Northwest were not especially well adapted to the periodic drought conditions of the Midwest. The challenge of developing varieties with specific adaptation to drought stress conditions is the variation in timing and severity of drought from one season to the next. Since drought tolerance is governed by many genes, and since the genes limiting yield potential under drought conditions may be different under dissimilar conditions (e.g., drought stress before, during, or after anthesis), the heritability of yield potential under these varied conditions will be lower than under more controlled moisture environments. Thus there is a great reduction in effectiveness of selection (Johnson and Frey, 1967). Whether or not it is possible to make substantial genetic gains for yield potential under a widely variable set of conditions is still in question (Quisenberry, 1982), but this is unlikely.

Choice of parents in a breeding program is one critical initial step. Plant breeders agree that it is most desirable and productive to find traits of interest in high-yielding, well-adapted varieties or inbred lines that are known to have superior combining ability, as compared to bringing in those traits from nonadapted introductions. Another generalization is that traits with relatively simple genetic control and high heritability will be more easily identified and manipulated in a breeding program than traits with complex inheritance and low heritability. For example, tolerance to low temperature in a number of crops is known, as reviewed by McDaniel (1982) and Marshall (1982). Likewise, the genetic nature of tolerance and heritability of plant response to some adverse soil conditions is fairly well understood, especially the reaction to acid soils (Devine, 1982). Within the germplasm collections at the international research centers and several key national programs, especially in the United States, there is a broad source of genetic variability for most traits of interest, although some of these collections have not been characterized in detail (Duke, 1982).

Both interspecific and intraspecific genetic diversity are important to agronomic and genetic research to increase system sustainability. Choices among potential component crops depend on how differently these species react to prevailing stress conditions. Parallel to this, the amount of genetic diversity within each species is one vital key to the potential for crop improvement. Wild relatives can be used to increase genetic variance for traits of interest. But this diversity must be translated into useful, heritable variance in order to contribute to increased gains in productivity. It is useful to know which traits contribute to stress tolerance and whether selection indices can be used to accelerate the process (Baker, 1986).

Decisions regarding which traits to include in a selection program depend on how well the resistance or tolerance to a given stress condition is understood as well as the resources or ability to repeat a testing environment or operate across multiple environments (the next section deals with testing). In the field, some conditions such as low or high temperature are likely to be quite uniform across a nursery or replicated test. Other stress factors such as drought or mineral problems are likely to be highly variable, even in the same field. For this reason, Martineau et al. (1979), Sullivan et al. (1977), and many others have searched for correlated traits that could be used to reduce plant-to-plant variability caused by diversity in the microenvironment, either in the field or the greenhouse. Such techniques have been used to measure heat stress, net photosynthesis, cold temperature shock, and tolerance to mineral element toxicity or deficiency, among other traits. These tests, as well as observations of phenotypic traits, often are more difficult in the field due to lack of uniformity and control of the environment (Marshall, 1982). Our ability to make genetic progress depends, then, on setting clear program objectives, determining which traits to incorporate, and designing an adequate field testing plan.

To develop varieties for sustainable systems, it would be desirable to identify genetic combinations that not only tolerate stress conditions but that also respond to favorable growing conditions. Results from experiments designed to select under contrasting conditions are nonconclusive. Gotoh and Osanai (1959) selected winter wheat under three fertility levels and found that lines selected under low fertility had superior yields and the widest range of adaptation. In spring wheat tested in multiple locations, Krull et al. (1966) found the same highest-yielding varieties in both highly productive and poor testing environments. Selection of oat lines under stress and nonstress conditions showed higher heritability for grain yield in the absence of stress; there was also a wider choice of useful varieties with a broader range of adaptation under nonstress conditions (Frey, 1964). Yet even within the same crop and research program there may be different results with different populations and sets of lines, as shown in Frey's oat breeding program in Iowa (Johnson and Frey, 1967; Vela-Cardenas and Frey, 1972). It is not apparent whether the same genes are contributing to yield under contrasting levels of stress.

In a recent paper using oat and maize examples, Atlin and Frey (1989) calculated the genetic correlation coefficients between yields in low-input and high-input environments. From the frequency of significant genotype by input level interactions reported in the literature, they suggest that alleles for specific adaptation to low-input or stress environments must be present in many species. In oat trials, they found heritabilities for yield to be about twice as high in high productivity environments. In a low-phosphorus environment, heritability for oat yields was about twice as great as in high-phosphorus conditions, while heritabilities were similar in high- and low-nitrogen conditions. Because of these results, they caution that separate breeding programs for high- and low-productivity conditions should not be initiated unless the breeder is sure that specific adaptation alleles do exist and that different varieties are justified for the contrasting conditions.

Using data from the literature, Atlin and Frey (1989) surveyed the results of testing of maize hybrids under different levels of nitrogen. Although results varied among the several studies used, they concluded: "Heritability for yield under low N conditions was sufficiently high to permit effective selection, and that yield at low N fertility was a character that was genetically distinct from yield under high N conditions. This suggests that maize hybrids can be selected for specific adaptation to reduced N fertility and that the most effective way to develop such hybrids is to select them under much lower N than normally used in the U.S. cornbelt."

These results are in contrast to the conventional wisdom in plant breeding that heritability for yield and yield components is always reduced by stress conditions that reduce yield (Blum, 1985; Bramel-Cox, 1988). In support of the conventional view, Allen et al. (1978) found higher estimates of expected gain in favorable environments than in unfavorable or intermediate environments in barley, wheat, oats, soybeans, and flax. This was due to higher heritabilities for yield in the more favorable environments. From these literature reports and experience in grain sorghum, Bramel-Cox (1988) concludes that the "continued selection of improved varieties under optimal conditions will result in a slower rate of gain under the lower yielding environments."

Heritability can be low due to low genetic variance and/or due to high environmental or genotype by environment variance. High environmental variance is not surprising under stress (M. E. Smith, personal communication), in part because it is difficult to generate a uniformly and consistently stressed environment. Making sure that every plant has optimal supply of all growth factors is the easiest route to uniformity in an inherently variable natural setting. This has been one justification for high inputs in breeding nurseries and trial plots. Low genetic variance may be due to many generations of selection under nonstressed conditions, resulting in the loss of genes or gene combinations that provide adaptation to stress conditions. Low heritability does not mean that genetic progress is impossible. It does mean that breeders need innovative techniques to help identify genetic differences under stress, such as better designs and replicated evaluation, selection for correlated traits, and collaborative work with physiologists to develop screening techniques.

There appears to be sufficient evidence that optimizing breeding for improved adaptation to environmental stress conditions that limit yield potential, such as low fertility, would not be possible if all selection and testing were carried out under near-optimal conditions. If energy and thus fertilizer inputs become limited in the future, we will need to find the best genetic combinations for moderate nutrient stress conditions. The development of new hybrids and varieties with better adaptation to a broader range of limiting factors will be an important component of future sustainable production strategies. Breeding may have a significant role in creating varieties and hybrids that can contribute to a more sustainable agriculture.

CHOICE OF TESTING ENVIRONMENTS

In a recent symposium on sustainable agriculture, Bramel-Cox (1988) stated that "an effective testing environment is characterized by a low error and genotype × environment variance, a high genetic variance, and a high correlation with the target environment." In general, this describes the high-input, controlled environments where much of our initial selection and testing has been done with most crops—on experiment stations and with farmers who use high inputs. Whether this is successful or not will depend on the prevalence of these favorable environments among the target farmers' fields (within a recommendation domain) and the overall adaptation of these new cultivars to a range of on-farm conditions. This improvement will depend on the correlation between yield performance under more favorable experiment station conditions and the more stressful on-farm conditions. Fehr (1987) and Jensen (1988) offer practical discussions about how to choose the number of environments as well as specific sites for testing.

In a relatively simplified example, Rosielle and Hamblin (1981) defined mean productivity as the average yield across sites, and stress tolerance as the difference among site means. In practicality, this is how most breeding programs conduct their testing. Bramel-Cox (1988) concluded in her review that "to date, selection has been for mean productivity which has resulted in little or no improvement under stress with a great improvement under nonstress." Thus, choice of testing environments is critical, and they need to conform as closely as possible to target environments, or "recommendation domains" as they are described by scientists in the international center network.

Multiple locations are useful for gaining information about new genetic combinations and their potential for sustained production under a range of conditions, yet this type of testing is expensive. One practical guideline for maximizing progress in a breeding program while minimizing cost is to concentrate in early generations on those plant traits that are simply inherited and least affected by the environment. If resistance to a specific pathogen is critical for varieties within the entire zone of application, or if seed color is important for consumer acceptance, these traits can be selected in early generations, perhaps in a single location under the most convenient conditions for the researcher—often in the controlled environment of the experiment station (Francis, 1986b).

This is analogous to what Boyd et al. (1976) consider selection for general adaptability in early generations, as a procedure for narrowing the amount of new material to be widely tested in later steps. Subsequent evaluation can be done in multiple locations for special adaptation to stress or marginal growing conditions—an adaptation often based on complex inheritance and highly influenced by the environmental variation from site to site and year to year. It is the magnitude of this environment-to-environment variation and the differential reactions of genotypes to differing conditions that complicate selection and testing. The same complication indicates genetic potential for specific adaptation to locations or stress conditions.

Breeders have assumed that the greater the number of environments used for testing experimental cultivars, the more will be known about their performance under varied conditions. Additional growing conditions can be created for testing by using different cultural conditions, locations, and years. Since time is of the essence, especially in highly competitive commercial breeding programs, more locations or manipulation of cultural practices within a location (for example, irrigated versus dryland, high versus low fertility, early versus late planting, high versus low density) have been used to substitute for additional years of testing.

In order to maximize information from a minimum number of locations, plant breeders have devised statistical methods to determine which of a number of locations is necessary to give cost-effective additional information. This involves checking the relative performance of a large group of genotypes that has been tested across a wide range of locations of interest to the breeder. Genotype by location interactions can be calculated for different combinations of locations to see which of these locations really are similar. Using this method, Horner and Frey (1957) determined that four locations were sufficient for testing oats in Iowa. Similarly, Konishi

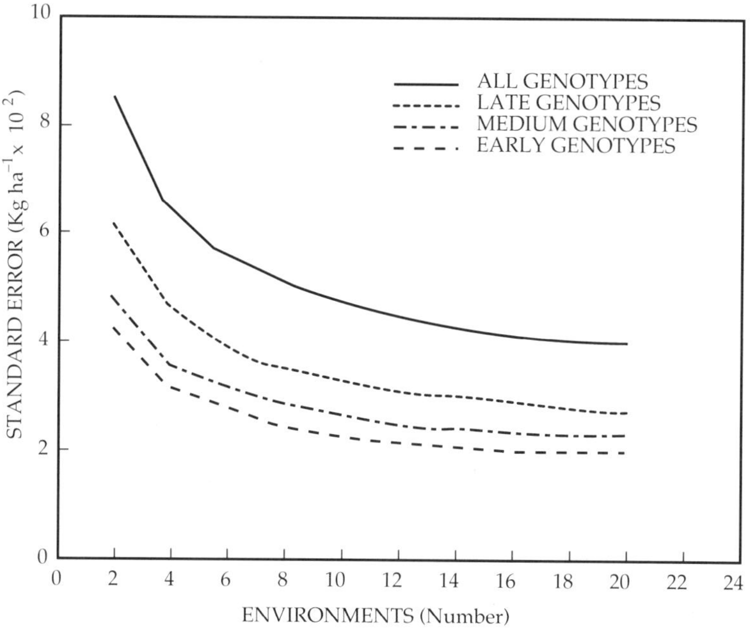

Figure 2.1 Expected standard error of a genotype mean yield at different maturity levels for various assumed numbers of environments when years = replications = 2. (Saeed et al., 1984.)

and Sugishima (1964) decided that three sites were needed for testing barley in Kyushu.

Other methods were summarized by Gotoh and Chang (1979) and by Fehr (1987). Using rank correlations among locations, Gotoh and Fujimori (1960) chose three types of environments in order to test soybeans in Hokkaido, and Guitard (1960) reduced the number of locations from ten to five for testing barley lines in Canada without losing much information. Finally, cluster analysis was used by Abou-El-Fittouh et al. (1969) to decide on five areas for testing cotton in the United States, and by Ghaderi et al. (1980) to limit wheat genotype testing to two locations in Michigan. These examples illustrate statistical ways to reduce locations needed for testing while maintaining a maximum amount of information about adaptation of potential new cultivars. Sustainable production systems depend on knowledge of performance under a wide range of conditions, but it is necessary to obtain this with minimum cost.

The potential for determining significant differences among new genotypes increases with a reduction in the variance of a genotype mean (Jones et al., 1960). Specific examples of how this applies to grain sorghum were presented by Saeed et al. (1984) using analysis of yield trials in Nebraska. Figure 2.1 shows the effect of increasing number of environments on the

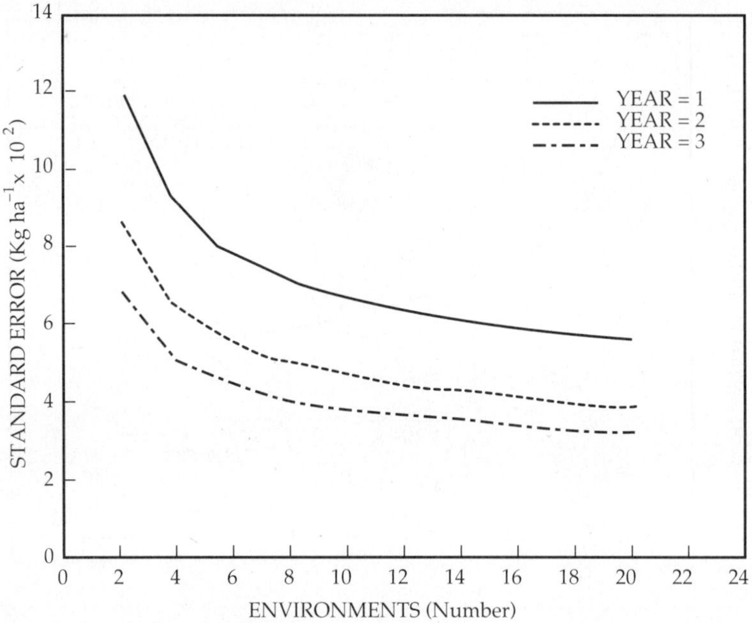

Figure 2.2 Expected standard error of a genotype mean yield for various assumed numbers of years and environments within years when replications = 2. (Saeed et al., 1984.)

standard error of a genotype mean for 11 early sorghum cultivars, 15 medium maturity cultivars, and 28 late cultivars. Grouping the cultivars by maturity reduced the standard error from the level achieved with all genotypes analyzed together. The effects of increasing number of years and number of environments are shown in Figure 2.2. Testing in ten environments in one year gave about the same level of error as testing in four environments in two years or two environments in three years. Finally, Figure 2.3 shows the effects of increasing replications within an environment. There is some advantage in using two replications, but the gain for adding one or two additional reps is small. These curves illustrate why number of environments is important, and how grouping genotypes, adding years, or adding replications can help improve the ability to detect differences among cultivars in a test.

The public plant breeder or commercial company is interested in the broadest possible adaptation of new genetic combinations, in order to effect the greatest impact on productivity of the crop in question or to gain market share. This interest extends to the individual crop producer only in terms of that adaptation applicable to a given farm over a series of years. Farmers lose specific adaptation and yield potential when breeders select and test for a large area. Thus, the testing over multiple locations or

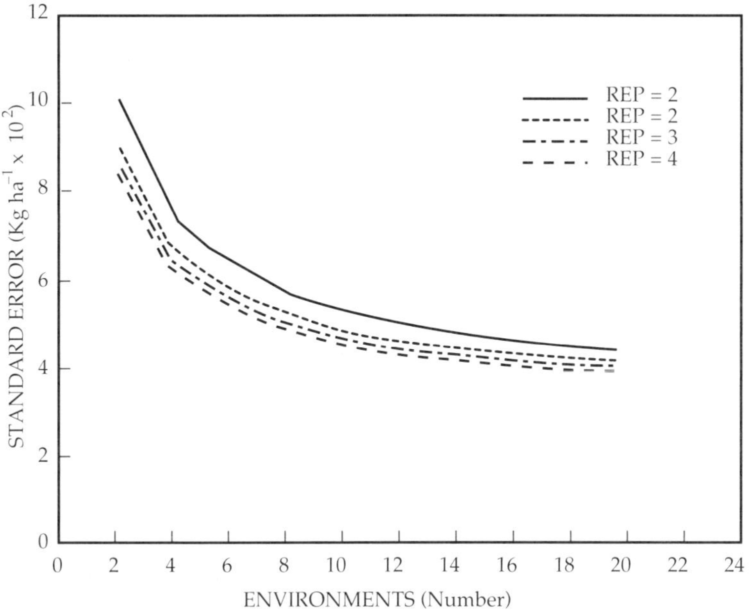

Figure 2.3 Expected standard error of a genotype mean yield for various assumed numbers of replications and environments when years = 2. (Saeed et al., 1984.)

cultural practices, a practical approach by the plant breeder to substitute for years, is meaningful to the producer only to the extent that those locations or practices simulate the variation that will occur over years on his or her farm. Sustainability of production will be determined in part by the ability of plant breeders to build in adaptation or yield stability to the range of expected conditions on each farm. Nontraditional systems such as intercropped species, overseeded legumes in an existing crop canopy, or agroforestry patterns may become important in the future. The book by Jensen (1988) dedicates an entire chapter to "Choosing Sites and Environments", and it should be consulted for more detailed practical insights and references.

YIELD STABILITY AND SYSTEM SUSTAINABILITY

The concept of yield stability or reliability across environments has captured the curiosity of plant breeders for more than three decades (Fehr, 1987; Jensen, 1988). Earlier described as "broad adaptation," this concept was first studied as the genotype by environment interactions from an analysis of variance of yields over locations and years (for example, Sprague and Federer, 1951). More interest and statistical alternatives were brought to the stability question through the regression analyses and reports of Finlay and Wilkinson (1963) and Eberhart and Russell (1966). These techniques described the performance of individual hybrids or varieties in terms of the mean performance of all varieties tested in each location. This showed the improvement of yields of individual genotypes in response to "improving environments." More recently, cluster analysis has been used to group genotypes by their reaction to a series of dissimilar environments (Ghaderi et al., 1980) in order to match varieties with testing locations and to get some indication of yield stability. Jensen (1988) dedicates a chapter to models and the use of the stability concept in breeding programs.

Stability is suggested to result from some type of genetic buffering by plants to changes in their environment (Fehr, 1987). For individual plants, this has been called developmental homeostasis and individual buffering (Allard and Bradshaw, 1964). For varieties or other collections of genotypes this was called genetic homeostasis and population buffering (Lerner, 1954), where stability of a group of plants is greater than that of individuals in the group. The multiline concept of blending a series of pure lines to introduce diversity into a variety was proposed by Jensen (1952). From these reports and others, it appears that heterozygous individuals have greater stability than homozygous individuals, and that heterogeneous mixtures of cultivars are more stable than homogeneous cultures. A logical extension of these observations is to explore genetically uniform cropping systems over time, and then compare them to systems that are more diverse.

Although there is still some debate among experts in biology (for example, Goodman, 1975; Loomis, 1984), it is generally agreed that more diverse natural ecosystems are more stable. There is a growing literature on multiple cropping that suggests a parallel in cultivated ecosystems (Francis, 1986a). Two field studies on different continents showed smaller variance in both grain yields and farmer incomes across environments from intercrops of two species compared to their component crops; these were bean/maize intercrops in Colombia (Francis and Sanders, 1978) and sorghum/pigeonpea intercrops in India (Rao and Willey, 1980). There is a need for more such biological and economic comparisons with long-term use of intercrops.

Crop rotations have been shown to consistently increase yields of all crops in the sequence, and the average advantage is estimated to be about 10% (Francis and Clegg, 1988). Even rotation of some dissimilar maize hybrids has shown a yield advantage over continuous culture of the same hybrid (Hicks and Peterson, 1981). There is much yet to be learned about the "rotation effect," but farmers currently use this biological phenomenon to advantage. It is thought that rotations of dissimilar crop species contribute fertility to the soil; break the cycles of some insects, pathogens, nematodes, or weed species; and otherwise affect the soil biological populations in some favorable way. Other possible benefits are increased organic matter, improved soil aeration, less soil compaction, better moisture retention due to improved structure, and increased nutrient availability. This is one important research frontier that will lead to better understanding of the biological events that contribute to improved productivity with diverse species.

In summary, the genetic implications for system stability and sustainability involve buffering of the reactions of crop genotypes in the field to the range of conditions that will occur in each location over a period of time. This begins with the genetic diversity present in an individual plant and in a variety, described as developmental and genetic homeostasis. Next is the diversity that is present in the cropping mixture in a given season, as in various intercropping patterns including an overlap of crops (relay cropping). Finally there is diversity over time in crop rotations of dissimilar species. The plant breeder is thus challenged to build in capacity for genetic contributions to increased sustainability of productivity at each of four levels: plant, variety or hybrid, cropping system (in one season), and rotation of species or mixtures of species in the field over time. Most work to date has concentrated on the individual plant and especially on the cultivar level, and the two more complex dimensions have received only limited attention. Cropping systems and rotations may hold the most promise for future genetic improvement. Cornell University maize scientists, for example, have initiated a breeding program to look specifically at sustainability in maize-based systems (T. Barker, personal communication). Although it is obvious that we will still be in the business of developing individual plants and collections of those plants (called cultivars or varieties), the challenge will be to improve cultivars to allow maximum genetic contribution to multiple cropping systems and to long-term rotations.

EXAMPLES FROM BREEDING FOR MULTIPLE CROPPING

One set of complex farming patterns where plant breeding has made a relatively minor contribution is in multiple cropping systems. These are defined as systems that include more than one crop in the field during a calendar year. Some of the variants include double and triple cropping (crops planted sequentially), relay cropping (overlapping of crop cycles), ratoon cropping (harvest more than once from a single planting), and many types of intercropping (two or more crops in the field at the same time for a significant part of their life cycles) (Francis, 1986a).

Although many of the better known examples of multiple cropping are used by small farmers in the tropics, there are significant applications of this principle in the temperate zone: double cropped wheat/soybeans, multiple species pasture mixtures, overseeded grasses and legumes used as cover crops. These complex systems are viewed by some experts as providing the diversity and flexibility needed to make food production more sustainable in the future.

Most of our recent crop improvement efforts have been directed at monoculture cropping. Thus, we have worked to improve the productivity of multiple cropping systems with existing cultivars. One notable exception is the development of short-cycle rice varieties in China to allow two and three crops per year. Beets (1982) described this report from Perkens (1969): "The development of an early rice variety in the year 1012 triggered a revolution in growing practices and made cultivation of a second crop possible (in the South). By the Ming period (1368–1644) cold tolerant varieties were developed which could be planted in midsummer after spring crops or early rice . . . . Further north, in Hunan, it was not until the seventeenth and eighteenth centuries that efforts were made to promote second crops." These principles have been used in recent years to provide short-cycle tropical varieties of rice that allow two and three crops per year in Southeast Asia.

Double cropping and relay cropping of wheat/soybean in the southeast United States have increased substantially over the past several decades. Yet these systems have depended on the wheat and soybean varieties already developed for monoculture systems. Overseeding of legumes and grass species into growing summer crops such as maize, grain sorghum, or soybeans has promise for capturing and retaining nutrients that would otherwise be surface eroded with soil or leached down through the soil profile. They provide some organic matter and nutrients during decomposition that can be used by the subsequent summer crop. All of the breeding efforts with these species have been in sole culture, and scientists are only beginning to screen available species and cultivars for their potential application in this new system. There have been conscious improvements of grasses and legumes for pasture and forage systems, and the competition models and selection techniques developed here could be useful for other multiple species systems (Gomm et al., 1976).

We can only speculate about the potential of plant breeding to improve crops or companion species for these multiple cropping systems. Experience has shown that selection for any new system or set of conditions can be successful if there is sufficient variability in the genetic raw material and a well-established testing scheme to sort out that variation in a consistent manner. If genetic variation is available for traits such as shade tolerance, ability to compete for other growth factors such as moisture and nutrients, and appropriate maturity to fit into a given niche in a new cropping system, there should be potential for substantial improvement through crop breeding. The magnitude of this potential will not be known until the system is tried. To date, there has been little research investment in this direction, although some of the international centers have put priority on developing varieties for more complex systems used by subsistence farmers (for example, drybean systems program in

CIAT, Colombia). This is a virtually unexplored area for future contributions of plant breeding research to increased productivity and sustainability of systems.

FUTURE CONTRIBUTIONS OF PLANT BREEDING

Contributions of plant breeding to future cropping systems may be more important than they have been in the past. This is speculative, because breeding hybrids of maize and grain sorghum and varieties of wheat, soybeans, and other crops have already accounted for at least half of the increased productivity of these crops (Fehr, 1984). As fossil fuel resources become scarcer and less accessible to agriculture, there will be a need to search for alternatives.

It is likely that systems will continue to evolve toward less tillage, greater erosion control (cover crop use), reduced water inputs from irrigation, more fine-tuning of fertility and less applied fertilizer, and fewer pesticides. During the past five years alone, U.S. farmers have reduced purchased inputs by about $25 billion per year, or by about 20%. Plant breeding could provide cost-effective substitutes for some fossil fuel–based inputs. Incorporation of drought tolerance, insect and pathogen resistance, weed tolerance, or ability to germinate and grow in cooler soils under a no-till environment demonstrate the types of potential that can be built into the seed.

Although plant breeders by nature are oriented toward the future, they sometimes exhibit a surprising conservatism in choice of parents and ways to handle their progeny. Jensen (1988) dedicates the last chapter of *Plant Breeding Methodology* to a list of ways to revitalize a breeding program that has been going for some time. Based on his practical experience of about five decades in plant breeding, this is a useful collection of suggestions for the beginning professional and the seasoned veteran. Too infrequently do we sit back and engage in creative speculation about what future cropping systems will look like and how breeding can contribute to them.

In one such attempt to predict future needs, this author drew up a list ten years ago, suggesting ten major changes in cropping systems for maize and grain sorghum (Francis, 1980). The list was based on survey responses from about 100 maize and sorghum breeders, from reading the literature, and from experience with the two crops both in the United States and overseas. The assumptions were future "increased demand, increased costs of energy and inputs, and greater concern about the importance of conservation." Here is the list of predicted changes for maize and sorghum:

1. Rotation with other crops.
2. Potentials of specific adaptation.
3. Multiple purpose crops.

4. Multiple cropping systems.
5. Perennial cereal crops.
6. Genetic engineering of corn and sorghum.
7. More variable hybrids.
8. Permanent sod culture.
9. Maximum return to limiting factors.
10. Maximum return per unit of resource used.

A complete list developed by any breeder today would probably differ from this one. Yet we can identify a number of key issues for the 1990s in this list drawn up in 1980. Some of the priorities on the list may become important at some future date. Priorities change with the development of new tools for the breeder; molecular genetics, for example, has become a well-funded driving force in the industry, to a greater extent than many would have predicted a decade ago.

Adaptation of crop cultivars to evolving cropping systems will require a long-term vision by breeders. We will have to be as accurate as possible in predicting the cropping patterns, specific crops needed, and resource constraints under which they will be grown at least one decade into the future. The current political and popular concern about nitrate and pesticide contamination of surface and ground waters is a powerful example of the broadening recognition by society as a whole of some of the unwanted and unpredicted consequences of high input agriculture. Improved nitrogen use efficiency and genetic pest tolerance can contribute to some of the solutions, in combination with modified cultural management of the total system. In any case, plant breeding will be central to the planning and implementation of more sustainable systems for the future.

Acknowledgments

The author thanks Dr. Paula Bramel-Cox (Kansas State University), Dr. Lenis Nelson, Dr. Wayne Youngquist, and Professor David Andrews (University of Nebraska), Dr. Mohammad Saeed (St. Cloud State University), Dr. Margaret Smith and Dr. Tom Barker (Cornell University), and Dr. Gary Atlin (Biotecnica Canada) for their considered review of the manuscript and relevant suggestions. JoAnn Collins prepared several drafts and the final manuscript.

REFERENCES

Abou-El-Fittouh, H. A. , J. O. Rawlings, and P. A. Miller. (1969). "Classification of environments to control genotype by environment interactions with an application to cotton," *Crop Sci.* **9**:135–40.

Allard, R. W. , and A. D. Bradshaw. (1964). "Implications of genotype-environmental interactions in applied plant breeding," *Crop Sci.* **4**:503–7.

Allen, F. L. , R. E. Comstock, and D. C. Rasmusson. (1978). "Optimal environments for yield testing," *Crop Sci.* **18**:747–51.

Altieri, M. A. , and M. Liebman. (1986). "Insect, weed, and plant disease management in multiple cropping systems." In *Multiple Cropping Systems* (Ed. C. A. Francis). pp. 183–218. Macmillan Publ. Co., New York.

Atlin, G. N. , and K. J. Frey. (1989). "Breeding crop varieties for low-input agriculture," *Amer. J. Alt. Agr.* 4 (in press).

Baker, R. J. (1986). *Selection Indices in Plant Breeding.* CRC Press, Boca Raton, Florida.

Beets, W. C. (1982). *Multiple cropping and Tropical Farming Systems.* Westview Press, Boulder, Colorado.

Bell, A. A. (1982). "Plant pest interaction with environmental stress and breeding for pest resistance: plant diseases." In *Breeding Plants for Less Favorable Environments* (Eds. M. N. Christensen and C. F. Lewis). pp. 335–64. John Wiley & Sons, New York.

Blum, A. (1973). "Components analysis of yield responses to drought of sorghum hybrids," *Exp. Agric.* **9**:159–67.

Blum, A. (1985). "Breeding crop varieties for stress environments," *CRC Crit. Rev. Plant Sci.* **2**:199–238.

Blum, A. (1988). *Plant Breeding for Stress Environments.* CRC Press, Boca Raton, Florida.

Bolaños, J., and G. O. Edmeades. (1988). "The importance of the anthesis-silking interval in breeding for drought tolerance in tropical maize," *Agron. Abstr.* Amer. Soc. Agron., Madison, Wisconsin. pp. 74–75.

Boyd, W. J. R., N. A. Goodchild, W. K. Waterhouse, and B. B. Singh. (1976). "An analysis of climatic environments for plant-breeding purposes," *Aust. J, Agric. Res.* **27**:19–33.

Brakke, J. P. , C. A. Francis, L. A. Nelson, and C. O. Gardner. (1983). "Genotype by cropping system interactions in maize grown in a short season environment," *Crop Sci.* **23**:868–70.

Bramel-Cox, P. (1988). "Resource efficient hybrids and varieties." In *Proc. Sustainable Agriculture in the Midwest* (Eds. C. A. Francis and J. W. King). pp. 21–23. Coop. Extension Service, Univ. Nebraska, Lincoln.

Briggle, L. W., and O. A. Vogel. (1968). "Breeding short-stature, disease resistant wheats in the United States," *Euphytica* Supplement **1**:107–30.

Brun, E. L. , and J. W. Dudley. (1989). "Nitrogen response in the USA and Argentina of corn populations with different proportions of flint and dent germplasm," *Crop Sci.* **29**:565–69.

Castleberry, R. M., C. W. Crum, and C. F. Krull. (1984). "Genetic yield improvement of U.S. cultivars under varying fertility and climatic conditions," *Crop Sci.* **24**:33–36.

Christiansen, M. N. , and C. F. Lewis, editors. (1982). *Breeding Plants for Less Favorable Environments.* John Wiley & Sons, New York.

Clark, R. B. (1982). "Plant response to mineral element toxicity and deficiency." In *Breeding Plants for Less Favorable Environments* (Eds. M. N. Christensen and C. F. Lewis). pp. 71–142. John Wiley & Sons, New York.

Clark, R. B. , J. W. Maranville, and H. J. Gorz. (1978). "Phosphorus efficiency of sorghum grown with limited phosphorus," In *Plant Nutrition 1978* (Eds. A.R. Ferguson, R. L. Bielski, and I. B. Ferguson). Proc. 8th Intl. Colloq. Plant Anal. Fert. Prob., Aukland, New Zealand. pp. 93–99.

Cox, T. S. , J. P. Shroyer, Liu Ben-Hui, R. G. Sears, and T. J. Martin, (1988). "Genetic improvement in agronomic traits of hard red winter wheat cultivars from 1919 to 1987," *Crop Sci.* **28**:756–60.

Devine, T. E. (1982). "Genetic fitting of crops to problem soils." In *Breeding Plants for Less Favorable Environments* (Eds. M. N. Christensen and C. F. Lewis). pp. 143–73. John Wiley & Sons, New York.

Dover, M. J. , and L. M. Talbot. (1988). "Feeding the earth: An agroecological solution," *Technology Review*, February/March, pp. 27–35.

Duke, J. A. (1982). "Plant germplasm resources for breeding of crops adapted to marginal environments." In *Breeding Plants for Less Favorable Environments* (Eds. M. N. Christiansen and C. F. Lewis). pp. 391–433. John Wiley & Sons, New York.

Eberhart, S. A. , and W. A. Russell. (1966). "Stability parameters for comparing varieties," *Crop Sci.* **6**:36–40.

Estrada, R. N. (1978). "Breeding frost-resistant potatoes for the tropical highlands," In *Plant Cold Hardiness and Freezing Stress: Mechanisms and Crop Implications* (Eds. P. H. Li and A. Sakai). pp. 333-41. Academic Press, New York.

Evans, L. T. (1976). "Physiological adaptation to performance as crop plants," Phil. Trans. Roy. Soc. London, B.**275**:71–83.

Evans, L. T. (1980). "The natural history of crop yield," *Amer. Sci.* **68**:388–97.

Fehr, W. R. , editor. (1984). *Genetic Contributions to Yield Gains of Five Major Crop Plants*. Crop Sci. Soc. Amer., Spec. Publ. No. 7, Madison, Wisconsin.

Fehr, W. R. (1987). *Principles of Cultivar Development, Vol. 1*. Macmillan Publ. Co., New York.

Finlay, K. W. , and G. N. Wilkinson. (1963). "The analysis of adaptation in a plant-breeding programme," *Australian J. Agr. Res.* **14**:742–54.

Francis, C. A. (1980). "Developing hybrids of corn and sorghum for future cropping systems," *Proc. Annual Corn Sorghum Res. Conf.* **35**:32–47.

Francis, C. A. (1981). "Development of plant genotypes for multiple cropping systems." In *Plant Breeding Symposium II, 1979* (Ed. K. J. Frey). pp. 179–231. Iowa State Univ. Press, Ames, Iowa.

Francis, C. A. , editor. (1986a). *Multiple cropping systems*. Macmillan Publ. Co., New York.

Francis, C. A. (1986b). "Variety development for multiple cropping systems," *CRC Crit. Rev. Plant Sci.* **3**:133–68.

Francis, C. A. , and M. D. Clegg. (1988). "Crop rotations in sustainable production systems." In *Proc. International Conf. Sustainable Agr. Systems* (Eds. C. Edwards et al.). Soil Water Cons. Soc. Amer., Ames, Iowa. (in press).

Francis, C. A. , C. A. Flor, and S. R. Temple. (1976). "Adapting varieties for intercropping systems in the tropics." In *Multiple Cropping* (Eds. R. I. Papendick, P. A. Sanchez, and G. B. Triplett). Spec. Publ. 27, Amer. Soc. Agron., Madison, Wisconsin.

Francis, C. A. , R. S. Moomaw, J. F. Rajewski, and Mohammad Saeed. (1986). "Grain sorghum hybrid interactions with tillage systems and planting dates," *Crop Sci.* **26**:191–193.

Francis, C. A. , and J. H. Sanders. (1978). "Economic analysis of bean and maize systems: Monoculture versus associated cropping," *Field Crops Res.* **1**:319–35.

Frey, K. J. (1964). "Adaptation reaction of oat strains selected under stress and nonstress environmental conditions," *Crop Sci.* **4**:55–58.

Furlani, A. M. C. , R. B. Clark, W. M. Ross, and J. W. Maranville. (1978). "Differential phosphorus uptake, distribution, and efficiency by sorghum inbred parents and their hybrids." In *Genetic Aspects of Plant Mineral Nutrition* (Eds. H. W. Gabelman and B. C. Loughman). pp. 287–98. Martinus Nijhoff Publ., Dordrecht, Netherlands.

Ghaderi, A., E. H. Everson, and C. E. Cress. (1980). "Classification of environments and genotypes in wheat," *Crop Sci.* **20**:707–10.

Gliessman, S. R. (1983). "Allelopathic interactions in crop/weed mixtures: applications for weed management," *J. Chem. Ecol.* **9**:991–99.

Gomm, F. B. , F. A. Sneva, and R. J. Lorenz. (1976). "Multiple cropping in the Western United States." In *Multiple Cropping* (Eds. R. I. Papendick, P. A. Sanchez, and G. B. Triplett). pp. 103–15. Amer. Soc. Agron. Spec. Publ. 27, Madison, Wisconsin.

Goodman, D. (1975). "The theory of diversity-stability relationships in ecology," *Quarterly Rev. Biol.* **50**:237–66.

Gorman, C. F. (1969). "Hoabinhian: A pebble complex with early plant associations in southeast Asia," *Science* **163**:671–73.

Gotoh, K., and T. T. Chang. (1979). "Crop adaptation." In *Plant Breeding Perspectives* (Eds. J. Sneep, A. J. T. Hendriksen, and O. Holbek). pp. 234–60. Centre Agr. Publ. and Doc., Wageningen, Netherlands.

Gotoh, K., and I. Fujimori. (1960). "Rank correlation analysis of local adaptability in soybeans," *Japanese J. Breed.* **10**: 272.

Gotoh, K., and S. Osanai. (1959). "Efficiency of selection for yield under different fertilizer levels in a wheat cross," *Japan. J. Breed.* **9**:173–78.

Grafius, J. E. (1981). "Breeding for winter hardiness." In *Analysis and Improvement of Plant Cold Hardiness* (Eds. C. R. Olien and M. N. Smith). CRC Press, West Palm Beach, Florida.

Guitard, A. A. (1960). "The use of diallel correlations for determining the relative locational performance of varieties of barley," *Can. J. Plant Sci.* **40**:645–51.

Hallauer, A. R. , and T. S. Colvin. (1985). "Corn hybrids response to four methods of tillage," *Agron. J.* **77**:547–50.

Heinrich, G. M. , C. A. Francis, J. D. Eastin, and Mohammad Saeed. (1985). "Mechanisms of yield stability in sorghum," *Crop Sci.* **25**:1109–13.

Hicks, D. R. , and R. H. Peterson. (1981). "Effect of corn variety and soybean rotation on corn yield," *Proc. 36th Ann. Corn and Sorghum Res. Conf.*, Amer. Seed Trade Assoc., Washington, DC, **36**:84–93.

Hoffer, G. N. (1926). "Some differences in the functioning of selfed lines of corn under varying nutritional conditions," *J. Amer. Soc. Agron.* **18**:322–34.

Horner, T. W. , and K. J. Frey. (1957). "Methods for determining natural areas for oat varietal recommendations," *Agron. J.* **49**:313–15.

Hurd, E. A. (1971). "Can we breed for drought resistance?" In *Drought Injury and Resistance in Crops* (Eds. K. L. Larson and J. D. Eastin). pp. 77–88. Crop Sci. Soc. Amer. Special Publ. no. 2, Madison, Wisconsin.

ICRISAT (International Crops Research Institute for the Semi-Arid Tropics). (1985). *Annual Report, 1984*. Patancheru P.O., Andhra Pradesh, India.

ICRISAT (International Crops Research Institute for the Semi-Arid Tropics). (1986). *Annual Report, 1985*. Patancheru P.O., Andhra Pradesh, India.

Jenkins, J. N. (1982). "Plant pest interactions with environmental stress and breeding for pest resistance: Insects." In *Breeding Plants for Less Favorable Environments* (Eds. M. N. Christensen and C. F. Lewis). pp. 365–74. John Wiley & Sons, New York.

Jensen, N. F. (1952). "Intravarietal diversification in oat breeding," *Agron. J.* **44**: 30–34.

Jensen, N. F. (1988). *Plant Breeding Methodology*. John Wiley and Sons, New York.

Johnson, G. R. , and K. J. Frey. (1967). "Heritabilities of quantitative attributes of oats (*Avena* sp.) at varying levels of environmental stress," *Crop Sci.* **7**:43–46.

Jones, G. L. , D. F. Matzinger, and W. K. Collins. (1960). "A comparison of flue-cured tobacco varieties repeated over locations and years with implications on optimum plot allocation," *Agron. J.* **52**:195–99.

Kerr, W. E. , and Posey, D. A. (1984). "Informacoes adicionais sobre a agricultura dos Kayapo," *Interciencia* **9(6)**:392–400.

Konishi, T., and H. Sugishima. (1964). "The nature of regional differences of barley varieties responsible for heading time," *Bull. Kyushu Agr. Exp. Station* **10**:1–10.

Krizek, D. T. (1982). "Plant response to atmospheric stress caused by waterlogging." In *Breeding Plants for Less Favorable Environments* (Eds. M. N. Christensen and C. F. Lewis). pp. 293–334. John Wiley & Sons, New York.

Krull, C. F. , I. Narvaez, N. E. Borlaug, J. Ortega, G. Vasquez, R. Rodriguez, and C. Meza. (1966). *Results of the Third Near East–American Spring Wheat Yield Nursery, 1963–1965*. Inter. Maize Wheat Impr. Center Res. Bull. 5, Mexico 6, D. F.

Larter, E. N. (1973). "A look at yield trends in Triticale." In *Wheat, Triticale, and Barley Seminar* (Ed. R. G. Anderson). pp. 215–20. CIMMYT, Mexico 6, D.F.

Lerner, I. M. (1954). *Genetic homeostasis*. Oliver and Boyd, London.

Loomis, R. S. (1984). "Traditional agriculture in America," *Ann. Rev. Ecol. Syst.* **15**:449–78.

Maranville, J. W. , R. B. Clark, and W. M. Ross. (1980). "Nitrogen efficiency in grain sorghum," *J. Plant Nutrition* **2**:577–89.

Marshall, H. G. (1982). "Breeding for tolerance to heat and cold." In *Breeding Plants for Less Favorable Environments* (Eds. M. N. Christensen and C. F. Lewis). pp. 47–70. John Wiley & Sons, New York.

Martineau, J. R. , J. H. Williams, and J. E. Specht. (1979). "Temperature tolerance in Soybeans. II. Evaluation of segregating populations for thermostability," *Crop Sci.* **19**:79–81.

Mason, H. L. (1983). *Evaluation of corn hybrids for no-till farming*. Unpublished M. S. Thesis, Library, Univ. of Missouri, Columbia, Missouri.

McDaniel, R. G. (1982). "The physiology of temperature effects on plants." In *Breeding Plants for Less Favorable Environments* (Eds. Christiansen, M. N., and C. F. Lewis). pp. 13–45. John Wiley & Sons, New York.

Mederski, H. J. , and D. L. Jeffers. (1973). "Yield responses of soybean varieties grown at two soil moisture stress levels," *Agron. J.* **65**:410–12.

Mock, J. J. (1982). "Breeding corn for no-till farming," *Proc. 37th Annual Corn Sorghum Res. Conf.* **37**:103–17.

Mock, J. J. , and D. C. Erbach. (1977). "Influence of conservation tillage environments on growth and productivity of corn," *Agron. J.* **69**:337–40.

Moll, R. H. , E. J. Kamprath, and W. A. Jackson. (1982). "Analysis and interpretation of factors which contribute to efficiency of nitrogen utilization," *Agron. J.* **74**:562–64.

Moll, R. H. , E. J. Kamprath, and W. A. Jackson. (1982). "The potential for genetic improvement in nitrogen use efficiency in maize," *Proc. 37th Ann. Corn Sorghum Res. Conf.* **37**:163–75.

Muruli, B. I. , and G. M. Paulsen. (1981). "Improvement of nitrogen use efficiency and its relationship to other traits in maize," *Maydica* **26**:63–73.

Newhouse, K. E. (1985). "Selection and performance of corn hybrids under different tillage systems," *Proc. 40th Ann. Corn Sorghum Res. Conf.*, Amer. Seed Trade Assoc., **40**:90–107

Newhouse, K. E. , and T. M. Crosbie. (1987). "Genotype by tillage interactions of S1 lines from two maize synthetics," *Crop Sci.* **27**:440–45.

Norlyn, J. D. (1980). "Breeding salt tolerant crop plants." In *Genetic Engineering of Osmoregulation: Impact on Plant Productivity for Food, Chemical, and Energy* (Eds. D. W. Raines, R. C. Valentine, and A. Hollaender). pp. 381– . Plenum Press, New York.

Perkens, D. H. (1969). *Agricultural Development in China*. Aldine Publ. Co., Hawthorne, New York.

Plucknett, D. L. , and N. J. H. Smith. (1986). "Historical perspectives on multiple cropping." In *Multiple Cropping Systems* (Ed. C. A. Francis). pp. 20–39. Macmillan Publ. Co. , New York.

Posey, D. (1983). "Indigenous knowledge and development: An ideological bridge to the future," *Cienc. Cult. Sao Paolo* **35**:877–94.

Quisenberry, J. E. (1982). "Breeding for drought resistance and plant water use efficiency." In *Breeding Plants for Less Favorable Environments* (Eds. M. N. Christiansen and C. F. Lewis). pp. 193–212. John Wiley & Sons, New York.

Rao, M. R. , and R. W. Willey. (1980). "Preliminary studies of intercropping combinations based on pigeonpea or sorghum," *Expl. Agric.* **16**:29–39.

Reinert, R. A. , H. E. Heggestad, and W.W. Heck. (1982). "Plant response and genetic modification of plants for tolerance to air pollutants." In *Breeding Plants for Less Favorable Environments* (Eds. M. N. Christensen and C. F. Lewis). pp. 259–92. John Wiley & Sons, New York.

Rosielle, A. A. , and J. Hamblin. (1981). "Theoretical aspects of selection for yield in stress and nonstress environments," *Crop Sci.* **21**:943–46.

Rush, C. W. , and E. Epstein. (1981). "Breeding and selection for salt tolerance by the incorporation of wild germplasm into domestic tomato," *J. Amer. Soc. Hort. Sci.* **106**:699–704.

Saeed, Mohammad, and C. A. Francis. (1983). "Yield stability in relation to maturity in grain sorghum," *Crop Sci.* **23**:683–87.

Saeed, Mohammad, C. A. Francis, and J. F. Rajewski. (1984). "Maturity effects on genotype × environment interactions in grain sorghum," *Agron. J.* **76**:55–58.

Saeed, Mohammad, C. A. Francis, and M. D. Clegg. (1986). "Yield component analysis in grain sorghum," *Crop Sci.* **26**:346–51.

Sasser, J. N. (1982). "Plant pest interactions with environmental stress and breeding for pest resistance: Nematodes." *Breeding Crops for Less Favorable Environments* (Eds. M. N. Christensen and C. F. Lewis). pp. 375–90. John Wiley & Sons, New York.

Shabana R., T. Bailey, and K. J. Frey. (1980). "Production traits of oats selected under low, medium, and high productivity," *Crop Sci.* **20**:739–44.

Smith, M. E. , and C. A. Francis. (1986). "Breeding for multiple cropping systems." In *Multiple Cropping Systems* (Ed. C. A. Francis). pp. 219–49. Macmillan Publ. Co., New York.

Specht, J. E. , J. H. Williams, and C. J. Weidenbenner. (1986). "Differential responses of soybean genotypes subjected to a seasonal soil water gradient," *Crop Sci.* **26**:922–34.

Sprague, G. F. , and W. T. Federer. (1951). "A comparison of variance components in corn yield trials," *Agron. J.* **43**:535–41.

Sullivan, C. Y. , N. V. Norcio, and J. D. Eastin. (1977). "Plant responses to high temperatures." In *Genetic Diversity in Plants* (Eds. A. Muhammed, R. Aksel, and R. C. von Borstel). Plenum Publ. Co., New York.

Suneson, C. A. , and H. G. Marshall. (1967). "Cold resistance in wild oats," *Crop Sci.* **7**:667–68.

Vela-Cardenas, M., and K. J. Frey. (1972). *Iowa State J. Sci.* **46**:381–94.

Wright, H. E. (1976). "The environmental setting for plant domestication in the Near East," *Science* **194**:385–89.

Zweifel, T. R. , J. W. Maranville, W. M. Ross, and R. B. Clark. (1987). "Nitrogen fertility and irrigation influence on grain sorghum nitrogen efficiency," *Agron. J.* **79**:419–22.

3 DESIGN OF PEST MANAGEMENT SYSTEMS FOR SUSTAINABLE AGRICULTURE

GEORGE W. BIRD

Department of Entomology,
and Department of Botany & Plant Pathology,
Michigan State University,
East Lansing, Michigan

TOM EDENS

Department of Resource Development and Entomology,
Michigan State University,
East Lansing, Michigan

FRANK DRUMMOND and ELEANOR GRODEN

Department of Entomology,
University of Maine,
Orono, Maine

As pests, insects and plant pathogens interact with humans in their quest for food, fiber, shelter, or space. Other pests function as vectors of disease-causing organisms or as nuisance organisms in relation to human comfort or welfare. The detrimental impact of pests can be associated with all

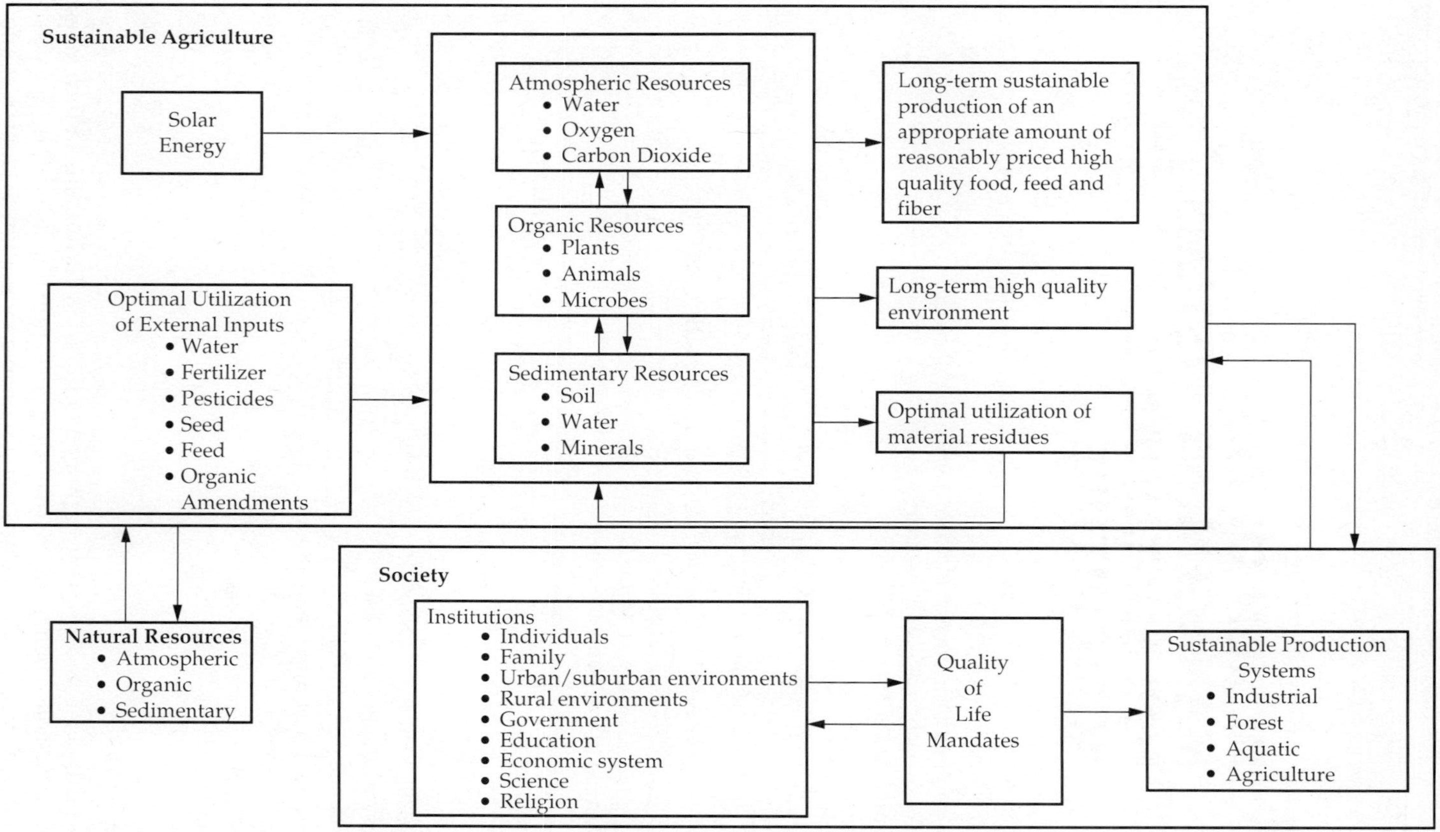

Figure 3.1 Conceptual model of sustainable agriculture in relation to society and natural resources.

components of society, including urban, suburban, and rural environments, as well as industrial, agricultural, forest, and aquatic systems. The objective of this chapter is to present an overview of the history, ecology, and management of detrimental insects, nematodes, bacteria, fungi, and viruses as they relate to the design of sustainable agricultural systems.

Agriculture is an essential system designed for the production of food, feed, and fiber and using solar energy and other atmospheric, organic, and sedimentary resources. A thorough understanding of the relationships between our institutions, production systems, and natural resources and the principles of ecology is necessary for the design and management of sustainable agriculture. A three-tier model of the interactions among institutions, production systems, and natural resources will be used in this chapter as the basis for developing a philosophy for the design of pest management systems for sustainable agriculture (Figure 3.1).

As organic resources, pests exist as individuals with population, community, and ecosystem components of the biosphere (Figure 3.2). Most organisms designated as pests are also beneficial components of energy-cycling food chains (Figure 3.3). At the beginning of each food chain is the primary producer (green plant) capturing solar energy. The energy

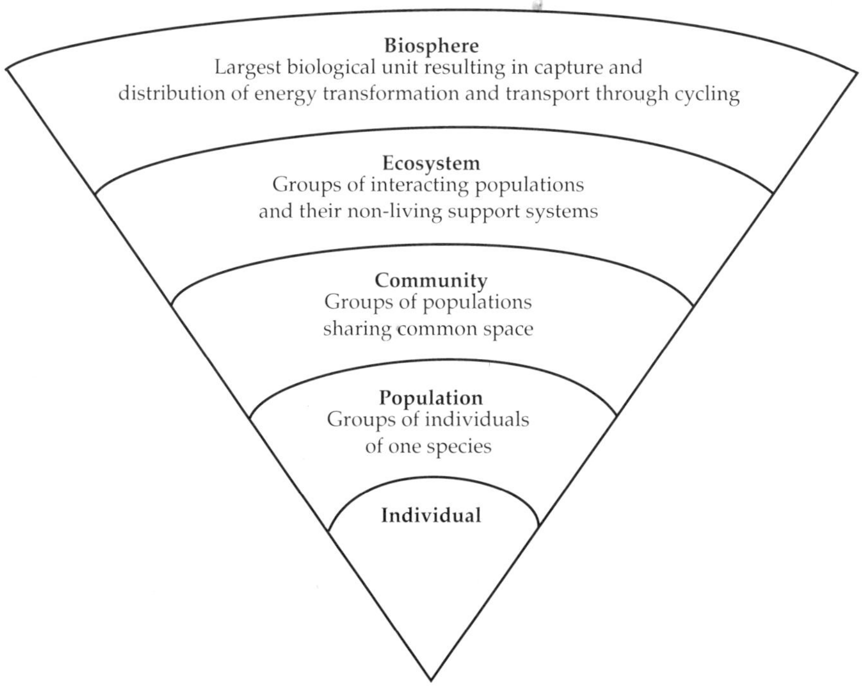

Figure 3.2 Tropic levels of organic resources and their nonliving environment.

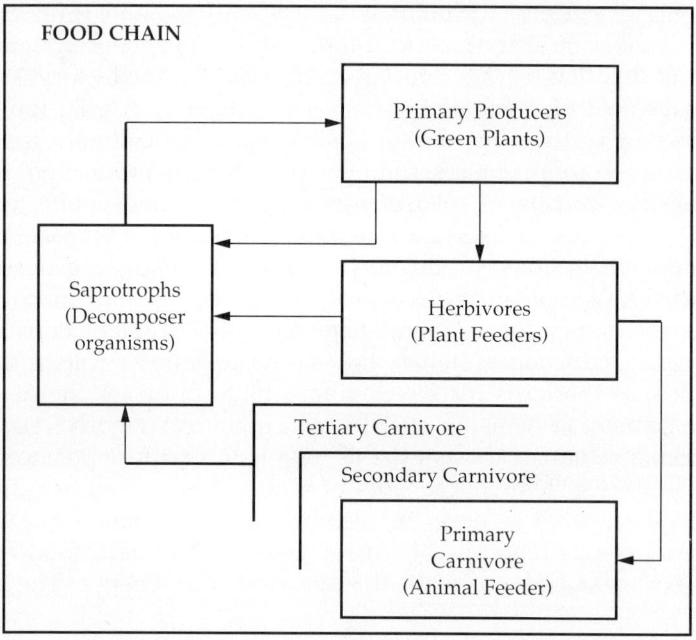

Figure 3.3 Illustration of energy flow through the organic resources of a food chain.

is transferred through the food chain to heterotrophs, which may be herbivores or carnivores, and finally to saprotrophic organisms. All of these ecological categories contain organisms that are designated as pests under specific environmental conditions. Diversity of feeding habitats has a highly significant impact on the potential of an organism to function as a pest and on the management procedures available for use in agriculture. Both herbivores and carnivores exist in food webs as monophagous or polyphagous organisms.

Sustainable agriculture can be defined broadly as the production and distribution of food, feed, or fiber in systems that are socially acceptable, agronomically viable, politically tractable, and environmentally sound over an intergenerational time horizon (Figure 3.1). The structure of a sustainable system, however, is not easily portrayed by standard parameters. Sustainability must be considered as a process toward or away from some current state. Indices describing sustainability must be dynamic and establish a standard for judging the temporal nature of the system.

The chapter is presented in eight sections: this introduction, a section on insects, pathogens and pest management, descriptions of the four eras of pest management, an introduction to systems science, and a prognosis. Ten case studies are used to illustrate the philosophies and procedures of the presynthetic pesticide, synthetic pesticide, integrated pest management, and sustainable agriculture eras of pest management.

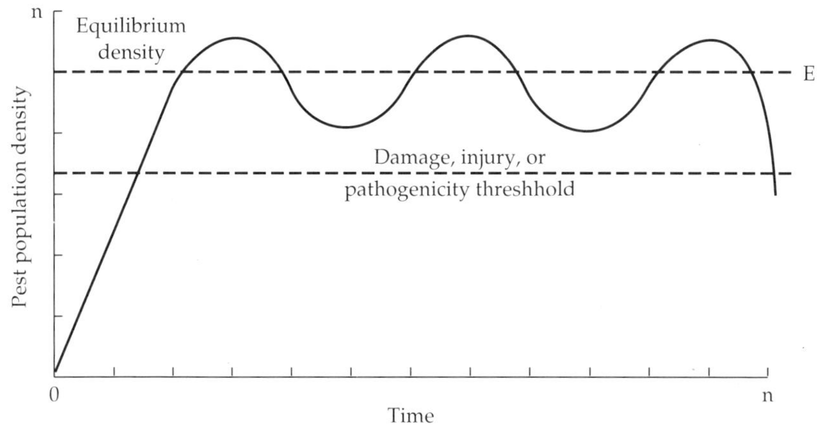

Figure 3.4 Pest population dynamics in relation to its equilibrium density and damage, injury or pathogenicity threshold.

INSECTS, PLANT PATHOGENS, AND PEST MANAGEMENT

Insects and plant pathogens that have significant detrimental impacts on agricultural production systems are considered pests. They are usually classified in one of the following four groups:

- **Key Pests**. Organisms that are commonly associated with a specific agricultural system and inflict severe commodity losses on the system (e.g., Colorado potato beetle, apple scab fungus, corn rootworm, soybean cyst nematode).
- **Minor Pests**. Organisms that are commonly associated with a system, but that cause relatively minor losses (e.g., European red mites).
- **Occasional Pests**. Organisms that periodically result in significant damage (e.g., Mexican bean beetle, armyworm, grasshoppers, late blight fungus).
- **Potential Pests**. Organisms that have the potential to be pests, but are either not present in the location of concern or are limited in some manner by the associated environment (e.g., corn cyst nematode).

In most situations the potential risk associated with a pest is related to its population density, host susceptibility, and environment. Pest population dynamics are limited by various factors such as temperature, water, light, food, shelter, space, parasites, predators, and disease. Under specific environmental conditions, the population density of an organism can increase to a level where the agricultural crop or animal is impacted detrimentally. This density is called the damage threshold, injury threshold, or pathogenicity threshold (Figure 3.4). The population density may also reach an equilibrium. These levels are distinctly different from the action and economic thresholds that will be discussed in the integrated pest management section of this chapter.

Insect and plant pathogen management consists of strategies and tactics called pest management. These are designed to improve quality of life by preventing organisms from becoming problems or by modifying an environment in a manner designed to alleviate pest problems. For the purpose of this chapter, agricultural pest management will be divided into presynthetic pesticide, synthetic pesticide, integrated pest management, and sustainable agriculture eras. This is an artificial classification, but it is useful for illustrating the shifts in philosophy and tools used in pest management during this evolutionary time frame of U.S. agriculture.

Four basic pest management strategies are available. These include excluding the organism from a specific area (problem avoidance), eradicating the target insect or plant pathogen or containing it in a specific area, controlling the target organism through population manipulation, and intentionally deciding to take no action. Seven general pest management tactics are available for use in the first three strategies. These include agroecosystem design, mechanical, biological, chemical, genetic, cultural, and regulatory procedures. They are used to manipulate the pest or the host, or to control the host-parasite relationship.

The history of pest management illustrates that change need not imply continuity and that scientific as well as social evolution is circular and cyclical rather than unidirectional and linear. We have all sensed a bit of deja vu while witnessing new breakthroughs and shifting trends. Just as a broad group of international development specialists embraced a "stages of growth" notion of economic development—in other words, ". . . that there is merely a 'time lag' between developed and underdeveloped countries" (Myrdal, 1972)—modern technologists continue to argue passionately (if not persuasively) for an orderly sort of technological determinism. Thus, we hear arguments (and excuses for inaction) based on probable technical fix scenarios that purportedly obviate the need for direct or indirect action. It is argued, for example, that resistance need not be cause for undue concern because we will be able to develop new pesticides in time. The technical solution to groundwater problems is to filter and purify water at the user end rather than to reduce or eliminate the sources of contamination. The current level of sustainability of temperate food systems is largely a result of the use and misuse of technology in the broadest sense.

We've all experienced the feeling of having been here before or heard the saying, "The more things change, the more they stay the same." Deep down, there is the feeling that we have not learned much from our prior experiences. The Irish potato famine and the Southern corn leaf blight epidemic serve as reminders of overdependence on narrow varietal germplasm. Yet we continue to encourage and reward, directly or indirectly, practices and behaviors that perpetuate and intensify our dependence on narrowly structured systems.

PRESYNTHETIC PESTICIDE ERA

Pest control procedures in the presynthetic presticide era were diverse. Sulfur was used for controlling insects and mites as early as 2500 B.C. Cultural procedures and various forms of habitat modification were part of early pest control programs. Biological control was used in citrus orchards in China in 307 A.D. Botanical pesticides such as pyrethrin and other toxicants including arsenic, mercury, Paris Green, and Bordeaux mixture were discovered and used in Europe during the late 18th and 19th centuries. The vedalia beetle was imported from Australia to the United States in 1888 to control cottony cushion scale of citrus in California. Some presynthetic pest control tactics were very successful, while others left much room for improvement.

Between 1850 and 1925, scientists working with agricultural pests identified new pest problems, developed improved pest management strategies, and discovered basic principles that have served as catalysts for important developments in other areas of science and technology, including human medicine. The Farmers Bulletin on root-knot nematode management, published by E.A. Bessey in 1911, illustrates that a truly integrated approach to pest management was available shortly after the turn of the century.

SYNTHETIC PESTICIDE ERA

The evolution of American agriculture and the U.S. food system is perceived primarily in the context of documented gains in increased substitution of capital for labor and the associated productivity of labor and expansion of total product. It is frequently overlooked that agricultural output and pest management are closely tied to the availability and cost of fossil fuel inputs and their derivatives (Edens and Koenig, 1980). The post–World War II chemotechnology revolution resulted in a vast array of inexpensive pesticides for use in agriculture. These included insecticides, acaricides, fungicides, herbicides, nematicides, bactericides, and rodenticides. Both weed science and the science of nematology evolved as a direct result of the development of this technology. During the past 40 years, numerous types of chemicals were formulated and used as pesticides in many different ways (Meister, 1986).

The economics of scale (size) evident in post–World War II American agriculture were made possible, and probably were necessitated, by cheap and abundant supplies of natural gas and petroleum. Some of the direct spinoffs of high-energy agriculture include increased specialization of the production process, reduced heterogeneity of cropping systems, and the associated decline in redundancy of the system. This resulted in a decrease in the resiliency of the system to perturbations and the movement toward larger production units.

Pest management practices are closely related to prevailing agricultural technology, which is, in turn, determined by the cost and availability of existing energy inputs. The availability of inexpensive high-energy technology led to the development of unique organic components of agriculture that would likely be vastly different in more labor-intensive and diversified systems. The concepts of agricultural pest management during the synthetic pesticide era will be illustrated through the use of four case studies: (1) U.S. agriculture, (2) Michigan agriculture, (3) Onion production, and (4) Colorado potato beetle.

Case Study 1: U.S. Agriculture

Energy inputs to the U.S. food system increased more than threefold from 1940 to 1970 (Steinhart and Steinhart, 1974) and fifteenfold from 1700 to 1986 (Pimentel and Pimentel, 1986). Labor-intensive practices on the farm have been replaced by energy-intensive mechanized production (Pimentel et al., 1973; Bender, 1984). From 1945 to 1972, human labor on farms decreased more than 40% (Edens and Haynes, 1982). Economies of scale favored the trend toward large specialized production systems, and this mode of production was reinforced by inexpensive transportation and food-processing facilities (Hightower, 1976; Berry, 1978). Shifts in agriculture resulted in regional specialization, with consequent changes in the structure of farms and agricultural research institutions (de Janvry and LeVeen, 1986). Edens and Koenig (1980) indicated that these changes dictate a need for critically evaluating the very systems of production and management that have been the basis of our agriculture.

In many crops, pesticides were used infrequently before 1945. In field corn production, for example, an average of only 0.1 kg of insecticide was applied per hectare in 1950. By 1983, the amount of insecticide applied per hectare increased to 3.0 kg—a thirtyfold increase (Pimentel and Pimentel, 1986). The development and use of synthetic organic insecticides resulted in a major change in agricultural production and philosophy (Flint and van den Bosch, 1983). Although firm-level cost:benefit ratios used to justify pesticide use in agriculture were such that every dollar invested in a chemical resulted in $4 to $5 of increased yields, externalities were not included in the analysis (Metcalf, 1982). In 1973, Jones reported that between 1927 and 1957, insect management research increasingly focused upon insecticide testing and sharply decreased in the area of insect ecology. Over 2 million metric tons of active ingredients of pesticides are currently applied annually worldwide (79% in the northern temperate climates). This is more than double the amount applied a decade ago (Hill, 1984). In effect, the evolution of farming system structures was brought on by the availability of inexpensive fossil fuels and their resulting technology.

The genetic diversity within many crops is limited (Wright, 1984; Kloppenburg and Kleinman, 1987). In the United States, 53% of the corn produced comes from three major lines, and 72% of the potatoes represent four cultivars (Merrill, 1976). The lack of genetic diversity is similar in cereal crops. As early as 1939, 85% of the commercial barley seed in the United Kingdom was derived from four individual plants (Potts, 1977). As farmers rely more on a few productive varieties, local and regional strains have been abandoned (Wright, 1984). A similar situation exists in dairy, livestock, and poultry husbandry (Cox and Atkins, 1979).

The structural changes accompanying the decrease of crop diversity have had a significant effect upon the farm as an ecosystem and increased the incidence of pest problems. A major difference between the agriculture of Europeans colonizing North America in the 1600s and native Indian agriculture was the greater degree of monoculture associated with the Euro-American agriculture. Monoculture reduces ecological diversity, and pest outbreaks were a common attribute of Euro-American agriculture. Native Indian fields were both small and diverse, and they were burnt regularly in an effort to manage diseases or insect pests (Cronon, 1983).

Polycultural cropping practices can produce food on an equal basis to monoculture and spread the risk of major economic loss (Gliessman, 1984). Risch (1981) suggests that concentration of a food resource, such as a monoculture, is attractive to herbivores and influences their behavior such that they are less likely to emigrate out of a monoculture than a diversified habitat resulting in severe pest outbreaks. This is less likely to occur in polycultures. Andow (1983) reviewed 150 studies to determine whether resource concentration or increased numbers of natural enemies is a major reason that increased on-farm crop diversity may result in a reduction in pest problems. He noted that polyphagous insects increased more than monophagous ones, and that more insect pests increased in perennial systems than in annual systems. From his analysis of the published literature, Andow concluded that the phenomenon of resource concentration has a major impact on reducing pest numbers, rather than an increase in natural enemies. In the 150 studies, Andow observed 36 predator species; 47% increased in response to increasing diversity, 17% decreased, 8% showed no change, and 28% showed a variable response. In the more diverse systems, 22 of 28 parasite species increased.

The distribution of natural enemies is not solely dependent on the distribution of target pests. Simplifying agroecosystems to large-scale monocultures has reduced the diversity of habitats and eliminated many habitats essential for natural enemies to complete their life cycles. To encourage the role of natural enemies in the suppression of pest populations, an understanding of factors influencing their distribution and abundance is essential. Pimentel (1981) concludes that a higher pest abundance in a monoculture brassica crop in relation to a mixed brassica stand is due to the camouflaging effect of weeds in mixed plantings.

Abiotic characteristics of habitats influence preferences and survival of natural enemies. Townes (1958) stated that moisture is an important requirement of Ichneumonid wasp parasitoids and, in many cases, its scarcity is a limiting factor in cultivated areas. Weseloh (1978, 1979a, 1979b) found that the spatial distributions of different gypsy moth parasitoids and hyperparasitoids varied with the microhabitats of the host and that, in some cases, these distributions are related to the relative humidity and temperature preferences of the parasitoids and hyperparasitoids. Merritt and Anderson (1977) also found relative humidity to be the determining factor in the abundance of some hymenopteran parasitoids in dung habitats. The fungal disease of the Colorado potato beetle, caused by *Beauveria bassiana* (Bals) Vuill., is limited by low relative humidity (Ferron, 1978), ultraviolet light (Toumanoff, 1933), and soil type and soil pH (Groden, 1988). Ground cover provided by weeds and intercropped plants can alter the microhabitat of crop plants and provide refuge for general predators (Altieri, 1987). Carabid (Dempster, 1969; Dempster and Coaker, 1974) and spider (Litsinger and Moody, 1976) populations have been effectively increased with ground cover.

Parasitoid distributions are strongly influenced by the availability of food resources for the adult parasitoids. Parasitoid success has been linked to the presence of a nectar or pollen source. The carbohydrate source has been shown to increase both the longevity and fecundity of certain species (Lewis, 1961; Lewis, 1963; Symes, 1975).

The effectiveness of some natural enemies in controlling pest populations depends on the proximity of habitats that support alternate hosts. The eggs of the grape leafhopper, *Erythroneura elegantula* Osborn, are parasitized by the mymarid wasp, *Anagrus epos* Girault. Though this host undergoes reproductive diapause during the fall and winter months, the parasitoid does not; instead, it continues to reproduce on a non-economically important leafhopper, *Dikella cruentata* (Gillette), on wild and cultivated blackberries, *Rubus* spp. (Doutt and Nakata, 1972). A flush of new growth of *Rubus* occurs in February, which stimulates *Dikella* females to oviposit. *Anagrus* responds to this rapid increase in the host population, and their numbers increase. The parasitoids developing in *Dikella* eggs peak in March, emerging in early April. This parasitoid emergence coincides with the initiation of the grape leafhopper oviposition in commercial vineyards. Vineyards within the vicinity of riparian habitats supporting *Rubus* spp. (5.6 km) have more Anagrus activity earlier in the season, resulting in effective control of the grape leafhopper. Establishment of *Rubus* spp. adjacent to orchards has been difficult, however, since the habitat requirements for grapes and *Rubus* spp. are different (Flaherty et al., 1985).

A similar relationship occurs with the tachinid parasitoid, *Lydella grisescens* Robineau-Desvoidy, introduced for the control of the European corn borer *Ostrinia nubilalis* Hubner (Hsiao and Holdaway, 1966). *Lydella* adults emerge in the spring when most of the corn borer larvae are too young to parasitize. The presence of older stalk borers, *Papaipema nebris* Guenter, in ragweed affords a suitable host for these parasitoids during their entire emergence period. The second generation of *Lydella* adults emerge in time to attack the first generation *O. nubilalis* in corn. Cover crops and weeds often supply important alternate hosts for natural enemies of pest insects and nematodes. The grass *Sorghum halepense*(L.) in vineyards serves as a reservoir for hosts of phytoseid mites that control pest mites on grapes (Flaherty et al., 1985). Certain weeds in the cotton agroecosystem sustain alternate hosts for natural enemies of *Heliothis* moths (Smith and Reynolds, 1972).

The resource concentration hypothesis has also been investigated with respect to plant pathogens (Burdon and Chilvers, 1982). Diversity in crop plants on a temporal scale (that is, crop rotation) has long been recognized as a means of reducing crop losses due to fungi, bacteria, and nematodes (Curl, 1963; Nusbaum and Ferris, 1973). Spatial scale diversity in regard to mixtures of crop species and incidence of disease tends to be highly correlated to the major mode of inoculum dispersal (Gibson and Jones, 1977). Some mechanisms of inoculum transmission (for example, autonomous, nematode-borne, and splash dispersal) tend to limit the dispersal to the immediate vicinity of the source of the pest. Such mechanisms favor within-plant transmissions, and thus a region of patchwork small fields would tend to have a lower overall disease incidence than a monocultural region where between-plant and within-plant transmissions would be equivalent (Burdon and Chilvers, 1982).

Other mechanisms of inoculum transmission include wind-dispersing fungal spores and aphid-borne viruses which spread inoculum over a wider area. In these cases, the relative importance of between-plant transmission is significant and crop diversity may not have a negative impact on disease incidence. In general, as with insect pest systems, good empirical studies that demonstrate relationships between crop loss due to disease and crop diversity are not abundant. Simulation studies of airborne dispersal (Waggoner, 1962; van der Plank, 1963; Zadoks and Kampmeijer, 1977) have shown that epidemics arising from primary infection foci due to inoculum arriving from a distance tend to start later and achieve less progress in a system of small fields. However, when disease is initiated from within the field or on the field borders from a reservoir host, large monoculture fields with their greater area to perimeter ratio may be superior in limiting the spread of inoculum.

In recent years, the dangers of the ease of disease spread resulting from genetic uniformity within a crop monoculture has been recognized (Day, 1973; Kloppenburg and Kleinman, 1987). The U.S. corn blight epidemic in 1970 underlines the danger of genetic uniformity in relation to disease

(Wilkes, 1977). Research on diversifying the basis of resistance in crop plants has become a major focus. One such strategy is the use of mixtures of varieties or isolines of the crop species. Little is understood, however, of the resulting selection pressure that these multilines may have on the virulence structure of the pathogen population. Will selected management practices produce super pathogenic populations consisting of high frequencies of many races? In the short term, there have been encouraging results. A fourfold reduction occurred in the number of spores of *Puccinia coronata* trapped over a 1:1 mixture of a susceptible and a resistant variety of oats compared to a monoculture of susceptible oats (Cournoyer, 1970). In a 1:1 mixture of corn, the amount of disease caused by *Helminthosporium maydis* was half that recorded in monoculture stands of the susceptible variety (Summer and Littrell, 1974). These and other studies (Berger, 1975; Burdon and Chilvers, 1982) suggest that multiline mixes can be used to reduce the incidence of disease and its rate of increase.

The examples presented show that diversity per se cannot be used in any kind of an analysis without regard to the specific pest or the specific makeup of the diversity or, in other words, the functional relationships among crop plants, pests, and natural enemies in the agroecosystem. There may not be any specific ecological principles that are universally useful in designing agroecosystems. Specific interactions may have to be evaluated on a case-by-case basis. For instance, increasing the diversity of a southern pine plantation with oak trees increases the disease incidence of fusiform rust of pine caused by *Cronartium fusiforme* (Wakely, 1954). This is because the oak is an alternate host for this obligate heteroecious rust fungus. Densities of biological control agents of insect pests may increase under certain crop mixtures (those plants providing alternate prey or nectar sources for various stages of the predators or parasites), but not under other crop mixtures. As some natural enemies orient to the odors of the pest's host plant for locating the host, crop diversification at a certain scale may interfere with the searching efficiency of the natural enemy (Chandler, 1968; Read et al., 1970). Way (1977) demonstrates how an increase in diversity of cotton agroecosystems can increase the abundance of the bollworm, *Heliothis armigera*.

The evolution of agricultural production systems in response to cheap fossil fuel energy and its products may have had more than just the direct effects of pesiticide mortality on natural enemies. Removal of pesticides alone will probably not ensure natural regulation of pest species. This is because the evolution of Euro-American agriculture broke the linkages of the functional diversity (Haynes et al., 1981). An example illustrating this concept will be presented in Case Study 3: Onion production.

Case Study 2: Michigan Agriculture

Michigan has a diverse agriculture that has changed significantly during the past 50 years. Agricultural land declined from 18 million acres in 1945 to less than 10 million acres in 1986, and from about 200,000 to 50,000 farms. Farm size increased from approximately 100 acres in 1945 to 175 acres in 1980. This is consistent with national trends (Perleman, 1976). If farm diversity is measured by the proportion of farms producing different commodities and by the geographic distribution of a commodity within a state, Michigan has a very diverse agriculture. This diversity is, however, declining. The percentage of Michigan farms that raise cattle fell from 88% in 1920 to only 34% in 1978. Concomitant with the decrease in livestock on farms was the increase in the use of synthetic fertilizer. In 1945, the ratio of synthetic nitrogen fertilizer to manure used in U.S. corn production was 0.7, and by 1983, this ratio increased to 3.7 (Pimentel and Pimentel, 1986). Michigan potato production declined in both the percentage of Michigan farms growing potatoes (80% in 1940 to 3% in 1985) and the total acreage (325,000 acres in 1940 to 50,000 acres in 1985). Although Michigan has 15 different commercial potato growing regions, 20% of the current potato acreage is concentrated in one county, and 56% is in 8 of 83 counties. Vegetables have always been a very important Michigan industry, but they have been limited to a small percentage of total farm acreage. There has been a 50% decline in farms growing vegetables commercially between 1940 and 1980, but regional specialization of vegetable crops is still evident. In 1980, 50% of Michigan tomato acreage was grown in one county and 70% of Michigan's onion crop was grown in six counties. A measure of relative crop diversity (calculated by deriving an average of the proportion of farms producing each crop commodity in Michigan from 1920 to 1986) indicates that as farms have become larger, the on-farm crop diversity has decreased.

The impact of 200,000 acres of soybeans or 8 million grapevines concentrated in two adjacent counties on the pest status of insects and pathogens is difficult to assess. However, research has shown that monoculture is indirectly related to insect diversity (Edens and Motyka, 1983). Ecosystem stability is usually considered as a direct result of diversity, because of the multitude of interpretations of stability and diversity, and the large spatial scale across which cropping patterns can exist. There is no definitive study that provides a basis for the principle that, in agroecosystems, diversity reduces economic loss due to specific pests (Dover and Talbot, 1987).

Case Study 3: Onion Production

The trend in onion production in Michigan has been toward large farms. Onions are grown in drained wetlands called muck or organic soils, and crop production is concentrated in 10- to 3,000-acre areas. Until recently (Carruthers, 1981; Whitfield, 1981; Groden, 1982; Drummond, 1982; Carruthers et al., 1985) natural enemies in Michigan's onion agroecosystem were not recognized as important in the control of the major insect pest, the onion maggot, *Delia antiqua* (Meigen). In 1980, the management of pests in the Michigan onion production system was in the crisis phase:

> After a variable number of years in the exploitation phase and the heavy use of insecticides, a series of events occurs. More frequent applications of pesticides and higher dosages are needed to obtain effective control. Insect populations often resurge rapidly after treatments, and the pest population gradually becomes tolerant to the pesticide. Another pesticide is substituted and the pest population becomes tolerant to it too (Luckmann and Metcalf, 1982).

To control onion maggot, granular soil insecticides were applied at planting, followed by 10 to 18 foliar insecticide treatments. Since the 1930s, there has been a constant replacement of insecticides, not because of more efficient chemicals, but because of the development of insecticide resistance in the onion maggot (Haynes et al., 1981). Other pests in the onion agroecosystem exhibit similar tendencies. The onion production system fits van den Bosch's (1978) concept of a "pesticide treadmill."

How were onions grown in Michigan before the synthetic pesticide era? Why is it that in the middle of a chemically intensive onion growing area, an organic onion grower using no pesticides can grow a crop? The answer to these questions is that the design of the chemically intensive monoculture does not provide the functional diversity or linkages that maximize the potential of biological and cultural control.

Aphaereta pallipes is a small braconid parasitoid of onion maggot larvae. Carruthers et al.(1985) found that herbicides and insecticides used in onion production cause mortality of these parasitoids. However, even with the elimination of sprays, this parasitoid is not abundant (Groden, 1982). *A. pallipes* has alternate hosts in fresh cow dung. When examining the spatial distribution of this parasite from an active cow pasture adjacent to an organically grown field of onions, it was noted that its relative abundance decreases into the onion field the farther one moves from the cow pasture. *A. pallipes* is also found in greater abundance in field borders compared to the open onion field. Laboratory studies show that the parasitoids live longer if they have access to a nectar source. *A. pallipes* abundance about the border habitat may reflect their necessity to frequent these areas for nectar from flowering plants.

Carruthers (1981) found that grassy field borders are important to the epidemiology of the fungus disease of onion maggot adults caused by *Entomphthora muscae*. When the pathogen matures in infected flies, it alters the flies' behavior such that they seek high resting places. The disease results in the death of the fly, leaving it firmly attached to a high resting place. At night, fungal spores are disseminated from the dead fly. The attachment location is advantageous to spore dispersal and increases the chance that spores will come in contract with a new infection count. Early in the season there are no suitable high attachment sites in onion fields. Grassy borders or other crops are the only habitats for attachment. The borders provide attachment sites for infected flies in areas frequented by healthy flies as they attempt to escape mid-day heat. Maintaining optimal attachment habitats maximizes the spread of the disease and the concomitant control of adult onion maggots.

Aleochara bilineata (rove beetle) is another natural enemy of the onion maggot (Whitfield, 1981; Groden, 1982). As an adult, it preys on onion maggot larvae, and it is a parasitoid of pupae as an immature. In an onion monoculture, the stages of onion maggot egg and larval activity and onion maggot pupation (there are three generations per year) represent brief windows. When forced to exist on the onion maggot as the sole host, there are periods of time when the rove beetle has no prey and disappears. This is very disruptive to the biology of this natural enemy. However, *A. bilineata* also preys upon and parasitizes other insects such as the cabbage maggot (*Delia brassicae*) and seed corn maggot (*Delia platura*). In Michigan, the simultaneous occurrence of these two maggot species is such that they occur in close to peak numbers when the critical onion maggot life stages are at a minimum. In a polyculture of onions and cole crops (which are grown in Michigan on muck soils, but generally in monocultures), there is a continuous food supply for both the adult and parasitoid stages of *A. bilineata*.

The third natural enemy linkage involves the tiger fly, *Coenosia tigrina*, a predatory fly of onion maggot flies. As an adult it captures many species of muscoid flies and is not particularly synchronized with the occurrence of onion maggot flies. Increasing the diversity of other maggot species, as mentioned for *A. bilineata*, should provide a continuous food source for *C. tigrina*, and optimize its predation on onion maggot. The larvae of *C. tigrina* are parasitoids of earthworms, which are found in extremely low densities in commercial onion fields in comparison to diversified fields (Edwards and Lofty, 1972; Drummond et al., 1988).

These examples of natural onion maggot controls are presented to illustrate how regional monocultures, mowing and spraying flowering field borders for aesthetic reasons, and pesticides break specific links in regulation of the population dynamics of the onion fly.

Case Study 4: Colorado Potato Beetle

Another example of a change in philosophy and practice of pest control accompanying the evolutionary change in the structure of U.S. agriculture is control of the Colorado potato beetle. This beetle is the major insect pest of potatoes in the north temperate United States, Canada, Europe, and Asia. It feeds on the foliage of potato plants in both the larval and adult stage, thereby significantly reducing yield. This insect is apparently native to Mexico, where its host plants are the solanaceous weeds *Solanum angustifolium* Mlll. and *S. rostratum* Dunal (Casagrande, 1988). The Colorado potato beetle first moved into Nebraska and Iowa, feeding on its native host plant *S. rostratum*, as settlers moved west planting potatoes. The beetle established itself as a pest on potatoes in 1859 and spread rapidly to the east, reaching the Atlantic seaboard by 1874. During the first years as a pest of potatoes, it decimated the potato crop, and as a result, the price of potatoes rose fourfold (Casagrande, 1985, 1988).

Before the widespread use of Paris Green in 1872, growers were encouraged to use an integrated approach (Riley, 1871). A multipronged strategy was followed:

1. Resistant and tolerant potato varieties were identified, and the planting of varieties attractive to potato beetle such as Russet was recommended in strips around fields (trap crops) of nonpreferred cultivars such as Peach Blow or Early Rose.
2. Hand picking and mechanical picking (a horse-drawn potato beetle picker) were developed in conjunction with reducing the acreage that could be grown under this type of management scheme.
3. Crop rotation (recommending that potatoes not be grown in any given field for more than one year at a time) was used, and individual potato fields were isolated from each other in the region.
4. The conservation of natural enemies was promoted (22 species preying on the Colorado potato beetle had been identified as early as 1872).

Despite these cultural control recommendations by entomologists and Riley's reference to them as the only "true remedy" (1871), Paris Green became the predominant control strategy by the middle 1870s (Casagrande, 1988). It is not hard to see why this came about, given that in the short run a chemical insecticide approach enabled farmers to manage larger acreages and therefore net more *private* profit from potatoes. For about 80 years, this approach appeared successful; but in the 1950s the Colorado potato beetle developed resistance to DDT (Gauthier et al., 1981). In the following 30 years, the Colorado potato beetle developed resistance to 14 additional insecticides, some after only one or two years of use against the pest (Forgash, 1985). In a few potato growing regions in the United States, the Colorado potato beetle is threatening the very existence of the potato production industry.

What appears to be a successful approach to Colorado potato beetle control has resulted in a change in farm structure. Growers and regions specialize in potato production, resulting in a centralized potato industry and short-term gains. The long-term stability of this system is highly suspect. Research on alternative management strategies for the Colorado potato beetle has been neglected as a result of concentrating on a single control tactic. The Colorado potato beetle situation is similar to the case of cotton in Texas (Adkisson et al., 1987). An integration of pest management approaches and tactics appears to be the best approach to long-term sustainable agricultural production. Indeed, the philosophy of integrated pest management evolved as a result of the synthetic pesticide era.

INTEGRATED PEST MANAGEMENT ERA

The publication of Rachel Carson's *Silent Spring* catalyzed a general awareness of potential human health and environmental risks associated with some uses of pesticides. As a result, the Federal Insecticide, Fungicide, and Rodenticide Act was amended in 1972. This changed the orientation of national pest control legislation from consumer protection for pesticide users to environmental protection. Concomitantly, the scientific community began to place increased research emphasis on the use of multiple pest control tactics and on potential environmental and human health hazards associated with pesticides. In addition to the highly significant benefits of the chemotechnology revolution to pest management, at least six unexpected consequences were identified: human health risks, environmental risks, development of pest resistance to pesticides, impacts on nontarget organisms, pest population resurgence, and the development of new pest problems. These factors resulted in the evolution of what has become known as integrated pest management, or IPM.

IPM is recognized as the development, use, and evaluation of pest control procedures that result in favorable socioeconomic and environmental consequences. In a 1979 Presidential Message to Congress, IPM was defined as "a systems approach to reduce pest damage to tolerable levels through a variety of techniques, including predators and parasites, genetically resistant hosts, natural environmental modifications and, when necessary and appropriate, chemical pesticides." Although the development and utilization of IPM is far from complete, it can be conceptualized as a process involving seven components (Figure 3.5).

Biological monitoring is one of the components of IPM. This is frequently referred to as "scouting." Biological monitoring consists of sampling procedures designed to estimate the stages and population densities of both pests and beneficial organisms. It also involves monitoring the stage of development and symptomatology of the associated agricultural commodity (crop or animal). Biological monitoring is a very knowledge-intensive procedure, and requires a highly trained scout. The system manager (farmer) or decision maker is responsible for the current state of the production system, and is hypothetically best suited for scouting. In reality, however, a private sector scout from a pest management association or private consulting firm is usually hired to do the biological monitoring. These individuals are frequently trained in biological monitoring by the state landgrant institutions.

Pest populations and the growth and development of crops are governed by environmental parameters such as air temperature, soil temper-

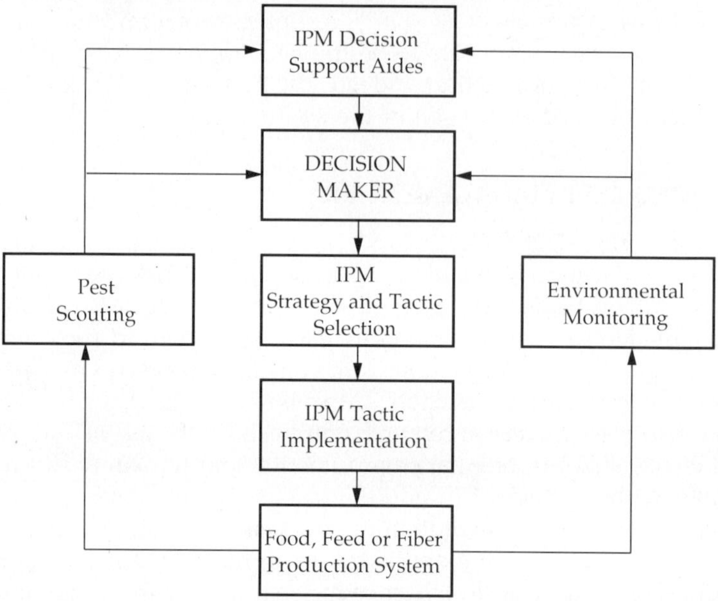

Figure 3.5 Conceptual model of the components and process of IPM (Integrated Pest Management).

ature, soil moisture, light intensity, and relative humidity. Weather monitoring information for IPM must be available on both a regional and local basis. This information must be readily available, user friendly, and as close to real-time as possible. Weather monitoring systems for use in IPM programs have improved greatly during the past decade. In addition to macroclimatological information, firm-level microclimatological data is frequently imperative for predicting diseases caused by fungi and bacteria. Dedicated microcomputers are currently available for use in environmental monitoring in a signficant number of agricultural systems.

Because of the complexity and the large number of potentially significant interactions between pests, beneficial organisms, crops, agricultural animals, and the environment, decision support aides are frequently important aspects of IPM systems. These may be relatively simple look-up tables or computerized systems. A decision support system may be based on simulation models from research data or developed as expert systems. Elementary aspects of artificial intelligence have been incorporated into a few systems.

The production system manager or designated representative is responsible for pest management decisions. This aspect of IPM is also a very knowledge-intensive process. An individual within the specific production enterprise may be assigned the IPM decision-making responsibility, or a private consultant may be hired for this activity. Individuals trained

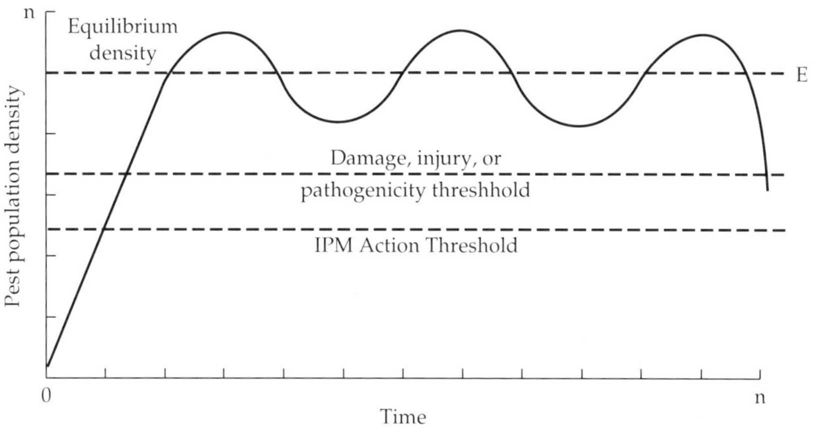

Figure 3.6 Relationships between pest population density, equilibrium density, damage threshold and action threshold.

only in pest scouting should never be delegated the responsibility of IPM strategy or tactic decision making.

The basis of IPM decision making is the threshold concept. The first threshold that needs to be considered is action (Figure 3.6). When the population density of a pest reaches an action threshold, or is predicted to reach this level in the near future, it is time to select appropriate IPM strategies and tactics for implementation. An economic threshold must be estimated when selecting IPM procedures. IPM procedures should usually be implemented when the marginal revenue derived from the management input is equal to or exceeds the marginal cost (Headley, 1975; Ferris, 1978; Elliott, 1980; Shoemaker, 1985; Teng, 1988). The economic threshold is a dynamic concept and depends on the cost and efficacy of the management input, production system economics, nature of the pest and population density, and other environmental parameters. Although IPM strives to reduce environmental and human health risks, the costs associated with these externalities are not usually available for incorporation into the economic threshold. Where the conditions of the economic threshold exist, an appropriate pest management procedure should be implemented. This will usually consist of manipulation of the pest, crop, or animal or regulation of the associated interaction (Figure 3.7).

After implementing an IPM procedure, it is necessary to continue the biological and environmental monitoring to determine if the desired pest management objectives were achieved or if there is need for additional action. IPM has been successfully implemented in a number of important agricultural systems. Although institutional, educational, and social constraints of the past decade have slowed its development, IPM has become a well-established concept. The principles of IPM appear to be ideally suited

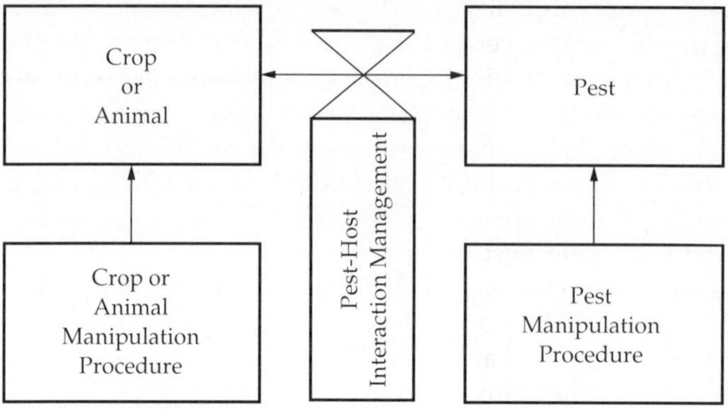

Figure 3.7 Pest management options.

for use in the pest management components of sustainable agricultural systems.

Reviews of IPM philosophy were published by Allen (1980), Frank (1981), Adkisson et al. (1987), Bird (1987b), Burn et al. (1987), Rajotte et al. (1987), and Pedigo (1989). Most IPM programs combine the approaches of efficiency and substitution (Hill, 1984). The efficiency approach is based on biological monitoring of insect pests and the estimated potential for crop loss. Monitoring methods include frequent examination of the crop and the crop environment using various sampling methods (Luckmann and Metcalf, 1982). When pest densities reach an action threshold, a control measure is applied. This is a more efficient use of pesticides than application on a predetermined calendar spray independent of the threat of crop loss due to the pest. Other ways to increase the efficiency of pesticides are better formulation and methods of application (for example, electrostatically charging the spray droplets for increased retention and decreased drift (Dover, 1985) and the use of disease or insect resistant plant varieties (Poincelot, 1986).

The substitution approach involves the replacement of synthetic pesticides with other pest management tactics (Dover, 1985). There has been much research emphasis in the past two decades on tools for substitution both synthetic and biological (Oelhaf, 1978; Matteson et al., 1984; Altieri, 1987). Use of biological control fits into this category. Hoy and Herzog (1985) edited a collection of papers that address the present status of biological control of insect pests, weeds, and plant pathogens in agricultural IPM systems. Much of the current focus of biological control research is in the area of integration of biological control with other management strategies such that the potential of biological control is optimized. An example of this, for which there are many, is increasing the tolerance of predaceous mites in orchard systems to insecticides, thereby giving the predator a competitive edge over the pest mite. This has been approached by genetic selection and may be accomplished in the future through genetic engineering (Beckendorf and Hoy, 1985). Biological control can utilize parasites such as parasitic wasps, certain flies, mites, and nematodes (Luck, 1981; Stehr, 1982; King et al., 1985). Examples of predators of insects are ladybird beetles, ground beetles, rove beetles, lacewing larvae, ants, spiders, birds, and many others (Whitcomb, 1981). Pathogens include bacteria, fungi, viruses, and protozoa (Splittstoesser, 1981; Fuxa, 1987). Introduction of sterile male or genetically unfit members of a pest population has also been used for specific pests (Lorimer, 1981). Biological control of plant-parasitic nematodes (Van Gundy, 1985; Jatala, 1986) and soil-borne and airborne pathogens (Baker and Cook, 1974; Blakeman and Fokkema, 1982; Baker, 1985; Hoitink and Fahy, 1986); is also being researched and utilized in current cropping systems.

Cultural controls often require a change in the production system design; however, sometimes this is not necessary. Planting date used to avoid damage of the Hessian fly in wheat is a good example of a substitution approach; host plant resistance is another (Roberts, 1981; Sailer, 1984; Poincelot, 1986). Other possibilities include safer botanically or microbially based pesticides, hormones, repellants, and oviposition deterrents (Pimental, 1981). Cultural practices have also been used effectively to manage plant disease (Palti, 1981; Jones, 1987; Robinson, 1987). Sanitation, crop rotation, multiple cropping, mulching, tillage, crop nutrition, moisture management, pruning, and host plant resistance are just a few of the many strategies that can be used.

Interactions exist between various cultural practices (Palti, 1981), biological control agents (Stimac and O'Neil, 1985), and a given cultural and biological control agent (Boethel and Eikenbary, 1986). These interactions can be positive, neutral, or negative. An example of a positive interaction is the action of sewage sludge used as a soil amendment (Habicht, 1975) in combination with a suppressive soil (Cook and Baker, 1983). The sewage sludge has a direct toxic effect, reducing the population densities of *Meloidogyne incognita*, but the sludge also increases the growth of competitive or antagonistic microbiota that attack *M. incognita* (Palti, 1981). An example of a negative interaction between a cultural control practice and biological control is described by Obrycki (1986). He shows that while the aphid-resistant potato, *Solanum berthaultii*, reduces aphid population growth due to granular trichomes, there are adverse effects on 11 species of aphidophagous insects.

Global agriculture and agribusiness appear to be changing rapidly. Although the potential benefits of IPM have become widely recognized, and adopted in a limited number of agricultural systems, resources for the design, research, education, and facilitation required for broader adoption have not developed as quickly. In some cases, production system managers and other decision makers have received mixed signals about IPM, and have not invested in the additional educational and support resources required by this knowledge-intensive system. Several of the important success stories in IPM have evolved from crisis and not from the planned change in procedures of education, facilitation, and persuasion. These lessons should be studied in detail as plans are made for future systems of sustainable agriculture. While IPM has been successful, it has not solved the problems of sustainability in agriculture. IPM is usually implemented as a strategy to deal with features of an agricultural system that are the causes of pest problems. IPM makes incremental adjustments to the system's trajectory (Flint and van den Bosch, 1983). Three case studies have been chosen to illustrate applied IPM.

Case Study 5: Alfalfa IPM

The California alfalfa IPM program is a suitable case history to start a survey of IPM programs. Alfalfa is a major crop in California, and represents one of the earliest examples of a holistic approach to pest management (Flint and van den Bosch, 1983). The first success was in managing the alfalfa caterpillar, *Colias eurytheme*. Biological monitoring based on sweep net sampling was used to make management decisions according to an action threshold. If the sweep net samples yielded catches greater than ten healthy caterpillars, a decision was made to apply an insecticide. If more than ten caterpillars per sweep resulted, but the caterpillars were parasitized by the parasitic wasp *Apanteles medicaginis*, or infected by a nuclear polyhydrosis virus, an insecticide was not applied since it had been shown that these natural mortality factors would bring the pest under control (Michelbacher and Smith, 1943).

The introduction of the spotted alfalfa aphid, *Therioaphis trifolii*, in 1954 threatened to destroy the alfalfa industry. Insecticide applications to control this aphid pest decimated ladybird (*Hippodamia* spp.) predators that could bring about natural control. A management strategy was developed to rectify this. First, an insecticide, demeton, was introduced that was not as toxic to parasites and predators as to the aphid. More efficient use of insecticides was made possible by the development of action thresholds based upon predator to pest ratios. Similar thresholds and sampling methodologies were developed for other insect pests of alfalfa. These included the pea aphid, grasshoppers, western yellow-striped armyworm, alfalfa weevil, Egyptian alfalfa weevil, spider mites, and beet armyworm. Second, the introduction of exotic hymenopteran parasites complemented predation of the ladybird beetles. Irrigation scheduling, designed to enhance the incidence of a fungus disease of the spotted alfalfa aphid, and strip harvesting, which provided refugia for natural enemies, were also integrated into the control of this aphid pest.

Control of the alfalfa weevil, *Hypera postica*, in the northeastern United States can be achieved through timed cutting and insecticides, combined with natural control of the weevil by the parasitoid *Bathyplectes curculionis*. A cutting of alfalfa late in the fall makes the field unattractive to colonizing adults. It is possible to time the first spring cutting of the crop so that a majority of weevil larvae are killed. An insecticide is used only if the phenology of the crop and the weevil larvae are not in synchrony. Insecticides can decimate the *B. curculionis* population in the alfalfa field, which often maintains the weevils below the economic threshold. A fungal pathogen *Erynia* sp. of the alfalfa weevil can have dramatic impact on population levels of this pest. Current research is attempting to determine ways of maximizing epizootics through cultural manipulations (Yeargan, 1985).

Another success story of classical biological control is importation of parasitoids for control of the introduced alfalfa pest, the alfalfa blotch leafminer, *Agronmyza frontella*. Hendrickson and Plummer (1983) reported that in Delaware these parasitoids brought levels of the alfalfa blotch leafminer well below economic injury levels. Yeargan (1985) states that resistant cultivars have been effectively used in alfalfa production for control of alfalfa stem nematode, bacteria wilt, anthracnose, and *Phytophthora* root rot.

Case Study 6: Apple IPM

Apples are a high-value crop and have a low tolerance for pest damage. Worldwide, apples have a very large complex of pests, making pest management difficult and complicated (Croft, 1982). Many states, however, have implemented apple IPM programs. Croft (1982) and Whalon and Croft (1984) reviewed the current status of apple IPM in the northern United States. Most of the apple IPM programs rely heavily upon monitoring pest and natural enemy densities. If pest densities surpass a specific action threshold, a decision is made to apply a recommended control strategy. Kovach and Tette (1988) conducted a survey of the New York apple IPM program, and concluded that the program was successful in terms of safe and efficient use of insecticides without a significant decrease in fruit quality. More than 80% of the apple growers surveyed used some aspect of IPM on their farm. The program in New York has reduced the total amount of pesticide applied. Growers following a comprehensive IPM program used 30% less insecticide, 47% less miticide, and 10% less fungicide than growers not using IPM.

Disease management has played a major role in apple IPM programs. Apple scab caused by *Venturia inaequalis* is a serious disease of apples. Symptoms occur as dark lesions on the leaves and fruit, and infected fruit exhibit uneven growth (Jones et al., 1980). Apple scab can be a severe problem for most commercial apple varieties, although resistant varieties have been developed and are commercially available. Biological and environmental monitoring can be used to reduce fungicide applications (Jones, 1986). Monitoring the maturity of asci in perithecia on overwintering apple leaves is performed in the early spring so that when weather conditions for infection are suitable and inoculum is present, protective fungicides are applied. Environmental monitoring of air temperature and hours of leaf wetness are the basis for predicting the probability of infection using Mills Tables (Mills and LaPlante, 1951) or more sophisticated disease epidemiology computer simulation predictors (Jones et al., 1980). Once infection has taken place, therapeutic systemic fungicides can be applied to control the pathogen.

Sampling insect pests in apple trees often includes the use of attractive traps based on visual cues or synthetic pheromones. There has been a high degree of success in the use of various types of traps for apple maggot, tarnish plantbug, European apple sawfly, spotted tentiform leafminer, San Jose scale, red banded leafroller, and codling moth (Coli et al., 1985). Traps can be dedicated to monitoring the presence or absence of a species in an orchard so that decisions can be made regarding when to apply the first and last insecticide sprays. Researchers have been able to determine correlations between cumulative trap catch and percentage of fruit damage, so that action thresholds can be established and used for decisions on whether or not to apply an insecticide, given the density of the pest (Prokopy and Hubbell, 1981).

Croft and Hoyt (1983) discussed an integrated mite control program of two-spotted spider mites, rust mites, and European red mites. This program relies on an economic threshold of 15 two-spotted spider mites per leaf, a predatory mite, *Amblyseius fallacis*, and selective acaracides that result in higher mortality of mite pest species than the mite predator species. During a growing season, pest mite populations can be managed in a variety of ways dependent upon the ratio of *A. fallacis* predator mite densities to pest mite densities. For instance, an *A. fallacis* to two-spotted spider mite ratio of 1 to 5 early in the season provides season-long biological control with no need for insecticide treatments (Croft, 1982). A ratio of 1:32 does not provide biological control, and pest mites surpass the economic injury threshold of 15 mites per leaf. A selective acaracide can be used to lower spider mite densities, while preserving predator mites, bringing the ratio down to a level where effective biological control can occur for the remainder of the season. This is a sophisticated program, and a complex sampling scheme is used, taking into account (1) the spatial distributions of mite species in the system, (2) varietal preference, and (3) the proper sampling of orchard, tree, and leaf combinations. Another strategy is the manipulation of apple rust mites to the detriment of the more serious two-spotted mite. The apple rust mite is useful as an alternate prey for *A. fallacis*, but also by feeding on apple foliage it conditions the foliage in an unknown manner such that it makes the foliage unsuitable for the two-spotted mite and suppresses population buildup of this pest (Croft and Hoying, 1977). The apple rust mite causes less damage than the two-spotted mite and is cheaper and much easier to control that the pesticide-resistant two-spotted mite.

Other novel approaches in apple IPM include the management of pesticide-resistant mite populations (Croft and Hoyt, 1983). Resistance management is an attempt to either slow down or reverse the rate of development of resistance in orchards by applying pesticides in a rotation or selecting classes of pesticides in which the pest is not already demonstrating a tolerance. By doing this, the tremendous genetic selection pressure that a single class of insecticides (all with the sample pharmacological mode of action) has on a population is reduced (Dover and Croft, 1986; Frisbie et al., 1986). Temperature-dependent development rate computer simulation models can be used to predict windows of time when growers can apply an insecticide while the pest density is high, but the specific pest's natural enemies densities are low so that little harm comes to the natural enemies (Drummond et al., 1985). A third approach is the use of a synthetic oviposition deterrent pheromone (ODP) applied to apple trees to repel apple maggot flies from laying their eggs in the fruit (Roitberg and Prokopy, 1986). Bees are important to apple production as pollinators, but quite often insecticides are needed to control pest species before bloom or after bloom when bees are still in the orchard. Ayers et al. (1984a, 1984b) have developed a creative approach to protecting honey bees under these circumstances. They have developed the concept of "diversionary plantings" on the periphery of the orchard. These diversionary plantings are more attractive to the bees than the flora inhabiting the orchard floor during peak insecticide application periods, thereby luring the bulk of the pollinator population away from poisoned flowers to highly attractive, nectar-bearing introduced plantings.

Case Study 7: Potato IPM

Potato production is an intensively managed system that can be impacted by numerous insects and pathogens. This system is an excellent example of successful implementation of IPM, especially through the production and planting of potato seed pieces certified as free of fungal, bacterial, viral, and nematode pathogens. This is a model program. For some potato pests, however, successful IPM procedures have not been developed to the level required for broad scale implementation.

The potato crops are plagued by many insects and plant pathogens, including the Colorado potato beetle. R.J. Wright (1984) demonstrated that crop rotation can greatly reduce the number of beetles colonizing a rotated field, resulting in less than economically damaging beetle densities. This practice is common in some areas and not feasible in others. Research is being conducted on host plant resistance and biological control (Casagrande, 1988), but has met with limited success except for a parasitic mite of the adult Colorado potato beetle which has been released in Rhode Island to help control this pest (Drummond, 1986). Synthetic insecticide applications based on damage thresholds are used to control the Colorado potato beetle, and in some cases less toxic, biologically based insecticides such as exotoxin, derived from the bacterium *Bacillus thuringiensis*, are now being used. Resistance management is receiving considerable attention. This may indeed be the only short-term strategy for saving the potato industry in regions where this pest is difficult to control and developing resistance at alarming rates, but the strategy of chemical control of the Colorado potato beetle is a flawed one, and continuing along this path can only lead to disaster (Casagrande, 1988).

Green peach aphid, *Myzuz persicae*, and late blight caused by *Phytopthora infestans* are two very destructive pests of the potato in the northern United States for which IPM programs have been developed and implemented in Pennsylvania. GPA-CAST is a computer model for green peach aphid management on potatoes (Smilowitz, 1981). Four delivery systems are used, including hand-held calculators, a dedicated minicomputer and mainframe computers, grower-owned microcomputers, and hand calculation procedures. Green peach aphids and the incidence of their natural enemies are sampled early in the season. To use GPA-CAST, the grower accesses the computer on a frequent basis and inputs green peach aphid densities per 50 leaves, maximum and minimum air temperatures since the last access, estimated market value of the crop, cost of control, and predicted maximum and minimum air temperatures for the next three days. Statistical models in GPA-CAST developed from three years of field data are used to provide the grower with information on future green peach aphid densities and their relation to estimated economic injury levels based upon market price, plant growth, and control costs. If green peach aphid densities are below the economic injury level, then the grower is advised not to spray and to access the program in a week. If green peach aphid densities are above the economic injury level, then a spray application is recommended to the grower for aphid control (see Smilowitz, 1981 for more detail).

Blitecast is a decision-making algorithm developed for control of late blight caused by *Phytophthora infestans*. It was this disease that resulted in the Irish potato famine in the mid-1840s (Large, 1940). Disease incidence has the potential to increase at a rapid rate, making it impossible to base management programs on symptomatology. Blitecast was developed to predict the potential for infection of potato by *P. infestans* based upon mathematical relationships between time in hours when the relative humidity is greater than 90% and the average air temperature during that time. This relationship between air temperature and relative humidity is used to assign daily disease severity levels. These reflect the probability of a late blight infection. When the weather has been especially conducive to late blight infection (moderate temperatures, high rainfall, and high relative humidity) Blitecast recommends that a fungicide be applied at specific intervals. To complement the weather-epidemiology aspect of late blight forecasting, other factors can be integrated into the program. These include host plant resistance to late blight, sanitation, and the use of therapeutic systemic fungicides. Blitecast can reduce fungicide applications as much as 60% compared to potato production under a weekly calendar fungicide application (Krause et al., 1975; Fry, 1981; Fry and Thurston, 1981; MacKenzie and Smilowitz, 1981).

INTRODUCTION TO SYSTEM SCIENCE

Systems science will very likely play an important role in the development of future technology for sustainable agriculture. During the past 15 years, the term *systems approach* has been used in pest management with increasing frequency (Haynes et al., 1973; Ruesink, 1976; Edens & Koening, 1980; Bird et al., 1985; Teng, 1988). Its meaning, however, appears to have taken on a number of different contexts, and is frequently a subject of controversy. The objective of this section is to describe the nature of systems science and show how it can be used for the identification of interactions among crop, pest, and pest management components of sustainable agriculture.

Systems science is the study of interactions among related entities. It is an engineering philosophy with its foundations in constructing models (both conceptual and mathematical) of real world systems to aid in evaluation and optimization of existing systems and design of new systems. This discipline evolved during the past 50 years to deal with complex military operations and space exploration (Churchman, 1968; Brooks, 1975; Checkland, 1981; Wilson, 1984; Beer, 1985; Miser and Quade, 1985). It has been shown to be of significant value in the design, evaluation, and management of agricultural and natural resource production systems (Feldman and Curry, 1982; Shoemaker and Onstad, 1983).

The first step in a systems approach is to identify the boundaries of the system. The process consists of dichotomization of the system (universe of concern) into the object of control and external environment (Figure 3.8). The objects of control are those aspects of the system that we wish to manage and therefore need to model. The external environment includes those aspects of the system that influence the object of control, but need not be modeled themselves in order to quantitatively describe the object of control. This separation does not imply that the object of control and external environment are mutually independent. On the contrary, it provides a framework for quantitative evaluation of the interdependence between the object of control and its surrounding environment. The system boundaries depend on the objectives of the analysis and the perspective of the analyst. For example, in a commercial onion production system in Michigan, an

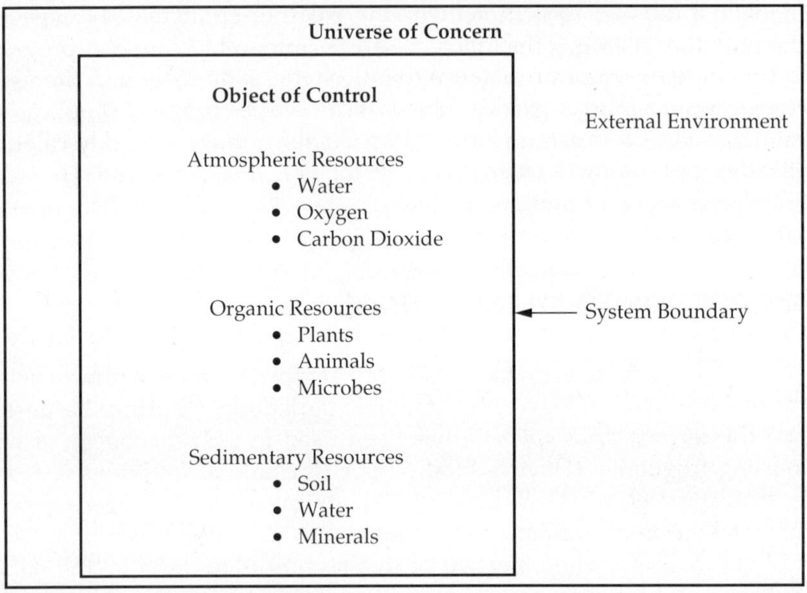

Figure 3.8 Dichotomization of the system (universe of control) into the object of control and external environment.

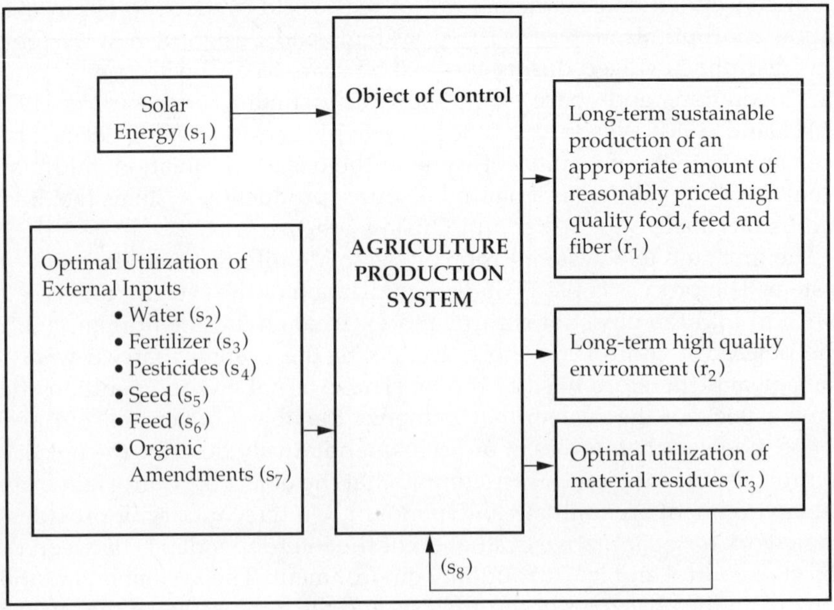

Figure 3.9 Conceptual model of eight stimuli (inputs) and three responses (outputs) of an agricultural production system.

entomologist might define the object of control to be (1) the onion plant, (2) the onion maggot and the onion thrips (insect pests of the onion plant), and (3) natural enemies of the onion maggot and onion thrips. The external environment would consist of (1) uncontrollable abiotic factors such as rainfall, wind, solar radiation, (2) controlled abiotic factors such as fertilization, irrigation, cultivation, (3) uncontrolled biotic factors such as crop pathogens, weeds, nematode pests, and (4) controlled institutional factors such as market prices, harvesting procedures, environmental regulations. The same system would have different system boundaries if it were defined by an economist.

The second and third steps in the process of describing a system consist of indentifying system stimuli and responses. In the case illustrated in Figure 3.9, the responses (r_1, r_2, and r_3) are synonymous with the system objective or proposed definition of sustainable agriculture. In this model, the first stimulus (s_1), solar energy is considered a noncontrollable variable, and may have to be monitored. For example, if the plant is part of the object of control, knowledge of daily solar radiation may be critical if modeling the daily growth rate is an objective. Monitoring the rate of solar incidence may be sufficient for a quantitative model of photosynthesis. Stimuli (s_2, s_3, s_4, s_5, s_6, and s_7) are designated as controllable external inputs. The optimal use of the material residues produced by the system (r_3) is considered as an additional system input (s_8).

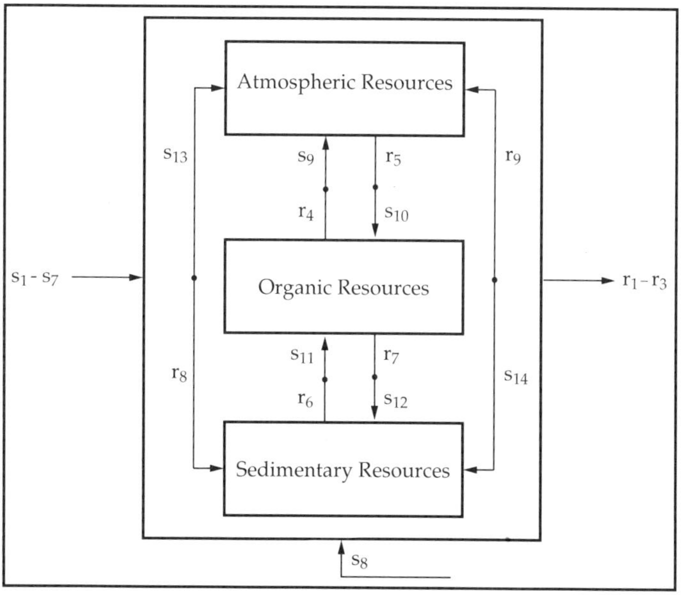

Figure 3.10 Conceptual model of the interactions among the atmospheric, organic, and sedimentary components of agriculture.

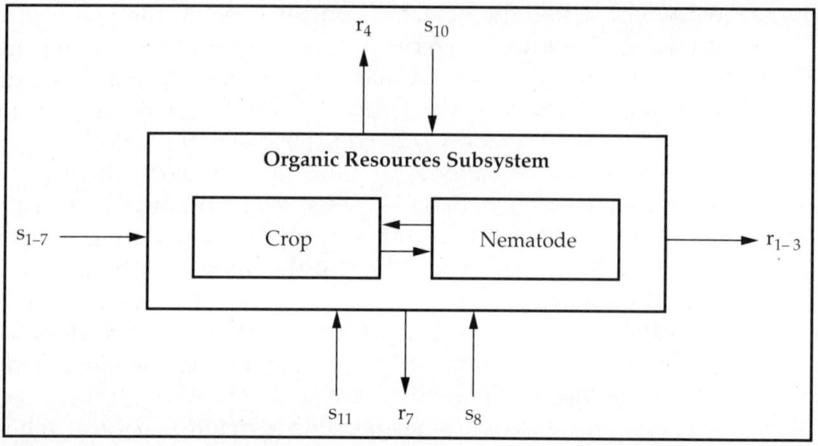

Figure 3.11 Conceptual model of the host-parasite relationships component of crop and nematode pathogen of an agricultural production system with the identification of ten stimuli and four responses of the system as described in Figure 10.

The procedures of systems science can be used to develop a framework for the use of quantitative procedures in the construction of models suitable for determining specific responses (r_1–r_9) resulting from stimuli s_1–s_{12} (Figure 3.10). Additional complexities of the system can be evaluated with precision through the use of crop and pest interaction models (Figure 3.11). Models can be static or dynamic. A static model generates responses by manipulating the current stimuli without taking into consideration any historical information about the system, while a dynamic model incorporates positive and/or negative feedback loops. This requires a more complete understanding of the current state of the system, and takes into consideration information about the historical development of the system. For example, a model of the potato cyst nematode (*Globodera rostochiensis*) linked to a model of a potato (*Solanum tuberosum*) root system can be used to predict nematode population densities and their impact on plant growth (Figure 3.12).

Insect and plant pathogen models have been linked to crop-plant models in order to evaluate various methods and timing of pest control strategies (Waggoner, 1962; Ruesink, 1976; Carruthers et al., 1986; Teng, 1988). Optimization schemes have been used in conjunction with both mathematical and simulation models in order to determine optimal control strategies (Shoemaker, 1985; Alocilja and Ritchie, 1988). An example is the continuation of the nematode model to include biological control (Figure 3.13).

A slightly different approach in utilizing pest-crop models forms the basis of On-line Pest Management (Haynes et al., 1973; Tummala and Haynes, 1977). Pest population dynamics and phenology models are used during the growing season in order to make short-term predictions. Real-

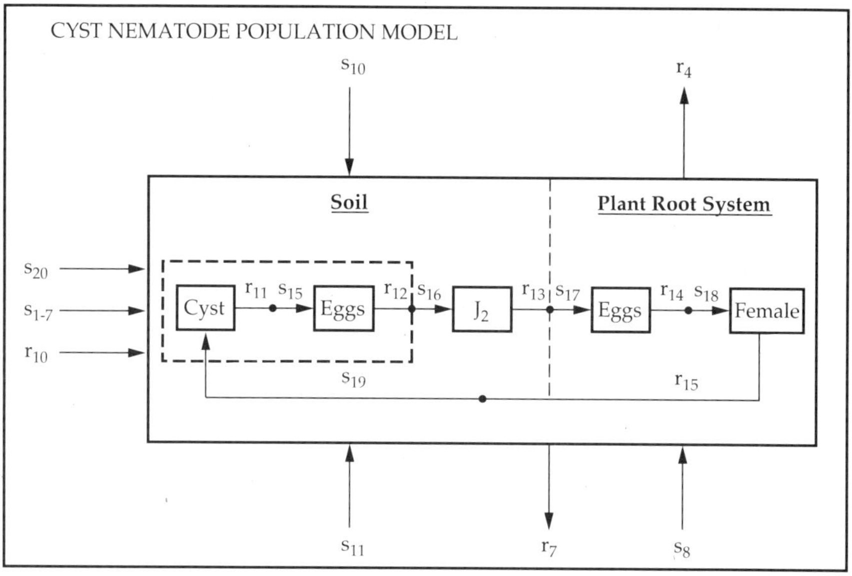

Figure 3.12 A conceptual model of the life cycle and life stages of *Globodera rostochiensis* (Potato cyst nematode) associated with soil and roots of *Solanum tuberosum.*

time weather data and information obtained from sampling growers fields (for example, initial spring pest densities and maturity of ascospores) are used as inputs to drive the models. Outputs from these models can be schedules for scouting or sampling fields based upon the predicted occurrence of pest stages, or decisions regarding control action that should be taken based upon predicted yield loss from pest densities, the presence of a vulnerable pest stage, or a window in time where a pesticide application will have a negligible effect on natural enemies. Models must be realistic and able to follow the general form of the system being modeled. Resolution is the opposite of generality and relates to the number of system attributes that a model attempts to reflect. Time scale is an example of resolution. A model designed to show yearly trends in pest populations would most likely lack the resolution necessary to predict changes in the system at one-minute intervals.

No modeling effort is complete without an appropriate validation. This phase is a critical and difficult task. Validation is easier if the parameters of the model can be measured and verified in the field or under controlled conditions through experimentation. The system structure is verified when the overall system behavior is in agreement with that of the simulated behavior. If the predicted population densities agree with the observed data at some satisfactory level of probability, the real world system is considered to be similar and the model can be accepted for use. If the

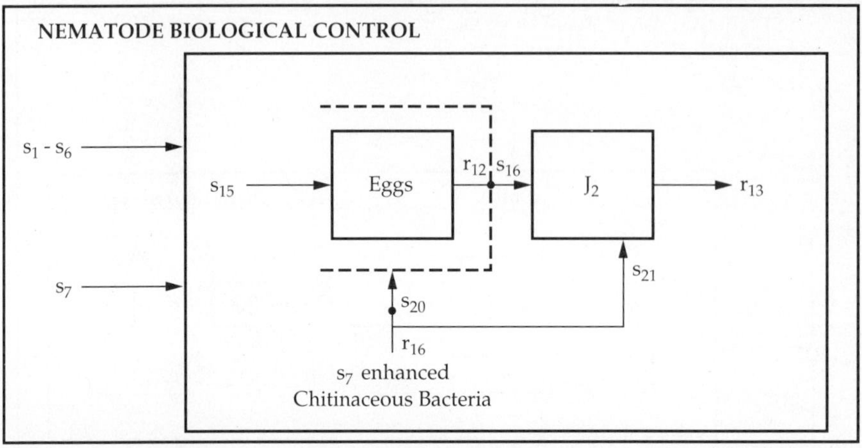

Figure 3.13 Conceptual model of nematode biological control resulting from the interaction between nematode eggs and chitinaceous bacteria, with identification of potential system stimuli (s_{20},s_{21}) and responses (r_{16}), in addition to those identified in Figures 3.9–3.12.

degree of agreement between the model and the measured data is inadequate, then the model must be modified. This refinement may involve changes in the perceived interactions within the system (system structure), equations, or values of constants.

Once a model is validated or accepted as a representation of the system under study, it can be used for various purposes. A sensitivity analysis of a specific parameter (for example, soil temperature) can be useful at this stage in evaluating the impact of this parameter on the system and its components. One can check for possible errors in the estimated values and their potential for control of the system by artificially modifying these parameter values.

Creating a model system to substitute for a real system of analysis is one possible objective of systems science. This type of modeling may be used to obtain information concerning the consequences of proposed actions or policies on the existing system structure. This use, including the redesign of agroecosystems toward a sustainable end, is demonstrated in the next case study. Through the use of quantitative models, new system structures are hypothesized and tested. This type of research is not practical without simulation. The cost of constructing and evaluating a series of agroecosystem designs on farms would be prohibitive. One of the greatest advantages of a model is that users can study the effects of hypothetical manipulations in the real system through simulation. Models can be used to enhance understanding of behavior of the system at very low costs and to design and test pest management and sustainable agriculture strategies. For instance, Shoemaker and Onstad (1983) used a model to analyze the

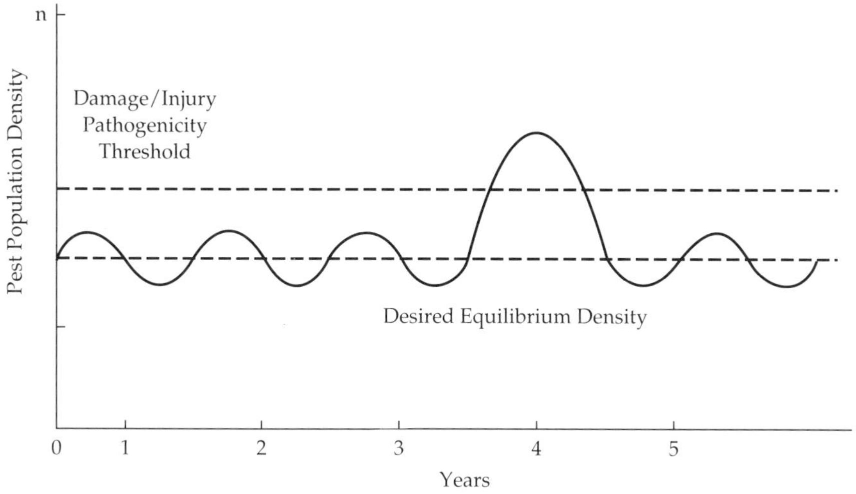

Figure 3.14 Population dynamics of a potentially major pest in a sustainable agricultural system in which the population density is maintained below the damage/injury/pathogenicity threshold in four out of five years.

integrated control of alfalfa weevil in central New York. They considered biological control by the parasitoid *Bathyplectes cuculionis*, insecticides, and cultural control by early harvesting and found that early harvesting is usually the most effective management procedure. To evaluate these management options experimentally in the field would have taken many years, and the cost would have been prohibitive.

SUSTAINABLE AGRICULTURE ERA

Nutrient cycling and pest management are important components of sustainable agriculture. The nature of pest management in future sustainable agriculture, however, is not well defined (Bird, 1987a). It is our prediction that the most important pest management strategy in sustainable agriculture will be pest problem avoidance through system design. The object will be to exclude pests (including major pests of current commercial agriculture) from the area of concern, or to maintain pest populations at a density below the damage/injury/pathogenicity threshold. For an agricultural system to be sustainable, we are suggesting that this pest management objective be achieved at least four out of every five years (Figure 3.14). This will mandate the use of system design as a management procedure. Because of the dynamics of pest populations, there may be years when appropriate short-term pest control tactics will be necessary to redirect the population dynamics of a pest and its associated natural control factors.

Moving toward a more sustainable agriculture will require the closure of cycles (energy conversion, soil, air, and water cycles, and waste) that have been broken or altered over the past 50 years, largely as a consequence of the availability of inexpensive liquid fossil fuels (Edens and Haynes, 1982). It will be necessary to design and implement production practices and consumption patterns consistent with a closed system agriculture—one in which cycling and recycling are fundamental. The underlying structure of the agricultural system of production and distribution is directly affected by the quality and quantity of energy inputs. Edens and Haynes (1982) suggest that as energy prices increase, a direct effect will be felt in the food transformation sector, which (through market forces) will result in an increased agricultural diversity at the state, regional, and farm levels. Hill (1982) stated that the average molecule of food is carried 1,300 miles in the United States before it is eaten and that only through crop diversification can agriculture remain economically and ecologically viable.

Sustainable agriculture can be viewed as the production and distribution of food, feed, or fiber in systems that are politically and socioeconomically viable and environmentally sound over an appropriately long period of time (Figure 3.1). To operationalize this generalized and somewhat philosophical concept of sustainable agriculture, it is necessary to establish criteria for identification of elements or activities that further sustainablity. Although it is difficult to achieve consensus on what promotes sustainable systems in a concrete or operational sense, and it is beyond the scope of this chapter to wrestle with this issue in great detail, it is necessary to introduce a protocol for future investigations. A change in the design and structure of agriculture to accommodate increased regional and farm-level diversity for plant protection has many implications. The plant protection disciplines will have to be involved in the redesign; otherwise, the system structure will likely be set and any solutions to pest problems will be short-term. For instance, many current monocultures evolved within the context of readily available, relatively cheap, toxic, and effective pesticides. If, due to pest resistance or other environmental issues, alternatives are required for crop protection (for example, biological control) the probability of this successful implementation will be low, because there is no accompanying change in the system structure. Therefore, system design is just as much a variable for plant protection as it is for plant nutrition, marketing, and a high-quality environment.

Paramount for a holistic approach is a systems philosophy. Sustainable cropping systems based on diversity do exist (Gliessman, 1984; Dover and Talbot, 1987). While we may learn important lessons by studying these systems, it would be a mistake to think that a Mexican polyculture system can be transferred to the United States, where the underlying sociological, economic, and ecological structure of the agroecosystem is so different. It would be repeating some of the same errors made in the course of the Green Revolution (A. Wright, 1984). The underlying theme must be centered around planned or functional diversity at all social, economic, and ecological levels (Dover and Talbot, 1987). Agriculturalists will have to embrace a systems philosophy and work at the level of the ecosystem in order to reduce current disciplinary isolation. Therefore, instead of attempting to determine which management practices will best close the cycles, given an existing agricultural system structure, the system structure must be treated as a variable and designed to facilitate adaptation to long-term change in the social, biological, and economic production environment. An example of this approach is contained in the following case studies from an entomological systems' perspective. The approach is only viable if scientists from other disciplines follow a similar course so that linkages between disciplinary components can be elucidated, quantified, and designed on an interdisciplinary level.

Three case studies have been selected to illustrate the pest management challenges related to sustainable agricultural systems.

Case Study 8: Sustainable Onion Pest Management

Within the context of a functionally diverse onion agroecosystem that optimizes mortality of the major insect pest (Figure 3.15), the design of the system was established as a major tool for managing the system. This design was derived from quantitative models describing the relationships among components in the system. From an understanding of these quantitative interactions, designs incorporating diseases, weeds, and marketing can be derived as long as the relationships that are used in the construction of these "free-body" models are structure independent or incorporate aspects of structure as a variable. For example, researching and quantifying the relationship between mycorrhizal infection of onion roots and the incidence of pink root disease caused by *Pyrenochaeta terrestris* in a commercial onion production system may yield dynamics that are very sensitive to fertilizer rates (especially phosphorus levels) and plant density and planting date. If the objective is to design an agroecosystem that, by design, aids in reducing pest numbers, then structurally dependent models will contribute only if small incremental adjustments to the system design are made. This philosophy is referred to as using design as a management option. Design is no longer a static environmental constant, but a dynamic management variable. How do we research complex interactions between components in a system, yet maintain the ability to integrate "free-body" models in search of design scenarios? Two approaches that are commonly used are (1) the factorial experimental design (discussed earlier under the systems science methodology) where rate functions of the significant components are researched in a multiway factorial type of design, and (2) the energy gradient design (Haynes et al., 1981). The

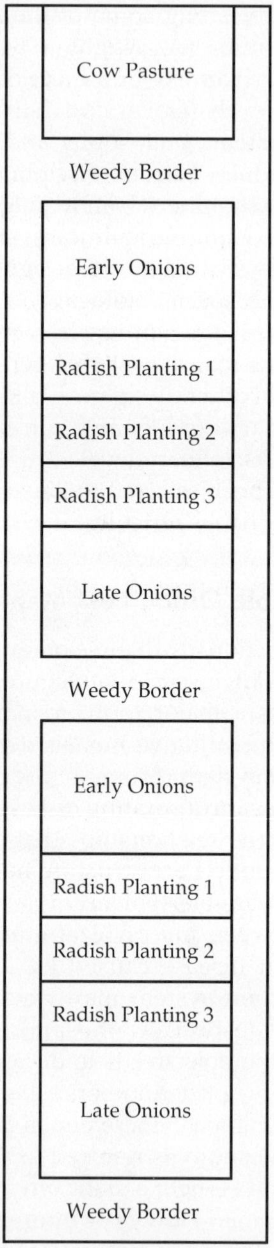

Figure 3.15 Sustainable agriculture planting for minimizing the impact of onion maggot and the need for the use of insecticides for control of this pest.

example of a systems approach to cyst nematode incorporated the factorial approach. This approach, however, can be unwieldy if, initially, the number of significant components impacting the dynamics of the object of concern (in the cyst nematode case) is large and only partly specified. One cannot always determine the key components if the structural design has eliminated some of them. This is the case described in the context of an onion agroecosystem. In this approach, key components affecting the dynamics of the object of concern (the onion maggot) are identified by quantifying the life system of the onion maggot across a gradient of onion agroecosystems representing different energy input levels. The systems analyzed included an old field ecosystem that was disked, planted with onions, and then left alone, a diversified "organic" production system, and, at the far end of the spectrum, a commercial, chemically intensive onion monoculture. This approach quickly allows identification of key components so that the effect of structure design becomes apparent. Conceptual models of this representative system structure can then be designed and tested either through simulation or in the field.

An alternative design of the onion agroecosystem stresses planned or functional diversity. The cow pasture and weedy borders provide alternate host and nectar for *A. pallipes* (Groden, 1982). The cow pasture also provides a rich resource for earthworms, thereby potentially maximizing the densities of the tiger fly predator of onion flies (Drummond et al., 1988). The long, narrow strips of onions minimize the distance from any point in the onion field and the weedy borders and cow pasture. This is important since *A. pallipes* numbers decline exponentially from weedy borders and cow pastures into the onion field (Groden, 1982). This is also true of onion flies infected with disease caused by *Entomopthora muscae* (Carruthers, 1981). Weedy field borders are not mowed so that attachment sites for diseased flies are provided. Narrow weedy borders maximize the probability of *E. muscae* spores encountering healthy flies by crowding together resting and attachment sites for healthy flies during mid-day. By mowing some of the weedy border, this crowding effect can be increased. The planting of radishes adjacent to onions provides an alternate host and thus a continuous food supply for the rove beetle *A. bilineata*. A number of plantings should be used in order to provide a season-long food resource for the cabbage maggot, and a number of different planting dates of onions should be incorporated into the design (Groden, 1982). Groden also showed that early planted onions adjacent to late planted onions serve as a highly attractive trap crop resulting in a concentration of the onion maggot population in the early planting. Because the later plantings go largely untouched, the early planting can be positioned near the radish interface so that the host pool for *A. bilineata* is concentrated, thereby making prey search more efficient.

A cultural practice that is not a design change but an important aspect of onion fly management is cull management (Drummond, 1982). In Michigan, after onions are harvested, an entire generation of onion flies emerges and searches for onions to colonize (Drummond, 1982). Any onions left in the field become an ideal food source for a third and final generation of flies. It is this generation that contributes most to the next spring generation, the most damaging to the onion crop. Many onion growers have developed the bad habit of sorting onions in the field during harvest, which results in thousands of cull onions per acre left in the field for the utilization by onion maggots. Plowing these cull onions down in the fall only makes matters worse. This practice stimulates sprouting of cull onions, which are more attractive to female onion flies looking for egg-laying sites, and at the same time onion fly survival is much higher on these culls than on unsprouted culls lying on the surface. Two management options are available. The first is to sort onions in storage sheds not in the field. The second option, which can be used in conjunction with the first (since even with the first strategy some onions end up in the field due to inefficient harvesting), is to sow a fall rye or oat cover crop immediately after harvest so that in a week the cover crop hides the cull onions in the field, making it difficult for the onion flies to find the culls. An organic onion grower in Michigan adds a twist to the fall cover cropping strategy: he chooses not to harvest a small section of onion rows, and then, while sowing the cover crop, cuts the tops off of the onions and leaves them on the ground. These cut tops are very attractive to the onion flies (more attractive than cull onions); however, the onion fly immatures cannot survive on them because they dry up before insect development can be completed. Thus, the cut tops serve to keep the onion flies from laying on the culls until the cover crop comes up, and then the searching efficiency of the female flies is drastically reduced. In addition to the cull-management strategies, crop rotation significantly reduces the number of flies colonizing an onion field in the spring (Mortinson et al., 1988).

As demonstrated in this previous case history, moving toward a balance with nature is fundamental to designing sustainable systems. Balancing the nutritional aspects of crop plants and livestock has long been known as an operating principle for organic farmers (Oelhaf, 1978). More than any other element used in fertilization, nitrogen affects the crop plants' susceptibility to disease and insect damage (Palti, 1981; Scriber, 1984). Most pronounced is the effect of nitrogen on the vigor and rate of plant growth. Vigorously growing crops are more susceptible (with some exceptions) to obligate parasites such as rusts, powdery mildews, downy mildews,and virus diseases.

Facultative parasites, such as *Alternaria* spp. and *Fusarium* spp., attack low-vigor crops more severely. There is also considerable evidence that shows increased plant nitrogen increases insect populations and the resulting damage caused by insect feeding. Scriber (1984) surveyed the entomological literature over the last 100 years and found 135 studies that clearly showed increased arthropod damage or population levels associated with an increase in nitrogen and 44 studies that showed a decrease in damage and arthropod population levels as nitrogen increased.

Case Study 9: Soil Ecosystem Equilibrium

Soil-borne pathogens causing vascular wilt diseases of plants can devastate crops and are difficult to manage. In a recent treatise on wilt diseases, Beckman (1987) reviews and synthesizes management strategies for vascular wilt diseases of tomato caused by *Fusarium oxysporum* f. sp. *lycopersici* and *Verticillium albo-atrum*. The main thrust of the strategy is biologically based, combining the innate resistance of the host plant with soil and crop management practices that provide a suboptimal environment for the development of the pathogens. The following measures make up the tools for this strategy (Beckman, 1987):

1. Use of resistant cultivars.
2. Management of soil fertility to decrease the growth, sporulation, and virulence of the pathogens.
    a. Addition of lime to obtain soil pH of at least 7.0.
    b. Avoidance of micronutrient soil amendments.
    c. Minimization of use of phosphorus and magnesium soil amendments.
    d. Utilization of nitrate rather than ammonium nitrogen.
    e. Application of fertilizers as bands close to tomato roots (do not broadcast fertilizer).
    f. The use of preplanting fallow schedules.
3. Avoidance of the use of diseased transplants or infected seed.
4. Prevention of the dissemination of the pathogen by eliminating movement of infested soil into disease-free areas.
5. Advantageous use and maintenance of suppressive soils.
6. Maintenance of pathogenic nematode species at low population densities.

The application of Beckman's strategy requires an in-depth understanding of the biological interactions that occur in the soil environment. For example, in the use of lime, maintaining a high pH often becomes a trade-off in an area where both *Fusarium* and *Verticillium* are present. Fusarium wilt is reduced, but Verticillium wilt is enhanced with the higher soil pH resulting from liming. Chemical controls also result in trade-offs. Chemical treatments can clearly be beneficial when control by other means has been lost (such as by poor management) and when it is economical to use them, but a reliance on chemical control will not move the grower toward a program aimed at developing and maintaining suppressive soils based upon diversity of the soil microflora and soil microfauna (Louvet et al., 1981).

Case Study 10: Livestock Health Maintenance

Another example of the importance of achieving a balance is in livestock production. Boehncke (1985) describes how the problems of veterinary medicine and the need and use of antibiotics are associated with structural changes in agriculture. When farms with different animal species became more highly specialized enterprises with higher stocking levels, rate incidence in diseases increased (Ekesbo, 1981; Niederstucke, 1982). The use of antibiotics to combat the increase in disease proved to be a sustainable "solution." Guillot et al. (1983) showed that over a four-year period of use, antibiotics put such selection pressure on the coli bacterial populations of calves that resistant strain increased from less than 5% of the total flora to almost 50%. Other negative effects of the high use of antibiotics include the residues that accumulate in milk, meat, and eggs and the development of resistant bacterial strains that may be hazardous to humans (Boehncke, 1985). Solutions to these problems do not lie in new drugs; rather, a redesign of the structure of livestock and poultry production systems that provide for welfare (behavioral, nutritional, and physical) of animals is required. Such changes necessitate moving away from an animal "factory" model form towards animal husbandry. The requirements of such a move are outlined by Boehncke (1985) and include the following components and characteristics:

1. Housing systems need to be free of congestion, ventilation increased, and livestock allowed movement and exercise to facilitate meeting the animals' behavioral needs (social, resting, and eating). The density of animals in a factory farm has been shown to both induce stress and create a suitable external environment for development of disease in livestock (Ekesbo, 1981; Sambraus, 1984).
2. Breeding goals need to be in tune with animal welfare and not just focused upon efficiency of feed conversion or maximizing profit. Boehncke reports that breeding in an effort to maximize a single characteristic has resulted in animals that are inherently unhealthy, for example, skeletal structure not capable of supporting body weight. An unhealthy or genetically unfit animal will be more susceptible to disease.
3. Proper nutrition is central to maintaining a healthy herd or flock. The use of concentrate feeds in order to obtain maximum yields does not outright kill dairy cows, but it does disturb the bacterial balance in the rumen, which increases the permeability of the rumen wall and allows toxic bacterial compounds to enter the liver via the blood stream. Pasture grazing can be detrimental to the herd if the pasture is poorly managed by intensive use of synthetic fertlizers. In addition, a high nitrate content of pasture grass can cause liver damage, and low diversity of pasture grasses can result in an unbalanced

potassium:sodium ratio in the diet, thereby predisposing the animal to disease.

PROGNOSIS

Sustainable agriculture can be viewed as a process or quest. Its objective can be stated in terms of temporal stability, inter and intra group equity, and ecological impact along all perceivable time horizons. The problem quickly encountered is how, from a scientific stance, does one measure whether or not a practice or process is sustainable or leads toward a more sustainable system. A potentially useful approach might be to focus on the process and to evaluate elements of the process in a dynamic way so as to assess their contribution to sustainability in terms of their impact on the length or duration of subsystem cycles. It has been argued elsewhere (Edens and Haynes, 1982) that sustainability is tantamount to the movement toward closure of cycles within subsystems in the sense that the long cycles apparent in both the production and marketing end of the North American food systems have, to a considerable extent, contributed to the formation and evolution of agricultural systems that do not appear to be sustainable.

As societies become more generally aware that we all live "downstream," both figuratively and literally, the significance of understanding the nature of our institutional systems in a quantitative way will be more apparent. In order to assess the contribution of any specific subsystem to the sustainability of the overall system, the relevant system boundaries must be specified. Spatially, these might consist of a nested set of regions: global, national, state, regional, and local. Accordingly, within these spatial areas, functional or operational delineations must be made. These might include a specification of subsystem nutrient, marketing, or materials cycles within the system. The Cornucopia Project undertaken by the Rodale Research Center is an example of such an evaluation approach in which the ratio of internal to external inputs (imports and exports across state boundaries) is a criterion for assessing subsystem contribution to system sustainability. Axinn and Axinn (1987) have suggested an analogous approach for material and energy flows. By focusing on cycles within the system, it is possible to establish subobjectives that are identified and chosen because of their contextual importance in the overall system and are appropriate for the application of conventional suboptimization techniques.

Depending on the level of generalization one chooses, sustainable agriculture can be defined at a rather high level of abstraction as an agriculture that " . . . does not deplete soils or people" (Berry, 1987), or more concretely and specifically as one that " . . . has the ability to evolve indefinitely toward greater human utility (including, for the foreseeable future, increased production), greater productivity (increased efficiency of resource use), and a more stable balance with an environment which is favorable to both humans and to the majority of other species" (Harwood, 1988). These definitions are complementary and differ largely as a result of their capacity to operationalize the concept of sustainability. One of the reasons we have focused our discussion on the closure of material, energy, and information cycles is that the relative openness of a cycle can be measured (ordinally if not cardinally), systems can be ranked or compared, and explicit criteria can be established to measure or evaluate success with respect to sustainability. This focus is not, however, intended to diminish the importance of less quantifiable philosophical criteria and objectives.

The earlier emphasis in this chapter on systems science and integrated pest management also stems from our belief that unless criteria can be established, at a level of specificity that allows ex poste evaluation, we will use too much of our time and effort playing word games. The systems approach facilitates the delineation of problems at a level of abstraction that supports analysis and evaluation and, at the same time, forces the researcher to remain cognizant of the larger contextual relationship between the subsystem being observed and/or manipulated and the larger system within which it exists.

Integrated Pest Management represents both a vision and a methodology. As James Kendrick Jr. succinctly put it, IPM " . . . is an ecological approach to maintaining plant health. It is an attitude evolving into a concept of controlling pest and disease damage to plants. It is based on an understanding of the entire ecological system to which the host that we are interested in keeping healthy belongs" (Kendrick, 1988). The systems approach facilitated our understanding and operational capacity to analyze these kinds of complex systems.

If we are to move towards a more sustainable agriculture, programs and techniques that have been developed under the rubric of IPM will be extremely useful. The issues of the past do not differ greatly from those of the future. Indeed, they all exhibit a deja vu quality. In 1973, Rachel Carson observed, "The entomologist, whose specialty is insects, is not so qualified by training, and is not psychologically disposed to look for undesirable side effects of his control program." To generalize this observation, one need only point to the writings of Barry Commoner in 1969: "We have become not less dependent on the balance of nature, but more dependent on it. Modern technology has so stressed the web of processes in the living environment at its most vulnerable points that there is little leeway left in the system." Commoner proceeds to point out two "simple" rules of environmental biology: "Everything has to go somewhere" and "Everything is connected to everything else." Indeed, this interconnectedness is the crux of our problem, and our willingness to place a high priority on understanding and mitigating associated impacts will dictate our future state.

Time Preference

A continuing issue that directly impacts our ability to understand and rectify unsustainable agricultural practices is the time horizon adopted by various key participants in the political decision-making process. The very use of the term *sustainable* implies a temporal dimension. It is the determination of this dimension that is critical. Economists and politicians use rather short market or term of office oriented time frames in their analyses. Geologists or anthropologists, on the other hand, are accustomed to using decades or centuries as their temporal frame of reference. It is perhaps more judicious and practical to speak of moving toward a more sustainable system rather than attaining some conceptual level of sustainability.

We must also be aware that political, economic, and social realities play a major part in determining for whom sustainability is sought and for how long. Societies or individuals who are on the edge of starvation are not likely to concern themselves with, or certainly take actions to provide for, the next generation. Indeed, even preparing for the next day or week may be too much to expect. Many of our global commons problems fall into this category: rain forests destroyed, woodlands removed, and waters irrevocably contaminated—often in the name of survival. Societies and individuals in stressed conditions need and deserve the attention and support of the wealthy nations in order to move toward a sustainable system. More saddening, however, is the plethora of unsustainable practices pursued in societies where life is not being lived on the edge. Mining soil, excessive erosion, depletion of aquifers, and pollution of waterways occur in the midst of plenty. Time horizons are shaped by economic forces influenced largely by private market incentives and, by their nature, are of a short duration.

Chemical Cycles

The major long-term impact of the chemical revolution has been the acceleration of chemical cycles at all levels of the system. At the agronomic level, we have witnessed a multifold increase in pesticide use. Estimates place the total value of pesticides used (purchased) in 1980 (excluding Russia and China) at nearly $10 billion (Braunholtz, 1981). It is further estimated that the costs associated with direct pesticide control were $2.8 billion and that those due to increased resistance exceeded $130 million (Pimentel et al., 1979). While direct costs associated with reliance on synthetic chemicals are alarming, a more insidious impact is discernible. It has been observed that while the use of pesticides is incrementally additive, it is not incrementally reversible (Edens and Koening, 1980). The development of resistance to pesticides is not known to be reversible in any meaningful way (Georghiou, 1986). Pesticides also destroy natural enemies, estimated in 1979 to amount to more than $150 million (Pimentel et al., 1979). As a result, removal of previously effective chemicals must be done in the absence of many of the key biological agents which at one time may have prevented the need for chemical control (Edens and Koening, 1982).

At the macro and global level, the speed-up of the chemical cycle is just beginning to be recognized. A variety of subtle signals have emerged: the lowering of the Ogalalla aquifer, the polar ozone void, possible CO_2 global warming, continuing excessive soil erosion throughout the world, increased incidence of surface and ground water nutrient and chemical contamination, more obvious signs of food chain magnification of heretofore unknown chemical contaminants. All are directly related to agricultural management practices and, more specifically, to pest management programs.

Tolerance as a Function of Measurement

In the future, criteria established for assessing the sustainability of our agricultural systems will be closely linked to technical and economic ability to measure the direct and indirect impacts of agriculture. Two decades ago, our ability to measure chemical contaminants was more often than not measured in parts per hundred or thousand. Current technology measures in parts per million, billion, and even trillion. During the era when measurement techniques only allowed reporting of parts per hundred or thousand, an actual level of anything up to 1,000 parts per million was reported as zero! Thus, in many instances, the levels of contamination have remained unchanged while our increased measurement ability has signaled to the public that environmental impacts are more severe. As the molecular age evolves, our technical capacity might well enable us to measure at the atomic and subatomic levels; only the economic cost of such technologies is likely to restrict our passion for minutae. Socially and politically, perceptions are as important as reality. In pest management, and more generally in agroecosystem management, we will be pressed to further reduce synthetic chemical and nutrient impacts on the environment. Emissions, if allowed above zero or the detectability threshold, will be severely restricted.

The sustainability of agricultural systems will be redefined according to social perceptions of acceptable impacts on environments, broadly defined to include human-ecological as well as biophysical systems. Perhaps the corollary is reflected in the basic concept of the conservation of matter or the notion of mass balance. In states or regions where a substance such as phosphorus is banned in common consumer products (soaps, detergents), the appearance of phosphates or nitrates in surface or ground water will be quickly traced to or assumed to emanate from agriculture. The existence of pesticides will be even more quickly associated with pest management practices. The burden of proof will rest with the agriculturalists, rather than the bureaucrats or environmentalists. Although, as scientists, we may argue that zero emissions are not a viable option in this era of extended measurement capacity, public opinion and social pressures will likely prevail.

Biotechnology

The media has given considerable attention to four major areas of technology: computers, robotics, telecommunications, and biotechnology. Although all of these affect agriculture directly or indirectly, the largest role will undoubtedly be placed out in the biotechnology arena. It is not clear at this juncture whether or not biotechnology will be used to facilitate a transition to a sustainable agriculture. Certainly it is equally plausible that these technologies will be used in an attempt to strengthen humankind's mastery over nature.

The issue over who owns the products of genetic engineering has not been fully resolved. Current trends in policy suggest that both transgenic plants and animals will be patentable and that segments of society will probably be excluded from their potential benefits. Although the benefits that might accrue from these technologies cannot be predicted, any more than the character of the technologies themselves, it is possible to sketch a few plausible scenarios. Resistance management will likely be greatly enhanced as the result of rapidly advancing technologies. If management strategies are pursued in the quest of furthering our technological control of nature, they will probably result in techniques oriented to increasing plant resistance or tolerance to new chemicals required to overcome resistance. One possibility is an expanded emphasis on research to increase the resistance of beneficial insects to selective chemicals so that natural enemies can be used in conjunction with chemicals to manage pest populations. If this general approach is taken in the biotechnology field, it will represent an extension of the chemical revolution rather than a transition to an era in which biological solutions are sought for potential insect and pathogen problems associatied with sustainable agricultural management.

CONCLUSION

When assessing the sustainability of an agricultural system, we are really posing a broader question. We are asking whether or not a society is in balance with its resources. There are few contemporary examples of such balance. Those that are identified are generally found in the cultural anthropology literature. Rather, the common situation is imbalance. In the case of North America, for example, natural resources are pressed to their limits. In other parts of the less industrialized world, there are similar pressures on natural resources, albeit for different reasons. In these cases, poverty places a high premium on today over tomorrow and on this generation over the next. Population pressure takes a similar toll on the environment as insatiable demands by the more affluent.

Sustainable agriculture transcends the agricultural sector. It is useless to argue for the sustainability of anything outside of the context of the total system within which it exists. This is one of the reasons we have stressed the importance of the systems approach. Insect and pathogen management is a small subsystem. The way in which this subsystem fits into the larger system in the context of a sustainable food system must be defined holistically. The determination and will to develop and nurture sustainable systems requires a vision for the future and a goal for the present. It is analagous to the common wisdom to "think globally, act locally." Our shared vision must be one of concern for the future. Our goals must relate to this vision. The technical capacity exists for us to rectify many, if not all, of our environmental concerns. The future of agricultural pest management is highly objective-dependent and presents several significant challenges. Unless there are major unlikely developments in the area of technological dominance of nature, agriculture will have to rely on an increasingly knowledge-dependent system of pest management. This must be designed to take advantage of multiple strategies and tactics for optimizing the long-term sustainability of agroecosystems. The question remaining is whether we can muster the courage to embrace a vision that requires some modicum of sacrifice and whether the global community stands prepared to share some of the transitional burdens on an equitable basis.

In development of future pest management systems, it is important to remember the wisdom imparted to Alice when she first met the Cheshire Cat in Lewis Carroll's *Alice in Wonderland*. Alice asked the Cat where she should go from here. The Cat replied that it depended a great deal on where she wanted to go. Alice replied that she didn't very much care, and the Cat informed her that it didn't make much difference then how she proceeded; if she traveled long enough and far enough, she was bound to get somewhere. Bertrand Russell once said, "Most people would agree that, although our age far surpasses all previous ages in knowledge, there has been no correlative increase in wisdom." He concluded: "With every increase of knowledge and skill, wisdom becomes more necessary, for every such increase augments our capacity for realizing our purposes are unwise. The world needs wisdom as it has never needed it before; and if knowledge continues to increase, the world will need wisdom in the future even more than it does now."

REFERENCES

Adkisson, P. L., R. E. Frisbie, and S. A. Joiner. (1987). A list of publications issued by the Consortium for Integrated Pest Management: 1979–1987. Texas A&M Univ., College Station. 135 pp.

Allen, G. E. (1980). "Integrated pest management," *BioSci.* **30**:665–701.

Alocilja, E. C., and J. T. Ritchie. 1988. "Upland rice simulation and its use in multicriteria optimization." IBSNAT, Univ. Hawaii. Research Report No. 1. 95 pp.

Altieri, M. (1987). *Agroecology: The Scientific Basis of Sustainable Agriculture,* 2nd ed. Westview Press, Boulder, CO. 160 pp.

Andow, D. (1983). "Effect of agricultural diversity on insect populations." In *Environmentally Sound Agriculture* (Ed. W. Lockeretz). pp. 91–115. Praeger, New York.

Axinn, G. H., and N. W. Axinn. (1987). "Farming systems research in its macropolicy dimensions." Proc. Farming Systems Research Symposium 1987. Farming Systems Research Paper Series, Paper No. 15. Univ. Ark. and Winrock International Inst. for Ag. Dev. pp. 461–73.

Ayers, G. S., R. A. Hoopingarner, and A. J. Howitt. (1984a). "Diversionary plantings for reduction of pesticide related bee mortality: Part 1. An introduction to the concept of diversionary plantings," *Am. Bee J.* **124(5)**:360–62.

Ayers, G. S., R. A. Hoppingarner, and A. J. Howitt. (1984b). "Diversionary plantings for reduction of pesticide related bee mortality: Part III. Initial attempts to divert bees from pesticide treated orchards," *Am. Bee J.* **124(7)**:514–16.

Baker, K. F., and R. J. Cook. (1974). *Biological Control of Plant Pathogens.* Freeman, San Francisco. 433 pp.

Baker, R. T. (1985). "Biological control of plant pathogens: Definition." In *Biological Control in Agricultural IPM Systems* (Eds. M. A. Hoy and D. C. Herzog). pp. 25–40. Academic Press, New York. 589 pp.

Beckendorf, S. K., and M. A. Hoy. (1985). "Genetic improvement of arthropod natural enemies through selection, hybridization, or genetic engineering techniques." In *Biological Control in Agricultural IPM Systems* (Eds. M. S. Hoy and D. C. Herzog). pp. 167–86. Academic Press, New York. 589 pp.

Beckman, C. H. (1987). *The Nature of Wilt Diseases of Plants.* APS Press, St. Paul, MN. 175 pp.

Beer, S. (1985). *Diagnosing the System for Organizations.* Wiley, New York. 152 pp.

Bender, M. (1984). "Industrial versus biological traction on the farm," In *Meeting the Expectations of the Land* (Eds. W. Jackson, W. Berry, and B. Colman). pp. 87–105. North Point Press, San Francisco, 250 pp.

Berger, R. D. (1975). "Disease incidence and infection rates of *Cercospora apii* in plant spacing plots," *Phytopathol.* **65**:485–87.

Berry, W. (1976). "Where cities and farms come together." In *Radical Agriculture* (Ed. R. Merrill). p. 16. Harper Colophon Books, New York. 549 pp.

Berry, W. (1987). "Whose head is the farmer using? Whose head is using the farmer?" In *Meeting the Expectations of the Land* (Eds. W. Jackson, W. Berry, and B. Coleman). pp. 19–30. North Point Press, San Francisco.

Berry, W. (1978). *The Unsettling of America Culture and Agriculture.* Avon, New York. 228 pp.

Bessey, E. A. (1911). *Root-knot and Its Control. USDA Bureau of Plant Industry Bull.* No. 217. 87 pp.

Bird, G. W. (1987a). "Alternate futures of agricultural pest management." *Amer. J. Alt. Agric,* **2**:25–29.

Bird, G. W. (1987b). "Role of Nematology in Integrated Pest Management Programs." In *Vistas in Nematology* (Eds. J. A. Veech and D. W. Dickson). pp. 114–21. Society of Nematologists, Hyattsville, MD. 509 pp.

Bird, G. W., R. L. Tummala, and S. H. Gage. (1985). The role of systems science and data management in nematology." In *An Advanced Treatise on Meloidogyne,* Volume II Methodology (Eds. K. R. Barker, C. C. Carter, and J. N. Sasser), pp. 159–73. Department of Plant Pathology, North Carolina State Univ. and the U.S. Agency for International Development. 223 pp.

Blakeman, J. P., and N. J. Fokkema. (1982). "Potential for biological control of plant diseases on the phylloplane," *Ann. R. Phyto.* **20**:167–92.

Boethel, D. J., and R. D. Eikenbary, Eds. (1986). *Interaction of Plant Resistance and Parasitoids and Predators of Insects.* Ellis Horwood, Ltd., West Sussex. 224 pp.

Boehncke, E. (1985). "The role of animals in a biological farming system". In *Sustainable Agriculture and Integrated Farming Systems, 1984 Conference Proceedings* (Eds. T. C. Edens, C. Fridgen, and S. L. Battenfield). pp. 22–33. Mich. State Univ. Press, East Lansing. 344 pp.

Braunholtz, J. T. (1986). "Crop protection: The role of the chemical industry in an uncertain future," *Philos. Trans. R. Soc. London, Ser. B* **295**:19–34. In National Research Council, Committee on Strategies for the Management of Pesticide Resistant Pest Populations, *Pesticide Resistance: Strategies and Tactics for Management,* National Academy Press, Washington, DC.

Brooks, F. P. Jr. (1975). *The Mythical Man-Month: Essays in Software Engineering.* Addison-Wesley, Reading, MA. 195 pp.

Burdon, J. J., and G. A. Chilvers. (1982). "Host density as a factor in plant disease ecology," *Ann. R. Phyto.* **20**:143–66.

Burn, A. J., T. H. Coaker, and P. C. Jepson. (1987). *Integrated Pest Management.* Academic Press, New York. 474 pp.

Carruthers, R. I. (1981). "The biology and ecology of *Entomophthora muscae* in the onion agroecosystem." Ph.D. dissertation, Michigan State Univ., East Lansing. 234 pp.

Carruthers, R. I., G. H. Whitfield, and D. L. Haynes. (1985). "The impact of pesticides on natural enemies of the onion maggot," *Entomophaga* **30**:151–61.

Carruthers, R. I., G. H. Whitfield, R. L. Tummala, and D. L. Haynes. (1986). "A systems approach to research and simulation of insect pest dynamics in the onion agroecosystem," *Ecol. Model.* **33**:101–21.

Carson, R. (1962). "Silent Spring. " Houghton Mifflin, Boston. 368 pp.

Carson, R. (1973). "Silent Spring." In *From Conservation to Ecology* by C. Pursell. Thomas Crowell, New York. 69 pp.

Casagrande, R. A. (1985). "The 'Iowa' potato beetle, *Leptinotarsa decemlineata,*" *Bull. Ent. Soc. Am.* **31(2)**:27–29.

Casagrande, R. A. (1988). "The Colorado potato beetle: 125 years of mismanagement," *Bull. Ent. Soc. Am.* **34(1)**:37–45.

Chandler, A. E. F. (1968). "Some factors influencing the occurrence and site of oviposition by aphidophagous Syrphyidae (Diptera)," *Ann. Appl. Biol.* **61**:435.

Checkland, P. (1981). *Systems Thinking, Systems Practice.* Wiley, New York. 330 pp.

Churchman, C. W. (1968). *The Systems Approach.* Dell, New York. 243 pp.

Coli, W. M., T. A. Green, T. A. Hosmer, and R. J. Prokopy. (1985). "Use of visual traps for monitoring insect pests in the Massachusetts apple IPM program," *Agric. Ecosystems and Environ.* **14:**251–65.

Commoner, Barry. (1969). "Can We Survive." In *From Conservation to Ecology* by C. Pursell. Thomas Crowell, New York. 69 pp.

Cook, R. J., and K. F. Baker. (1983). *The Nature and Practice of Biological Control of Plant Pathogens.* American Phytopathological Society, St. Paul, MN. 539 pp.

Cournoyer, B. M. (1970). "Crown Rust Epiphytology with Emphasis on the Quality and Periodicity of Spore Dispersal from Heterogeneous Oat Cultivar—Rust Race Populations." Ph.D. dissertation, Iowa State Univ. Ames. 244 pp.

Cox, G. W., and M. D. Atkins. (1979). *Agricultural Ecology: An Analysis of World Food Production Systems.* W. H. Freeman, San Francisco. 721 pp.

Croft, B. A. (1982). "Apple pest management." In *Introduction to Insect Pest Management,* 2nd ed. (Eds. R. L. Metcalf and W. H. Luckmann). pp. 465–98. Wiley, New York. 577 pp.

Croft, B. A., and S. A. Hoying. (1977). "Competitive displacement of *Panonychus ulmi* (Koch) by *Aculus schlechtendali* Nalepa in apple orchards," *Can. Entom.* **109**:1025–35.

Croft, B. A., and S. C. Hoyt, eds. (1983). *Integrated Management of Insect Pests of Pome and Stone Fruits.* Wiley, New York.

Cronon, W. (1983). *Changes in the Land, Indians, Colonists and the Ecology of New England.* Hill and Wang, New York. 241 pp.

Curl, E. A. (1963). "Control of plant diseases by crop rotation," *Biol. Rev.* **29**:413–79.

Day, P. R. (1973). "Genetic variability of crops," *Ann. R. Phyto.* **11**:293–312.

Dempster, J. P. (1969). "Some effects of weed control on the numbers of small cabbage white (*Pieris rapae* L). on brussel sprouts," *J. Appl. Ecol.* **6**:339.

Dempster, J. P., and T. H. Coaker. (1974). "Diversification of crop ecosystems as a means of controlling pests." In *Biology of Pest and Disease Control* (Eds. D.P. Jones and M.E. Solomon). Wiley, New York.

Doutt, R. L., and J. Nakata. (1972). "The *Rubus* leafhopper and its egg parasitoid: An endemic biotic system useful in grape pest management," *Environ. Entomol.* **2(3)**:381–86.

Dover, M. J. (1985). *A Better Mousetrap: Improving Pest Management for Agriculture.* World Resources Institute, Washington, DC. 84 pp.

Dover, M. J. , and B. A. Croft. (1986). "Pesticide resistance and public policy," *BioSci.* **36(2)**:78–85.

Dover, M., and L. M. Talbot. (1987). *To Feed the Earth: Agro-Ecology for Sustainable Development.* World Resources Institute, Washington, DC. 88 pp.

Drummond, F. A. (1982). "Post harvest biology of the onion maggot, *Hylemya antiqua* (Meigen)." M.S. thesis, Michigan State Univ., East Lansing. 353 pp.

Drummond, F. A. (1986). The biology of *Chrysomelobia labidomerae* Eickwort, and its potential to control the Colorado potato beetle." Ph.D. dissertation, Univ. Rhode Island, Kingston. 200 pp.

Drummond, F. A., E. Groden, D. L. Haynes, and T. C. Edens. (1988). "Some aspects of the biology of a predaceous Anthomyiid fly, *Coenosia tigrina.*" *Great Lakes Entomol.* **22**: 11-18.

Drummond, F. A., R. G. Van Driesche, and P. A. Logan. (1985). "Model for the temperature-dependent emergence of overwintering *Phyllonorycter crataegella* (Clemens) (Lepidoptera: Gracillariidae), and its parasitoid, *Sympiesis marylandensis* Girault (Hymenoptera: Eulophidae)," *Environ. Entomol.* **14**:305–11.

Edens, T. C., and D. L. Haynes. (1982). "Closed system agriculture: Resource constraints, management options, and design alternatives," *Ann. R. Phyto.* **20:** 363–95.

Edens, T. C., and H. G. Koening. (1980). "Agroecosystem management in a resource-limited world." *BioSci.* **30(10)**:697–701.

Edens, T. C., and G. Motyka. (1983). "A comparison of heterogeneity and abundance of pests and beneficials across a spectrum of pesticide and cultural controls." IPM Technical Report No. XX. Michigan State Univ., East Lansing.

Edwards, C. A. and J. R. Lofty. (1972). *Biology of Earthworms*. Bookworm Publ. Co., London. 283 pp.

Ekesbo, I. (1981). "Some aspects of sow health and housing," In *The Welfare of Pigs*. (Ed. W. Sybesma), pp. 250–64. Martinus Nijhoff, The Hague. 334 pp.

Elliott, A. P. (1980). "Ecology of *Pratylenchus penetrans* associated with navy beans (*Phaseolus vulgaris* L)." Ph.D. dissertation, Michigan State Univ., East Lansing. 351 pp.

Feldman, R. M., and G. L. Curry. (1982). "Operations research for agricultural pest management." *Operat. Res.* **30(4)**:601–18.

Ferris, H. (1978). "Nematode economic thresholds: Derivation, requirements, and theoretical considerations," *J. Nematol.* **10**:341–50.

Ferron, P. (1978). "Biological control of insect pests by entomogenous fungi," *Ann. Rev. Eutomol.* **23**:409–42.

Flaherty, D. L., L. T. Wilson, V. M. Stern, and H. Kido. (1985). "Biological control in San Joaquin valley vineyards." In *Biological Control in Agricultural IPM Systems*. (Eds. M. A. Hoy and D. C. Herzog). pp. 501–20. Academic Press, Orlando, FL. 589 pp.

Flint, M. L., and R. van den Bosch. (1983). *Introduction to Integrated Pest Management*. Plenum Press, New York. 240 pp.

Forgash, A. J. (1985). "Insecticide resistance in the Colorado potato beetle," *Proceedings of the Symposium on the Colorado Potato Beetle* (Eds. D. N. Ferro and R. H. Voss). Mass. Agr. Exp. Stn. Res. Bull. 704, Amherst.

Frank, J. R. (1981). "Integrated pest management: Present and future." Proc. of Symposium, sponsored by Weed Control and Pest Management Working Group of the American Society for Horticultural Science. *Hort. Sci.* **16**:500–16.

Frisbie, R. E., P. Weddle, and T. J. Dennehy. (1986). "The role of cooperative extension and agricultural consultants in pesticide resistance management." In *Pesticide Resistance: Strategies and Tactics for Management*, National Research Council, pp. 410–20. National Academy Press, Washington DC.

Fry, W. E. (1981). "Management of late blight." In *Advances in Potato Pest Management* (Eds. J. H. Lashomb and R. A. Casagrande). pp. 244-60.

Fry, W. E., and H. D. Thurston. (1981). "The relationship of plant pathology to integrated pest management," *BioSci.* **30(10)**:665–69.

Fuxa, J. R. (1987). "Ecological considerations for the use of entomophathogens," *Ann. Rev. Entomol.* **32:**225–52.

Gauthier, N. L., R. N. Hofmaster, and N. Semel. (1981). "History of Colorado potato beetle control." In *Advances in Potato Pest Management* (Eds. J. H. Lashomb and R. Casagrande). pp. 13–33. Hutchinson Ross, Stroudsburg PA. 289pp.

Georghiou, G. P. (1986). "The magnitude of the resistance problem." In *Pesticide Resistance: Strategies and Tactics for Management*, National Research Council, pp. 14–37. National Academy Press, Washington, DC.

Gibson, J. A. S., and T. Jones. (1977). "Monoculture as the origin of major forest pests and diseases." In *Origins of Pest Parasite, Disease, and Weed Problems,* (Eds. J. M. Cherret and G. R. Sogar). pp. 139–61. *Blackwell Sci. Publ.*, London.

Gleissman, S. R. (1984). "An agroecological approach to sustainable agriculture," In *Meeting the Expectations of the Land* (Eds. W. Jackson, W. Berry, B. Colman), pp. 160–71. North Point Press, San Francisco. 250 pp.

Groden, E. (1982). "The interactions of root maggots and two parasitoids, *Aleochara bilineata* (Gyll.) and *Aphaereta pillipes* (Say)." M.S. thesis, Michigan State Univ., East Lansing. 152 pp.

Groden, E. (1988). "Natural mortality of the Colorado potato beetle." Ph.D. dissertation, Michigan State Univ., East Lansing. 200 pp.

Guillot, J. F., J. P. LaFont, and E. Chaslus-Dancla. (1983). "Antibiotherapie on medecine veterinaire et antibioresistance en pathologie animale," *Rec. Med. Vet.* **159**:581–91.

Habicht, W. A. (1975). "The nematicial effect of various rates of raw and composted sewage sludge as soil organic amendments on a root-knot nematode." *Plant Dis. Rep.* **59**:631–34.

Harwood, R. (1988.). "Developing sustainable agriculture in the third world: Lessons from Indo-U.S. experience (draft)." Presented at the Winrock International Colloquium of Future U.S. Development Assistance: Food, Hunger, and Agricultural Issues, Petit Jean Mountain, Morrilton, AR. February 17–19, 1988.

Haynes, D. L. (1973). "Population management of the cereal leaf beetle," In *Insects: Studies in Population Management* (Eds. P. W. Geier, L. R. Clark, D. J. Anderson and H. A. Nix). pp. 232–40. Ent. Soc. Aus. (Memoirs I), Canberra.

Haynes, D. L., R. K. Brandenbury, and P. D. Fisher. (1973). "Environmental monitoring network for pest management systems," *Environ. Entomol.* **2**:889–899.

Haynes, D. L., R. L. Tummala, and T. C. Ellis. (1981). "Ecosystem management for pest control," *BioSci.* **30(10)**:690–96.

Headley, J. C. (1975). "The economics of pest management." In *Introduction to Insect Pest Management* (Eds. R. L. Metcalf and W. H. Luckman). pp. 75–98. Wiley, New York. 587 pp.

Hendrickson, R. M., and J. A. Plummer. (1983). "Biological control of alfalfa blotch leafminer in Delaware," *J. Econ. Entom.* **76**:757–61.

Hightower, J. (1976). "Hard tomatoes, hard times: The failure of the land grant college complex." In *Radical Agriculture* (Ed. R. Merrill). pp. 87–110. Harper Colophon Books, New York. 459 pp.

Hill, S. B. (1984). "Controlling pests ecologically,"*Soil Assoc. Quart. Rev.* pp. 13–15.

Hill, S. B., and P. Ott. (1982). *Basic Techniques in Ecological Farming*. Birkhauser, Basel, Switzerland. 365 pp.

Hoitink, H. A. J., and P. C. Fahy. (1986). "Basis for the control of soilborne plant pathogens with composts." *Ann R. Phyto.* **24**:93–114.

Hoy, M. A., and D. C. Herzog. (1985). *Biological Control in Agricultural IPM System.* Academic Press, New York. 589 pp.

Hsiao, T. H., and F. G. Holdaway. (1966). "Seasonal history and host synchronization of *Lydella grisescens* (Diptera: Tachinidae) in Minnesota," *Ann. Entomol. Soc.* **59**:125–33.

de Janvry, A., and E. P. LeVeen. (1986). "Historical forces that have shaped world agriculture: A structural perspective." In *New Directions for Agriculture and Agricultural Research* (Ed. K. A. Dahlberg). pp. 83–106. Rowman and Allanheld, Totowa, NJ. 436 pp.

Jatala, P. (1986). "Biological control of plant-parasitic nematodes," *Ann. R. Phyto.* **24**:453–89.

Jones, A. L. (1986). "Role of wet periods in predicting foliar diseases," In *Plant Disease Epidemiology: Population Dynamics and Management* (Eds. K. J. Leonard and W. E. Fry). pp. 87–100. Macmillan, New York. 372 pp.

Jones, A. L., S. L. Lillevik, P. D. Fisher, and T. C. Stebbins. (1980). "A microcomputer-based instrument to predict primary apple scab infection periods," *Plant Dis. Rep.* **64**:69–72.

Jones, D. G. (1987). *Plant Pathology: Principles and Practice.* Open Univ. Press, Stratford. 191 pp.

Jones, D. P. (1973). "Agricultural entomology." In *History of Entomology* (Eds. R. F. Smith, T.E. Mittler, and C. N. Smith). pp. 307–32. Annual Reviews, Palo Alto, CA.

Kendrick, J. Jr. (1988). IPM Technical Report No. 46, Michigan State Univ.

King, E. G., K. R. Hopper, and J. E. Powell. (1985). "Analysis of system for biological control of crop arthropod pests in the United States by augmentation of predators and parasites," In *Biological Control in Agricultural IPM System.* (Eds. M. A. Hoy and D. C. Herzog). pp. 201–25. Academic Press, New York. 589 pp.

Kloppenburg, J., and D. L. Kleinman. (1987). "The plant germplasm controversy," *BioSci.* **37(3)**:190–98.

Kovach, J. K., and J. P. Tette. (1988). "A survey of the use of IPM by New York apple producers," *Agric. Ecosystems and Environ.* **20**:101–08.

Krause, R. A., L. B. Massie, and R. A. Hyre. (1975). "Blitecast: A computerized forecast of potato late blight," *Plant Dis. Rep.* **59**:95–98.

Large, E. C. (1940). *The Advance of the Fungi.* Dover, New York. 488 pp.

Lewis, K. (1961). "Influence of food on fecundity and longevity of *Itoplectis conquisitor* (Say) (Hymenoptera:Ichnewmonidae)," *Can. Entom.* **95**:202–07.

Lewis, K. (1963). "Effects of pollens on fecundity and longevity of adult *Scambus buolianae* (Htg.) (Hymenoptera:Ichneumonidae)," *Can. Entom.* **93**:202–07.

Litsinger, J. A., and K. Moody. (1976). "Integrated pest management in multiple cropping systems," In *Multiple Cropping*, Spec. Publ. 27 (Eds. R. I. Papendick, P. A. Sanchez, and G. B. Triplett). Amer. Soc. Agron., Madison, WI.

Lorimer, N. (1981). "Genetic means for controlling agricultural insects," In *CRC Handbook of Pest Management in Agriculture*, vol. II (Ed. D. Pimentel). pp. 299–314. CRC Press, Boca Raton, FL. 501 pp.

Louvet, J., C. Alabouvette, and F. Rouxel. (1981). "Microbiological suppressiveness of some soils to Fusarium wilts," In *Fusarium: Diseases, Biology, and Taxonomy* (Eds. P. E. Nelson, T. A. Tousson, and R. J. Cook). pp. 261–75. Penn. State Univ. Press, University Park.

Luck, R. I. (1981). "Parasitic insects introduced as biological control agents for arthropod pests," In *CRC Handbook of Pest Management in Agriculture*, vol. II (Ed. D. Pimentel). pp. 125–284. CRC Press, Boca Raton, FL. 501 pp.

Luckmann, W. H., and R. L. Metcalf. (1982). "The pest-management concept," In *Introduction to Insect Pest Management*, 2nd ed. (Eds. R. L. Metcalf and W. H. Luckmann). pp. 1-32. Wiley, New York. 577 pp.

MacKenzie, D. R., and Z. Smilowitz. (1981). "Automated pest management," In *Advances in Potato Pest Management* (Eds. J. H. Lashomb and R. A. Casagrande). pp. 261–67. Hutchinson Ross, Stroudsburg, PA. 288 pp.

Matteson, P. C., M. A. Altieri, and W. C. Gagne. (1984). "Modification of small farmer practices for better pest management," *Ann. Rev. Entomol.* **29**:383–402.

Meister, R. T. (1986). *Farm Chemicals Handbook*. Meister, Willoughby, OH.

Merrill, R. (1976). "Toward a self-sustaining agriculture," In *Radical Agriculture* (Ed. R. Merrill). pp. 284-327. Harper Colophon Books, New York. 459 pp.

Merritt, R. W., and J. R. Anderson. (1977). "The effects of different pasture and rangeland ecosystems on the annual dynamics of insects in cattle droppings," *Hilgardia* **45**:35–71.

Metcalf, R. L. (1982). "Insecticides in pest management." In *Introduction to Insect Pest Management*, 2nd ed. (Eds. R. L. Metcalf and W. H. Luckmann). pp. 217–78. Wiley, New York. 577 pp.

Michelbacher, A. E., and R. F. Smith. (1943). "Some natural factors limiting abundance of the alfalfa butterfly," *Hilgardia* **15**:369–97.

Mills, W. D., and A. A. LaPlante. (1951). "Diseases and insects in the orchard," *Cornell Ext. Bull.* **711**:21–27.

Miser, H. J., and E. S. Quade. (1985). *Handbook of Systems Analysis: Overview of Uses, Procedures, Application and Practice*, Vol. 1. Wiley, New York. 346 pp.

Mortinson, T. E., J. P. Nyrop, and C. J. Eckenroad. (1988). 'Dispersal of the onion fly (Diptera: Anthomyiidae) and larval damage in rotated onion field," *J. Econ. Entom.* **81**:509–14.

Myrdal, G. (1972). *Against the Stream: Critical Essays on Economics*. Pantheon Books, Random House, New York. 336 pp.

Niederstucke, F. H. (1982). "On the economy of health promoting measures in pig production," *Deutsche Wechenschrift* **89**:370–73.

Nusbaum, C. J., and H. Ferris. (1973). "The role of cropping systems in nematode population management," *Ann. R. Phyto.* **11**:423-40.

Obrycki, J. J. (1986). "The influence of foliar pubescence on entomophagous species," In *Interaction of Plant Resistance and Parasitoids and Predators of Insects* (Eds. D. J. Boethel and R. D. Eikenbary). pp. 61–83. Ellis Horwood, West Sussex. 244 pp.

Oelhaf, R. C. (1978). *Organic Agriculture*. Allanheld, Osmun, Totowa, NJ. 271 pp.

Palti, J. (1981). *Cultural Practices and Infectious Crop Diseases*. Springer-Verlag, New York. 243 pp.

Pedigo, L. P. (1989). *Entomology and Pest Management*. Macmillan, New York. 646 pp.

Perleman, M. (1976). "Efficiency in agriculture: the economics of energy," In *Radical Agriculture* (Ed. R. Merrill). pp. 64–86. Harper Colophon Books, New York. 459 pp.

Pimentel, D. (1981). *CRC Handbook of Pest Management in Agriculture*, vol. III. CRC Press, Boca Raton, FL.

Pimentel, D., L. E. Hurd, A. C. Bellotti, M. J. Forster, I. N. Oka, O. D. Sholes, and R. J. Whitman. (1973). "Food production and the energy crisis," *Science* **182**:443–49.

Pimentel, D., and S. Pimentel. (1986). Energy and other natural resources used by agriculture and society. In *New Directions for Agriculture and Agricultural Research* (Ed. K. A. Dahlberg). pp. 259–90. Rowman and Allanheld, Totowa, NJ. 436 pp.

Poincelot, R. P. (1986). *Toward a More Sustainable Agriculture*. AVI, Westoport, CT. 241 pp.

Potts, G. R. (1977). "Some effects of increasing the monocultures of cereals." In *Origins of Pest, Parasite, Disease, and Weed Problems* (Eds. J. M. Cherret and G. R. Sogar). pp. 183–204. Blackwell Sci. Publ., London.

Prokopy, R. J., and G. L. Hubbell. (1981). "Susceptibility of apple to injury by tarnished plant bug adults," *Environ. Entomol.* **10**:977–79.

Rajotte, E. G., R. F. Kazmierczak, G. W. Norton, M. T. Lambur, and W. A. Allen. (1987). *The Natioinal Evaluation of Extension's Integrated Pest Management (IPM) Programs*. VCES, Virginia. 123 pp.

Read, D. P., P. P. Feeny, and R. B. Root. (1970). "Habitat selection by the aphid parasite *Diaeretiela rapae* (Hymenoptera: Braconidae) and hyperparasite *Charips brassicae* (Hymenoptera:Cynipidae)," *Can. Entom.* **102**:156–57.

Riley, C. V. (1871). *Third annual report on the noxious beneficial, and other insects of the state of Missouri.* Horace Wilcox, Jefferson City, MO. 176 pp.

Risch, S. J. (1981). "Insect herbivore abundance in tropical monocultures and polycultures: An experimental test of two hypotheses," *Ecology* **62(5)**-:1325–40.

Roberts, D. (1981). "Using regulatory programs, biological methods, cultural practices, and resistant varieties to control diseases of plants," In *CRC Handbook of Pest Management in Agriculture*, vol. II (Ed. D. Pimentel). pp. 69–78. CRC Press, Boca Raton, FL. 501 pp.

Robinson, R. A. (1987). *Host Management in Crop Pathosystems*. Macmillan, New York. 263 pp.

Roitberg, B. D., and R. J. Prokopy. (1986). "Insects that mark host plants: An ecological evolutionary perspective on host-marking chemicals," *BioSci.* **37(6)**:400–06.

Ruesink, W. (1976). "Status of the systems approach to pest management," *Ann. Rev. Entomol.* **21**:27–46.

Sailer, R. (1981). "Extent of biological and cultural control of insect pests of crops," In *CRC Handbook of Pest Management in Agriculture*, vol. II (Ed. D. Pimentel). pp. 57–67. CRC Press, Boca Raton, FL. 501 pp.

Sambraus, H. H. (1984). "Advantages and disadvantages of modern housing systems for cattle," *Tieraerztl. Umsch.* **39**:399–404.

Scriber, J. M. (1984). "Nitrogen nutrition of plants and insect invasion," In *Nitrogen in Crop Production* (Ed. D. Hauck). pp. 441–60. *Am. Soc. Agron.*, Madison, WI.

Shoemaker, C. A. (1985). "Integrating population dynamics with economic thresholds in alfalfa pest management," In *Integrated Pest Management on Major Agri-*

cultural Systems, (Eds. R. E. Frisbe and P. L. Adkisson). pp. 567–86. Texas Agric. Expt. Sta. MU-1616.

Shoemaker, C. A. and D. W. Onstad. (1983). "Optimization analysis of the integration of biological, cultural, and chemical control of alfalfa weevil (Coleoptera: Curculionidae)," *Environ. Entomol.* **12**:286–95.

Smilowitz, Z. (1981). "GPA-CAST: A computerized model for green peach aphid management on potatoes," In *Advances in Potato Pest Management* (Eds. J. H. Lashomb and R. A. Casagrande). pp. 193–203. Hutchinson Ross, Stroudsburg,PA. 288 pp.

Smith, R. F., and H. T. Reynolds. (1972). "Effects of manipulation of cotton agroecosystems on insect pest populations," In *The Careless Technology* (Eds. M. T. Farvar and J. P. Milton). Natural History Press, New York.

Splittstoesser, C. M. (1981). "The use of pathogens in insect control," In *CRC Handbook of Pest Management in Agriculture,* vol. II (Ed. D. Pimentel). pp. 285–98. CRC Press, Boca Raton, FL. 501 pp.

Stehr, F. W. (1982). "Parasitoids and Predators in Pest Management." In *Introduction to Insect Pest Management,* 2nd ed. (Eds. R. L. Metcalf and W. H. Luckmann). pp. 135–74. Wiley, New York. 577 pp.

Steinhart, C., and J. Steinhart. (1974). *The Fires of Culture: Energy Yesterday and Tomorrow.* Wadsworth, Belmont, CA. 273 pp.

Stimac, J. L., and R. J. O'Neil. (1985). "Integrating influences of natural enemies into models of crop/pest system," In *Biological Control in Agricultural IPM Systems.* (Eds. M. A. Hoy and D. C. Herzog). pp. 323–43. Academic Press, New York. 589 pp.

Summer, D. R., and R. H. Littrell. (1974). "Influence of tillage, planting date, inoculum survival, and mixed populations on epidemiology of southern corn leaf blight," *Phytopatho.* **64**:168–73.

Symes, P. A. (1975). "The effect of flowers on the fecundity and longevity of two native parasites of the European pine shoot moth in Ontario," *Environ. Entomol.* **4**:337-46.

Teng, P. S. (1988). "Pests and pest loss models." IBSNAT, Univ. Hawaii. *Agrotech. Transfer* **8**:1–10.

Toumanoff, C. (1933). "Actions des champignons entomophytes sur la pyrale du mais (*Pyrausta nubilalis* Hubner)," *Ann. Parasitol. Humaine Comparee.* **11**:129–43.

Townes, H. (1958). "Some biological characteristics of the Ichneumonidae (Hymenoptera) in relation to biological control," *J. Econ. Entom.* **51**:650–52.

Tummala, R. L., and D. L. Haynes. (1977). "On-line pest management systems," *Environ. Entomol.* **6**:339–49.

van den Bosch, R. V. (1978). *The Pesticide Conspiracy.* Doubleday, Garden City, New Jersey. 226 pp.

van der Plank, J. E. (1963). *Plant Diseases: Epidemics and Control.* Academic Press, New York. 349 pp.

Van Gundy, S. D. (1985). "Biological control of nematodes: Status and prospects in agricultural IPM systems," In *Biological Control in Agricultural IPM Systems* (Eds. M. A. Hoy and D. C. Herzog). pp. 467–77. Academic Press, New York. 589 pp.

Waggoner, P. E. (1962). "Weather, space, time, and chance of infection," *Phytopathol.* **52**:1100–08.

Wakeley, P. C. (1954). *Planting the southern pines*. USDA Agric. Managr. No. 18. 233 pp.

Way, M. F. (1977). "Pest and disease status in mixed stands versus monocultures: The relevance of ecosystem stability," In *Origins of Pest, Parasite, Disease, and Weed Problems* (Eds. J. M. Cherret and G. R. Sogar). pp. 127–38. Blackwell Sci. Publ., London.

Weseloh, R. M. (1978). "Seasonal and spatial mortality patterns of *Apanteles melanoscelus* predators and gypsy moth hyperparasites," *Environ. Entomol.* **7(5)**:662–65.

Weseloh, R. M. (1979a). "Competition among gypsy moth hyperparasites attacking *Apanteles melanoscelus*, and influence of temperature on their field activity," *Environ. Entomol.* **8(1)**:86–90.

Weseloh, R. M. (1979b). "Comparative behavioral responses of three *Brachymeria* species and other gypsy moth parasitoids to humidity and temperature," *Environ. Entomol.* **8(4)**:670–75.

Whalon, M. E., and B. A. Croft. (1984). "Apple IPM implementation in North America," *Ann. Rev. Entomol.* **29**:435–70.

Whitcomb, W. H. (1981). "The use of predators in insect control," In *CRC Handbook of Pest Management in Agriculture*, vol. II (Ed. D. Pimentel). pp. 105–23. CRC Press, Boca Raton, FL. 501 pp.

Whitfield, G. H. (1981). "Spatial and temporal population dynamics of the onion maggot, *Hylemya antiqua*, in Michigan." Ph.D. dissertation, Michigan State Univ., East Lansing. 379 pp.

Wilkes, G. (1977). "The world's crop plant germplasm—an endangered resource," *Bull. Atom. Sci.* 33:8–16.

Wilson, B. (1984). *Systems Concept, Methodologies and Applications*. Wiley, New York. 339 pp.

Wright, A. (1984). "Innocents abroad: American agricultural research in Mexico," In *Meeting the Expectations of the Land* (Eds. Wes Jackson, W. Berry, and B. Colman). pp. 135–51. North Point Press, San Francisco. 250 pp.

Wright, R. J. (1984). "Evaluation of crop rotation for control of the Colorado potato beetle (Coleoptera: Chrysomelidae) in commercial potato fields on Long Island." *J. Econ. Entom.* **77**:1254–59.

Yeargan, K. V. (1985). "Alfalfa status and current limits to biological control." In *Agricultural IPM Systems* (Eds. M. Hog and D. Herzog). pp. 521–36. Academic Press, New York.

Zadoks, J. C. and P. Kampmeijer. (1977). "The role of crop populations and their deployment, illustrated by means of a simulator, EPIMUL 76," *Ann. N.Y. Acad. Sci.* **287**:164–90.

4 SUSTAINABLE WEED MANAGEMENT PRACTICES

MATT LIEBMAN

Department of Plant and Soil Sciences,
University of Maine,
Orono, Maine

RHONDA R. JANKE

Rodale Research Center,
Kutztown, Pennsylvania

Crop rotation
Intercropping systems
Allelopathic cover crops
Weed-suppressive crop varieties
Crop density and spatial arrangement
Soil fertility management
Manipulation of soil temperature and moisture
Insect pests and pathogens of weeds
Use of livestock
Cultivation and flame weeding
An integrated approach to sustainable weed management

INTRODUCTION

Weed management is a critical component of any farming system. Its importance is emphasized by the quantities of resources that farmers devote to reducing negative impacts of weeds on crop production and by the losses that farmers suffer from weeds they cannot kill or effectively suppress. Currently, U.S. farmers spend more than $6.2 billion annually to control weeds on crop and pasture land, including an estimated $3.6 billion for use of nearly 200 million kilograms of herbicides (Shaw, 1982; Pimentel and Levitan, 1986). About 50% of all tillage in the United States is performed specifically for weed control (McWhorter and Chandler, 1982), and one to three cultivation operations are common in many row crop production systems (Zimdahl, 1981). These mechanical weed control operations are accomplished with 4.5 million tractors that use up to 75 billion liters of crude oil annually (Wiese and Chandler, 1979). Despite applications of herbicides, use of cultivation, and other weed management technologies, weeds still deprive U.S. farmers of about 10% of their potential crop yields, an annual monetary loss of $7 to $12 billion (Chandler, 1981; Shaw, 1982). Any consideration of new and possibly more sustainable farming systems must deal with the ubiquitous and substantial threat posed by weeds.

Over the last several decades, weed management in North American agriculture has become increasingly dependent on herbicide applications, particularly as more farmers have shifted toward reduced or minimum tillage systems (Koskinen and McWhorter, 1986). Reliance on herbicides has also been affected by increases in farm size and decreases in crop diversity within farms (USDA, 1973); these factors make timely cultivation more difficult. A large number of farmers and researchers view herbicides as key ingredients for effective weed management and profitable farming, even when non-herbicide tactics such as cultivation and crop competition are brought to bear against weeds in an integrated manner (Baldwin and Santelmann, 1980; Hill, 1982; McWhorter and Shaw, 1982; Shaw, 1982; Aldrich, 1984).

Despite the current emphasis on herbicides in North American agriculture, several factors have recently led to a reappraisal of their use. First, discovery of herbicides in drinking water supplies has created concern among farmers and nonfarmers over unintended environmental and human health effects (Berteau and Spath, 1986; Cohen et al., 1986; Hallberg, 1986; Knudson, 1986; "Rice growers," 1987; Williams et al., 1988). Second, financial hardships currently facing many farmers have led to considerations that farm profitability might be increased by reducing use of purchased inputs (such as herbicides) if effective, but less costly, alternative production techniques were available (Papendick, 1987; Francis and King, 1988). Finally, farmers and researchers have become increasingly aware of full-time farms that operate profitably with little or no use of herbicides (USDA, 1980; Lockeretz et al., 1981; Culik et al., 1983; Thompson and Thompson, 1984; Sinclair, 1985, 1987a, 1987b). Weed management on these farms is dependent on a variety of careful biological and physical manipulations of crops, weeds, and their environment.

Based on experiences of farmers and results from research efforts, we believe a number of biological and physical practices offer opportunities for reducing heavy reliance on agricultural herbicides and for potentially improving farm profitability and environmental quality. These practices involve crop rotation, intercropping, allelopathic cover crops, weed-suppressive crop varieties, crop density and spatial arrangement, soil fertility management, manipulation of soil temperature and moisture characteristics, insect pests and pathogens of weeds, control of livestock grazing regimes, and specialized tillage and cultivation equipment. Many of these tactics can be used together in a complementary manner. Herbicides can be employed as tools to aid in the transition to alternative methods and to improve the effectiveness of biological and physical tactics, but smaller quantities should be required for effective weed control. In some cases use of these tactics may make herbicides unnecessary. Sustainable weed management thus involves the use of a variety of biological and physical weed management tactics in an integrated manner, and the use of herbicides only when these materials are environmentally appropriate and economically necessary.

The appropriateness of biological and physical weed management tactics will vary according to soil conditions, climate, crop and weed species, and other factors. However, to be effective, any of these tactics will require experience, fine-tuning, commitment to "getting the system to work," attention to appropriate timing, and some knowledge of the dynamic relationships among crops, weeds, and their environment. Sustainable weed management is thus thought- and information-intensive.

In the following pages we describe components of sustainable weed management systems. Our intent is to offer examples of successful approaches and to suggest other approaches that might be useful if refined by farmers and researchers. We do not believe that sustainable weed management strategies necessarily exclude the use of herbicides, but given the large amount of information about chemical weed control that is readily available, we have chosen to emphasize ecological and cultural approaches. Because we believe that many farmers have a wealth of experience and practical knowledge concerning biological and physical methods of weed control, we have intentionally included examples of on-farm weed management practices from popular literature sources. For further information concerning sustainable weed management, the reader is referred to Zimdahl (1980), Walker and Buchanan (1982), Radosevich and Holt (1984), Rosenthal et al. (1984), Aldrich (1984), and Altieri and Liebman (1988).

CROP ROTATION

Use of crop rotation establishes the framework for sustainable weed management. Crop rotation will not eliminate interference from weeds, but it can limit build-up of weed populations and prevent major shifts in weed species composition. Crops tend to be affected by particular weed species that possess similar growth habits and thrive under the same cultural conditions as the crop (Sumner, 1982). By growing sequences of crops that differ in planting and maturation dates, competitive characteristics, and soil management requirements, growth and reproduction of a given weed species can be disrupted. Crop rotation in conventional farming systems now implies rotating both crops and herbicides, but significant effects on weed growth can still be attributed directly to tillage and soil disturbance patterns and to interference from crops (Walker and Buchanan, 1982; Aldrich, 1984).

Aldrich (1984, citing MacHoughton, 1973) reported that corn-wheat-soybean rotations had much more stable weed community dynamics than did monocultures of the respective crops. In an experiment conducted in Illinois in which standard cultivation practices were used but no herbicides were applied, weed seed composition of the soil was measured after six years of continuous (monoculture) or rotational cropping. Continuous corn favored build-up of giant foxtail (*Setaria faberii* Herrm.) seeds; continuous wheat favored build-up of giant foxtail, crabgrass (*Digitaria sanguinalis* [L.] Scop.), and fall panicum (*Panicum dichotomiflorum* Michx.) seeds; and continuous soybeans favored build-up of giant foxtail and velvetleaf (*Abutilon theophrasti* Medic) seeds. In contrast, planting a rotation sequence of the three crops resulted in relatively small increases in giant foxtail, crabgrass, and fall panicum seeds and a decrease in velvetleaf seeds. Aldrich (1984) concluded that the monocultures provided maximum opportunities for the best-adapted weed species to increase, while the rotation limited these opportunities.

Forcella and Lindstrom (1988a), working in Minnesota, compared weed population dynamics in continuous corn and corn-soybean rotation systems grown with ridge and conventional tillage. No herbicides were used in a portion of the study, and dominant weed species were green foxtail (*Setaria viridis* [L.] Beauv.), yellow foxtail (*Setaria glauca* [L.] Beauv.), redroot pigweed (*Amaranthus retroflexus* L.), and common lambsquarters (*Chenopodium album* L.). Weed seed production was lower for both conventional and ridge tillage systems when crops were grown in the rotation. With conventional tillage, weeds produced an average of 25 seeds/m^2 in continuous corn, but only 4 seeds/m^2 in corn or soybeans grown in rotation. With ridge tillage, weeds produced an average of 915 seeds/m^2 in continuous corn, 12 seeds/m^2 in corn grown in rotation, and 0 seeds/m^2 in soybeans grown in rotation. These trends in weed seed production were paralleled by numbers of viable weed seeds buried in the soil of each treatment. The researchers noted that greater crop yield loss to weed interference was associated with greater numbers of buried viable weed seeds.

Yields of corn and soybeans grown with ridge tillage are generally equal to those grown with conventional tillage on light and medium textured soils, but in contrast to conventional tillage, ridge tillage greatly promotes soil conservation. Forcella and Lindstrom (1988b) examined the effects of herbicides on crop yields in ridge tilled continuous corn and corn-soybean rotation systems and found that after seven to eight years of good to excellent weed control with herbicides, crop yield losses in the absence of herbicides were 10% to 27% in continuous corn, compared to 0% for corn and 6% to 10% for soybeans in corn-soybean rotations. The researchers concluded that, at least for ridge tillage systems, withholding herbicides for one year or herbicide failure within a single cropping season would not be expected to affect crop yields appreciably in corn-soybean rotations, whereas continuous corn production requires herbicide application and high levels of herbicide efficacy each year.

In the corn-soybean-wheat and corn-soybean studies noted above, no green manure or hay crop was included in the rotation sequences. However, the value of including forage grasses and legumes within rotations should not be overlooked. In addition to their ability to maintain good soil structure and fertility, forage grasses and legumes, once established, are very effective in suppressing growth of annual weeds. This is a consequence of leaving the soil surface undisturbed, providing dense crop canopy cover and root development, and mowing, which has a much more severe effect on the growth of annual weeds than on forage grasses and legumes. Although certain perennial weeds such as quackgrass (*Agropyron repens* [L.] Beauv.) are well adapted to meadow environments (Sumner, 1982), tillage and competition from subsequent crops can help suppress growth of these species (for example, see Mortimer, 1983).

Data from a long-term cropping systems experiment at the Rodale Research Center (Kutztown, PA) indicated that weed growth was reduced 30% to 67% in corn following a sequence of small grain + red clover-red clover hay than in corn following a sequence of corn-soybeans (Peters et al., 1988a). No herbicides were used in these treatments and dominant weed species were annuals such as common lambsquarters, redroot pigweed, and giant foxtail.

From an ecological perspective, rotation sequences that include clean-cultivated annual crops (for example, corn or soybeans), densely planted and highly competitive grain crops (for example, barley, oats, or wheat), and mowed and untilled perennial crops (for example, alfalfa or forage grass/clover mixtures) are desirable, because they create an unstable and often unfavorable environment for survival and reproduction of annual and perennial weeds. Further improvements in weed management may be gained by including crops within the rotation sequence that have early spring, late spring, summer, and fall planting dates, thus challenging weed populations that have different optimal seasons for growth and establishment (Dotzenko et al., 1969). Densely planted, rapidly growing, short duration "smother crops" (such as buckwheat, sorghum-sudangrass, or Japanese millet) can also be considered as weed management options within crop rotation sequences, particularly when followed by harrowing and other forms of cultivation. Crops with midsummer maturation times (for example, in the northeastern United States: winter wheat, rye, lettuce, peas, or crimson clover planted as a green manure) offer opportunities for effective harrowing of perennial weeds before planting late season crops (for example, brassicas, overwintered spinach, or winter cover crops).

Despite the ecological advantages of using long, diversified crop rotations, short-term economics may favor use of cash-crop monocultures or short rotations of cash crops with very similar ecological characteristics (Stonehouse et al., 1988). One possibility for providing farmers with some of the ecological benefits of crop diversification without decreasing cash or crop opportunities is intercropping.

INTERCROPPING SYSTEMS

Intercropping involves growing two or more crops in mixture, as contrasted to sole cropping, in which a single crop is grown alone. Intercrop component crops may be planted simultaneously or on different dates; the latter practice is often called relay cropping. Forage legume species intercropped with small grains or row crops are sometimes referred to as seeding with a "nurse crop," and may also be called living mulches. Intercropping is prevalent in tropical areas of the world where human labor and draught animals are more abundant than agricultural machinery; these systems also have potential for increasing the biological and economic efficiency of temperate, mechanized farming systems (Horwith, 1985). As the following examples indicate, intercropping systems can have important, beneficial effects on weed management.

More than 20% of the agricultural herbicides applied each year in the United States are used to treat soybean fields (Eichers, 1981; Pimentel and Levitan, 1986). Intercropping soybean with wheat or barley may greatly decrease the need to use these herbicides, while providing acceptable crop yields and higher profits. Soybean/cereal intercropping systems use nonsynchronous planting dates for the two crops: wheat or barley is planted first, then interseeded with soybean. Both winter and spring cultivars of barley or wheat have been used in soybean/cereal intercropping systems. Soybean/grain intercropping systems have been used successfully in many states, including Illinois, Indiana, Kansas, Mississippi, Missouri, Nebraska, Ohio,Pennsylvania, and Tennessee (Jeffers and Triplett, 1979; Volak and Janke, 1987). In regions with growing seasons too short for double cropping (planting crops sequentially with no period of overlap), soybean/grain intercropping enables farmers to grow two cash crops per year on the same unit of land (Jeffers and Triplett, 1979; Peters et al. 1988b).

Reinbott et al. (1987) investigated the yield characteristics of soybean/winter wheat intercrops in Missouri. Wheat was planted in October, and soybean was interseeded with a no-till planter into immature wheat in April or May of the following year. Wheat grew above the younger soybean crop and grain was harvested with a combine. The soybean crop then grew to maturity amid the wheat stubble. Physical damage from field operations and interspecific interference reduced yields of intercropped wheat by 19% and intercropped soybean by 27%, as compared to conventional sole crops. However, combined yield of the intercropped species was approximately 50% higher than could be obtained from growing the two crops separately on the same total land area. Intercropped soybeans produced 28% more than double cropped soybeans (planted immediately after wheat harvest).

Recent studies have investigated the effects of soybean/cereal intercropping on weed management. Prostko and Ilnicki (1988) reported that, in New Jersey, intercropping soybeans into barley without herbicides gave acceptable control of fall panicum, redroot pigweed, and common ragweed (*Ambrosia artemisiifolia* L.), as compared to sole cropped soybeans treated with metolachlor and linuron. All interseeded soybeans yielded less than sole cropped soybeans, but the interseeded treatments without herbicide yielded as much as or more than interseeded treatments receiving herbicides.

Peters et al. (1988b) found that, in Pennsylvania, net economic returns from soybean/small grain intercrops grown without herbicides were greater than those from soybean sole crops grown with herbicides. Increased profitability of the intercropping systems reflected increased cash value from two crops (cereal + soybeans vs. soybeans alone) and lack of expenses for soybean herbicides. Depending on the year, intercropped soybeans produced between 83% and 114% of the sole cropped soybean yields. More weed growth was found in the intercropped soybeans, but weeds had little or no effect on yields. Double cropping was not investigated in this study.

Limitations to the soybean intercropping system should be noted. The system appears to work well under adequate rainfall and/or irrigation. However, on droughty soils, or in years of low rainfall, soybeans may fail to establish (for a review, see Volak and Janke, 1987). If the small grain crop is fertilized for maximum yield, the soybean crop will be suppressed. A lodged cereal crop can severely suppress interseeded soybeans. The timing of the boot stage and precise harvest date of the small grain are also important. Poor results were obtained when soybeans were drilled into oats (Volak and Janke, 1987) as compared to wheat and barley, probably because the oat crop was harvested two to four weeks later than the other cereals. A cereal that matures early, such as winter barley, will optimize soybean growth, but may be injured by drill traffic at the time of soybean planting if grain heads have become visible (Peters, 1988).

More than 52% of the agricultural herbicides applied in the United States are used to treat corn fields (Pimentel and Levitan, 1986). Recent research efforts suggest that intercropping corn (a summer annual species) with subterranean clover (a winter annual species) may have potential for decreasing the need to use corn herbicides in areas of the country where this clover species can overwinter successfully.

In experiments conducted in New Jersey, subterranean clover was found to grow primarily during spring and fall months, and to lay quiescent as a mulch between corn rows during the summer. Enache and Ilnicki (1987, 1988, 1989) compared treatments in which corn was grown alone using conventional tillage to treatments in which corn was planted without tillage into subterranean clover. The entire plot area was treated with herbicides in the conventional treatment, but herbicides were applied only in a band over the corn row in the intercrop treatment. A broad spectrum of weed species were present in the experiment: ivyleaf morningglory (*Ipomea hederacea* [L.] Jacq.), common lambsquarters, redroot pigweed, common ragweed, velvetleaf, horseweed (*Conyza canadensis* L.), jimsonweed (*Datura stramonium* L.), and fall panicum.

Depending on the year, weed biomass was 55% to 91% lower in the intercrop treatment than in the conventional treatment. Corn grain yields in the corn/clover intercrop were 9% to 23% lower than in the conventional treatment; corn silage yields were 22% to 35% lower in the intercrop. However, because of reduced input costs, economic returns from the intercrop system compared favorably with the conventional system (Enache and Ilnicki, 1988). Further work with this system appears promising and the investigators are expanding their studies (A. Enache, personal communication). Lanini et al. (1988) have initiated integrated studies concerning weed and insect pest management, soil nutrient dynamics, crop performance, and economic feasibility for corn/subterranean clover intercrops in California.

Mixtures of clover with small grain crops (such as barley, wheat, or oats) are another form of intercropping that can provide weed management benefits. Clover species can be sown with the cereal or, in the case of winter cereals, broadcast seeded onto frozen ground the following spring, forming an understory that has little or no effect on grain yield if water and nutrients are not limiting. Following grain harvest, the clover continues to grow, fixing atmospheric nitrogen and providing soil cover to limit erosion. Farmers commonly use this intercropping practice to establish legumes for hay, especially red clover and alfalfa. A tradeoff is made, however, between obtaining a cereal harvest plus one cutting of hay during the establishment year versus two or three cuttings of hay from a clear-seeded legume. Usually herbicides are used if hay crops are seeded without a small grain. The grain crop helps to suppress weeds in the intercrop mixture.

Dyke and Barnard (1976) found in three years of experiments in England that intercropping red clover with barley reduced growth of quackgrass by 42% to 62%, as compared to barley sole crop treatments. The researchers suggested that interplanting clovers with grain crops can be useful for slowing the spread of quackgrass if cultivation or herbicide spraying is delayed after grain harvest.

Studies by Vrabel et al. (1980), Scott et al. (1987), and others have demonstrated that clovers and other forage legumes can be seeded into row crops such as corn that have been growing for several weeks, without any detrimental effect on crop yields. Early season weed control in these systems is provided by cultivation, sometimes supplemented with rapidly degraded herbicides applied at the time of planting of the row crop. More research is needed concerning the effect of the interseeded legumes on late season weed growth.

Hartwig (1987) has developed no-tillage crop production systems in Pennsylvania using established sods of crownvetch; the perennial legume is maintained from year to year rather than being repeatedly plowed down and reseeded. Crownvetch sods help suppress weed establishment and growth, but are prevented themselves from interfering with crops by low, nonlethal doses of herbicides. Similar approaches are being explored with a variety of species in different areas of the country. In some studies, mowing has been used instead of herbicides to regulate growth of the established legume sods. Other researchers are examining the feasibility of using strip tillage or antitranspirants such as waxes to suppress the live mulch. It is clear that perennial legume/row crop intercrops have many potential benefits, including weed control and excellent soil erosion control, but these systems require a high level of sophisticated management to permit adequate crop yields.

Intercropping with clover has been shown to have important weed management effects in pastures. In experiments conducted in southeast Texas, Evers (1983) found that interseeding arrowleaf or subterranean clovers into pastures of bermudagrass or bahiagrass essentially eliminated weeds from the pastures. Like subterranean clover, arrowleaf clover is an annual species that grows during the cool fall and winter seasons and reseeds itself naturally. The pasture grasses are perennials that grow during the warm spring and summer seasons. Less weed dry matter was produced in plots of pasture grasses interseeded with the clovers than in plots treated with simazine, which also provided very effective weed control. This effect was attributed to interference from the fall-germinating clovers toward winter- and spring-germinating weeds. Evers (1983) concluded that intercropping cool-season clovers with warm-season forage grasses can extend the grazing season, provide high-quality forage, reduce the need for N fertilizer, and eliminate the cost of a herbicide and its application.

ALLELOPATHIC COVER CROPS

The use of reduced and minimum tillage systems has generally been associated with increased requirements for the use of synthetic herbicides. Recent research suggests, however, that chemical compounds released or derived from certain cover crop species can substitute for purchased herbicides in some systems, or they can be combined with reduced herbicide rates in others. Management systems are now being developed in which cash crops are grown without tillage through cover crops that lie as mulch on the soil surface. Through an interaction called allelopathy, chemicals from the cover crop residues inhibit germination and growth of many weeds, particularly annual broadleaf species (Barnes and Putnam, 1983, 1986; Liebl and Worsham, 1983; Putnam and DeFrank, 1983; Shilling et al., 1985, 1986; Barnes et al., 1987). Additionally, because many weed species are specifically adapted for establishment in disturbed soil conditions, eliminating tillage can, by itself, reduce weed populations by eliminating environmental cues necessary for germination (Roberts and Potter, 1980; Putnam and DeFrank, 1983; Shilling et al., 1985). Early season competition between the established mulch crop and emerging weeds and the physical shading effect of the dead mulch are also important factors.

In Michigan, Putnam et al. (1983) grew cover crops of rye, wheat, sorghum, barley, and oats to a height of 40–50 cm, dessicated the cover crops with a herbicide (glyphosate or paraquat) or natural winter freezing, and planted a variety of vegetable crops into mulches of the cover crop residues. Vegetables with large seeds, such as corn, cucumber, pea, and snapbean, generally germinated and grew well in the presence of the mulches. In contrast, vegetables with small seeds, such as cabbage, lettuce, tomato, and carrot, tolerated residues of only certain mulch species, or consistently performed poorly under mulched conditions.

The effect of the mulches on weed growth was dramatic. For example, when snap beans were no-till planted into rye mulch, weed growth was reduced 80%, as compared to conventionally tilled plots without the cover crop (Putnam and DeFrank, 1983). Rye reduced emergence of ragweed, green foxtail, redroot pigweed, and purslane (*Portulaca oleracea* L.) by 43%, 80%, 95%, and 100%, respectively (Putnam et al., 1983). However, it had no effect on emergence of yellow foxtail.

Shilling et al. (1985) reported that when soybeans and sunflowers were planted into desiccated rye, elimination of tillage and presence of the rye mulch reduced aboveground biomass of common lambsquarters, redroot pigweed, and common ragweed 99%, 96%, and 92%, respectively. Planting corn without tillage into a desiccated wheat cover crop reduced growth of a mixture of morningglory species (*Ipomea lacunosa* L. and *I. purpurea* L.) 79%, compared to a nonmulched, tilled treatment. No deleterious effects of the mulches on crop growth were observed.

Brusko (1987) described management of a commercial farm in Illinois in which soybeans were planted without tillage into rye when the cover crop reached the flowering stage. The cover crop was then shredded with a flail mower. Growth of broadleaf weeds was very slight, and only small amounts of herbicides were required for control of volunteer rye and grass weed species. Yield of soybeans with no-till rye mulch management was as good as or better than that with conventional management. Combined cost of the rye cover crop and supplemental grass herbicide was lower than cost of the full spectrum of herbicides required for the conventionally managed soybean crop.

Putnam et al. (1983) found that when sorghum, wheat, or rye cover crops were used as mulches for weed control in Michigan apple and cherry orchards, growth of the trees was equal to or better than tree growth in treatments in which weed control was accomplished with repeated applications of paraquat or tillage. The cover crops were planted in the fall and killed with one herbicide application in May. When alive, the cover crops competed with weeds for light, water, and nutrients; after dessication, the cover crops provided allelopathic weed control for up to 60 days.

In a study conducted at the Rodale Research Center in 1988, 85% to 95% weed suppression was obtained six weeks after corn planting from winter rye, winter wheat, and hairy vetch, which were flail mowed in early June. In a second study, corn that was no-till planted into mown hairy vetch was equal in yield to corn planted into conventionally plowed vetch and cultivated for weed control (Janke and Peters, 1989).

The potential of the no-till, allelopathic cover crop systems to conserve soil resources, while decreasing requirements for purchased herbicides, strongly suggests these systems can be important for enhancing agricultural sustainability. Many management details of these systems remain to be worked out, however. Because cover crops can sequester nutrients, deplete early season soil moisture, and increase late season soil moisture, particular attention needs to be directed toward developing a better understanding of nutrient and water relations in these systems (Wagger and Mengel, 1988).

WEED-SUPPRESSIVE CROP VARIETIES

Crop varieties can differ in their ability to suppress and tolerate weeds (Zimdahl, 1980; Minotti and Sweet, 1981; Walker and Buchanan, 1982). These differences have generally been attributed to differential ability to compete for light, water, and nutients, but allelopathic crop-weed interactions might also be involved (Putnam and Duke, 1978; Minotti and Sweet, 1981). The ability to suppress weed growth allelopathically has been noted not only for small grain crops such as rye (Barnes and Putnam, 1986), wheat (Steinsiek et al., 1982), and oats (Fay and Duke, 1977), but also such crops as sunflower (Leather, 1983) and sweet potato (Harrison and Peterson, 1986). U.S. crop breeding programs have placed relatively little emphasis on the development of superior varieties for growth under weed-infested conditions, but the studies discussed below indicate that such efforts might be extremely useful in the development of sustainable weed management programs.

Guneyli et al. (1969) measured the yield performance and weed-suppression ability of six sorghum hybrids and six parental inbreds under weedy and weed-free conditions in Nebraska. Dominant weed species were tall waterhemp (*Acnida altissima* Riddell), smooth pigweed (*Amaranthus hybridus* L.), and green foxtail. Weed biomass produced in association with the different sorghum genotypes ranged from 4.0 to 2.6 tons/ha. Hybrid RS 609 produced the highest yields under both weed-free and weed-infested conditions and permitted only 4% more weed growth than the most weed-suppressive sorghum. Competitive advantage of sorghum over weeds was attributed to rapid germination, emergence, and root and shoot growth during early stages of sorghum development.

Similarly, a cultivar of tall fescue (*Festuca arundinacea* Schreb.) with a high leaf area expansion rate suppressed velvetleaf growth 14% to 24% when compared to an isogenic line with a lower rate of leaf area expansion (Forcella, 1987).

Downy borme (*Bromus tectorum* L.) is the most prevalent winter annual grass weed in winter wheat fields of the central Great Plains region of the United States (Challaiah et al., 1986). Use of weed-suppressive crop varieties is particularly relevant to this crop-weed association because similarity of the crop and weed makes a selective herbicide control program very difficult.

Challaiah et al. (1986) measured grain yield and weed-suppression ability of ten varieties of winter wheat grown with downy brome in Nebraska. Centura wheat, which produced the highest grain yield under weed-infested conditions, provided 91% of the yield of the highest yielding variety under weed-free conditions. Downy brome produced only 1300 kg/ha of biomass in association with Centura wheat, but produced 1890 kg/ha of biomass in association with the wheat variety that had the highest yield under weed-free conditions. Thus, a wheat variety that produced a relatively high grain yield could also substantially suppress weed growth. Wheat height was the parameter best correlated with reduction in downy brome yield.

In a study comparing interference between 20 winter wheat varieties and summer annual weeds, taller wheat varieties tended to intercept more light, have higher grain yields, and be more effective in suppressing weed growth (Wicks et al., 1986). The two shorter wheat varieties that were exceptions to these correlations had the highest tiller numbers observed in the experiment.

Although crop height is often associated with superior weed suppression ability, this is not always the case. Bridges and Chandler (1988) examined the weed suppression ability of three cotton varieties with average heights at maturity of 66, 122, and 168 cm. No differences were noted among cultivars in their ability to suppress growth of johnsongrass (*Sorghum halepense* [L.] Pers.).

Minotti and Sweet (1981) noted large differences among potato varieties in their ability to tolerate the presence of weeds and to suppress weed growth. Under weed-free conditions, Katahdin produced approximately 20% more yield than Green Mountain, but under weed-infested conditions, Katahdin produced approximately 25% less than Green Mountain. Presence of common lambsquarters and redroot pigweed had almost no effect on tuber yields of Green Mountain potato, but decreased tuber yields of Katahdin potato by almost 50%.

Dry weight of yellow nutsedge (*Cyperus esculentus* L.) grown in association with Green Mountain potato was only 10% of its dry weight in association with Katahdin potato. Differences in the ability of the potato varieties to suppress weeds were found to be highly correlated with patterns of canopy development and light interception. Potato varieties that were superior competitors rapidly established dense canopies and prevented light from reaching the weeds. Below-ground interactions between weeds and potatoes were found to have little importance in determining the outcome of competition (Sweet et al., 1974).

The studies noted above suggest that varieties with the ability to suppress weeds while providing reasonable yields may already exist, or could be developed, for many crops. As noted by Walker and Buchanan (1982), use of genetic variability has provided marked success in protection of crops from insect pests and pathogens. It seems likely that attention from crop breeders to crop-weed interactions could also decrease yield losses while reducing requirements for herbicides.

CROP DENSITY AND SPATIAL ARRANGEMENT

Experiments conducted with a variety of crops, including soybeans, peas, peanuts, barley, wheat, flax, corn, sorghum, and sugarcane, indicate that under weed-infested conditions, increases in crop seeding rate generally result in decreased weed growth and higher crop yield (Felton, 1976; Zimdahl, 1980; Walker and Buchanan, 1982; Ghafar and Watson, 1983; Lawson and Topham, 1985). This phenomenon can be interpreted as a reflection of earlier and more complete capture of available water, light, and nutrients with increasing crop density (Minotti and Sweet, 1981).

Planting crops in narrow rather than wide rows can also decrease weed growth and increase crop yields, if early season weed growth is adequately controlled. This is illustrated by experiments conducted by Teasdale and Frank (1983) with snap beans and annual and perennial weed species. The investigators varied distances between bean rows from 15 to 91 cm, while maintaining a constant number of bean plants per unit area. Thus, the design of the experiment allowed effects of varying crop spatial arrangement to be determined independently of effects of varying crop density.

When weeds were allowed to emerge with the crop, narrow row spacings (15 to 36 cm) suppressed weed growth by 18%, compared to conventional wide row spacing (91 cm). In contrast, when weeds were controlled for the first half of the growing season, beans in narrow rows suppressed weed growth by 82%, compared to beans in wide rows. Crop yields from narrow row treatments were an average of 23% higher than yields from wide row treatments. Increases in crop yield and decreases in weed growth that were associated with narrow crop rows were attributed to earlier, denser crop canopy cover and increased shading of associated weeds (Teasdale and Frank, 1983).

If herbicides are necessary for effective weed control, quantities required might be substantially lower in higher density, narrow row crops. Research in Minnesota indicated that when crops were planted in 25 or 38 cm rows, rather than 76 cm rows, recommended herbicide rates for corn, soybean, and sunflower crops could be reduced 67% to 75%, while maintaining adequate weed control and high crop yields ("Narrow-spaced", 1988; Forcella and Westgate, 1988). The narrow row crops developed canopy cover faster, intercepted more sunlight, and created more shade on the soil surface than did crops grown in conventional wider rows. Herbicide use in high density, narrow row cropping systems might be decreased further by use of allelopathic cover crops, weed-suppressive crop varieties, cultivation with a rotary hoe before crop emergence, or interrow cultivation (if row spacing and operator skill permit).

SOIL FERTILITY MANAGEMENT

Conventional agricultural systems developed in the era of cheap and abundant energy have relied heavily on the use of synthetic fertilizers to meet crop nutrient requirements. These materials are frequently applied in excess of crop demand (Hallberg, 1987; Papendick et al., 1987). Recent experiences in Pennsylvania and Maine indicate that livestock producers using large quantities of nutrients in the form of concentrate feeds may also apply animal manures at rates in excess of crop demand. Because crop and weed growth can be highly responsive to nutrients, excessive applications of synthetic fertilizers and manure raise questions related to weed management.

In a review of studies concerning effects of fertilizers on crop-weed interactions, Alkamper (1976) noted that weeds are often capable of absorbing nutrients faster and in relatively larger amounts than crop plants. Thus, in the presence of a high weed population density, fertilizer application may stimulate weed growth so greatly that crop plants can be overgrown and suppressed.

This possibility is clearly illustrated in results of experiments conducted by Carlson and Hill (1985). Effects of nitrogen (N) fertilizer on interactions between wild oat (*Avena fatua* L.) and spring wheat were studied in California. Under weed-free conditions, increasing rates of N fertilizer increased wheat grain yields. In contrast, under weed-infested conditions, increasing rates of N decreased wheat yields and increased reproductive yield of the weed. The weed was so responsive to soil fertility conditions that N increased wheat yields only when weed density was below 1.6% of the total (crop + weed) plant density. These results emphasize the importance of carefully avoiding overfertilization when weeds are more responsive than crops to fertilizer.

Even when fertilizer application does not cause weeds to depress crop yields below levels obtained without fertilizer, weeds may negate the value of fertilizer. Appleby et al. (1976) found that N increased yield of winter wheat under weed-free conditions, but had no significant effect on crop yield when Italian ryegrass (*Lolium multiflorum* Lam.) was present as a weed. As compared to weed-free wheat plots, the grass weed created proportionally larger reductions of wheat yields as the N fertilizer rate increased. This result was interpreted as a reflection of increased interference from the weed toward the crop at higher fertilizer rates.

In some cases, fertilizer application can alleviate interference effects of weeds on crops and/or increase crop suppression of weed growth. Janke (1987) conducted greenhouse and field studies on interference between alfalfa and quackgrass and found that increased potassium (K) fertilization favored alfalfa growth; low rates of K fertilization favored growth of the weed. Staniforth (1957) reported that application of N increased corn yields two to three times more than growth of annual grass weeds. Nieto and Staniforth (1961) found that reductions of corn yields due to grass weed interference averaged 1254, 878, and 627 kg/ha with N fertilizer applications of 0, 78, and 157 kg/ha, respectively. In situations such as these, potential economic benefits from improved crop nutrition with added fertilizer should be weighed against costs of intensified weed control.

Whatever the relative responses of weeds and crops to fertilizers, it seems critical to place crop nutrients where they are most available to crops and least available to weeds. This argues against single, high-dose, broadcast fertilizer applications and for slower release, split, banded, and sidedressed applications. Little (1987) noted that weed problems in Pacific Northwest wheat fields diminished when crop planting and fertilization patterns were shifted to clumped, paired rows with fertilizer placed between paired rows. (The planting pattern had seed rows with a distance of 13 cm between pair members and 38 cm between pairs.) Weeds growing in the wide unplanted areas were effectively deprived of nutrients, and crop yields increased.

Exploiting differences between species in their means of acquiring nutrients can be an important component of weed management through rotations. For example, legumes such as field pea that use atmospheric N can avoid competition with weeds for soil N, and can be highly competitive against weeds for light (Liebman, 1986). When a legume crop is grown with little or no N fertilizer, after a nonlegume crop that has depleted residual soil N, weed populations can be starved for N and more effectively controlled with crop competition, cultivation, and other methods.

MANIPULATION OF SOIL TEMPERATURE AND MOISTURE

In certain geographic areas, soil temperature and moisture can be manipulated to dramatically reduce negative effects of weeds on crops. Soil solarization, a method of heating moist soil by covering it with plastic tarps to trap solar radiation (Horowitz et al., 1983), has been used to reduce weed populations in the southern United States and California (Egley, 1983; Standifer et al., 1984; Stapleton and DeVay, 1986), and may be effective as far north as Connecticut (R. DeGregorio, personal communication).

Soil solarization causes large increases in soil temperature, increases in CO_2/O_2 concentrations in the soil atmosphere, increases in moisture content of upper soil levels, changes in soil chemical properties, and changes in population levels of soil microorganisms (Egley, 1983; Rubin and Benjamin, 1984; Stapleton and DeVay, 1986). These and possibly other factors are believed responsible for reducing populations of viable seeds of annual weeds buried in the soil. When applied over moist soil and maintained in place for several weeks of hot sunny weather, clear polyethylene sheeting is more effective for weed suppression than black polyethylene (Horowitz et al., 1983; Standifer et al., 1984). In temperate areas, the method seems best adapted as a means of summer soil preparation for late summer and fall crop production. Cost of the tarping may limit its use to high value crops.

The effects of solarization on weeds are demonstrated by results of research conducted by Egley (1983) in Mississippi. Solarization for one week in midsummer reduced numbers of viable buried seeds of prickly sida (*Sida spinosa* L.), common cocklebur (*Xanthium strumarium* L.), velvetleaf, and spurred anoda (*Anoda cristata [L.]* Schlect.) by 93%, 85%, 36%, and 32%, respectively. Solarization treatments for one to four weeks reduced emergence of prickly sida, pigweeds (*Amaranthus* spp.), spurred anoda, morningglories (*Ipomoea* spp.), horse purslane (*Trianthema portulacastrum* L.), and several grass species for 14 weeks following tarp removal by 64% to 94%, as compared to an untarped control treatment. Emergence of purple nutsedge (*Cyperus rotundus* L.), however, was not affected or was increased by solarization treatments.

Use of subsurface drip irrigation to deprive weeds of water while placing water in the crop root zone has been shown to be successful in a study conducted in California with dry-season irrigated tomatoes (Grattan et al., 1988), and is almost certainly applicable wherever irrigation is the major source of water for plant growth. Dominant weed species in the study conducted by Grattan and his colleagues were barnyardgrass (*Echinochloa crus-galli* [L.] Beauv.) and redroot pigweed. Three irrigation methods (furrow, sprinkler, and subsurface drip) were compared for plots that did or did not receive herbicides (napropamide + pebulate). The drip irrigation system delivered water into the crop row, 25 cm below the soil surface. The tomato crop in all treatments was direct seeded, sprinkler irrigated until the crop was 23 cm tall, and cultivated once before the different irrigation treatments were started.

Grattan et al. (1988) reported that weed growth in fields receiving subsurface drip irrigation was several orders of magnitude lower than in fields receiving sprinkler or furrow irrigation. Weed growth in drip irrigated fields was equally low whether or not herbicides were applied. Yields of red tomatoes were inversely related to weed growth, with drip irrigated plots producing highest yields.

Several factors may affect grower adoption of this approach to weed management, including need for careful monitoring of water delivery, higher initial costs because of the specialized materials involved, and need to match subsequent crops and field operations to the buried irrigation equipment. However, the system appears to have excellent potential as an ecologically sound weed control method, and individuals who use the system claim that higher start-up costs are offset in subsequent years by reduced field traffic demands and labor savings (Grattan et al., 1988). McLeod and Swezey (1980) noted that subsurface drip irrigation is also used effectively for weed control in California orchards.

McWhorter (1972) reported that flooding soil with five to ten cm of water for two weeks was extremely effective in killing freshly planted (simulating recently tilled) johnsongrass rhizomes. A delay in flooding until culms emerged greatly reduced control. McWhorter noted that flooding is practical on thousands of acres of level land in the lower Mississippi River valley that are infested with johnsongrass, and he suggested that flooding in early spring would permit crop production during the summer of the same year. In rice production systems, flooding is an important means of controlling weed species that are unable to establish and survive in water, but species adapted to flooded conditions can become more abundant (Barrett and Seaman, 1980; Walker and Buchanan, 1982).

INSECT PESTS AND PATHOGENS OF WEEDS

Insects and pathogens have had limited roles as biological weed control agents in most agricultural systems, but important exceptions do exist. Recent successful development of several weed-specific microbial herbicides suggests that an expansion of this approach may become more common.

Insects appear to be more successful as weed control agents in perennial, rather than annual, agricultural systems. The greater environmental stability of perennial systems allows insect populations to be introduced in relatively low numbers, then grow and persist until the host plant (the weed) is reduced to a low level of abundance. Weed control by insects in perennial systems can be relatively inexpensive and long term (Charudattan and DeLoach, 1988). In North America, success with insects as weed control agents has generally occurred on pasture and rangeland. Notable examples include control of St. Johnswort (*Hypericum perforatum* L.) by a beetle (*Chrysolina quadrigemina*), and control of musk thistle (*Carduus nutans* L.) by weevils (*Rhinocyllus conicus* Froelich and *Trichosirocalus horridus* [Panzer] Colonelli) (Andres, 1982; Aldrich, 1984).

The frequent habitat disturbances and relatively rapid growth of weeds associated with annual cropping systems make it extremely difficult to synchronize population dynamics of insect control agents with crops and weeds in such a way that economically effective weed control is achieved (Bernays, 1985). To overcome this problem biologically, insects might be mass-cultured and released into crop fields. However, insectary rearings involve careful control of environmental conditions and considerable expense, factors that limit the rapid and widescale acceptance of this approach.

An alternative approach for using insects as weed control agents in annual cropping systems would be to manipulate environmental conditions to enhance the impacts of indigenous insects feeding on target weed species. Brust and House (1988) have recently shown that certain native beetle and rodent species are more prevalent in no-till soybean cropping systems than in conventional tillage systems and that more weed seeds are eaten by these animals in no-till systems. Further research is needed to better understand factors affecting consumption of weeds by indigenous animals, particularly those feeding on weed seeds and seedlings.

Microbes, especially fungi, appear to have real potential as weed control agents in both perennial and annual cropping systems (Charudattan and Walker, 1982). Pathogens of weeds are particularly useful because of the possibility of high host specificity and because they are usually much less expensive to produce than insects (Charudattan and DeLoach, 1988).

A commercial preparation of the fungus *Phytophthora palmivora* Butler (Devine™) is available for postemergence control of citrus strangler vine (*Morrenia odorata* Lindl.), and its use has proven effective in Florida (Quimby and Walker, 1982; Charudattan, 1985). The pathogen infects and wilts seedlings or established vines in two to ten weeks after application, and complete control in an infested area may be achieved in 18 months (Charudattan, 1985).

The fungal pathogen *Colleotrichum gloeosporioides* (Penz.) Sacc. f. sp. *aeschynomene* (Collego™) is now sold for effective postemergence control of northern jointvetch (*Aeschynomene virginica* [L.] B. S. P.), a weed that infests rice fields in Arkansas and the southeastern United States (Quimby and Walker, 1982; Charudattan, 1985). Infection by the pathogen usually occurs seven to ten days after application, and death of the weed may occur four to five weeks later (Charudattan, 1985).

Phatak et al. (1987) reported that yellow and purple nutsedge (*Cyperus esculentus* L. and *C. rotundus* L.) are susceptible to a rust fungus, *Puccinia canaliculata* (Schw.) Lagerh. Release of the fungus early in the spring reduced yellow nutsedge stand by 46% and tuber formation by 66%, and it completely inhibited flowering. Combined application of the rust fungus and paraquat allowed 99% control of yellow nutsedge, compared with 60% control with rust and 10% with paraquat alone. Commercial formulations of the fungus are not yet available, although the wide distribution of *Cyperus* spp. might justify the development of such a product.

The fungus *Alternaria cassiae* is currently under commercial development as a microbial herbicide for control of sicklepod (*Cassia obtusifolia* L.), coffee senna (*Cassia occidentalis* L.), and showy crotalaria (*Crotalaria spectabilis* Roth), important weeds in soybean fields in the southeastern United States (Charudattan, 1985; Bannon, 1988). Field efficacy has been demonstrated for pathogens of other soybean weeds, as well as weeds of corn, cotton, and rangelands (Charudattan, 1985). A large number of pathogens currently under study as weed control agents are listed by Charudattan and DeLoach (1988).

USE OF LIVESTOCK

Careful management of livestock grazing regimes can have important effects on pasture species composition and productivity; it can also be a key component in increasing the profitability of livestock production enterprises (Murphy, 1987; see also Chapter 8). Short duration, high-intensity grazing of immature pasture plants, followed by suitable recovery periods and regrazing, can promote an increase in the relative abundance of nutritious legume and grass species, which are well adapted to repeated grazing, and a decrease of noxious weeds, which are somewhat palatable to livestock when immature and generally poorly adapted to repeated grazing. Murphy (1987) contends that the weakening of weed species by rotational-intensive grazing is exacerbated by increased competition from legumes and grasses, as the latter species begin to flourish. Removal of weeds from intensively managed pastures can be hastened by mowing or, for particularly recalcitrant species, spot application of systemic herbicides.

One particularly attractive aspect of using intensive pasture management as a primary method of forage production is that it greatly decreases or eliminates the need for corn and other silage crops and thus the need for weed control in these crops. One profitable full-time dairy in Maine produces high milk yields from intensively managed pastures, hay, and a small amount of purchased concentrate; no corn is grown on the farm, and no silage or haylage is made.

Because goats prefer to eat a large amount of brush relative to herbaceous species, they can be extremely effective in eliminating brush from pastures, and are virtually noncompetitive with cattle or sheep for available forage when sufficient brush is available (Scifres, 1981; Wood, 1987). Wood (1987) observed almost total elimination of brush species and a large increase in grass production within two years of grazing with goats on brushy Vermont pastureland. Sheep and cattle were much less effective in promoting these changes in vegetation composition. Wood suggested that (1) mixed grazing with goats and cattle or goats and sheep should be the most efficient first step in restoring brush-infested land for livestock production, (2) the number of goats should be gradually reduced as the brush declines to prevent competitive foraging, and (3) maintaining a few goats with cattle and sheep could protect renovated pasture from reinvasion by brush and some weedy forbs.

Sheep can serve as biological control agents for leafy spurge (*Euphorbia esula* L.). In a three-month grazing study (Landgraf et al., 1984), no definite preference for or avoidance of leafy spurge was detected. A maximum intake of 40% to 50% of the diet did not significantly affect weight gain of ewes when compared to ewes in pastures free of leafy spurge.

McLeod and Swezey (1980) noted the use of weeder geese in orchards, vineyards, and fields of garlic, tomatoes, cucumbers, cotton, and mint in California and Oregon. Use of weeder geese has also been reported for strawberry and potato fields. Crops must be of sufficient size so as to not be eaten by the birds, and stocking rate should be five or six birds per hectare. Fencing, protection from predators, adequate water, and a small amount of supplemental grain are necessary ("Weed control," 1981).

McLeod and Swezey (1980) noted that hogs have been used for control of perennial weeds between cropping seasons in Washington. When fenced into fields at a rate of 25 animals per hectare, hogs have removed roots of weed species such as bindweed (*Convolvulus arvensis* L.) to a depth of more than 75 cm.

CULTIVATION AND FLAME WEEDING

Well-timed cultivation to uproot, bury, or cut off weeds is one of the most important and effective forms of weed control (Kepner et al., 1978). On a per-acre basis, cultivation operations can cost an estimated $1.74 to $6.89. In contrast, use of pre- and postemergence herbicides may cost $10 to $20 for a single chemical, and often more than one chemical is needed if a broad spectrum of weed species is present in the field (Johnson, 1985; Micka, 1985; Roeth and Selley, 1987).

The most serious problem with cultivation can be inadequate control of weeds in the crop row. However, applications of narrow bands of herbicides directly over the crop row paired with cultivation between rows can provide excellent weed control at low cost (Johnson, 1985). For example, Ludwig (1982) reported that herbicide use and associated costs were reduced by two-thirds on his Illinois farm when herbicides were banded over corn and soybean rows, rather than applied over the entire field.

Early control of weeds, including within crop rows, can be at least partially achieved with a rotary hoe cultivator (Roeth and Selley, 1987). This implement can be pulled at relatively high speeds (10 km/h or more), has low traction power requirements, and can be used with crops such as corn and large-seeded grain legumes (soybean, dry bean, lupin, etc.) from before crop emergence up until crops are a few centimeters tall. The whirling, curved tines of the rotary hoe throw small weeds out of the upper layers of the soil to dry out and die. The rotary hoe is particularly effective on germinated but unemerged weed seedlings. Some damage may occur to emerged crop plants, but increasing the sowing rate slightly can compensate for such losses. Relatively dry conditions are necessary for rotary hoeing to be effective. In a manner analogous to the use of the rotary hoe, flexible tine harrows are being used for postplanting weed control in Canadian Maritime small grain and potato fields; weeds are "combed" out of the fields.

While the crop is small, shields can be used for crop protection when between-row weeds are removed with cultivating implements such as sweeps, shovels, and knives. Shields can be used to allow cultivation of young crops at quite narrow row spacings. Once the crop is well established, a variety of cultivation implements (for example, "spiders," torsion weeders, and hoe ridgers) are available to throw soil into crop rows and bury small weeds (Terpstra and Kouwenhoven, 1981).

Farmers producing small grains and other crops that are difficult to cultivate can often improve weed control by delaying planting after initial seedbed preparation, allowing a cohort of weeds to germinate and emerge, and then cultivating immediately before sowing the crop (Wookey, 1985). This method is sometimes called the "false seedbed technique."

Listing is a specialized tillage system in which crops are planted into furrows formed at the time of planting. Early cultivations remove weeds from between crop rows, but limit movement of soil back toward young crops within the furrows. After crops are established, later cultivations throw dirt into the furrows, burying small weeds. Cramer (1987) reported the experiences of a Nebraska farmer who, by using this method, was able to eliminate use of herbicides in first-year corn and use just an 18 cm band application of atrazine and butylate in second-year corn.

Ridge tillage systems involve constructing soil ridges, truncating the ridge tops immediately before sowing crops such as corn and soybean, and excavating soil from furrow areas and depositing it on ridge crests after crops are well established. Ridge tillage not only provides soil conservation benefits, but also removes an appreciable proportion of weed seeds from the crop row at the time of planting (Forcella and Lindstrom, 1988b). This displacement can negate or reduce interference of weeds toward crops, depending on the size of the original weed seed population and the proportion displaced. For supplemental weed control within crop rows, ridge tops can be rotary hoed after planting or treated with band applications of herbicides. Because relatively large amounts of crop residues are maintained on surfaces of the furrows and sides of the ridges in this system, specialized heavy duty cultivation equipment is required. The cost of this equipment may be offset by reduction or elimination of requirements for preplanting tillage operations and machinery.

Geier and Vogtmann (1987) tested the weed control value of a multiple row brush hoe on commercial and research farms in Germany. The implement is rear- or mid-mounted on a tractor and is PTO driven. Flexible brushes rotate around a shaft mounted parallel to the ground and penetrate the soil to a depth of 5 cm. The working widths of the brushes between the crop rows can be varied to any distance from 16 cm upwards. Crops are protected within shielding tunnels that are 6 to 20 cm wide and 60 to 80 cm long. Reported advantages of the tool include the ability to work very early and close to the crop row and better kill of weeds under relatively wet conditions. Brush bristles not only uproot weeds but also wipe soil off the roots, permitting rapid drying by sun and wind.

Flame weeding has been used to control agricultural weeds in North America and Europe to a minor extent for decades (Kepner et al., 1978; Daar, 1987). It is most effective when weeds are small (<3 cm), and should be used only when crops have not yet emerged or are well established. Combination of between-row cultivation and within-row flaming was shown to be effective for weed control in corn, when flaming was done at the "match" stage (corn 3 to 4 cm tall) (Geier and Vogtmann, 1987). Crops such as cotton, sorghum, and soybean can also be flame weeded effectively (Kepner et al., 1978; Daar, 1987), but the energy-intensiveness of this method suggests that it might be used most economically for high-value crops planted on limited acreages.

Ott and Desvaux (cited in Daar, 1987), for example, developed methods for flame weeding carrots and onions. For carrots, seedbeds should be prepared and irrigated at least ten days before the crop is sown; this stimulates weeds to germinate. Carrots are then drilled into the seed bed. Shortly before the carrot seedlings emerge, the seedbed is scorched with a flamer; this flaming usually occurs five or six days after carrots are sown. The timing of the field operations encourages a large proportion of weed seeds to germinate before the crop has emerged, and allows weeds to be flamed without injuring the crop. When this method was followed on commercial French farms, weed populations were reduced by 50% to 80%, as compared to unflamed controls. Ott and Desvaux noted that weeds between crop rows that survive flaming can be removed by subsequent cultivation operations; one rapid hand weeding may be required to eliminate the few weeds within crop rows that escape flaming and cultivation.

Ott and Desvaux recommended use of flame weeding on well established onion plants. When used at the crop's four-leaf stage, one flame weeding gave yields only 4% lower than those obtained by hand weeding.

Flaming before the four-leaf stage damaged the crop and gave lower yields.

AN INTEGRATED APPROACH TO SUSTAINABLE WEED MANAGEMENT

Because weed communities respond dynamically to changes in management practices, sustainable weed management requires a combination of approaches to function effectively. In this chapter we have described some time-honored methods of weed control, such as cultivation, crop rotation, and use of weed suppressive crop varieties. We have also described methods that are relatively new for North American farmers, such as no-till planting into allelopathic cover crops and soil solarization. Other weed control methods, such as biological control with insects or pathogens, could be extremely important if given adequate research attention, but they currently have limited value in annual row crop systems. Choosing methods to combine into an integrated weed management system must reflect the particular situation of each farm. Selection of weed management methods will depend on the kind of information, financial, labor, seed, and equipment resources that are available; the time and labor constraints that are involved for different operations; the kind and quantities of crops that must be produced; and the soil and pest management practices that are used.

The following examples illustrate how several weed management methods can be used in combination on working farms. The examples are not provided to be copied blindly and imposed on other farms. Rather, they demonstrate the flexibility and variety of techniques required for weed management that uses little or no herbicides.

An Illinois farmer combines crop rotation, cover crops, cultivation, and drip irrigation under plastic mulch to control weeds in melons (Cramer, 1988). Melons are rotated with wheat and legumes, generally hairy vetch or field peas. Vetch in the rotation serves as a smother crop and "chokes" out weeds prior to planting. The vetch is disked, black plastic and the drip irrigation system are laid down, and melons are transplanted into slots in the plastic. A cultivator controls weeds between the plastic and throws soil over the edge of the plastic to help conserve moisture. After fields are walked over twice to pull weeds growing through slots in the plastic, the drip irrigation system ensures that melons (and not weeds) will have enough water during dry periods that are typical of July and August in the Midwest.

Samson (1987) described a "4C" approach (crop rotation, cultivation, cover cropping, chemical banding) for reducing herbicide use and conserving soil in corn-soybean-wheat rotations grown in Ontario and Quebec. In the first year of the rotation, a 15–19 cm band of herbicide is applied over corn at planting time. Fertilizer is banded near the crop row, and a substantial proportion of the corn's nitrogen requirement is provided by plowed down clover that had been intercropped with wheat in the preceding phase of the rotation. Two cultivations clear weeds between rows; a postemergent herbicide can be banded during a cultivation operation if necessary. Between 11 and 17 kg/ha of ryegrass are interseeded at the time of the second cultivation (when corn is approximately 15–25 cm tall) using a seed box on a modified cultivator. The ryegrass suppresses weed growth after corn harvest and in the spring of the subsequent year.

In the second year of the rotation, the ryegrass cover crop is plowed down and soybeans are planted in 53 cm rows. The between-row distance is close enough to provide shading of weeds, but wide enough to permit between-row cultivation operations. The soybeans are rotary hoed twice and cultivated once or twice. No herbicides are applied, and fertilizers are banded in small quantities if they are required. Winter wheat is aerially interseeded into the maturing soybeans at 168 kg/ha before 25% leaf drop in the soybeans.

In the third year of the rotation, red clover is frost seeded into the wheat at 8 to 11 kg/ha in early spring. The broadcast seed is worked into the ground by freezing and thawing the soil. Winter wheat acts as an excellent competitor to many annual weeds, and interseeded clover can help suppress weed establishment in the wheat crop and weed growth after wheat harvest. If weeds are a problem, the field can be cultivated after wheat harvest and seeded to oilseed radish, a rapidly growing, cold-tolerant cover crop that "smothers" weeds. Nitrogen fertilizer is applied to the wheat crop as a topdressing in the spring, when the crop has a well-established root system and can rapidly absorb the fertilizer.

Samson (1987) compared the "4C" corn-soybean-wheat rotation with cover crops to continuous corn and found that over a 36-month period, the rotation/cover crop system had 32 months of canopy cover, as compared to 15 months of canopy cover in the monoculture. By banding herbicides in corn, using only a rotary hoe and cultivation in soybeans, and not applying herbicide on wheat, the rotation/cover crop system uses 1/15 the herbicides of a continuous corn system. Fertilizer use is also reduced to less than one-third that of continuous corn if banding techniques are used.

These examples emphasize not only the efficiencies gained from using combinations of weed management methods, but also the potential importance of restructuring farming operations and systems. For example, for cover cropping to be successful, it must be compatible with rates, dates, and types of herbicide applications, if these materials are used. Farming systems that include different phases of rotation sequences each year need to be designed so that cultivation and other operations can be performed in a timely manner and so that areas sown to feed and cash crops are compatible with the producer's income and production goals.

Major restructuring of farming systems and changes in weed management practices can take place when emphasis is placed on the broader objectives of the producer, rather than adherence to conventional production technologies. For example, a central Maine dairy farmer has eliminated the need to use herbicides or cultivation in corn silage production by eliminating the need to produce corn silage. He satisfies his overall objective of producing milk abundantly and profitably by using rotational-intensive grazing management to produce hay and nutritious pasture. His cows produce between 7,700 and 8,200 kg milk/year, more than 1,000 kilograms above the state average, yet he feeds less than half the amount of purchased concentrates used by conventional dairy farmers in the state. His weed control methods involve cows eating immature stages of the weeds, interference from vigorous clover and grass swards, and occasional clipping to prevent thistles from seeding. These methods require minimal management time and expense.

In addition to restructuring, the concept of tolerable threshold levels for weeds has an important place in sustainable weed management. Weed threshold research can help determine whether or not various management practices are successful in keeping weeds below economically damaging levels. This research may reassure those farmers accustomed to seeing extremely clean fields that a certain level of weed infestation is tolerable, especially toward the end of the growing season. Threshold information can also be used by farmers earlier in the growing season to decide whether or not to apply postemergence herbicides if other control methods are not effective (Marra and Carlson, 1983). This option is especially important to farmers who are beginning the transition to reduced herbicide use, and can serve as a back-up method when new practices are attempted. Unfortunately, because the economic impacts of weeds on crops are dependent on a large number of variables, including soil type, weather, crop cultivar, weed species and genotype, relative times of crop and weed emergence, crop and weed densities, costs of weed control, and crop prices (Marra and Carlson, 1983; Aldrich, 1987; Auld and Tisdell, 1987; Kropff, 1988; Oliver, 1988), economic weed thresholds are not readily available for most crops in most areas.

A final point concerning sustainable weed management involves the integration of research efforts with on-farm practices. As new weed management techniques are developed by researchers, it is important that their usefulness be examined under real production conditions and constraints. It is also important for farmers to communicate their own innovations and observations to researchers. Successful development of sustainable weed management requires an iterative, back-and-forth flow of information between farmers and researchers. Farmers and researchers will need to work together to fully take advantage of new information- and thought-intensive technologies that are now coming to the forefront.

Acknowledgments

We are grateful for information and publications provided to us by A. Enache, W. T. Lanini, J. A. Liebman, A. R. Putnam, and A. D. Worsham.

We thank R. Andrews and E. Dyck for useful comments and criticisms. This chapter is contribution 1387 of the Maine Agricultural Experiment Station and a contribution of the Rodale Research Center.

REFERENCES

Aldrich, R. J. (1984). *Weed-Crop Ecology: Principles in Weed Management*. Breton, North Scituate, MA.

Aldrich, R. J. (1987). "Predicting crop yield reductions from weeds," *Weed Technology* **1**:199–206.

Alkamper, J. (1976). "Influence of weed infestation on effect of fertilizer dressings," *Pflanzenschutz-Nach.* **29**:191–235.

Altieri, M. A., and M. Liebman (Eds.). (1988). *Weed Management in Agroecosystems: Ecological Approaches*. CRC Press, Boca Raton, FL.

Andres, L. A. (1982). "Integrating weed biological control agents into a pest management program," *Weed Sci.* **30** (suppl.): 25–30.

Appleby, A. P., P. D. Olson, and D. R. Colbert. (1976). "Winter wheat yield reduction from interference by Italian ryegrass," *Agron. J.* **68**:463–66.

Auld, B. A. and C. A. Tisdell. (1987). "Economic thresholds and response to uncertainty in weed control," *Agric. Sys.* **25**:219–27.

Baldwin, F. L. , and P. W. Santelmann. (1980). "Weed science in integrated pest management," *BioSci.* **30(10)**:675–78.

Bannon, J. S. (1988). "CASST™ herbicide (*Alternaria cassiae)*: A case history of a mycoherbicide," *Am. J. Alt. Agric.* **3(2 and 3)**:73–76.

Barnes, J. P. and A. R. Putnam. (1983). "Rye residues contribute weed suppression in no-tillage cropping systems," *J. Chem. Ecol.* **9(8)**:1045–57.

Barnes, J. P. , and A. R. Putnam. (1986). "Allelopathic activity of rye (*Secale cereale* L.)." In *The Science of Allelopathy*, (Eds. A. Putnam and C.S. Tang). pp. 271-286. Wiley, New York.

Barnes, J. P., A. R. Putnam, B. A. Burke, and A. J. Aasen. (1987). "Isolation and characterization of allelochemcials in rye herbage," *Phytochem.* **26(5)**:1385–90.

Barrett, S. C. H. , and D. E. Seaman. (1980). "The weed flora of California rice fields," *Aquat. Bot.* **9**:351–76.

Bernays, E. A. (1985). "Arthropods for weed control in IPM systems," In *Biological Control in Agricultural IPM Systems* (Eds. M. A. Hoy and D. C. Herzog). pp. 373–88. Academic Press, Orlando, FL.

Berteau, P. E., and D. P. Spath. (1986). "The toxicological and epidemiological effects of pesticide contamination in California groundwater." In *Evaluation of Pesticides in Groundwater* (Eds. W. Y. Garner, R. C. Honeycutt, and H. N. Nigg). pp. 423–35. American Chemical Society, Symposium Series 315.

Bridges, D. C., and J. M. Chandler. (1988). "Influence of cultivar height on competitiveness of cotton (*Gossypium hirsutum)* with johnsongrass (*Sorghum halepense*)," *Weed Sci.* **36**:616–20.

Brusko, M. (1987). "Ten dollar weed control in no-till beans," *New Farm* **9(1)**:10–11.

Brust, G. E., and G. J. House. (1988). "Weed seed destruction by arthropods and rodents in low-input soybean agroecosystems," *Am. J. Alt. Agric.* **3(1)**:19–25.

Carlson, H. L., and J. E. Hill. (1985). "Wild oat (*Avena fatua*) competition with spring wheat: Effects of nitrogen fertilization," *Weed Sci.* **34(1)**: 29–33.

Challaiah, O. C. Burnside, G. A. Wicks, and V. A. Johnson. (1986). "Competition between winter wheat (*Triticum aestivum*) cultivars and downy brome (*Bromus tectorum*)," *Weed Sci.* **34**:689–93.

Chandler, J. M. (1981). "Estimated losses of crops to weeds." In *CRC Handbook of Pest Management in Agriculture, Vol. 1* (Ed. D. Pimentel). pp. 95–109. CRC Press, Boca Raton, Fl.

Charudattan, R. (1985). "The use of natural and genetically altered strains of pathogens for weed control." In *Biological Control in Agricultural IPM Systems* (Eds. M. A. Hoy and D. C. Herzog). pp. 347–72. Academic Press, Orlando, FL.

Charudattan, R., and C. J. DeLoach, Jr. (1988). "Management of pathogens and insects for weed control in agroecosystems." In *Weed Management in Agroecosystems: Ecological Approaches* (Eds. M. A. Altieri and M. Liebman). pp. 245–64. CRC Press, Boca Raton, FL.

Charudattan, R., and H. L. Walker, Jr., Eds. (1982). *Biological Control of Weeds with Plant Pathogens*, Wiley, New York.

Cohen, S. Z., C. Eiden, and M. N. Lorber. (1986). "Monitoring ground water for pesticides." In *Evaluation of Pesticides in Ground Water* (Eds. W. Y. Garner, R. C. Honeycutt, and H. N. Nigg). pp. 170–96. American Chemical Society, Symposium Series 315.

Cramer, C. (1987). "Listing leaves herbicides behind," *New Farm* **9(3)**:38–40.

Cramer, C. (1988). "Building a drought-proof farm," *New Farm* **10(7)**:10–15.

Culik, M. N., J. C. McAllister, M. C. Palada, and S. L. Rieger. (1983). *The Kutztown farm report: A study of a low-input crop/livestock farm.* Regenerative Agriculture Association, Rodale Research Center, Kutztown, PA.

Daar, S. (1987). "Update: Flame weeding on European farms," *IPM Practitioner* **9(3)**:1–4.

Dotzenko, A. D., M. Ozkan, and K. R. Storer. (1969). "Influence of crop sequence, nitrogen fertilizer and herbicides on weed seed populations in sugar beet fields," *Agron. J.* **61**:34–37.

Dyke, G. V., and A. J. Barnard. (1976). "Suppression of couch grass by Italian ryegrass and broad red clover undersown in barley and field beans," *J. Agric. Sci.* (Camb.) **87**:123–26.

Egley, G. H. (1983). "Weed seed and seedling reductions by soil solarization with transparent polyethylene sheets," *Weed Sci.* **31**:404–09.

Eichers, T. R. (1981). "Use of pesticides by farmers." In *CRC Handbook of Pest Management in Agriculture, Vol. 2*, (Ed. D. Pimentel). pp. 3–25. CRC Press, Boca Raton, FL.

Enache, A., and R. D. Ilnicki. (1987). "Tillage and mulch systems study in field corn," *Proc. Northeast. Weed Sci. Soc.* **41**:13–17.

Enache, A., and R. D. Ilnicki. (1988). "Subterranean clover: A new approach to weed control," *Proc. Northeast. Weed Sci. Soc.* **42**:34.

Enache, A., and R. D. Ilnicki. (1989). "Weed control by subterranean clover (*Trifolium subterraneum*) used as a living mulch," *Weed Technology* (in press).

Evers, G. W. (1983). "Weed control on warm season perennial grass pastures with clovers," *Crop Sci.* **23(1)**:170–71.

Fay, P. K., and W. B. Duke. (1977). "An assessment of allelopathic potential in *Avena* germ plasm," *Weed Sci.* **25**:224–28.

Felton, W. L. (1976). "The influence of row spacing and plant population on the effect of weed competition in soybeans (*Glycine max*)," *Aust. J. Exp. Agr. & Anim. Husb.* **16**:926–31.

Forcella, F. (1987). "Tolerance of weed competition associated with high leaf area expansion rate in tall fescue," *Crop Sci.* **27**:146–47.

Forcella, F., and M. J. Lindstrom. (1988a). "Weed seed populations in ridge and conventional tillage," *Weed Sci.* **36**:500–503.

Forcella, F., and M. J. Lindstrom. (1988b). "Movement and germination of weed seeds in ridge-till crop production systems," *Weed Sci.* **36**:56–59.

Forcella, F., and M. E. Westgate. (1988). "Reduced herbicide rates for narrow-row crop production," *Agron. Abstr. 1988*: 125.

Francis, C. A., and J. W. King. (1988). "Cropping systems based on farm-derived, renewable resources," *Agric. Sys.* **27**:67–77.

Geier, B., and H. Vogtmann. (1987). "The multiple row brush hoe—a new tool for mechanical weed control," *International Federation of Organic Agriculture Movements (IFOAM) Bulletin* **1**:4–6.

Ghafar, Z., and A. K. Watson. (1983). "Effect of corn (*Zea mays*) population on the growth of yellow nutsedge (*Cyperus esculentus*)," *Weed Sci.* **31**:588–92.

Grattan, S. R., L. J. Schwankl, and W. T. Lanini. (1988). "Weed control by subsurface drip irrigation," *Calif. Agric.* **42(3)**:22–24.

Guneyli, E., O. C. Burnside, and P. T. Nordquist. (1969). "Influence of seedling characteristics on weed competitive ability of sorghum hybrids and inbred lines," *Crop Sci.* **9**:713–16.

Hallberg, G. R. (1986). "From hoes to herbicides: Agriculture and groundwater quality," *J. Soil Water Conserv.* **41(6)**:357–64.

Hallberg, G. R. (1987). "Agricultural chemicals in ground water: Extent and implications," *Am. J. Alt. Agric.* **2(1)**:3–15.

Harrison, H. F., Jr., and J. K. Peterson. (1986). "Allelopathic effects of sweet potatoes (*Ipomea batatas*) on yellow nutsedge (*Cyperus esculentus*) and alfalfa (*Medicago sativa*)," *Weed Sci.* **34**:623–27.

Hartwig, N. L. (1987). "Cropping practices using crownvetch in conservation tillage." In *The Role of Legumes in Conservation Tillage Systems* (Ed. J. F. Power). pp. 109-110. Soil Conservation Society of America, Ankeny, Iowa.

Hill, G. D. (1982). "Herbicide technology for integrated weed management systems," *Weed Sci.* **30** (suppl.):35–39.

Horowitz, M., Y. Regev, and G. Herzlinger. (1983). "Solarization for weed control," *Weed Sci.* **31**:170–79.

Horwith, B. (1985). "A role for intercropping in modern agriculture," *BioSci.* **35(5)**:286–91.

Janke, R. R. (1987). "Weed Management in Established Alfalfa Based on Cultural Control of Crop-Weed Interactions." Ph. D. dissertation, Cornell University, Ithaca, New York.

Janke, R. R., and S. Peters. (1989). "Cover crops, crop rotation and weed control in sustainable cropping systems," *Proceedings of the Third Penn–Jersey Tillage Conference, February 22, 1989.* Bethlehem, PA.

Jeffers, D. L., and G. B. Triplett, Jr. (1979). "Management needed for relay intercropping soybeans and wheat," *Ohio Rep.* **58**:67–69.

Johnson, R. R. (1985). "A new look at cultivation: Tillage can boost yields and cut erosion," *Crops & Soils* **37(8)**:12–16.

Kepner, R. A., R. Bainer, and E. L. Barger. (1978). *Principles of Farm Machinery, 3rd Ed. AVI, Westport, CT.*

Knudson, T. J. (1986). "Experts say chemicals peril farm belt's water," *New York Times,* September 6, 1986, A-8.

Koskinen, W. C., and C. G. McWhorter. (1986). "Weed control in conservation tillage," *J. Soil Water Conserv.* **41**:365–70.

Kropff, M. J. (1988). "Modelling the effects of weeds on crop production," *Weed Res.* **28**:465–71.

Landgraf, B. K., F. K. Fay, and K. M. Havstad. (1984). "Utilization of leafy spurge (*Euphorbia esula*) by sheep," *Weed Sci.* **32**:348–52.

Lanini, W. T., D. Pittenger, F. Munoz, W. L. Graves, and N. C. Toscano. (1988). "The use of legume living mulches for managing pests in vegetable production systems." Sustainable agriculture project research progress report, Univ. California, Davis.

Lawson, H. M., and P. B. Topham. (1985). "Competition between annual weeds and vining peas grown at a range of population densities: Effects on the weeds," *Weed Res.* **25**:221–29.

Leather, G. R. (1983). "Sunflowers (*Helianthus annuus*) are allelopathic to weeds," *Weed Sci.* **31**:37–42.

Liebl, R. A., and A. D. Worsham. (1983). "Inhibition of pitted morningglory (*Ipomea lacunosa* L.) and certain other weed species by phytotoxic components of wheat (*Triticum aestivum* L.) straw," *J. Chem. Ecol.* **9(8)**:1027–43.

Liebman, M. (1986). "Ecological Suppression of Weeds in Intercropping Systems: Experiments with Barley, Pea, and Mustard.," Ph. D. dissertation, Univ. of California, Berkeley.

Little, C. E. (1987). *Green Fields Forever: The Conservation Tillage Revolution in America.* Island Press, Washington, DC.

Lockeretz, W., G. Shearer, and D. H. Kohl. (1981). "Organic farming in the corn belt," *Science* **211**:540–47.

Ludwig, D. L. (1982). "Integrated approach to weed control in the cornbelt, U.S.A." In *Basic Techniques in Ecological Farming: The Maintenance of Soil Fertility* (Eds. S. Hill and P. Ott). pp. 171–77. Birkhauser Verlag, Basel.

MacHoughton, J. (1973). "Ecological Changes in Weed Populations as a Result of Crop Rotations and Herbicides." Ph. D. dissertation, Univ. of Illinois, Urbana.

Marra, M. C., and G. C. Carlson. (1983). "An economic threshold model for weeds in soybeans (*Glycine max*)," *Weed Sci.* **31**:604–09.

McLeod, E. J., and S. L. Swezey. (1980). "Survey of weed problems and management technologies," Research leaflet, Univ. of California Appropriate Technology Program, Davis.

McWhorter, C. G. (1972). "Flooding for johnsongrass control. " *Weed Sci.* **20(3)**:238-241.

McWhorter, C. G., and J. M. Chandler. (1982). "Conventional weed control technology." In *Biological Control of Weeds with Plant Pathogens* (Eds. R. Charudattan and H. L. Walker). pp. 5–24. Wiley, N.Y.

McWhorter, C. G., and W. C. Shaw. (1982). "Research needs for integrated weed management systems," *Weed Sci.* **30** (supplement):40–45.

Micka, E. S. (1985). *Maine Farm Planning Guide.* Maine Coop. Exten. Service, Univ. of Maine, Orono.

Minotti, P. L., and R. D. Sweet. (1981). "Role of crop competition in limiting losses from weeds." In *CRC Handbook of Pest Management in Agriculture, Vol. 2* (Ed. D. Pimentel). pp. 351-367. CRC Press, Boca Raton, FL.

Mortimer, A. M. (1983). "On weed demography." In *Recent Advances in Weed Research* (Ed. W. W. Fletcher). pp. 3-40. Commonwealth Agricultural Bureaux, Farnham Royal, England.

Murphy, B. (1987). *Greener Pastures on Your Side of the Fence: Better Farming with Voisin Grazing Management*, Arriba Publishing, Colchester, VT.

Nieto, J. N., and D. W. Staniforth. (1961). "Corn-foxtail competition under various production conditions," *Agron. J.* **53**:1-5.

Oliver, L. R. (1988). "Principles of weed threshold research," *Weed Tech.* **2**:398–403.

Papendick, R. I. (1987). "Why consider alternative production systems?" *Am. J. Alt. Agric.* **2(2)**: 83–86.

Papendick, R. I., L. F. Elliot, and J. F. Power. (1987). "Alternative production systems to reduce nitrate in ground water," *Am. J. Alt. Agric.* **2(1)**:19–24.

Peters, S. (1988). "Soybean interseeding: Summary of on-farm research results 1987, Lancaster Co., PA," Rodale Research Center report, Kutztown, PA.

Peters, S., R. Andrews, and R. Janke. (1988a). "Rodale's farming systems experiment 1981–1987," Rodale Research Center (RRC) Bulletin 88/1, RRC, Kutztown, PA.

Peters, S., R. Andrews, and R. Janke. (1986b). "Comparing soybeans intercropped with small grains to a soybean monoculture within a cropping systems experiment," Rodale Research Center report, Kutztown, PA.

Phatak, S. C., M. B. Callaway, and C. S. Vavrina. (1987). "Biological control and its integration in weed management systems for purple and yellow nutsedge (*Cyperus rotundus* and *C. esculentus*)," *Weed Tech.* **1**:84–91.

Pimentel, D., and L. Levitan. (1986). "Pesticides: Amounts applied and amounts reaching pests," *BioSci.* **36(2)**:86–91.

Prostko, E., and R. D. Ilnicki. (1988). "Residual weed control in intercropped soybeans," *Proc. Northeast. Weed Sci.* **42**:22–23.

Putnam, A. R., and J. DeFrank. (1983). "Use of phytotoxic plant residues for selective weed control," *Crop Protection* **2(2)**:173–81.

Putnam, A. R., J. DeFrank, and J. P. Barnes. (1983). "Exploitation of allelopathy for weed control in annual and perennial cropping systems," *J. Chem Ecol* **9(8)**: 1001–10.

Putnam, A. R., and W. B. Duke. (1978). "Allelopathy in agroecosystems," *Ann. R. Phyto.* **16**:431–51.

Quimby, P. C. Jr., and H. L. Walker. (1982). "Pathogens as mechanisms for integrated weed management," *Weed Sci.* **30** (suppl.) 30–34.

Radosevich, S. R., and J. S. Holt. (1984). *Weed Ecology: Implications for Vegetation Management*, Wiley, N.Y.

Reinbott, T. M., Z. R. Helsel, D. G. Helsel, M. R. Gebhardt, and H. C. Minor. (1987). "Intercropping soybean into standing green wheat,"*Agron. J.* **79**:886–91.

Roberts, H. A., and M. E. Potter. (1980). "Emergence patterns of weed seedlings in relation to cultivation and rainfall," *Weed Res.* **20**:377–86.

Roeth, F. W., and R. Selley. (1987). "Weed control alternatives and costs." In *Sustainable Agriculture: Wise and Profitable Use of Our Resources in Nebraska.* pp. 179–81. Nebraska Coop. Exten. Service, Univ. of Nebraska, Lincoln.

Rosenthal, S. S., D. M. Maddox, and K. Brunetti. (1984). *Biological Methods of Weed Control*, Thomson Publications, Fresno, CA.

Rubin, B., and A. Benjamin. (1984). "Solar heating of the soil: Involvement of environmental factors in the weed control process," *Weed Sci.* **32**:138–42.

Samson, R. (1987). "4C making continuous corn a poor alternative," *Resource Efficient Agricultural Production (R.E. A.P.)-Canada* **1(2)**:1, 6-7.

Scifres, C. J. (1981). "Selective grazing as a weed control method." In *CRC Handbook of Pest Management in Agriculture, Vol. 2,* (Ed. D. Pimentel). pp. 369–76. CRC Press, Boca Raton, FL.

Scott, T. W., J. Mt. Pleasant, R. F. Burt, and D. J. Otis. (1987). "Contributions of ground cover, dry matter, and nitrogen from intercrops and cover crops in a corn polyculture system," *Agron. J.* **79**:792–98.

Shaw, W. C. (1982). "Integrated weed management systems technology for pest management," *Weed Sci.* **30** (suppl.):2–12.

Shilling, D. G., L. A. Jones, A. D. Worsham, C. E. Parker, and R. F. Wilson. (1986). "Isolation and identification of some phytotoxic compounds from aqueous extracts of rye (*Secale cereale* L.)," *J. Ag. & Food Chem.* **34**:633–38.

Shilling, D. G., R. A. Liebl, and A. D. Worsham. (1985). "Rye (*Secale cereale* L.) and wheat (*Triticum aestivum* L.) mulch: The suppression of certain broadleaved weeds and the isolation and identification of phytotoxins." In *The Chemistry of Allelopathy: Biochemical Interactions Among Plants* (Ed. A. C. Thompson). pp. 247–71. American Chemical Society, Washington, D.C.

Sinclair, W. (1985). "Organic farmers still harvesting profits: Bucking no-till trend and shunning chemicals, they go against the government grain," *Washington Post,* September 1, 1985, A9.

Sinclair, W. (1987a). "Farm achieves natural balance: Earthworms, songbirds, profit abound after chemicals forsaken," *Washington Post,* March 1, 1987, A3.

Sinclair, W. (1987b). "Organic farming is blossoming: Pollution and costs turn growers away from use of chemicals," *Washington Post,* November 23, 1987, A3.

Standifer, L. C., P. W. Wilson, and R. Porche-Sorbet. (1984). "Effects of solarization on soil weed seed populations," *Weed Sci.* **32**:569–73.

Staniforth, D. W. (1957). "Effects of annual grass weeds on the yield of corn," *Agron. J.* **49**:551–55.

Stapleton, J. J. and J. E. DeVay. (1986). "Soil solarization: A non-chemical approach for management of plant pathogens and pests," *Crop Protection* **5(3)**:190–98.

Steinsiek, J. W., L. R. Oliver, F. C. Collins. (1982). "Allelopathic potential of wheat (*Triticum aestivum*) straw on selected weed species," *Weed Sci.* **30**:495–97.

Stonehouse, D. P., B. D. Kay, J. K. Baffoe, and D. L. Johnston-Drury. (1988). "Economic choices of crop sequences on cash-cropping farms with alternative crop yield trends," *J. Soil Water Conserv.* **43(3)**:266–70.

Sumner, D. R. (1982). "Crop rotation and plant productivity." In *CRC Handbook of Agricultural Productivity, Vol. 1, Plant Productivity,* (Ed. M. Rechcigl, Jr.). pp. 273–313. CRC Press, Boca Raton, FL.

Sweet, R. D., C. P. Yip, and J. B. Sieczka. (1974). "Crop varieties: Can they suppress weeds?" *N.Y. Food Life Sci. Q.* **7(3)**:3-5.

Teasdale, J. R., and J. R. Frank. (1983). "Effect of row spacing on weed competition with snap beans (*Phaseolus vulgaris*)," *Weed Sci.* **31**:81–85.

Terpstra, R. and J. K. Kouwenhoven. (1981). "Inter-row and intra-row weed control with a hoe-ridger," *J. Agri. Engineering Res.* **26**:127–34.

Thompson, D., and S. Thompson. (1984). "Farming without chemicals," *EPA J.* **10(5)**:33–34.

USDA. (1973). *Monoculture in agriculture: Extent, causes, and problems. Report of the task force on spatial heterogeneity in agricultural landscapes and enterprises,* USDA, Washington, D.C.

USDA. (1980). *Report and recommendations on organic farming,* USDA, Washington, D.C.

Volak, B., and R. Janke. (1987). "Soybean interseeding technical report," Rodale Research Center (RRC) Bulletin 88/32, RRC, Kutztown, PA.

Vrabel, T. E., P. L. Minotti, and R. D. Sweet. (1980). "Seeded legumes as living mulches in sweet corn," *Proc. Northeast. Weed Sci. Soc.* **34**:171–75.

Wagger, M. G., and D. B. Mengel. (1988). "The role of nonleguminous cover crops in efficient use of water and nitrogen." In *Cropping Strategies for Efficient Use of Water and Nitrogen* (Ed. W.L. Hargrove). pp. 115–27. Amer. Soc. of Agron., Crop Science Society of America, and Soil Science Society of America, Madison, WI.

Walker, R. H., and G. A. Buchanan. (1982). "Crop manipulation in integrated weed management systems," *Weed Sci.* **30** (suppl.):17–24.

Wicks, G. A., R. E. Ramsel, P. T. Nordquist, J. W. Schmidt, and Challaiah. (1986). "Impact of wheat cultivars on establishment and suppression of summer annual weeds," *Agron. J.* **78**:59–62.

Wiese, A. F., and J. M. Chandler. (1979). "Weeds." In *Introduction to Crop Protection,* (Ed. W. B. Ennis, Jr.). pp. 232–38. Amer. Soc. of Agron., Madison, WI.

Williams, W. M., P. W. Holden, D. W. Parsons, and M. N. Lorber. (1988). "Pesticides in ground water data base, 1988 interim report," U.S. Environmental Protection Agency, Office of Pesticide Programs, Environmental Fate and Effects Division, Washington, DC.

Wood, G. M. (1987). "Animals for biological brush control," *Agron. J.* **79**:319–21.

Wookey, C. B. (1985). "Weed control practices on an organic farm," *1985 British Crop Protection Conference—Weeds,* pp. 577–82.

Zimdahl, R. L. (1980). *Weed-Crop Competition: A Review*. International Plant Protection Center, Corvallis, OR.

Zimdahl, R. L. (1981). "Extent of mechanical, cultural, and other nonchemical methods of weed control." In *CRC Handbook of Pest Control in Agriculture, vol. 2* (Ed. D. Pimentel). pp. 73–83. CRC Press, Boca Raton, FL.

5 SUSTAINABLE SOIL FERTILITY PRACTICES

LARRY D. KING

Department of Soil Science,
North Carolina State University,
Raleigh, North Carolina

INTRODUCTION

The term *sustainable soil fertility* implies that plant nutrients will be available forever. One way to assure sustainable fertility is to make sure that all the nutrients taken up by plants are returned to the soil so they can be used again by the plants. In this manner a nutrient cycle is established. In actual soil-plant systems, the cycle is not completely closed because nutrients are continuously added to or lost from the cycle. Managing the

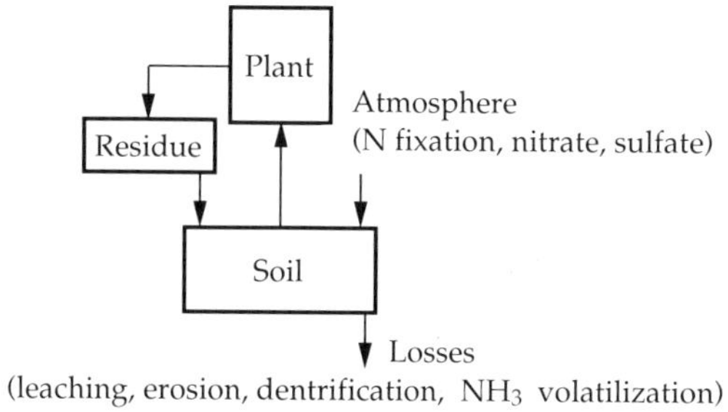

Figure 5.1 Nutrient cycling in a natural system.

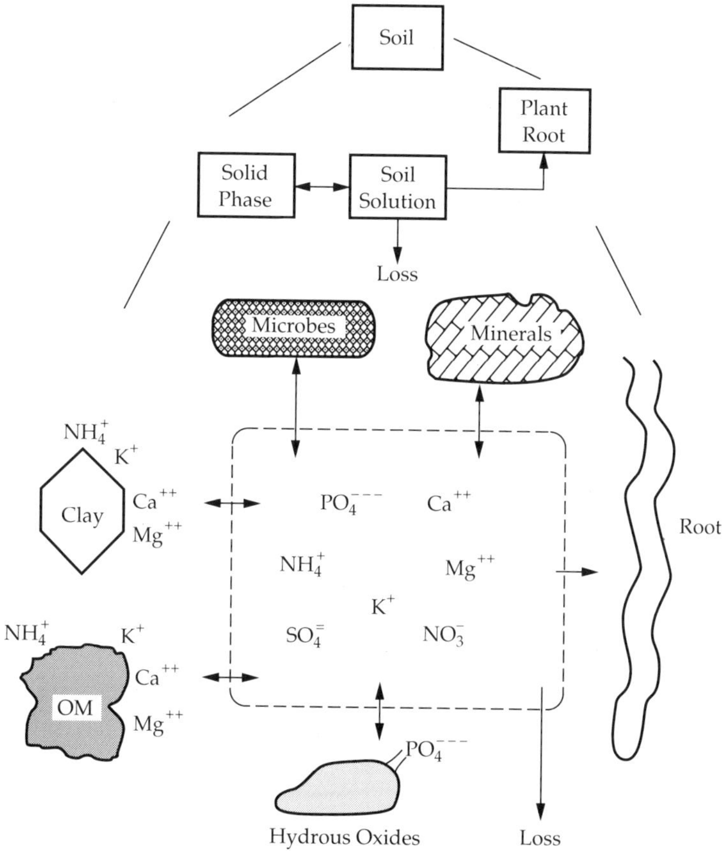

Figure 5.2 Nutrient dynamics among soil components.

cycle to minimize losses and to supply necessary inputs is the key to fertility management in sustainable agricultural systems.

Cycling can be viewed at several levels. On a field level in a natural system, nutrients move from soil into plants and are returned to the soil via residue as plants die (Figure 5.1). Although most of the nutrients remain in the cycle, losses and additions occur. For example, nitrogen (N) is added through biological N fixation and in precipitation but is lost through soil erosion, leaching, denitrification (microbial conversion of nitrate to nitrous oxides or N_2), and ammonia (NH_3) volatilization. One may look more closely at this model by focusing on nutrient dynamics in the soil (Figure 5.2). Here nutrients move between the solution and solid phases of the soil with the plant root acting as a nutrient sink. In the soil, positively charged ions move between the soil solution and clay or organic matter (OM). Phosphorus (P) sorbed on oxides of iron (Fe) and aluminum (Al) in acid soil or present as calcium (Ca) phosphates in calcareous soils cycles less readily than do the cations. However, sources of P are in equilibrium with P in the soil solution. Microorganisms absorb nutrients from solution as they decompose organic matter, and these nutrients are returned to solution as the organisms die and are decomposed by other microorganisms. Weathering of minerals supplies nutrients to the soil solution. The plant root is the nutrient sink in this system.

The natural cycle shown in Figure 5.1 is often considered the one after which agricultural systems should be patterned. However, the big difference between natural and agricultural systems is the relatively large export of nutrients through crop harvest (Figure 5.3). If the agricultural system is to continue, these nutrients must be replaced. In conventional agricultural systems, these nutrients are replaced with fertilizer and manure.

The nutrient cycle can be expanded to include an entire farm (Figure 5.4). On a farm that produces both crops and animals, nutrients are removed from the fields via crop harvest and may leave the farm either as grain, hay, or as animal products. A large fraction of nutrients consumed

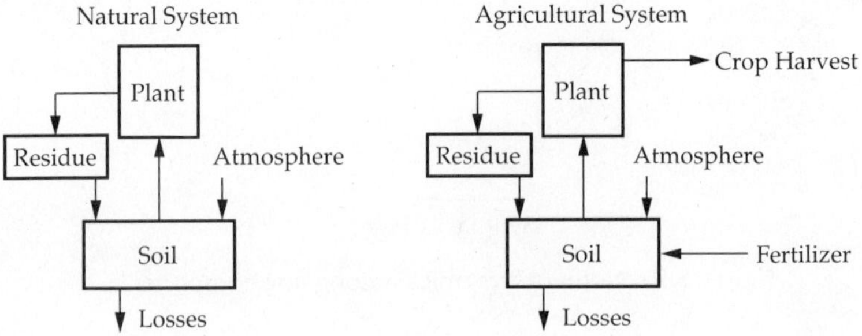

Figure 5.3 Comparison of nutrient cycling in natural and agricultural systems.

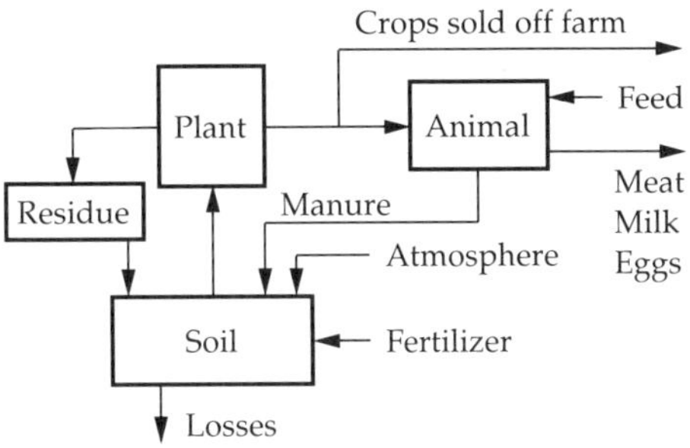

Figure 5.4 Nutrient flow on a farm.

by animals is returned to the soil in manure. Nutrients leaving the farm are replaced with fertilizer and purchased feed.

The cycle can be expanded further to include nutrient cycling in a region (Figure 5.5). Harvested crops and/or animal products leave the farm and are processed before being sold to consumers in the city. Some of the nutrients in these products eventually are deposited in surface water or landfills. Nutrients may be returned to the farm if the farmer uses food processing by-products, sewage sludge, or effluent as a nutrient source. Use of these materials is not true cycling because the actual nutrients removed from a farm are not the same ones returned since many farms contribute crops/livestock to the processing step. However, the effect is the same: nutrients are returned to the farm to replace those that are removed.

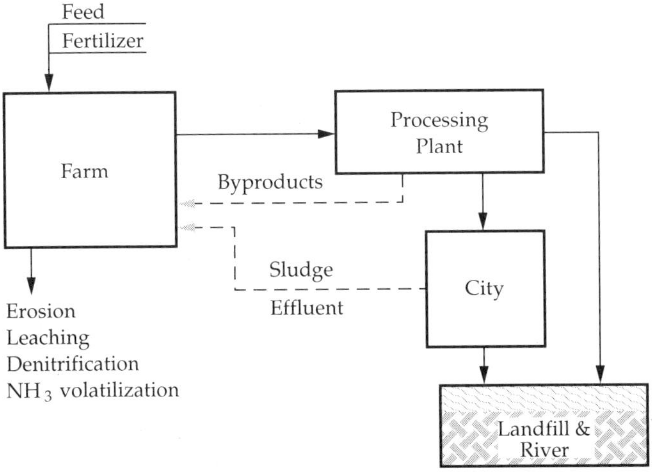

Figure 5.5 Nutrient flow in a region.

As one moves up from cycling of nutrients among clay, OM, microorganisms, oxides, and the soil solution (Figure 5.2) to regional nutrient cycling (Figure 5.5), losses from the cycles increase. For example, in the farm cycle (Figure 5.4) the method of manure management will affect loss of nutrients. Leaching and volatilization losses of nutrients from manure stored uncovered, NH_3 volatilization from surface-applied manure, and leaching losses due to excessive rates of manure application are examples of losses due to poor manure management. Efficient cycling on a regional basis is rarely achieved, or if it is achieved, the nutrients are returned only to farms near the sources of by-products and sludge or effluent. Consequently, the sale of products must be considered a loss from the cycle for most farms. The loss can be considerable from a grain farm or much lower from a livestock farm. On swine and poultry farms, the net nutrient budget may be positive because of the high input of nutrients from purchased feed.

Sustainability can be improved by minimizing nutrient loss other than loss due to sale of products, maximizing utilization of nutrients currently present in the soil, and maximizing biological N fixation—and all of these efforts must be *economically* sustainable. Finally, one must consider sources of nutrients to replace those lost in products leaving the farm. Some or all of the required N may be supplied by biological N fixation, but P, potassium (K), Ca, magnesium (Mg), and other nutrients must be supplied by fertilizer, wastes, or purchased feed.

REDUCING LOSSES

If the key to sustainable agriculture is nutrient cycling, then losses from the cycle must be minimized to reduce the need for externally supplied nutrients. Improper manure management, soil erosion, denitrification, and leaching are examples of mechanisms by which nutrients can be lost from the cycle.

Manure

Production and Composition. Use of manure for crop production is an ancient practice and one of the most obvious methods of recycling nutrients. The nutrient content of manure varies with the type of animal, diet, and method of manure management, so the data in Table 5.1 are, by necessity, general and should not be used to determine loading rates for a specific manure. Many state and private laboratories provide manure analysis just as they do soil and plant analysis, and this service should be used to determine nutrient content of a specific manure before determining application rates.

TABLE 5.1 Manure Production and Nutrient Content (Adapted from Barker, 1980).

Animal	Animal Weight kg	Fresh Manure kg/yr	Nutrient in Fresh Manure		
			N* kg/1000kg	P kg/1000kg	K kg/1000kg
Dairy	290	17,000	6	1.5	6
Beef	200	11,000	6	1.5	4
Swine	20	1,400	6	1.5	2.5
Caged layer	1	45	14	4.5	5
Dry litter:					
broiler	0.5	7	26	9	14
turkey	3	22	18	7	6

* Generally about half of the N in fresh manure is lost before being applied to the field.

Approximately 160 million metric tons of manure was produced in the United States in 1979, of which 90% was returned to the land. Approximately 40% of the manure was produced by confined animals, of which 75% was returned to the land. The other 60% was produced by grazing animals and was returned directly to the land.

Pastures. Although direct return of manure by grazing animals relieves the farmer of the problem of collection, storage, and application to crops, it is an inefficient mechanism of pasture fertilization. Inefficiency results from the high rate of application because of the small area covered by the individual excreta, the concentration of excreta near watering sites and along fences, and the large portion of the pasture not receiving excreta. Petersen et al. (1956) reported that nutrient application rates for individual excreta from grazing cattle were (in kg/hectare) 850 N, 170 P, and 410 K for feces and 450 N, 7 P, and 400 K for urine. A typical recommendation for annual application of commercial fertilizer to a bermudagrass pasture is 270 N, 27 P, 100 K. Since plants cannot use these high rates of excreta-applied nutrients efficiently, much of the N and some of the K is carried below the root zone and lost from the nutrient cycle in the pasture. Krantz et al. (1944) and Sears (1951) reported that the effect of nutrients supplied by excreta was evident in pastures for only about three months. Although losses will vary with soil and climate, the high application rate does result in significant loss of nutrients from the pasture cycle.

Nitrogen loss is particularly high because it is lost by several mechanisms. Much of the N in feces and urine is in the ammonia form or is quickly converted to ammonia. Since the excreta remains on the soil surface (with the exception of the fraction of the urine that moves below the surface layer of soil), the potential for ammonia volatilization is high.

Nitrogen converted to nitrate can be lost from the cycle by being leached below the root zone, or it may be lost by denitrification. Moist feces deposits provide an excellent environment for denitrification. It occurs when nitrate is subjected to an anaerobic environment and is subsequently reduced to nitrous oxides or N_2. Anaerobic environments can result from excessive soil moisture or, as with feces deposits, a large source of readily available organic matter. The organic matter stimulates rapid microbial growth, oxygen concentration is reduced via microbial respiration, and anaerobic conditions develop. In the feces deposit, N is converted to nitrate in the aerobic surface layers of the deposit, and if this nitrate moves into anaerobic zones in the center of the deposit or at the deposit-soil interface, it will be denitrified and lost from the cycle. A sketch of N pathways in feces deposited in a pasture is shown in Figure 5.6.

Potassium can be lost from the cycle by leaching if soils have low cation exchange capacity. Since P does not move readily in soil, it will remain in the root zone and not be lost from the cycle.

In addition to excreta being deposited at high rates, it is deposited nonuniformly and consequently much of the pasture receives no nutrient input from excreta. At a stocking rate of 2.5 dairy cattle per hectare, Petersen et al. (1956) estimated that ten years would be required for 95% of the pasture to be covered by at least one excretion (feces or urine). Nutrient distribution can be improved by pulling a lightweight harrow over the pasture to spread the manure droppings.

The practice of intensive rotational grazing (for example, the Voisin system [Voisin, 1959], also see Chapter 8) should increase the efficiency of nutrient cycling in pastures. Manure distribution is improved because the higher stocking density results in more excreta per unit area.

Confined Animals. When animals are confined, manure can be managed to more efficiently use the nutrients. Management strategy should be to minimize the loss of nutrients during collection, storage, and application and to apply manure uniformly to maximize nutrient utilization by crops. As in a pasture system, N is the nutrient that is most readily lost from manure in confinement systems. Therefore, manure should be collected as quickly as possible and properly stored or applied to fields to reduce NH_3 volatilization. Safley et al. (1986) reported a 23% loss of N from dairy manure from the time of defecation onto the barnlot until removal from storage (stored as liquid in above-ground tanks or earthen lagoons). Most of the loss (NH_3 volatilization) occurred during the 24-hour period between barnlot cleanings. Nitrogen loss during storage was negligible. Potassium loss from defecation to removal was 10% (probably from loss of urine in the barnlot), but P loss was essentially zero. The method of loading liquid manure storage tanks/lagoons affects N loss. Muck et al. (1984) reported a 3% to 8% total N loss from bottom-loaded facilities and a 29% to 39% loss from top-loaded facilities.

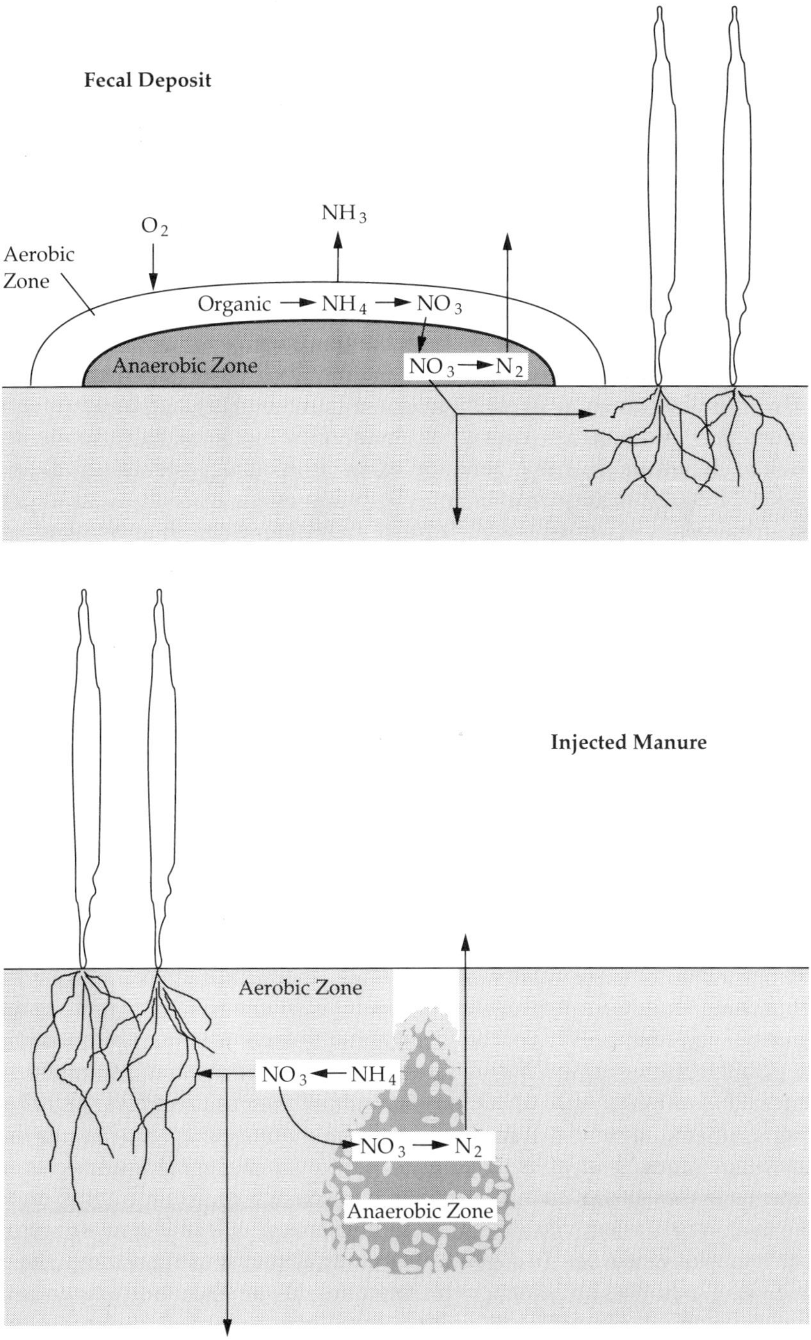

Figure 5.6 Pathways of N in feces deposited in a pasture and in injected manure.

In dry manure systems, nutrient retention is enhanced by using sufficient bedding to absorb the urine. However, N loss is still substantial in dry systems. Muck and Richards (1983) reported a 40% to 60% loss of total N in free-stall dairy barns. A practice used in the past was spreading superphosphate fertilizer in the barn so that ammonium in the manure would react with the superphosphate to form ammonium phosphate.

Some confined animal systems use relatively large quantities of water to remove manure to a storage facility. For example, water is used to flush manure in some swine and dairy barns. Although water is recycled in these systems, lagoons are required to retain the large volume of water needed. The use of mechanical aeration in lagoons to reduce odor by keeping the surface layer aerobic enhances N loss by two mechanisms: NH_3 volatilization due to the violent mixing of the lagoon surface and denitrification as nitrate formed in the aerobic surface layer moves down into the anaerobic lower layer. For example, Burns et al. (1985) reported a total N concentration range of 200 to 300 mg/L over a six-year period for effluent from a nonaerated lagoon receiving swine manure and wash water. When a surface aerator was added, N concentration dropped to 150 mg/L. Barker et al. (1980) reported 80% loss of total N in an aerated lagoon.

Composting is sometimes advocated as a means of stabilizing manure prior to applying it to fields. Composting reduces odor and improves the physical condition of manure so that it can be handled more easily. If the thermophilic process is used, temperatures of 45 to 70°C are reached and weed seed and pathogens are killed. The big disadvantage of composting is the loss of N in the process and the low plant availability of N remaining. Pratt and Castellanos (1981) found that availability of N in composted dairy manure averaged only half of that in fresh manure.

Manure may be applied to crop land as a solid from dry storage, as a slurry from liquid storage tanks/lagoons or as an effluent from a lagoon. Nitrogen loss by NH_3 volatilization is likely from each application method. Immediate incorporation of dry manure or injection of liquid manure is essential to reduce NH_3 volatilization. In a four-year study, Beauchamp et al. (1982) suface-applied liquid dairy manure to supply an average of 327 kg total N/hectare^{-1} (ha) of which 176 kg/ha^{-1} was in the ammonium form. They reported an average loss of 29% of the ammonium-N (48 kg N/ha^{-1}) via volatilization over a six-day period. Lauer et al. (1976) reported ammonium-N losses of 55% to 75% from surface-applied solid dairy manure (mean loss of 100 kg N/ha^{-1}). Although injection of liquid manure reduces ammonia loss, it does increase the potential for denitrification. The aerobic zone around the injection trench and the development of an anaerobic zone in the manure result in a nitrification-denitrification sequence (Figure 5.6). Comfort et al. (1988) found that high nitrate concentrations, readily oxidizable carbon (C), and high moisture associated with manure injection created an environment favoring denitrification. Rice et al. (1988) measured denitrification associated with injected fermentation waste and estimated that over half of the N added was lost by denitrification.

Since up to 90% of the N in lagoon effluent may be in the ammonium form, the potential for NH_3 volatilization during irrigation is significant. By comparing the N concentration in a swine lagoon with the concentration in samples collected after irrigation, Westerman et al. (1983) found a 15% loss of N during the irrigation process.

After techniques have been developed to minimize N loss during application, rates should be used to supply adequate but not excessive nutrients to crops. One example of excess application is the practice of applying high rates of manure to fields near the manure storage facility to minimize the time and distance involved in manure application. In a study of dairy farms in the Piedmont of North Carolina, nitrate-N in the upper two meters of soil in fields nearest manure storage facilities was found to range from 200 to 700 kg N/ha^{-1} (L.D. King, unpublished data). Most of this nitrate was below the root zone and thus unavailable to crops. Efficiency of nutrient use can be improved by applying manure close to the time crops will need the nutrients, for example, as close to planting as possible. Fall application for spring-planted crops usually results in significant nutrient loss from the cycle.

Nutrient Availability. Proper manure management requires that not only the nutrient content of the manure be known but also the availability of those nutrients (particularly N) so manure can be applied at rates such that crop nutrient needs are not exceeded. To determine total nutrient content, *representative* samples of manure from storage facilities should be submitted to state or private laboratories for analysis. Nutrients in inorganic forms in manures and other organic by-products are immediately available to plants. However, organic forms of nutrients must be converted to inorganic forms through decomposition by microorganisms before they are available to plants. Consequently, much research has been conducted to predict the availability of nutrients in organic by-products or to estimate their relative efficiency as compared to commercial fertilizer. Chemical tests (Chescheir et al., 1985), laboratory incubation (King and Vick, 1978; King, 1984), use of a portable N meter (Chescheir et al., 1985), greenhouse studies (Castellanos and Pratt, 1981), and field studies (Safley et al., 1986) are methods by which N availability or relative efficiency has been measured. These studies have shown that availability of manures varies considerably and is related primarily to the type of manure and the type of soil to which the manure is applied. Of course, availability under field conditions also will be influenced by methods of storage and application.

In view of the many variables that can affect nutrient availability, the practical approach would be for a farmer to use nutrient analysis of the manure (or other waste), soil and plant analysis, and crop yield records to develop an estimate of nutrient availability for the manure used on the farm.

Refeeding. Refeeding animal manure is a shortcut in the normal nutrient cycle. Crinkenberger and Goode (1978) reported that ensiling a mixture of 30% broiler litter (feces, bedding, feathers) and 70% corn produces a feed that provides adequate crude protein, Ca, and P for most beef cattle. Stacking poultry litter, allowing it to heat, and then adding 20% ground corn results in a good wintering ration for beef cattle.

Erosion

While the mechanisms of nutrient losses from manure are subtle, one can readily observe runoff water and sediment movement during a rainstorm, or dust being carried away by high winds, and realize that nutrients are being removed by these losses. Since the magnitude of loss is affected by factors such as cropping system, conservation measures, and slope (Table 5.2), much of this potential loss can be reduced through management. Obviously, the magnitude of nutrient loss is directly related to the amount of soil loss, but it is also a function of the nutrient content of the soil. From a crop production standpoint, the short-term importance of the nutrient loss depends less on the amount of nutrient loss than on how available the nutrients would have been to crops if the nutrients had remained on site. For example, in some erosion studies the total amount of P lost via erosion is reported (Olness et al., 1975), but in others an estimate of plant available P (for example, the amount extractable with a soil test extractant) is reported (Alberts et al., 1978). Therefore, in interpreting nutrient losses by erosion, the decision must be made as to whether the concern is the immediate effect on the nutrient cycling (loss of "available" nutrients) or the long-term effect (loss of total nutrients).

Since erosion studies like those shown in Table 5.2 are very expensive to conduct, measuring erosion and nutrient loss for all possible situations is not feasible. However, nutrient loss from any site can be estimated by first using the Universal Soil Loss Equation (USDA, 1976) to calculate soil loss and then using either total nutrient content of the soil or some measure of available nutrients to estimate the nutrient loss from erosion. This procedure overestimates loss from most fields because it does not consider redeposition of sediment within a field. Thus, it should be considered a worst-case estimate for most sites.

Denitrification

Environmental factors significantly affecting denitrification in soils are the nature and amount of organic matter present, degree of aeration, moisture

TABLE 5.2 Nutrient Losses from Runoff and Erosion

			Annual Loss		
Location	Crop	Site	N kg ha^{-1}	P kg ha^{-1}	Reference
Iowa	Corn	2–18% slope:			
		contour cropped	57	2[a]	Alberts et al., 1978
		terraced	5	<1[a]	
Missouri	Corn/wheat/clover	—	29	9	Miller and Krusekopf, 1932
North Carolina	Grass	35–40% slope	3	<1	Kilmer et al., 1974
Oklahoma	Cotton	<0.5% slope	6	5[b]	Olness et al., 1975
	Wheat	"	4	3[b]	
	Alfalfa	"	3	2[b]	
	Pasture	3% slope			
		continuous grazing	8	5[b]	
		rotational grazing	1	1[b]	

[a] $NaHCO_3$—extractable P

[b] Total P

status, pH, and temperature (Alexander, 1961). Some of these factors can be affected by management. Low pH reduces denitrification, but the threshold ($\approx$ 5.5) below which denitrification is minimal is also below that required for good growth of most crops. The anaerobic conditions required for denitrification can result from excess soil moisture, which hinders oxygen (0_2) movement into the soil. Climate determines the amount of moisture that a soil receives. Landscape position, slope, vegetative cover, and soil structure and texture dictate the rate at which that moisture moves through the soil and thus these factors affect the potential for denitrification. Soil moisture status can sometimes be modified by management (surface or subsurface drainage, irrigation scheduling to minimize saturated conditions), but in many situations, little can be done to control soil moisture. As previously mentioned, anaerobic conditions also may develop as a result of 0_2 depletion by high rates of microbial respiration resulting from addition of large quantities of readily available C. Examples of N loss by denitrification stimulated by organic matter additions are shown in Table 5.3. These studies show that up to 60% of the N in organic materials may be lost from the cycle due to denitrification. Stimulation of denitrification by organic matter poses a dilemma: Manures, other organic by-products, and green manure crops are important nutrient sources for sustainable agricultural systems, yet their use contributes to N loss by denitrification. Although addition of organic matter may stimulate denitrification initially, in the long term it also increases aeration by improving soil structure and thus produces a soil condition in which denitrification is less likely.

In addition to methods of controlling soil moisture in a field, several management options are available to reduce N loss by denitrification. One is to minimize the period that nitrate is exposed to a denitrifying environment. Organic nutrient sources should be applied as closely as practical to the time the crop will require N. If the material must be applied much earlier than the crop will need nutrients (for example, in the fall for a spring crop), then a cover crop can be used to accumulate N, store it during the winter, and then release it for use by subsequent crops through decomposition.

The method of application affects N loss by denitrification. As mentioned earlier, application methods that result in high concentrations of organic material in localized zones (for example, manure defecated in a pasture or liquid manure injected into soil) stimulate denitrification (Table 5.3). The detrimental effects of these concentrated zones can be reduced by mixing the organic material more uniformly with the soil, for example, surface applying the liquid manure and then mixing it with soil with a disc.

Leaching

The magnitude of nutrient loss via leaching is controlled by climate, soil chemical and physical properties, type of crop, and the specific nutrient

TABLE 5.3 Loss of N by Denitrification* from Soils Amended with Organic Materials.

Organic Material	Type of Study	Application Method	N Added (kg/ha^{-1})	N Denitrified (% of Applied)	Time Period (Days)	Reference
Cell waste from fermentation	Field	Injected	740	60	73	Rice et al. 1988
	Lab	Injected	280	12	14	
		Incorporated	280	4	10	
Fermentation waste	Lab	Incorporated:				
		wafers	1,000	31	224	King and Vick, 1978
		powder	1,000	61	224	
Cattle manure	Greenhouse	Incorporated	1,470	48	39	Guenzi et al., 1978
Sewage sludge	Lab	Incorporated	780	15	154	King, 1973
		Surface	780	20	154	
Subterranean clover	Lysimeter	Incorporated	78	25	365	Muller, 1987

* Data from Rice et al. were direct measurements of denitrification. In the other studies, the applied N that was not present in the soil or crop at the end of the experiment was assumed to have been lost by denitrification.

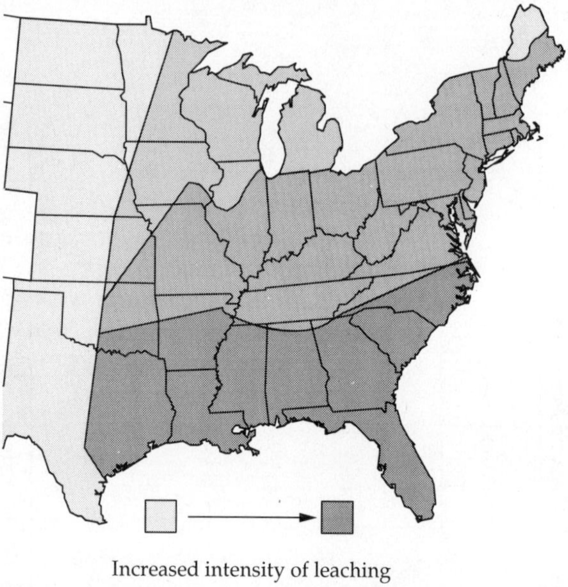

Figure 5.7 Relative intensity of leaching during the winter months in the eastern United States (from Nelson and Unland, 1955).

and its concentration in the soil. The numerous possible combinations of these factors result in a wide range of leaching losses. Water is the vehicle by which nutrients are carried out of the root zone and thus are lost from the cycle. Type and distribution of precipitation influence the rate of this leaching loss of nutrients. Leaching losses are highest in the southeastern United States and decline as one moves north or west (Figure 5.7).

Thomas (1970) has summarized climate and soil data to show the differences in nutrient losses among the regions. Generally, in the Northeast leaching is negligible from December through February because of frozen soil. However, losses during the summer are more likely than in any other part of the United States. High evaporation in the Southeast essentially prevents moisture from moving below the root zone in the summer, so leaching losses are negligible during this period. However, since soils in this region are seldom frozen, significant leaching losses occur during the winter and early spring. In the Midwest, low precipitation in winter plus frozen soil prevent losses during winter. High rainfall in the spring and early summer make this period the time of maximum leaching loss. Because of low rainfall and high evaporation in the Great Plains, leaching losses are minimal.

The effect of climate is modified by soil properties and the interaction of specific nutrients with these properties. Soil permeability, which is influenced by structure and texture, affects the rate at which water moves through the soil. Sandy soils or clayey soils with porous structure allow relatively rapid water movement, whereas clayey soils with less porous structure retain moisture. The extent to which nutrients move with water is influenced by their reaction (or lack of reaction) with soil. Clay and organic matter are negatively charged and thus retard the movement of cations (for example, K^+, Ca^{++}, Mg^{++}) through soil. The higher this net negative charge (cation exchange capacity or CEC), the greater the retention of cations. Extremes in leaching of cations are represented by a loss of 2 kg K $ha^{-1}yr^{-1}$ from a silt loam soil in Illinois and a loss of 140 kg (out of a total of 150 kg of total K present) from a coarse sandy soil in a greenhouse study in Florida (Tisdale and Nelson, 1966). Because nitrate ions are negatively charged, the negative charges on soil particles enhance nitrate leaching. An exception is in the red subsoils of the southeastern United States where positive charges associated with Al and Fe oxides retain significant amounts of nitrate (Westerman et al., 1987). Phosphorus reacts with Al and Fe compounds in acid soils, and with Ca in alkaline soils, to form insoluble compounds. Consequently, leaching losses are negligible except in very sandy soils or in organic soils that are inherently low in Al and Fe.

Vegetation affects leaching losses by two mechanisms. Vegetation removes moisture from soil as a result of transpiration from leaf surfaces. With a closed canopy (for example, a sod or a row crop that completely shades the soil surface), transpiration is the main mechanism of evaporative moisture loss from soil. This evaporative loss decreases soil moisture movement below the root zone. Secondly, plants remove nutrients from soil solution and prevent them from moving out in drainage water.

An obvious method of reducing leaching losses is to avoid suppling crops with excess nutrients; that is, one should supply nutrients at a rate equal to the rate of uptake by the crop. One means of achieving this balance is to apply commercial fertilizer several times during the growing season rather than applying it all prior to planting the crop. This "split application" of fertilizer is common on very permeable soils. Another approach is to use nutrient sources that release nutrients slowly during the growing season. The most obvious examples are manures and green manure crops that release nutrients as they decompose in the soil. However, a green manure crop that produces more N than the subsequent crop needs, or manure applied at rates supplying N in excess of crop needs, will result in nitrate leaching out of the root zone.

Rate of release from cover crops can be slowed by leaving the residue on the surface rather than incorporating it. For example, incorporating hairy vetch prior to planting corn resulted in 120 kg ha^{-1} soil nitrate N at the time of corn emergence, whereas leaving the vetch residue on the surface resulted in 78 kg ha^{-1} (Hubbard, 1986). Corn yields were comparable with both methods of residue management.

Some slow-release commercial N fertilizer sources have been developed, but because of their relatively high cost, they are used only with high-value crops. One disadvantage of slow-release materials is that they may continue to release nutrients after crop harvest and thus increase the risk of loss by leaching during the winter (King et al., 1977). If appreciable nutrients are present after harvest, cover crops can be used during the noncropping season to take up these nutrients, store them, and then release them to subsequent crops as the cover crop decomposes.

An added advantage of minimizing leaching losses is that the potential for loss of N by denitrification is also minimized.

Summary of Losses

The key to increasing sustainability of agricultural systems is to develop management practices that reduce nutrient losses from the system. Since the magnitude of each loss is influenced by factors such as farm management, climate, soil type, topography, and cropping system, it is not possible to present a table showing typical losses from erosion, leaching, N volatilization, and crop removal. The magnitude of these losses is site-specific. With adequate data, the potential for loss via the various mechanisms mentioned above can be estimated. As an example, estimated N losses measured in the Coastal Plain of North Carolina are shown in Table 5.4. The greatest N removal from the cycle was in the harvested grain, but leaching losses on the well-drained soil and denitrification losses on the poorly drained soil were appreciable. Erosion losses measured may be exaggerated above what actually would be lost from a field because measurements were made on small plots, but on nearly level fields like the experimental sites, significant redeposition of sediment can occur within the field.

TABLE 5.4 Nitrogen Budget for Continuous Corn Grown in the Coastal Plain of North Carolina. (Adapted from Gambrell et al. 1975).

	Soils	
	Well Drained (Aquic Paleudult) kg $ha^{-1}yr^{-1}$	Poorly Drained (Typic Umbraquult) kg $ha^{-1}yr^{-1}$
Inputs		
Fertilizer N	160	196
Losses		
Harvested grain	92	92
Runoff/erosion*	22	29
Leaching	46	16
Denitrification	0	60

* Measurement of erosion from small plots probably overestimates losses from nearly level fields because much of the sediment may be redeposited in the field.
Source: Gambrell et al, 1975.

MAXIMIZING AVAILABILITY OF SOIL NUTRIENTS

Once the losses of nutrients from the cycle have been minimized, then methods should be developed to use the remaining nutrients as efficiently as possible. Efficient use of these nutrients reduces the need for nutrient inputs and thus increases sustainability. Factors influencing the quantity of nutrients that a soil can supply to plants include inherent fertility, residual fertility from previous fertilizer applications, and microbial activity.

Inherent Soil Fertility

The degree to which a soil can supply nutrients on a sustained basis is dependent on its inherent fertility. Inherent fertility is a function of the chemical makeup of the parent material from which the soil was formed, the climate to which the parent material has been subjected, the length of time soil-forming processes have been acting, and the vegetation that has evolved in response to the soil properties and climate. Soils in the Piedmont area of the southeastern United States have developed from rock that is low in most plant nutrients, the rock has been subjected to soil-forming processes for several million years, and the processes have been enhanced by the relatively high temperature and rainfall in the area. Consequently, the soils are highly weathered, low in nutrient content due to leaching, and thus have low inherent fertility. In contrast, many of the soils in the northern United States have developed from parent material exposed by the relatively recent action of glaciers ($\approx$ 12,000 years ago). Because these soils are younger and have developed in a colder climate, their native nutrient status is higher than that of soils further south. Local conditions such as topography and atypical parent material can modify the soil-forming trends noted above. The effect of location on total nutrient concentrations in soils is shown in Table 5.5. Generally, nutrient content increases as one moves north or west.

TABLE 5.5 Total Amounts of Nutrients in Selected Soils in the United States.

Nutrient	*Florida* Norfolk Fine Sand mg kg^{-1}	*Virginia* Sassafras Sandy Loam mg kg^{-1}	*New York* Ontario Loam mg kg^{-1}	*Iowa* Marshall Silt Loam mg kg^{-1}
N	200	200	1600	1700
P	220	98	440	525
K	1,300	13,000	16,000	19,000
Ca	70	3,000	6,000	6,000
Mg	140	2,600	5,300	6,000
S	90	130	145	220

Source: Marbut, 1935.

TABLE 5.6 Insoluble Forms of Nutrients in Mineral Soils.

Nutrient	Form
N	Proteins, amino acids
P	Apatite, secondary Ca, Fe, and Al phosphates
K	Feldspars and micas, illitic clays
Ca	Feldspars, hornblende, calcite, dolomite
Mg	Mica, hornblende, dolomite, serpentine, certain clays
S	Pyrite, gypsum, organic forms

The nutrients shown in Table 5.5 occur predominately in insoluble form—either in organic matter or in primary or secondary minerals (Table 5.6)—and thus are not available to plants. However, the soluble forms of the nutrients in the soil solution are in equilibrium with these insoluble forms (Figure 5.2), so the total nutrient content is an estimate of the long-term nutrient-supplying capacity of the soil. The speed with which the insoluble forms can supply nutrients is rather rapid for organic matter (for example, mineralization of organic N as organic matter decomposes) and relatively slow for primary minerals (for example, dissolution of K from feldspars).

Residual Fertility from Long-term Use of Fertilizer

Fields that have received fertilizer for a long time often contain P and K concentrations that far exceed the critical concentration required for maximum crop production. This accumulation of nutrients was discussed with soil scientists/agronomists from Alabama, Iowa, Michigan, Minnesota, Missouri, Nebraska, and Vermont (Cramer et al., 1985). All agreed that many fields in their respective states had K and particularly P concentrations that exceeded the concentration required for maximum crop production. Two common reasons were given for the high concentrations: (a) continued use of a fertilizer program developed for a soil that was originally infertile and not adjusting rates down as unused nutrients accumulated, and (b) some soil-testing laboratories recommend fertilizer rates that are higher than needed for maximum yield.

Soil test data from 1950 and 1987 from the North Carolina Department of Agriculture were analyzed to determine the extent of excess P and K in the Coastal Plain soils of North Carolina (Figure 5.8). Only data from corn land were analyzed to avoid areas that have received very high rates of fertilizer (for example, tobacco and vegetables—although some of the fields sampled may have had tobacco in a rotation).

With Mehlich 3 (M3) extractant (1987 data), the critical soil test fertility index falls about the midpoint of the medium range for P (McCollum, 1987) and K (Eugene Kamprath, personal communication) for corn and

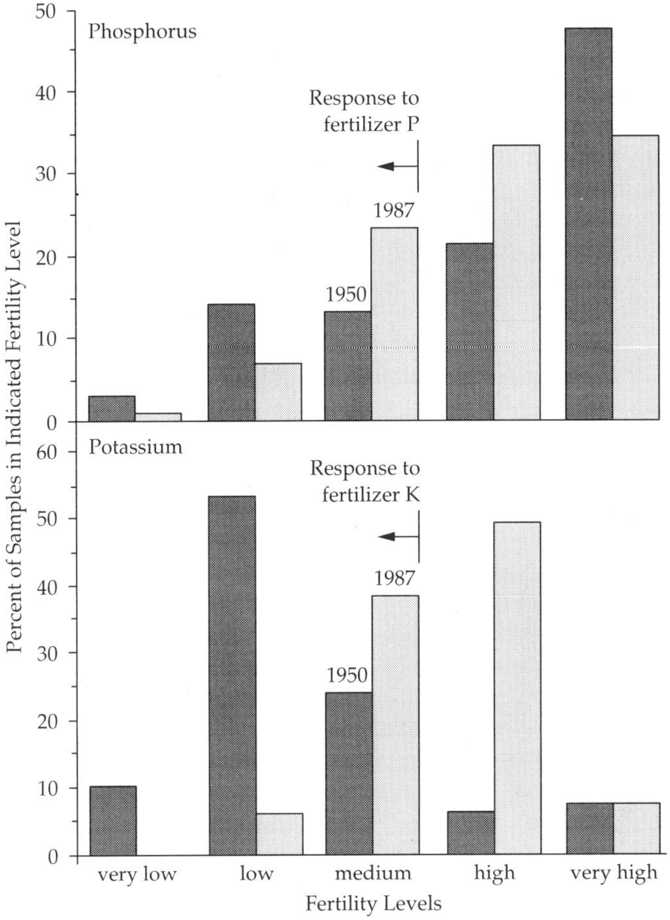

Figure 5.8 Distribution of P and K soil test fertility levels in soils from corn fields in the Coastal Plain of North Carolina in 1950 (3,731 samples) (Welch and Nelson, 1951) and 1987 (40,400 samples)(Tucker, 1988). Extractant in 1950 was 0.05 N HCl and in 1987 was Mehlich 3 (Mehlich, 1984).

soybeans. A conservative approach was used and the breakpoint for fertilizer response was assumed to be between the medium and high ranges, as indicated in Figure 5.8. In 1950, 49% of the samples were in the very high range and a total of 70% were within a range such that response to P fertilizer would not be expected. During the next 37 years, the percentage of soils dropped in the very high range but increased in the medium and high ranges. The net result was that about the same percentage of the soils (68%) would not be expected to respond to fertilizer P in 1987 as in 1950.

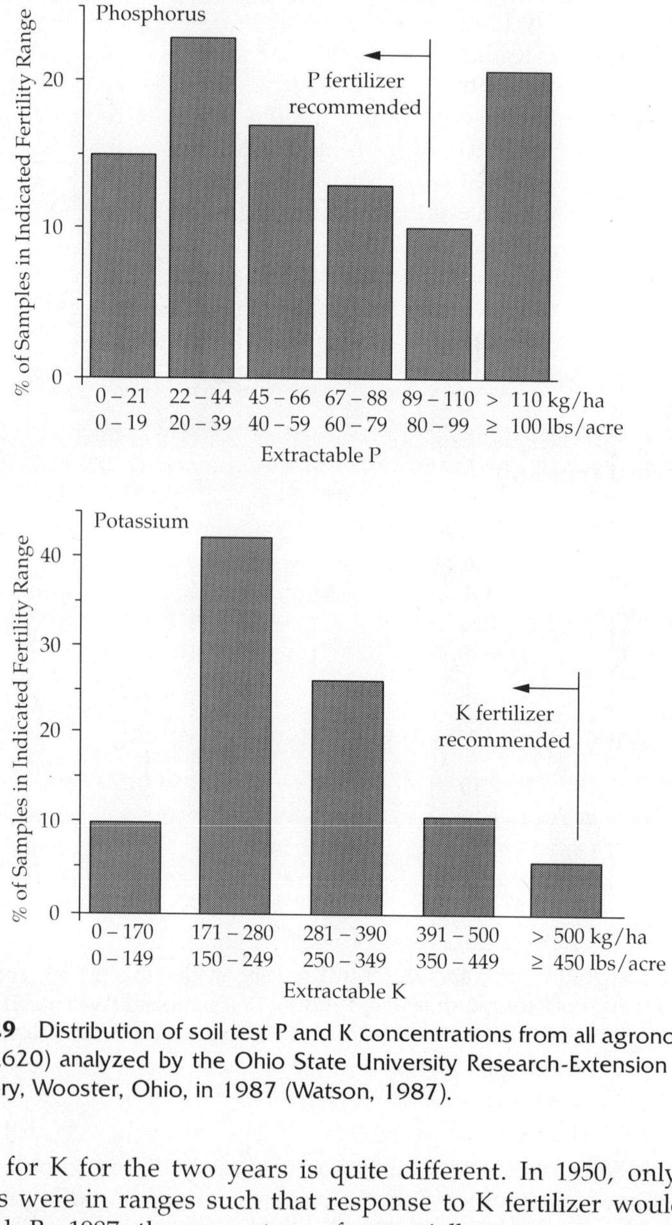

Figure 5.9 Distribution of soil test P and K concentrations from all agronomic samples (29,620) analyzed by the Ohio State University Research-Extension Analytical Laboratory, Wooster, Ohio, in 1987 (Watson, 1987).

Data for K for the two years is quite different. In 1950, only 13% of the soils were in ranges such that response to K fertilizer would not be expected. By 1987, the percentage of potentially nonresponsive soils had increased to 56%.

Similar data from Ohio for 1987 showed less reserve fertility than was found in the North Carolina soils (Figure 5.9). For a yield goal of 9,400 kg corn grain ha^{-1} (150 bu acre^{-1}), the extractable P concentration above which no P fertilizer would be recommended is 95 kg ha^{-1}. The corresponding K concentration is dependent on CEC and is 550 kg ha^{-1} for a

CEC of approximately 15 milli-equilivants per 100 g. Thus, 30% of the soils would not require P fertilizer, but only 5% would not require K fertilizer.

This reserve of nutrients can be utilized as the sole source of nutrients for crop production for some time. McCollum (1987) reported on a 30-year study with P on a Portsmouth soil (fine-loamy, siliceous, thermic Typic Umbraquults) in the Coastal Plain of North Carolina. Initially the soil had a M3-extractable concentration of 40 g/m^{-3} (midpoint of the medium level in Figure 5.8). Phosphorus was added incrementally over a period of eight years or as one large application the first year. After the initial eight-year period, the decline in extractable P and the effect of extractable P concentration on the yield of corn and soybeans grown in rotation was determined for 22 years. The yield limiting M3-extractable P concentration was determined to be 38 g/m^{-3} for both crops. From the data on decline of extractable P with time, McCollum calculated the time to reach yield-limiting concentration or the period one could grow crops without using additional P fertilizer. Using his decline curve and the 1987 P data in Figure 5.9, one can estimate that 23% of the soils tested in 1987 would not respond to P fertilizer application for three years, 33% would not respond for ten years, and 35% would not respond for at least 12 years. These estimates indicate how sustainability can be increased by utilizing residual fertility and eliminating the need for fertilizer inputs for several years.

Effect of Microbial Activity on Nutrient Availability

Soil organic matter influences soil productivity through its effect on physical and chemical properties. Organic matter stabilizes soil aggregates and thus affects physical properties such as moisture storage capacity, aeration, infiltration, erosion, and the power required for tillage operations. Chemical properties affected by organic matter include cation exchange capacity and storage of nutrients in organic forms.

The gross effects of organic matter on soil productivity are rather obvious. However, the role of the "living" organic matter (soil microorganisms) is less obvious. Although on a weight basis they make up only about 3% of the total soil organic matter, soil microbes are the processors through which plant and animal residue must pass to be transformed into the products that affect soil properties. As shown in Figure 5.10, residue from plants is partially decomposed by microorganisms and the resulting by-products are humus (well-decomposed plant or animal material), dead microbial cells, and nutrients not used by the microorganisms.

The size of the population (usually referred to as microbial biomass) is a function of the amount and type of organic matter added to soil. Application of organic matter that is readily decomposable (for example, high in amino acids, mono- and oligosaccharides) results in higher populations than does addition of material high in cellulose and lignin. Although microbial biomass can be estimated by several techniques, the activity of

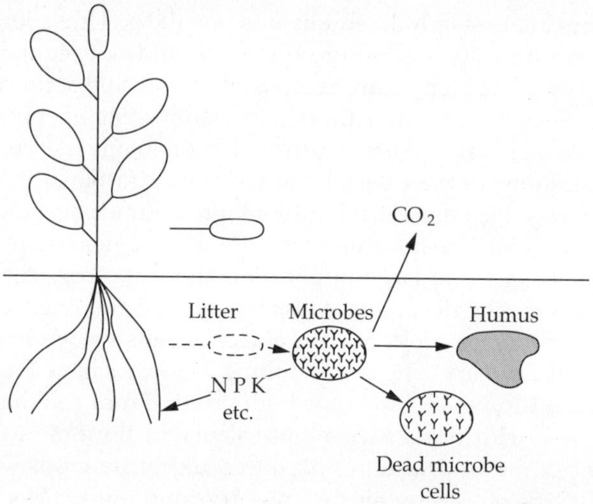

Figure 5.10 Conversion of plant material to soil organic matter.

the microbes is really a better indicator of their effect on soil properties than is a measure of their total weight. This activity can be estimated by determining the concentration of various microbially produced enzymes. Measurements of enzyme concentrations and microbial respiration rates provide an estimate of the overall microbial population and its activity (Nannipieri et al., 1978). Studies have shown soil enzymatic activity to be higher with crop rotations or green manure crops than with monocropping (Gauger, 1987; Khan, 1970). Verstraete and Voets (1977) found winter wheat yields positively correlated with soil phosphatase activity, but Gauger (1987) found no relationship between activity and extractable soil nutrients or nutrient concentrations in corn.

The activity of the microbes influences nutrient availability directly and indirectly. A direct effect is the breakdown of organic matter and subsequent release of nutrients not used in cell building and maintenance processes. These "extra" nutrients are available to plants. Also, since the microbial biomass itself is a relatively labile fraction of the soil organic matter, nutrients in biomass become available as dead microbial cells are attacked by other microbes.

Indirect effects result from the interaction of microbial by-products with soil constituents and nutrients. Examples of effect of by-products on P availability are shown in Figure 5.11: (A) organic acids released during decomposition and carbonic acid formed by dissolution of CO_2 evolved during respiration help solubilize insoluble Ca phosphates, (B) humate ions are able to displace phosphate ions sorbed on oxides of Fe and Al, (C) humus coatings may prevent phosphate ions from coming into contact with Fe/Al oxides and thus reduce P fixation by these oxides, and

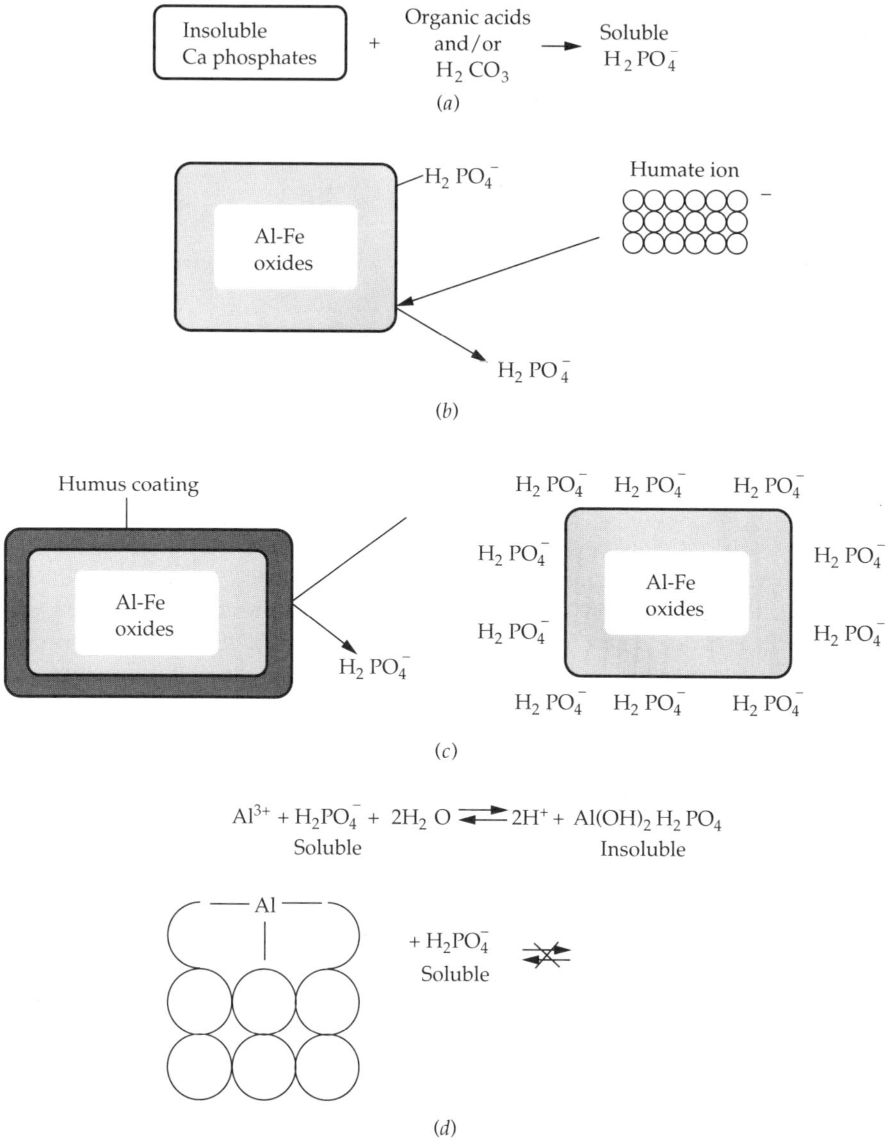

Figure 5.11 Effect of microbiological activity on P availability to plants.

(D) organic compounds chelate soluble Al and reduce the formation of insoluble Al phosphates. Another indirect effect is the enhancement of uptake of metal nutrients such as Fe, Cu, and Zn by chelation as soluble, low molecular weight chelates. Chelation reduces metal tie-up by soil constituents, but since the chelates are soluble, they can move to plant roots where the metal can be absorbed by the roots.

TABLE 5.7 Average Fixation of N by Legumes.

Legume	N Fixed kg ha^{-1}	Legume	N Fixed kg ha^{-1}
Alfalfa	217	Lespedeza	95
Ladino clover	200	Vetch	90
Sweet clover	133	Peas	72
Red clover	128	Soybeans	65
Kudzu	120	Winter peas	56
White clover	115	Peanuts	44
Cowpeas	100	Beans	45

Source: Tisdale and Nelson, 1966.

NUTRIENT INPUTS

Suspending or reducing fertilizer application for a few years and "mining" the nutrient reserves may not be a long-term sustainable practice, but it is a viable option that will reduce input costs for several years. Reducing these concentrations also will reduce nutrient losses from erosion, surface runoff, and leaching. However, once the nutrient status has declined to a concentration that limits economical yield, external sources of nutrients will be required to balance the cycle. These sources include biologically fixed N, commercial fertilizer, unprocessed nutrient sources, recycled wastes, and purchased feed.

Biologically Fixed N

Biological N fixation can supply much or all of the N required in some cropping systems. For example, soybeans and forage legumes can supply all of their N needs, forage legumes in rotations or a legume cover crop can supply N to subsequent crops, and grass yields are increased in legume-grass mixtures. Data on *average* quantity of N fixed by various legumes are shown in Table 5.7. The word *average* is stressed because, in addition to being influenced by the specific legume, the quantity of N fixed also depends on the growth of the legume. In a study in North Carolina, crimson clover planted in a conventionally prepared seedbed producd 7,200 kg of biomass containing 200 kg of N ha^{-1} but overseeded into soybeans produced only 4,500 kg of biomass and contained 130 kg of N ha^{-1} (L.D. King, unpublished data). A detailed discussion of legumes is presented in Chapter 6.

Commercial Fertilizer

Sustainability of commercial fertilizer supply is dependent on the reserve of raw materials and the energy required for manufacture. Most N fertilizer is manufactured by combining N_2 from the atmosphere with H_2 using the Haber-Bosch process:

$$\underset{\text{air}}{N_2} + \underset{\text{natural gas}}{CH_4} \xrightarrow[\text{pressure \& catalyst}]{\text{heat}} NH_3\ (82\%\ N)$$

The NH_3 may be used directly or processed to produce other types of N fertilizer (urea, ammonium nitrate, etc.). The energy consumed in production of fertilizer N is 18,000 kcal kg^{-1} of N (Pimentel et al., 1973), so the availability of N fertilizer is dependent mainly on energy supply. Sustainability of P and K fertilizer supplies is dependent on the world reserves of rock phosphate and KCl deposits. The life expectancy of these raw materials is influenced by the current rate of use, the projected annual rate of increase in use, and the increased cost of mining as the easily mined reserves are depleted. Known plus estimated phosphate rock reserves are estimated to last 650 years if use continues at the 1981 annual rate (CAST, 1988). However, if calculations are made based on the 1981 rate with a projected 3.6% annual increase suggested by the U.S. Bureau of Mines and Geological Survey, these reserves would last only 88 years. Similarly, K reserves are estimated to last 3,640 years if used at 1974 rates but only 107 years if the annual growth rate is 5%. Thus, the increase (or decrease) in nutrient use in the future will have a dramatic effect on the life of the reserves. Several factors should reduce the annual growth in the United States. As previously mentioned, excessive rates of fertilizer are being used and many soils have such high fertility reserves that no fertilizer is necessary for several years. The concern with water quality has caused many states to ban P in detergents and to require an additional P removal step in waste water treatment plants, thus reducing the nonfertilizer use of P and increasing the P content of sewage sludge.

The energy required to produce P and K fertilizer is much less than that required in N fertilizer production: 3,000 kcal kg^{-1} P and 2,300 kcal kg^{-1} K (Pimentel et al., 1973), so their availability and cost are less affected by energy availability and cost as compared to N.

Unprocessed Sources of Nutrients

Unprocessed sources of P and K are sometimes used as nutrient sources in low-input systems. Rock phosphate is the raw material for all P fertilizers, but can be used directly as a P source. An example of rock phosphate is francolite, a carbonate fluorapatite: $Ca_{10}F_2(PO_4)_6 \cdot XCaCO_3$. It contains 13% to 17% P and 3% to 4% F. Manufacturing P fertilizers consists of reacting the rock with acids to make the P in the rock more water-soluble. Sulfuric acid is used to produce ordinary superphosphate (OSP); phosphoric acid is used to produce concentrated superphosphate (CSP):

$$\text{rock P} + H_2SO_4 \rightarrow Ca(H_2PO_4)_2 + CaSO_4 + HF \quad (\text{OSP}, 9\% \text{ P})$$

$$\text{rock P} + H_3PO_4 \rightarrow Ca(H_2PO_4)_2 + HF \quad (\text{CSP}, 20\% \text{ P})$$

$$(\text{rock P}, 13\text{–}17\% \text{ P})$$

In 1975, rock phosphate used directly for fertilizer was 5% of world consumption. In 1951, rock phosphate accounted for 16% of the P used as fertilizer in the United States, but by 1974 that percentage had dropped to 0.2.

The availability of P in rock phosphate is affected by the mineralogy of the rock and consequently is affected by the source of the rock. The effect of the source of rock on plant growth is shown in Table 5.8.

Since availability of P in most rock phosphate is increased by acidity, rock phosphate is usually more effective in acid soils than in neutral to alkaline soils. However, this beneficial effect of low pH must be balanced with the detrimental effect of low pH and other factors affecting crop growth. Rock phosphate is more effective in soils with severe to moderate P deficiencies than in soils with relatively high concentration of available P. This results from the fact that the P in rock phosphate is in equilibrium with P in the soil solution (which is in equilibrium with other sources of P in the soil). Thus, the lower the P concentration in solution, the greater the movement of rock P into solution. The quantity of Ca in the soil solution affects rock phosphate effectiveness in the same manner by controlling the solubility of Ca from the rock (Figure 5.12).

Because of the low P solubility, rock phosphate cannot maintain sufficient P in soil solution to supply crops with high P requirements. It is more effective on crops with large root systems and on crops with slow growth rates (for example, perennials rather than annuals). Sweet clover has been shown to respond equally well to rock phosphate or CSP fertilizer. However, most studies comparing rock phosphate with OSP or CSP have shown that twice as much P is required from the rock source as compared

TABLE 5.8 Effect of Source of Rock Phosphate on Corn Yield in a Greenhouse Study (Adapted from Khasawneh and Doll, 1978).

Rock Phosphate Source	Corn Yield g/pot^{-1}
North Carolina	41
Florida	25
Tennessee	13
Missouri	5
Concentrated superphosphate	53
Control	5

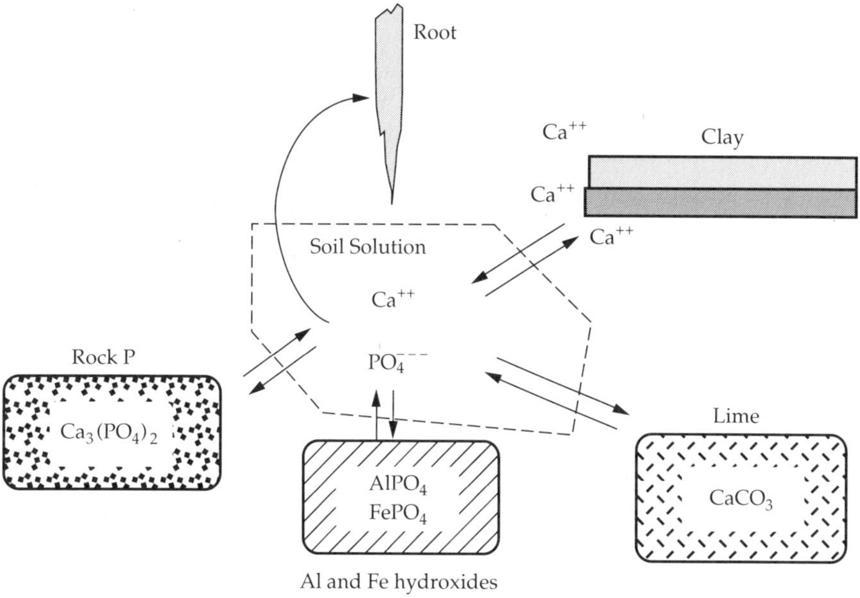

Figure 5.12 Equilibrium of P and Ca in rock phosphate with P and Ca in soil solution.

to OSP and CSP sources to produce comparable yields (Khasawneh and Doll, 1978). This was true even in 30-year studies.

Ground rock sources of K have been used in some low-input systems. Granite meal is a by-product of crushing and screening quarried granite. It contains 3% to 4% K in biotite and feldspar. Greensand contains the mineral glauconite, $KFeSi_2O_6 \cdot nH_2O$, which is a product of alteration of biotite by seawater. Greensand typically contains 7% K. Greenhouse studies comparing granite meal, greensand, and KCl fertilizer for clover production showed no response to greensand but showed comparable responses to granite meal and KCl (Graham and Albrecht, 1952). However, in a field study comparing granite meal and KCl for Coastal bermudagrass production, Jackson and Burton (1958) found that yields were comparable only when granite meal was applied at a rate to supply eight times as much K as was supplied by KCl.

Certain industrial by-products may be used as alternative sources of K. Wood ashes, or potash, were the original materials used to supply K and are still used by some gardeners. Many large wood processing industries use wood-fired boilers and must dispose of large quantities of fly ash. Fly ash from a wood siding manufacturing plant in North Carolina contained 2.7% K and significantly increased available soil K as determined from a laboratory incubation study (unpublished data, L.D. King). Fly ash from coal-fired boilers also contains K, but studies have shown decreased crop yields due to boron toxicity (Holliday et al., 1958; Martens et al.,

1970; Adriano et al., 1978). Cement kiln dust (a by-product of cement manufacturing) contains from 2% to 21% K and has been shown to be comparable to KCl and K_2SO_4 as a K source for crop production (Kulich, 1974; van Lierop et al., 1982). Fly ash and kiln dust also have significant liming potential. Fly ash from the wood siding manufacturing plant was 30% as effective as dolomitic limestone in raising soil pH (unpublished data, L.D. King). Golden (1972) reported a calcium carbonate equivalent of 89% for kiln dust.

Recycled Wastes

Agricultural use of municipal and food processing wastes is a method of returning nutrients to the farm cycle (Figure 5.5). Median concentrations of nutrients and heavy metals in samples of sludge and other wastes from the southern United States are shown in Table 5.9. A large fraction of these wastes is being applied to agricultural land, and that practice is increasing. However, the overall impact of these nutrients on U.S. agriculture is not great. An estimate was made of the cropland required to receive all municipal sewage sludge expected to be produced in the United States in 1985 at rates to supply 112 kg available N ha^{-1} (CAST, 1976). Approximately 0.5% of U.S. cropland would be required. If a P concentration of 1.6% were assumed and sludge were applied to supply 25 kg of total P ha^{-1}, then about 4% of the U.S. cropland would be required. These estimates show that sludges cannot supply a large fraction of the national need for nutrients, but they are a valuable source of nutrients to cropland within economical hauling distance of the wastewater treatment plant.

TABLE 5.9 Median Nutrient and Heavy Metal Concentrations in Wastes from the Southern United States. (Adapted from King, 1986).

	Municipal	Textile	Fermentation	Wood Processing
		% of Dry Weight		
N	2.6	2.8	3.5	0.4
P	1.6	0.9	0.2	0.1
K	0.2	0.2	0.1	0.1
		mg kg^{-1} of Dry Weight		
Pb	335	135	6	36
Zn	1,750	940	40	73
Cu	475	416	13	58
Ni	37	40	18	60
Cd	11	4	<1	<1
Cr	380	1,820	10	30

Source: King, 1986.

When interest in sludge utilization increased in the 1960s, the question of potential heavy metal toxicities to crops, animals, and humans was raised. Concern about heavy metals has decreased since that time for several reasons. Industrial pretreatment ordinances have eliminated much of the metal input from industries. Research conducted during the last 20 years has shown that heavy metal accumulation in plants is minimal if sludges with very high concentrations of metals are avoided, sludges are applied at rates to supply only the nutrient needs of the crops, crops that accumulate metals in their marketable portions are avoided (for example, tobacco), and soil pH is maintained at 6.0 or greater.

Purchased Feed

One other external source of nutrients is purchased animal feed. The importance of this source would be lowest on a beef cattle farm ($\approx$ 10% purchased feed), intermediate on a dairy farm ($\approx$ 40%) and highest on poultry and swine farms ($\approx$ 90–100%). In fact, on many poultry and swine farms this high nutrient input causes environmental problems because inadequate cropland is available on which to apply the manure.

SUMMARY

From a plant nutrient standpoint, the key to sustainable agriculture is nutrient cycling. In an agricultural system, one of the largest losses from the cycle is the harvested crop. This loss is minimized if crops are fed to animals and only animal products leave the farm. Management practices can be used to minimize nutrient losses from manure and losses due to erosion, leaching, and denitrification. Once these losses have been minimized, inputs must be obtained to offset the losses. Currently, commercial fertilizer is the main source of nutrient inputs. However, past inputs in the form of residual fertility may offset the need for new inputs for a short or an extended period. Legumes in rotations or used as cover crops can supply some or all of the N required for subsequent crops. Some nutrients leaving in the harvested crop or animal product may be recycled via use of sewage sludge or food processing wastes. Greater awareness of the potentials of nutrient cycling in temperate agricultural systems is needed. Enhanced nutrient cycling has strong economic and environmental implications for a more sustainable agriculture.

REFERENCES

Adriano, D. C., T. A. Woodford, and T. G. Ciravalo. (1978). "Growth and elemental composition of corn and bean seedlings as influenced by soil applications of coal ash," *J. Environ. Qual.* **3**: 416–21.

Alberts, E. E., G. E. Schuman, and R. E. Burwell. (1978). "Seasonal runoff losses of nitrogen and phosphorus from Missouri Valley loess watersheds," *J. Environ. Qual.* **7**: 203–08.

Alexander, M. (1961). *Introduction to soil microbiology*, Wiley, New York.

Barker, J. C. (1980). "Livestock manure production rates and approximate fertilizer content." North Carolina Agricultural Extension Service leaflet 198 (revised).

Barker, J. C., F. J. Humenik, M. R. Overcash, R. Phillips, and G. D. Wetherill. (1980). "Performance of aerated lagoon–land treatment systems for swine manure and chick hatchery wastes." International Symposium on Livestock Waste. American Society of Agricultural Engineers, St. Joseph, MI, pp. 217–20.

Beauchamp, E. G., G. E. Kidd, and G. Thurtell. (1982). "Ammonia volatilization from liquid dairy cattle manure in the field," *Can. J. Soil Sci.* **62**: 11–19.

Burns, J. C., P. W. Westerman, L. D. King, G. A. Cummings, M. R. Overcash, and L. Goode. (1985). "Swine lagoon effluent applied to 'Coastal' bermudagrass: I. Forage yield, quality, and element removal," *J. Environ. Qual.* **14**: 9-14.

CAST. (1988). "Long-term viability of U.S. Agriculture." Council for Agricultural Science and Technology Report No. 114.

Castellanos, J. Z., and P. F. Pratt. (1981). "Mineralization of manure nitrogen—correlation with laboratory indexes," *Soil Sci. Soc. Am. J.* **45**: 354–57.

Chescheir, G. M. III, P.W. Westerman, and L.M. Safley, Jr. (1985). "Rapid methods for determining nutrients in livestock manure," *Trans. Am. Soc. Agric. Eng.* **28**: 1817–24.

Comfort, S. D., K. A. Kelling, D. R. Keeney, and J. C. Converse. (1988). "The fate of nitrogen from injected liquid manure in a silt loam soil," *J. Environ. Qual.* **17**: 317–22.

Cramer, C., G. DeVault, M. Brusko, F. Zahradnik, and L. J. Ayers (eds.). (1985). *The Farmer's Fertilizer Handbook*, Regenerative Agriculture Assn. Emmaus, PA.

Crickenberger, R. G., and L. Goode. (1978). "Guidelines for feeding broiler litter to beef cattle." North Carolina Agricultural Extension Service, 5-78-2M, AG-61.

Frink, C. R. (1969). "Frontiers of plant science." Connecticut Agricultural Experiment Station, New Haven.

Gambrell, R. P. , J. W. Gilliam, and S. B. Weed. (1975). "Nitrogen losses from soils of the North Carolina Coastal Plain", *J. of Environ. Qual.*, **4**: 317-323.

Gauger, R. E. (1987). "The effect of green manure crops on microbial biomass and soil enzymes." Master of science thesis, North Carolina State Univ., Raleigh.

Golden, L. E. (1972). "The effect of kiln dust on yield and nutrition of sugarcane," *LA Agric.* **15(4)**: 12–15.

Graham, E. R., and W. A. Albrecht. (1952). "Potassium-bearing minerals as soil treatments." Univ. Missouri Agricultural Experiment Station Bulletin 510.

Groffman, P. M., P. F. Hendrix, and D. A. Crossley, Jr. (1987). "Nitrogen dynamics in conventional and no-tillage agroecosystems with inorganic fertilizer or legume nitrogen inputs," *Plant & Soil*, **97**: 315–32.

Guenzi, W. D., W. E. Beard, F. S. Watanabe, S. R. Olsen, and L. K. Porter. (1978). "Nitrification and denitrification in cattle manure-amended soil," *J. Environ. Qual.* **7**: 196–202.

Holliday, R., D. R. Hodgson, W. N. Townsend, and J. W. Wood. (1958). "Plant growth on fly ash," *Nature* **181**: 1079–80.

Hubbard, N.L. (1986). "Management of a hairy vetch cover crop as a nitrogen source and mulch for corn." Master of science thesis, North Carolina State Univ., Raleigh.

Jackson, J. E., and G. W. Burton. (1958). "An evaluation of granite meal as a source of potassium for Coastal bermudagrass," *Agron. J.* **50**: 307–8.

Khan, S. U. (1970). "Enzymatic activities in a gray wooded soil as influenced by cropping systems and fertilizers," *Soil Biol. Biochem.* **3**: 309–15."

Khasawneh, F. E., and E. C. Doll. (1978). "The use of phosphate rock for direct application to soils." In *Advances in Agronomy,* (Ed. N. C. Brady). pp. 159–206.

Kilmer, V. J., J. W. Gilliam, J. F. Lutz, R. T. Joyce, and C. D. Eklund. (1974). "Nutrient losses from fertilizer grassed watersheds in western North Carolina," *J. Environ. Qual.* **3**: 214–19.

King, L. D. (1973). "Mineralization and gaseous loss of nitrogen in soil-applied liquid sewage sludge," *J. Environ. Qual.* **2**: 356–58.

King, L. D. (1984). "Availability of nitrogen in municipal, industrial, and animal wastes," *J. Environ. Qual.* **13**: 609–12.

King, L. D. (1986). "Introduction to the agricultural use of municipal and industrial sludges." In L. D. King (Ed) *Agricultural use of municipal and industrial sludges in the southern United States,* Southern Cooperative Series Bulletin 314, North Carolina State University, Raleigh, NC.

King, L. D., and R. L. Vick, Jr. (1978). "Mineralization of nitrogen in fermentation residue from citric acid production," *J. Environ. Qual.* **3**: 315–18.

Kulich, J. (1974). "Untraditional sources of water-soluble K_20 for nutrition of vegetables and cereals," *Acta Fytotech.* **29**: 75–82.

Ladd, J. N., and M. Amato. (1986). "The fate of nitrogen from legume and fertilizer sources in soils successively cropped with wheat under field conditions," *Soil Biol. Biochem.* **18**: 417–25.

Lauer, D. A., D. R. Bouldin, and S. D. Klausner. (1976). "Ammonia volatilization from dairy manure spread on the soil surface," *J. Environ. Qual.* **5**: 134–41.

Marbut, C. F. (1935). "Soils of the United States," *Atlas of American Agriculture, part 3.* U.S. Department of Agriculture, Washington, DC.

Martens, D. C., M. G. Schnappinger, Jr., and L. W. Zelazny. (1970). "The plant availability of potassium in fly ash," *Soil Sci. Soc. Am. Proc.* **34**: 453–56.

McCollum, R. E. (1987). "The buildup and decline in soil phosphorus: Thirty-year trends on Portsmouth soil," *Soil Sci. Soc. N.C. Proc.* **30**: 28–64.

Mehlich, A. (1984). "Mehlich 3 soil test extractant: A modification of Mehlich 2 extractant," *Comm. Soil Sci. Plant Anal.* **15**: 1409–16.

Miller, M. F., and H. H. Krusekopf. (1932). "The influence of systems of cropping and methods of culture on surface runoff and soil erosion," *Research Bulletin* 177, Missouri Agricultural Experiment Station.

Muck, R. E. and B. K. Richards. (1983). "Losses of manurial nitrogen in free-stall barns," *Agric. Wastes,* **7**: 65–79.

Muck, R. E., R. W. Guest, and B. K. Richards. (1984). "Effects of manure storage design on nitrogen conservation," *Agric. Wastes* **10**: 205–20.

Muller, M. M. (1987). "Leaching of subterranean clover-derived N from a loam soil," *Plant & Soil* **102**: 185–91.

Nannipieri, P., R. L. Johnson, and E. A. Paul. (1978). "Criteria for measurement of microbial growth and activity in soil," *Soil Biol. Biochem.* **10**: 223–29.

Nelson, L. B., and R. E. Unland. (1955). "Factors that influence loss of fall applied fertilizers and their probable importance in different sections of the United States," *Soil Sci. Soc. Am. Proc.*, **19**: 492–96.

Olness, A., S. J. Smith, E. D. Rhoades, and R. G. Menzel. (1975). "Nutrient and sediment discharge from agricultural watersheds in Oklahoma," *J. Environ. Qual.* **4**: 331–36.

Peterson, R. G., W. W. Woodhouse, Jr., and H. L. Lucas. (1956). "The distribution of excreta by freely grazing cattle and its effect on pasture fertility: I. Excreta distribution. II. Effect of returned excreta on the residual concentration of some fertilizer elements," *Agron. J.* **48**: 440–49.

Pimentel, D. (1973). "Food production and the energy crisis," *Science* **182**: 443-48.

Pratt, P. F., and J. Z. Castellanos. (1981). "Available nitrogen from animal manures," *Calif. Agric.* **35(7&8)**: 24.

Rice, C. W., P. E. Sierzega, J. M. Tiedje, and L. W. Jacobs. (1988). "Stimulated denitrification in the microenvironment of a biodegradable organic waste injected into soil," *Soil Sci. Soc. Am. J.* **52**: 102–8.

Ryan, J. A., D. R. Keeney, and L. M. Walsh. (1973). "Nitrogen transformations and availability of an anaerobically digested sewage sludge in soil," *J. Environ. Qual.* **2**: 489–92.

Safley, L. M., Jr., P. W. Westerman, and J. C. Barker. (1986). "Fresh dairy manure characteristics and barnlot nutrient losses," *Agric. Wastes* **17**: 203–15.

Safley, L. M., Jr., P. W. Westerman, J. C. Barker, L. D. King and D. T. Bowman. (1986). "Slurry dairy manure as a corn nutrient source," *Agric. Wastes* **18**: 123–36.

Thomas, G. W. (1970). "Soil and climatic factors which affect nutrient mobility." In *Nutrient mobility in soils: accumulation and losses*, (Ed. O.P. Engelstad). pp. 1–20. Soil Science Society of America, Madison, WI.

Tisdale, S. L., and W. L. Nelson. (1966). *Soil fertility and fertilizers*, Macmillan, New York.

Tucker, M. R. (1988). "Soil test summary from 7/1/86–6/30-/87." Agronomic Division, North Carolina Department of Agriculture, Raleigh.

USDA. (1976). "The universal soil loss equation with factor values for North Carolina." USDA, Soil Conservation Service, Raleigh, NC.

van Lierop, W., T. S. Tram, and S. Morissette. (1982). "Evaluation of cement kiln flue dust as a potassium and sulfate fertilizer," *Comm. Soil Sci. Plant Anal.* **13**: 157–73.

Verstraete, W., and J. P. Voets. (1977). "Soil microbial and biochemical characteristics in relation to soil management and fertility," *Soil Biol. Biochem.* **9**:253–58.

Watson, M. E. (1987). "Soil test summary—1987." Ohio Agricultural Research and Development Center, Wooster, OH.

Welch, C. D., and W. L. Nelson. (1951). "Fertility status of North Carolina soils," North Carolina Department of Agriculture, Raleigh.

Westerman, P. W., L. D. King, J. C. Burns, and M. R. Overcash. (1983). "Swine manure and lagoon effluent applied to fescue," EPA-600/S2-83-078. Robert S. Kerr Environmental Research Laboratory, U.S. Environmental Protection Agency, Ada, OK.

Westerman, P. W., L. D. King, J. C. Burns, G. A. Cummings, and M. R. Overcash. (1987). "Swine manure and lagoon effluent applied to a temperate forage mixture: II. Rainfall runoff and soil chemical properties," *J. Environ. Qual.* **16**: 106–12.

6 LEGUMES AND CROP ROTATIONS

JAMES F. POWER
Department of Agronomy, ARS/USDA
University of Nebraska,
Lincoln, Nebraska

INTRODUCTION

Throughout the recorded history of agriculture, the use of legumes in cropping systems has been a standard practice and a prime method by which fixed dinitrogen was added to the soil-plant ecosystem. References to the use of legumes in cropping systems are found in ancient Chinese, Greek, and Egyptian writings (MacRae and Mehuys, 1985). Legume-based rotations were likewise dominant in American agriculture until the last few decades. In areas where grazing livestock enterprises are an integral part of the farming systems, legume-based rotations are still widely used because such systems provide the needed balance between grain and forage crops.

With the development of the Haber-Bosch process for synthetic N fixation, especially during World War II, agriculture gained a practical method for economically producing N fertilizers on a wide scale. Technology was quickly developed to produce fertilizer N from natural gas, and distribution systems were established. As a consequence, in the last three decades, fertilizer N has become widely available and is now a major source of N in American agriculture.

With the increased use of fertilizer N, U.S. farmers have concurrently drastically reduced the use of legumes in cropping systems. This is illus-

TABLE 6.1 Crop Yields, Legume Seed Production (with and without Alfalfa) and Fertilizer N Usage in the United States, 1959 through 1979[a].

Year	Average Crop Yield (Corn, Wheat, Soybean, Sorghum) ($kg\ ha^{-1}$)	Legume Seed: Including Alfalfa[b] ($kg \times 10^6$)	Legume Seed: Excluding Alfalfa ($kg \times 10^6$)	N Fertilizer Applied ($t \times 10^6$)
1959	2,274	180	123	2.4
1960	2,531	172	110	2.7
1961	2,621	151	94	3.0
1962	2,710	145	91	3.5
1963	2,822	170	98	3.5
1964	2,598	160	97	3.9
1965	2,912	136	81	4.1
1966	2,934	128	75	4.7
1967	2,979	104	53	5.4
1968	3,024	100	48	6.1
1969	3,270	103	56	6.3
1970	3,002	116	54	6.7
1971	3,494	99	45	7.2
1972	3,606	78	30	7.1
1973	3,360	77	30	7.4
1974	2,733	81	30	8.2
1975	3,230	74	31	7.7
1976	3,248	68	32	9.3
1977	3,360	71	27	9.5
1978	3,606	65	23	8.9
1979	3,853	67	22	9.5

[a] Corn (*Zea mays* L.), wheat, (*Triticum aestivum* L.), soybean [*Glycine max* (L.) Merr.], sorghum [*Surghum bicolor* (L.) Moench], alfalfa (*Medicago sativa* L.), red clover (*Trifolium pratense* L.), sweet clover (*Melilotus* sp.), lespedeza (*Lespedeza* sp.), crimson clover (*Trifolium incarnatum* L.), hairy vetch (*Vicia villosa* Roth), white clover (*Trifolium repens* L.), ladino clover (*Trifolium repens* L.).

[b] Includes alfalfa, red clover, sweet clover, lespedeza, crimson clover, hairy vetch, white clover, and ladino clover.

Source: Power and Doran (1984).

trated in Table 6.1, in which data from both fertilizer N production and legume seed production (excluding alfalfa and soybeans) in the United States for two decades are presented. (The U.S. Department of Agriculture ceased reporting legume seed production statistics in 1981.) Legume seed production was used because valid statistical information on legume acreages is limited. Data are presented both including and excluding alfalfa because much alfalfa is produced primarily for pasture and hay rather than in crop rotations, so alfalfa production has remained relatively constant. The information in Table 6.1 strongly supports the conclusion that the use of legumes as an N source for grain crops has largely been replaced by

TABLE 6.2 Estimates of Area Harvested and N Fixed by Grain and Forage Legumes in the United States.

Legume Crop	Area Harvested (ha × 10^3)	N_2 Fixed (t × 10^3/yr)
Grain legumes		
Soybean [*Glycine max* (L.) Merr.]	28,000	2,352
Dry edible beans (*Phaseolus* sp.)	716	36
Peanut (*Arachis hypogaea* L.)	613	29
Dry edible peas (*Pisum sativum* L.)	53	4
Others	33	2
	29,415	2,423
Forage legumes		
Alfalfa (*Medicago sativa* L.)	10,873	2,196
White clover (*Trifolium pratense* L.)	5,431	755
Red clover (*Trifolium pratense* L.)	4,075	566
Sweet clover (*Melilotus* sp.)	1,128	157
Vetch (*Vicia* sp.)	760	65
Trefoil (*Lotus corniculatus* L.)	686	81
Crimson clover (*Trifolium incarnatum* L.)	328	46
	23,281	3,866
Total	52,696	6,289

Source: Phillips and Dejong (1984).

fertilizer N. Data in Table 6.2 show recent estimates of the quantity of N biologically fixed by legumes in U.S. agriculture (Phillips and DeJong, 1984). Soybean and alfalfa account for almost 75% of the total 6.3 million tons annual estimate. Currently, about 10 million tons of fertilizer N are used each year in the United States.

As a result of these changes, monocultures of grain crops are often practiced (Power and Follet, 1987). However, the energy crises of the 1970s demonstrated the vulnerability of monocultures that depend greatly on inputs of fossil fuels. Rapid and drastic increases in fuel prices occurred, with natural gas prices increasing tenfold from 1974 to 1982 (Beck, 1982). With finite supply and unstable world conditions, such fluctuations can occur again, thereby greatly affecting fertilizer prices and subsequent permanence and profitability of monocultures. Also, we see evidence of environmental degradation resulting from intensive chemical use in agriculture. It is therefore prudent to look again at the attributes of legumes to determine how they might best be utilized in modern crop production systems (Power, 1987).

As a consequence of these problems, the objectives of this chapter include (1) describing the process of biological N fixation, (2) estimating the quantities of N fixed by legumes, (3) describing the effects of legumes on various soil properties, (4) evaluating the effects of legumes on the environment, and (5) discussing the potential for utilizing legumes in different types of modern cropping systems.

LEGUMES AND BIOLOGICAL N_2 FIXATION

The process of dinitrogen fixation by the *Rhizobium* bacteria found in nodules of legumes has been recognized for a century. In 1888, Beijerinck discovered that the rhizobia organisms were capable of dinitrogen fixation. Likewise, Hellriegel (1886) recognized that the inclusion of legumes in a cropping system added N to the soil and improved N availability. Throughout this century, the mechanisms and controls involved in dinitrogen fixation have been elucidated (Havelka et al., 1982).

In the fixation process, a symbiotic relationship is established between the host plant and the rhizobia in the nodules. The host plant provides fixed carbon (C) as an energy source for the bacteria, and also provides for the development of the nodule with its unique mechanisms to protect the N-fixation process from an oxidative environment. In turn, the rhizobia organisms, through the catalytic action of the nitrogenase enzymes they produce, convert atmospheric N_2 gas into ammonia, which is readily converted to amino acids and carried in the xylem to growing tissue.

Dinitrogen fixation in legume nodules is an energy-intensive process. The host plant must provide energy for root hair infection, nodule development, and bacteroid and nodule maintenance, in addition to that required for the stoichiometry associated with nitrogenase activity. Six electrons must be transferred per molecule N_2 fixed, requiring the reduction of 21 molecules ATP per molecule N_2 fixed (4kg carbohydrates per kg N_2) under ideal conditions (Hardy and Havelka, 1975). However, many nodules also evolve H_2 gas in the process of dinitrogen fixation, reducing the efficiency of the electron transfer process. Adding this to the above stoichiometric requirement plus maintenance requirements for nodule and bacteroids results in a requirement of 8 to 17 kg carbohydrates per kg N_2 fixed (Hardy et al., 1977). Consequently, it is estimated that frequently 10% to 40% of the C fixed in legumes by photosynthesis may be used in N_2 fixation (Havelka et al., 1984). This drain on photosynthate reduces the rate of dry-matter accumulation in legumes accordingly. Emerich and Evans (1984) showed that *Bradyrhizobium japonicum* species capable of oxidizing H_2 increased dry weights of soybean by over 20% when compared to species without this characteristic (Table 6.3).

Because of the high energy requirement for the N_2 fixation process, any factor that affects the rate of photosynthesis likewise affects N_2 fixation. Gibson (1978) gives a thorough review of the effects of a number of environmental factors upon the N_2 fixation process. Adverse soil temperature,

TABLE 6.3 Dry Weight Accumulation in Soybeans When Nodulated by *B. japonicum* Species with and without H_2-Oxidizing Capacity.

Growth Period (days)	Dry-Matter Accumulation in Plants Nodulated by *B. japonicum*: Without H_2-Oxidizing Capacity (grams/culture)	With H_2-Oxidizing Capacity (grams/culture)	Difference (grams/cultures)	Increase (%)
15	0.87	0.87	0	0
20	1.34	1.39	0.05	3.7
25	2.06	2.23	0.17	8.1
30	3.17	3.56	0.39	12.4
35	4.88	5.70	0.82	16.8
40	7.51	9.12	1.61	21.5

Source: Emerich and Evans (1984).

water stress, and low soil pH are common factors that control the quantity of N_2 fixed by a legume. Elevated soil NO_3-N levels can likewise reduce N_2 fixation. Sears (1953) estimated that the fraction of the total N in soybean accounted for by biological fixation decreased from 63% to 21% as soil N availability increased. More recently, Johnson et al. (1975) showed the fraction of total N in soybean derived from N_2 fixation declined, from 46% to 10% as fertilizer N rate increased from 0 to 448 kg N ha^{-1} (Table 6.4.)

A close relationship also exists between legume species and associated rhizobia species or even strain. Certain combinations of bacterial species or strains with a given legume are usually more efficient in N_2 fixation than are others. Considerable research is in progress to improve rhizobia capabilities, both through identification of mutant strains and by genetic transfer (Phillips and DeJong, 1984). These relationships are discussed in more detail by Burton (1979). Additional information concerning the N_2 fixation process in legumes is discussed by Silver and Hardy (1976).

TABLE 6.4 Percent of Total N in Soybean Derived from Soil, Fertilizer, and Atmospheric N as Affected by Fertilizer N Rate.

N Fertilizer (kg N ha^{-1})	Percent of Total N in Soybean Derived from: Soil	Fertilizer	Atmosphere
0	54	0	46
112	50	23	27
224	49	41	10
448	33	57	10

Source: Johnson et al. (1975).

Quantity of N_2 fixed in normal agricultural soils varies widely because of the factors mentioned above. Evans and Barber (1977) indicated that perennials such as alfalfa are most effective in N_2 fixation, with much smaller amounts usually fixed by annuals. Frequently, N_2 fixation by annual seed legumes such as soybean can be less than the quantity of N removed in the harvested grain, resulting in a net reduction in N balance for the production system (Heichel, 1987).

Quantity of N added to the soil by growing legumes is also highly dependent on the manner in which the legume is managed. Removal of forage for hay or fodder, of course, greatly reduces the quantity of N returned to the soil. As is the case with seed legumes, often less N is present in the roots and stubble of freshly harvested alfalfa than was biologically fixed by the whole crop. If the alfalfa is killed at this point and used as a green manure, there may be a net decrease in soil N (Haas et al., 1976; Heichel, 1987). However, if the alfalfa is killed after some regrowth has occurred, a positive N balance can be obtained.

With all these variables affecting potential N input by legumes, agronomists must develop a system for N credits from legumes when making fertilizer recommendations. This is generally done on a state-by-state basis because of the variability involved. Kurtz et al. (1984) summarized some of these recommendations (Table 6.5). Likewise, Meisinger (1984) also reviewed legume credits for fertilizer N recommendations.

LEGUMES AND SOIL PROPERTIES

Producing legumes in some type of sequence with grain crops can profoundly affect numerous soil properties (MacRae and Mehuys, 1985). Probably one of the most impressive large-scale examples of the effects of legumes on soil properties is the vastly improved productivity of cropland in eastern, southern, and southwestern Australia since World War II. This has resulted from extensive use of a ley-farming system along with correction of mineral deficiencies in these ancient, weathered, infertile soils. In these systems, clover-based pastures have been rotated with grain crops on a fixed basis. Number of years in each crop is relatively site-specific, varying with soils, climate, management options, and other factors. This ley-farming system has resulted in vastly increased production of both grain and livestock. Changes in agricultural practices in Australian agriculture and their effects on productivity are well described by Puckridge and French (1983).

Including a legume in a cropping system usually reduces the C:N ratio of crop residues added to the soil. This, in turn, can greatly alter microbiological activity, N transformations, nutrient availability, and plant growth (Doran et al., 1987b). Likewise, changes in microbiological activity and root growth can affect soil aggregation, thereby affecting soil water-aeration regimes

TABLE 6.5 Reductions in N Fertilizer Requirements for Corn after Legumes Compared to Continuous Corn.

	Fertilizer N Reduction (kg ha^{-1}) Previous Crop				
	After Alfalfa or Sweet Clover				Soybean
	First Year			Second Year	
Location	Good	Average	Poor		
Illinois	112	56	0	34	45
Indiana	78–90	68	34		
Iowa	157	112	22		45
Michigan		60			
Minnesota	112	56		56	22
Missouri	56				34
Ohio	79–180	101	56	22	
Wisconsin	90	45	22		1.12 kg^{-1} of yield

Source: Kurtz et al. (1984).

and subsequent biological activity in the soil (Tisdall and Oades, 1982). The effects of legumes on soil properties are discussed from the aspects of effects on soil physical conditions, soil erodibility, soil water and temperature regimes, and soil nutrient availability. Much of our information on effects of legumes on soil properties was obtained prior to 1960, when legumes were still a major component in most cropping systems. These have been discussed by Allison (1968), Lutz (1954), Elson (1941), and others. Some of these results have been summarized by Hoyt and Hargrove (1986).

Soil Physical Conditions

Many of the effects of legumes on soil physical properties are expressed through their effects upon soil aggregation and aggregate stability. Tisdall and Oades (1982) indicated that aggregation is influenced by three types of agents: (1) transitory materials, such as polysaccharides, that are usually products of microbial activity, (2) temporary effects through binding action of fungal hyphae and plant roots, and (3) persistent effects resulting from action of polyvalent cations and strongly sorbed organic polymers. They concluded that total quantity of soil organic matter present has a major influence on aggregation and aggregate stability. Therefore, use of legumes in a cropping system could affect aggregation through changes in soil organic-matter content and microbial activity. Because of the relatively narrow C:N ratio of legume residues, microbial biomass may be temporarily increased, increasing aggregation due to hyphal binding. On the other hand, grass roots are usually more fibrous than those of legumes, so aggregation resulting from root binding may be greater under grasses than under legumes.

Strickling (1950) illustrated these effects of cropping systems on soil organic matter and aggregation (Table 6.6). In general, he found that water-stable aggregates (greater than 0.25 mm) were closely related to soil organic-matter content. Aggregation in soil in continuous bluegrass was much greater than that for any other treatment. For cultivated soils, aggregation was greatest for a rotation containing two years of alfalfa-grass hay. Continuous ryegrass was intermediate and continuous corn was very low in aggregation; however, lowest values were recorded for corn/soybean hay. In general, including soybean in a rotation often decreased aggregation.

This problem of soybean often decreasing soil aggregation and degrading soil physical properties has been the subject of numerous studies because this degradation of surface structure by soybean often increases soil erosion potential (Laflen and Moldenhauer, 1979). Usually this effect cannot be predicted based on aggregation studies because measurement

TABLE 6.6 Average Soil Aggregation and Organic-Matter Content in the Surface of a Celina Silt Loam.

Rotations	Average Aggregation in 1949 (%)	Soil Organic Matter Content (%)
Continuous bluegrass	40.6	3.35
Corn-wheat-alfalfa grass hay-alfalfa grass hay	33.0	2.76
Soybean for seed-wheat (sweet clover for green manure)	31.7	2.74
Corn-soybean for seed-wheat-alfalfa grass hay	24.4	2.46
Corn (ryegrass)	23.4	2.17
Corn-wheat (sweet clover for green manure)-corn-soybean for seed	22.6	2.39
Corn-soybean for seed	22.2	2.17
Continuous soybean for seed	21.8	2.03
Corn-wheat (sweet clover for green manure)	21.5	2.03
Corn-wheat (sweet clover for green manure)-soybean for seed	18.8	2.04
Continuous corn	17.5	2.11
Corn-soybean hay	11.4	1.85

Source: Strickling (1950).

of water-stable aggregates does not take into consideration the effects of raindrop impact upon aggregate stability. Gantzer et al. (1987) found slightly greater splash from soils covered with soybean residues than with corn residues, but they found no difference between crops with respect to soil strength. For both types of crop residues, shear strength and aggregate size were maximum 14 days after initiating incubation studies, corresponding to the time of peak microbial activity. As indicated by Tisdall and Oades (1982), these results suggest that the changes in these soil properties were microbially mediated.

Earthworms as well as soil microorganisms are involved in the decomposition of crop residues and subsequent effects upon aggregation. Mackay and Kladivko (1985) found 8 and 16 earthworms m^{-2} in plowed and no-till soil, respectively, when in corn, compared to populations of 62 and 141 for soybean and 470 for a clover/grass pasture. Thus, the presence of leguminous crop residues, as well as absence of tillage, enhances earthworm populations. They measured a corresponding effect upon water-stable aggregates in surface soils, with much greater aggregation in soils with earthworms than without them (Table 6.7). Differences between effects of corn and soybean residues were minor in regard to aggregation.

Soil Erodibility

The effects of legumes on soil erodibility have been mentioned in the previous discussion. This subject is reviewed in more detail in Bruce et al. (1987). Legumes are effective in aiding in soil erosion control, not only by providing surface cover for the soil but also by improving soil physical conditions, thereby enhancing infiltration and reducing runoff. However, as is often the case with seed legumes such as soybeans, erosion can actually be greater for the legume than for a nonlegume such as corn, because smaller amounts of crop residue are produced and those present decompose much more rapidly. These points are discussed by Hoyt and Hargrove (1986).

Legumes used in rotation with grain crops can alter such soil properties as aggregation, water intake rate and permeability, and soil strength,

TABLE 6.7 Effects of Earthworms on Percentage Water-stable Aggregates (over 0.21 mm) in a Raub Silt Loam as Affected by Crop Residues.

	% Water Stable Aggregate		
Earthworms per Pot	No Residues	Soybean Residues	Corn Residues
0	44	56	48
30	60	75	71

Source: Mackay and Kladivko (1985).

thereby altering the resistance of soils to the detachment and transport of soil particles by the erosion process. Such effects have long been recognized from the early work of a number of investigators (Johnston et al., 1943). Frequently, these changes in soil properties resulting from using legumes in rotation reduce soil erosion potential and help maintain soil productivity.

Any changes in surface soil properties that reduce water infiltration, such as crusting, compaction, and reduced aggregation, are likely to increase potential for soil erosion. For example, the information in Table 6.6 on effects of cropping systems on aggregation would be expected to be reflected in erodibility (Strickling, 1950). Many investigators have verified this conclusion. The work of Johnston et al. (1943) cited above is a typical example. Likewise, Olmstead (1947) and Wilson and Browning (1946) also discuss the effects of cropping practices on soil properties and erodibility. Chepil (1955) later showed that aggregation and crop residues were important in the control of erosion by wind.

The major effect of legumes on reducing soil erosion results from the fact that sod legumes, compared to row crops, provide greater ground cover and soil protection. Thus, in many of the studies referred to earlier, soil erosion from legume-based rotations was less than from continuous grain crop rotations, primarily because greater soil cover was provided a greater portion of the year. This factor is discussed by Hoyt and Hargrove (1986). While soil in legume-based rotations is often better aggregated and has a higher infiltration rate than bare cultivated soils, legume-based rotations seldom provide better erosion protection than no-till systems because both systems maintain good surface cover (Hall et al., 1984).

Soil Water and Temperature Regimes

Legumes in a crop rotation affect the physical soil environment by altering soil water and temperature regimes. Temperature effects result from the influence of legumes on soil surface cover and from influence of altered soil water regimes on soil temperature regimes. The insulating effect of legume residues on the soil surface is no different from that of nonlegume residues. For example, Utomo et al. (1987) found that soil temperatures under no-till hairy vetch residues and corn stover were 1.5° and 1.2°C lower, respectively, than for clean, cultivated corn. The main effect of legumes on reducing temperature and potential evaporation rates results from the fact that legume-based cropping systems often provide more ground cover than occurs under clean cultivation. However, a living mulch of legumes reduces soil water content, thereby reducing the heat sink in the soil. Thus, the drier soil under legumes may heat up faster and compensate somewhat for the insulating effect, resulting in little net effect of legume cover on soil temperatures.

The effects of legume cover crops on soil water were discussed by Hargrove and Frye (1987). When used as a cover crop, legumes utilize stored soil water during the noncrop period of the grain crop with which the cover crop is associated. This can have a positive, negative, or neutral effect on the following grain crop. In poorly drained soils, or when excessive precipitation is received during the noncrop period, use of a legume cover crop reduces soil water content, thereby reducing the adverse effect of the excess water on crop growth. Likewise, the cover crop also reduces the likelihood of nutrients and pesticides leaching into ground water. For drier climates, however, legume cover crops can reduce soil water content to such an extent that the following grain crop suffers. For example, Koerner and Power (1987) showed that under eastern Nebraska conditions hairy vetch, if not properly managed as a winter cover crop, reduced soil water storage and increased competition, reducing yields of the following corn crop.

In the wheat-producing regions of the northwestern United States, various legumes are frequently grown in different types of rotation with winter wheat. Elliott et al. (1987) showed that soil water storage at wheat seeding time varied with the legume used (Table 6.8). Water storage was increased most with a spring pea rotation, and least with red clover or hairy vetch in rotation. Legume dry-matter production and amount of N_2 fixed by the legume generally increased as soil water storage increased, except for spring pea. These results indicate that legume species differ

TABLE 6.8 Total Dry-Matter Production, Fixed N, and Soil Profile Water (to 90 cm) prior to Tillage, 1985.

Legume	Rotation	Legume Dry Matter (kg ha^{-1})	N_2 Fixed (kg ha^{-1})	Soil Profile Water (mm)
Red clover	Spring barley-spring wheat/red clover-winter wheat	4,060	59	163
Hairy vetch	Spring barley/winter wheat/hairy vetch-winter wheat	4,570	108	163
Austrian winter pea	Spring barley/winter wheat/Austrian winter pea-winter wheat	6,430	113	178
Spring pea	Spring barley-winter wheat-spring pea-winter wheat	3,870	75	196

Source: Elliott et al. (1987).

TABLE 6.9 Soil Water Content (to 1.8 m) at Spring Wheat Seeding and Spring Wheat Yields following Wheat or Crested Wheatgrass/Alfalfa.

	Years after Crested Wheatgrass/Alfalfa					
Preceding Crop	1	2	3	4	5	6
	Soil Water Content (mm)					
Wheat	389	371	422	356	343	351
Crested wheatgrass/alfalfa	284	300	338	333	295	320
	Spring Wheat Yield (kg ha^{-1})					
Wheat	1,140	820	1,250	380	1,020	150
Crested wheatgrass/alfalfa	480	670	1,400	370	1,030	150

Source: Haas et al. (1976).

significantly in their water requirements as well as in N_2 fixation. Ramig (1987) also showed that water use efficiency by spring pea was dependent upon tillage method, with greatest efficiency from fall tillage.

In drier regions, use of legumes in crop rotations is often restricted because of water availability. Haas et al. (1976) showed that deep-rooted legumes such as alfalfa or sweet clover, when grown in rotation with wheat in North Dakota, frequently depleted soil water reserves to 2 m or greater (Table 6.9). As a consequence, the following grain crops had no subsoil reserve of soil water, and yields for the first several years after plowing up the sod suffered accordingly. Brown (1964) came to a similar conclusion after summarizing long-term data from legume-based rotations at a number of locations throughout the Great Plains.

Soil Nutrient Availability

Numerous authors have published on the effects of legumes upon soil nutrient availability and uptake by subsequent crops. Most such publications have focused on N availability because of the frequent enhancement of soil N resulting from N_2 fixation by legumes. Heichel (1987) recently summarized this topic. As expected, quantity of N_2 fixed by legumes varies widely with species, location, management, and other factors. Heichel's summary of N_2-fixation data appears in Table 6.10. While values range widely because of different conditions, it is interesting to note that values for seed and pulse legumes are comparable to those for forage legumes.

A number of factors affect the amount of biologically fixed N_2 resulting from legume production. Because the N_2-fixation rate is essentially driven by photosynthesis, any factor affecting season-long photosynthesis may affect biological fixation. Primary among these variables would be temperature and water availability. Zachariassen and Power (1987) showed that optimum temperature for N_2 fixation varied with species. For example,

TABLE 6.10 Seasonal Total of N_2 Fixation by Forage and Seed or Pulse Legumes, Measured by the Difference or Isotope Dilution Methods.

Species	N_2 Fixation (kg N ha^{-1} Growing $Season^{-1}$)	Location	Measurement Method
Forage Legumes:			
Alfalfa	212	Lexington, KY	Difference
Alfalfa	114–223	Rosemount, MN	Isotope dilution
Alfalfa-orchardgrass sward	15–136	Lucas Co., IA	Isotope dilution
Alsike clover	21	New Jersey	Difference
Birdsfoot trefoil	49–112	Rosemount, MN	Isotope dilution
Crimson clover	64	New Jersey	Difference
Hairy vetch	111	New Jersey	Difference
Ladino clover	165–188	Maryland	Difference
Red clover	68–113	Rosemount, MN	Isotope dilution
Subterranean clover	58–183	Hopland, CA	Isotope dilution
Subterranean clover-soft chess sward	21–103	Hopland, CA	Isotope dilution
Sweet clover	4	New Jersey	Difference
White clover	128	Lexington, KY	Difference
Seed and Pulse Legumes:			
Chickpea	24–84	Alberta	Isotope dilution
Common bean	12–121	Alberta	Isotope dilution
	2–110	Alberta	Difference
Faba bean	178–251	Alberta	Isotope dilution
Field pea	174–196	Alberta	Isotope dilution
Lentil	167–189	Alberta	Isotope dilution
Soybean	13–75	Iowa	Difference
	22–80	Minnesota	Difference
	55–160	Arkansas	Difference
	262–310	Washington	Difference
	55–129	Minnesota	Difference
	104	Nebraska	Difference

Source: Heichel (1987).

greatest fixation rate was 10°C for hairy vetch and faba bean, 20°C for soybean, and 30°C for lespedeza.

Another very important variable affecting N_2-fixation rate is soil inorganic N level (Heichel, 1987). Generally, N_2 fixation is reduced as the soil inorganic N level increases. Consequently, many management practices such as fertilization and tillage can affect the amount of biological N fixation that occurs with a legume crop.

The availability of N immobilized in legume residue to following crops also varies widely, depending on a number of factors. Hesterman et al. (1987) found that from 18% to 70% of the N in legume residues was utilized

TABLE 6.11 Effect of Legume Treatment on Use Efficiency of N by Corn from Incorporated Legume Residue at Two Minnesota Locations.

Legume Cultivar Management	N in Legume Used by Next Corn, %			
	Becker		Rosemount	
	Whole Plant	Grain	Whole Plant	Grain
Saranac AR alfalfa, 1-cut	37	27	70	46
Saranac AR alfalfa, 3-cut	66	46	56	32
MN Root N alfalfa, 1-cut	27	20	49	27
MN Root N alfalfa, 3-cut	44	30	39	29
Nodulated soybean	18	13	19	12
Non-nodulated soybean	39	30	50	44
$LSD_{(0.05)}$	20	16	NS	NS

Source: Hesterman et al. (1987).

by the following corn crop (Table 6.11). Legume species or even cultivar had a major influence on the quantity of legume N taken up by the corn. Likewise, legume management practices (such as frequency of cutting, date of plowing, degree of nodulation) also influence legume N recovery. In addition, soil water availability and temperature are important factors. Ladd and Amato (1986) showed that fertilizer practices also affect legume N recovery. Power et al. (1986) showed that a large percentage of the N in soybean residues was utilized by the next crop. Thus, recovery of legume N by following grain crops can vary widely and is highly dependent on management, climate, and soils.

While it is recognized that legumes may also affect availability of soil P, cations, and minor elements, a discussion of these effects is beyond the scope of this chapter. Frequently, deep-rooted legumes utilize and recycle subsoil calcium and other nutrients, affecting availability and utilization of these nutrients by the following grain crops.

LEGUMES AND THE ENVIRONMENT

One of the primary reasons frequently given for the need to develop sustainable agricultural production systems is to protect the environment from degradation by agricultural chemicals, soil erosion, and other agents. The logic here is that, by making greater use of legumes in cropping systems, the need for synthetic N fertilizers and other agrichemicals would be greatly reduced, thereby reducing input of soluble N and pesticides into the environment. Other effects of legumes that may improve the environment include reduction of soil erosion and sedimentation and the potential to remove residual nitrates from the soil profile, thereby reducing potential for nitrate contamination of ground water.

The effects of legumes upon the N cycle were discussed from various viewpoints earlier in this chapter. While data in Table 6.1 show that fertilizers have largely replaced legumes as a source of N for crop production, legumes still contribute significantly to American agriculture, adding over 6 million tons of N to our soils annually (Table 6.2). Estimates of quantities of N_2 biologically fixed by different legumes under varying management systems (see Tables 6.5, 6.8, and 6.10) illustrate the fact that legumes have great potential for supplying a significant part of the N required for modern agriculture. One of the major research challenges ahead of us is to develop economically feasible management systems that enhance this objective.

Definitive data on the influence of legumes on the environment are generally scarce. However, several conclusions can be drawn from the available data. It is clear that legumes, when used as a cover crop for grain monocultures, not only provide N for the following grain crop, but can also aid in soil erosion control and utilize residual soil nitrates, thereby reducing ground water pollution potential (Power, 1987).

A number of legumes can be used for cover crops. For continuous corn or sorghum monocultures in the southeastern United States, favorite species include crimson clover and hairy vetch (Knight, 1987; Neely et al., 1987; Utomo et al., 1987). Bigflower vetch also appears to have some merit. In more northern areas, only hairy vetch is sufficiently winter-hardy to perform satisfactorily (Koerner and Power, 1987). For use with other grain crops, various *Trifolium* and *Medicago* species may be considered as a cover crop. Likewise, annual species of peas and lentils can be used (Power, 1987), especially in more northern latitudes or during cooler periods of the growing season. Typically, a legume cover crop reduces the N fertilizer requirement for the following grain crop by 50 to 95 kg N ha^{-1} (Table 6.12).

TABLE 6.12 Estimates of the Nitrogen Fertilizer Equivalent Contributed by Winter Legumes to the Nitrogen Requirement of No-till Corn, Grain Sorghum, and Cotton.

Location	Crop	Cover Crop	Fertilizer N Equivalent (kg ha^{-1})
Kentucky	Corn	Hairy vetch	95
		Bigflower vetch	50
		Crimson clover	83
Georgia	Grain sorghum	Hairy vetch	90
		Common vetch	59
		Subterranean clover	57
Alabama	Cotton	Hairy vetch	68
		Crimson clover	68

Source: Hargrove and Frye (1987).

TABLE 6.13 Average Microbial Biomass, Microbial Numbers, and Dehydrogenase Enzyme Activity in the Surface 0- to 7.5-cm Soil Layer of the Conversion Project in June and July 1982.

Management System Crop, by Year		Microbial Biomass C	Microbial Numbers/ cm^3 Soil		Dehydogenase Enzyme Activity
1981	1982	($kg\ ha^{-1}$)	Fungi X 10^4	Bacteria X 10^6	(mg formazan/ 6 cm^{-3} d^{-1})
Grain-Forage Rotation + Manure					
Oat + red clover	Red clover	831 a[a]	22 a	96 a	0.82 a
Corn	Soybean	681 bc	18 ab	39 b	0.54 b
Corn	Wheat + red clover[b]	792 a	20 a	64 ab	0.84 a
Legume-Cash Grain Rotation					
Oat + red clover	Corn	637 c	20 a	63 ab	0.58 b
Soybean	Oat + red clover	666 bc	18 ab	45 b	0.76 a
Corn	Soybean	662 c	19 ab	32 b	0.52 b
Conventional Cash Grain					
Corn	Corn	587 c	15 c	42 b	0.42 b
Soybean	Corn	627 c	18 ab	47 b	0.53 b
Corn	Soybean	611 c	17 bc	36 b	0.53 b

[a] Means for each variable followed by the same letter do not differ significantly at $P < 0.05$.
[b] Beef cattle manure (9000 $kg\ ha^{-1}$) was applied on May 4, 1981.
Source: Doran et al. (1987a).

Koerner and Power (1987) found that a hairy vetch cover crop provided all the N required by dryland corn in eastern Nebraska.

The fact that legume N may replace some or all fertilizer N in a cropping system usually results in reduced amounts of nitrate leached from the soil. However, there may be exceptions. Hargrove (1986) found higher soil nitrate levels at the end of the growing seasons in plots with legume cover crops than in those without. In soil column studies, Muller (1987) found that 17% of the ^{15}N in subterranean clover (*Trifolium subterraneum L.*) residues was leached from fallow soil. When legume stubble and residues are incorporated into a warm, moist soil, the narrow C:N ratio of these materials enhances rapid mineralization of the N in these residues, so temporary accumulations of soil nitrate are not uncommon under these conditions. Consequently nitrate may leach from such soils. For example, Adams and Pattinson (1985) in New Zealand reported 90 kg N ha^{-1} leached after returning pea (*Pisum* spp) residues to the soil.

The question of effects of commercial fertilizers and pesticides upon soil biology and the microbiology of soil N transformations was studied extensively by Doran et al. (1987a). They were able to detect no direct adverse effects of agricultural chemicals upon soil microbial populations, their activity, or on N cycling (Table 6.13). This agrees with the conclusions of Goring and Laskowski (1982). However, Doran and associates found that cropping systems, particularly those containing legumes such as hairy vetch or red clover, greatly enhanced microbial biomass and pool sizes of organic and potentially mineralizable soil N. Thus, this larger pool of potentially available N in legume-based cropping systems offers the producer more management alternatives, but requires more enlightened management decisions than are required when fertilizer N is used.

Frequently, crops grown in rotation may yield 10% to 40% more than when grown as a monoculture even with sufficient N (Table 6.14). In this example, the yield for well-fertilized continuous corn was exceeded by that for corn following soybean or wheat receiving only 55 to 110 kg N ha^{-1}; by corn following three-cut alfalfa with 0 to 55 kg N ha^{-1}; and by corn after alfalfa cut only once receiving no fertilizer. Maximum yields for corn in rotation with any crop were nearly 2.5 Mg ha^{-1} greater than maximum yield for continuous corn. This 2.5 Mg ha^{-1} improvement in yield could have resulted from a combination of several factors other than N availability—fewer insect, disease, and weed problems (Power, 1987); elimination of phytotoxicity; enhancement of growth-promoting substances (Ries et al., 1977); and improvement in soil physical properties.

An example of a short-term legume crop rotation would be the corn-soybean rotation commonly used in the Midwestern United States. Likewise, other rotations such as corn-alfalfa, wheat-peas, wheat-lentils, and others would be additional examples. In many such rotations, a cereal

TABLE 6.14 Corn Grain as Affected by Previous Crop and Fertilizer N Rate.

Corn Grain Yield, Mg Ha^{-1}					
N Rate (kg ha^{-1})	Previous Crop				
	Corn	Soybean	Wheat	Alfalfa (3-cut)	Alfalfa (1-cut)
0	3.1	3.6	3.6	4.1	6.8
55	4.1	5.6	6.2	7.0	8.7
110	6.3	7.6	8.0	7.7	8.9
165	6.5	8.6	8.0	7.9	8.6
220	6.3	8.8	9.0	8.4	8.8

Source: Hesterman et al. (1986).

crop is rotated with a seed legume in a two-year sequence. Such a rotation usually provides for the "rotation effect" in which yield of the cereal is greater than that of continuous cereal regardless of the fertilizer rate used. This is illustrated by data in Table 6.15 in which yields of corn grown with 45 kg N ha^{-1} in rotation with soybean, wheat, or a wheat-alfalfa system were greater than those for continuous corn, even at 225 kg N ha^{-1}. Also, corn yields for all three rotations were approximately equal, suggesting that the legume rotations exhibited little or no advantage over a straight cereal rotation. These results suggest that this rotation effect is related to factors such as disease, weed, and insect problems and improved soil structure, in addition to improved N availability.

From the information given in this section, it appears that inclusion of legumes in a cropping system enhances environmental protection by several means, as compared with a monoculture system that uses commercial fertilizers as a primary source of N. With legume-based cropping systems, soil cover is provided a major part of the year, thereby reducing erosion potential. There is less fertilizer N input, reducing potential for

TABLE 6.15 Effect of Previous Crop on Corn Grain Yield.

Corn Grain Yield, Mg Ha^{-1}				
Nitrogen Rate (kg ha^{-1})	Previous Crop			
	Corn	Soybean	Wheat	Wheat-Alfalfa
0	4.45	6.89	6.85	7.25
45	5.80	8.14	8.00	8.00
90	6.46	8.70	8.78	8.80
135	6.85	8.97	8.78	9.05
180	7.36	9.37	8.98	9.24
225	7.53	9.37	9.20	9.11

Source: Randall (1981).

fertilizer N to escape into the water or atmosphere. Soil organic matter and biological activity is often enhanced (Table 6.16), which may improve soil tilth and water relations. However, under certain conditions, N may be leached from decomposing legume residues. Consequently, good management is required to achieve the objectives of environmental protection while maintaining an economically viable farming enterprise.

LEGUME CROPPING SYSTEMS

Historically, one of the primary reasons for growing legumes has been to provide N for the following crops. The widespread advent of commercial fertilizer N, especially anhydrous ammonia, in the United States has provided an economical alternate source of N for crop production. At current prices, few farmers can afford to forego a year or two of grain production in order to grow legumes as a source of N. Even if they harvest and sell legume hay from the land devoted to legume production, seldom will an economic analysis indicate that legumes are competitive with commercial fertilizers when legumes are valued as an N source only. At present, American farmers can obtain N as anhydrous ammonia for about \$0.25 kg^{-1}, so 200 kg N ha^{-1} will cost only \$50. In 1985, American farmers spent over \$7 billion for fertilizer and lime to produce over \$70 billion in crops (Berry and Hargett, 1987). How long these price structures and this economic situation will continue is unknown and highly speculative. The economic implications associated with legume-based rotations were discussed by Domanico et al. (1986).

The economic benefits derived from other uses of legumes in cropping systems are extremely difficult to quantify: benefits such as enhanced soil organic matter levels, reduced runoff and erosion, and reduced groundwater contamination. This difficulty in quantifying such benefits results in inexact information on the relative economic returns from legumes versus the use of commercial N. We recognize that there are other significant benefits from legumes, but the value of these benefits is often subjective and variable.

Many of the attributes of legume-based rotations have been described in earlier sections of this chapter. For example, data in Tables 6.5 and 6.10 address the N contribution of legumes used in rotation to the N requirements of following grain crops. Tables 6.6, 6.7, 6.9, 6.13, and 6.16 provide information on the effects of legumes in rotation on various physical, chemical, and biological properties of soils. Collectively, the information presented shows that, in legume-based crop rotations, a major part of the N required by the cropping system can be provided by the legume, and the soil cover resulting from these cropping systems provides important protection from soil erosion and thereby helps maintain soil productivity.

TABLE 6.16 Effect of Long-term Green Manuring on Organic Matter and Nitrogen Content of Soils.

Study	Length of Study (yr)	Green Manure Used	Rotation (X = Green Manure)	Soil Type	Initial Soil O.M (%)	Initial Soil N (%)	Changes at End of Study	
							Organic Matter	Soil N
Potting studies								
Prince et al. (1941)	40	Vetch	Corn-X-oats-X-oats-wheat-timothy	Loam	3.9	0.197	Decrease	Decrease
Leuken et al. (1962)	150-day incubation	Legumes and nonlegumes	No crop	Silty clay	10.3	0.491	Increase	N.R.[a]
				Clay loam	2.8	0.131	Increase	N.R.
De Haan (1977)	10	Legumes and nonlegumes	No crop, annual incorporation	Sand	3.9	0.156	Increase	Increase
				Clay	3.0	0.224	Increase	Increase
Field studies								
Poyser et al. (1957)	25	Legumes and nonlegumes	X-wheat-corn-wheat	Clay	7.8	0.370	Decrease	Increase
Mann (1959)	18	Legumes and nonlegumes	X-kale[b] or cabbage-barley	Sandy loam	1.48	0.090	Increase	Increase
Chater and Gasser (1970)	30	Trefoil	X-barley[c] potatoes[d]	Sandy loam	1.48	0.090	Same	Increase
[cont. of Mann (1959)]		Ryegrass	X-barley	Sandy loam	1.48	0.090	Same	Decrease

[a] N.R. = not reported.
[b] *Brassica oleracea.*
[c] *Hordeum vulgare.*
[d] *Solanum tuberosum.*

Source: MacRae and Mehuys (1985).

For purposes of this discussion, the use of legumes is divided into three categories: (1) long-term use of legumes in crop rotations (over 1 year), (2) short-term use of legumes (1 year or less), and (3) multiple or double cropping (cover crops). Because few, if any, woody leguminous or N_2-fixing species (*Acacia* or *Prosopis*) are routinely used in cropping systems in the United States, such systems are omitted from this discussion.

Use of legumes in crop rotations not only reduces or eliminates fertilizer N requirement and soil erosion potential; it has the added benefit of providing the "rotation effect" to a monoculture. The rotation effect includes all increases in yield of the grain crop when grown in rotation with other species as compared to grain crop production in a monoculture. Baldock et al. (1981) divided the rotation effect into N effects and all other effects. Russelle et al. (1987) indicated that all other effects may account for as much as 50% of the increase in yield resulting from growing a crop in rotation compared to growing it as a monoculture.

As mentioned earlier, grain legumes grown in rotation with cereal can contribute little or can even result in a depletion of the N balance of the soil. Frequently, more N is removed in the harvest of seed legumes than is biologically fixed by that crop. Consequently, such rotations would increase soil organic matter only through the increased total biomass production that resulted from the rotation effect. This has been shown to occur in the Morrow plots in Illinois (Odell et al., 1982), where a corn-soybean rotation since 1967 (corn-oat, 1876 to 1966) increased total soil N 680 kg ha^{-1} over that of continuous corn (both with no erosion). However, for erodible situations, soil erosion losses from a corn-soybean rotation may not be greatly different or may even be greater than from continuous corn (Bruce et al., 1987).

The merits of legumes used as cover crops or as intercrops have been reviewed by Power (1987). Again, several of the tables in this chapter document the beneficial effects normally observed. In addition to exhibiting the rotation effect, legume cover crops usually improve the N nutrition of the following cereal crop (Table 6.12). Shurley (1987) concluded that the use of legume cover crops for continuous corn in Kentucky was economically profitable, even without giving any consideration to benefits arising from enhanced soil erosion control or water quality. Tisdall and Oades (1982) concluded that additional organic matter returned to the soil, whether from increased plant growth resulting from the rotation effect or from other sources, builds up the readily mineralizable and more recalcitrant pools of soil N, thereby increasing both available soil N supply and aggregate formation and stability.

CONCLUSIONS

Legumes and crop rotations have historically been and remain an important part of American agriculture. In earlier generations, legumes and rotations were usually necessary as a cropping system to provide the feed and forage required by draft animals and the livestock needed for a largely self-sufficient enterprise. A degree of self-sufficiency was required because of a lack of the processing and transportation technology and the infrastructure present in today's agriculture. For these same reasons, self-sufficiency is still needed in many less developed areas of the world.

With the advent of agricultural chemicals, complete mechanization, and modern developments in processing and transport of agricultural products, coupled with encouragement from government programs within this last generation, the self-sufficient, diversified farm has been replaced in some regions by the large monoculture enterprises that depend heavily on purchased inputs. Few, if any, livestock are involved. Consequently, the use of legumes and rotations has decreased considerably. In view of the dependence of such an enterprise upon inputs of fossil fuels, and because of the susceptibility of monocultures to soil erosion losses and degradation of water quality, the long-term sustainability of monoculture systems is frequently questioned. Thus, we have need to evaluate rotations and legumes in light of modern agriculture. Except in enterprises heavily involved with grazing livestock, it is not likely that the old rotations of earlier generations will be economically viable again, but there are other types of rotations and management systems where there is a potential for greater use of legumes. Some of these situations have been described and discussed in this chapter.

An example of new legume management systems is the common short rotation of corn with soybean that has developed in the last generation of American farmers. When compared to continuous corn, there are many merits to this rotation in terms of corn yields, pest control, N fertilizer requirements, and N-use efficiency. However, a major problem with this rotation is greater soil erosion potential. This problem can be corrected with the use of reduced and no-tillage options or with legume cover crops. Compared to the older, long-term rotations involving corn, oats, and perennial legumes, the corn-soybean rotation provides greater income and does not require a livestock enterprise. Other short-term rotations (such as wheat-peas, wheat-lentils, corn-alfalfa) also have their respective merits and deficiencies.

There has been considerable interest in recent years in the use of legume cover crops as a means of reducing production costs and providing soil and water protection. Many experiments have demonstrated the relative merits of a number of legume species when used as cover crops in different grain monocultures. Some of the most promising are hairy vetch (or crimson clover, where winterkill is not severe) as a cover crop for continuous corn or sorghum, especially when used with a ridge-till system. This practice reduces or eliminates fertilizer N inputs, provides erosion protection, and helps maintain quality of surface and ground water. Again, much remains to be learned about how best to manage these green manures for varying soils and climates.

The above discussion illustrates a few examples of how rotations and legumes can be used in modern cropping systems. The economic impact of these practices varies widely, depending on relative costs of inputs, soils, climate, effects on grain crops, pest control, and other factors. Thus, practices that are profitable in one place or situation might not be profitable under other conditions. A major research effort is required to identify and define factors that affect profitability of these practices, both in the short term and in the long term. Also, much new information is needed to provide inputs needed to develop government agricultural programs that will promote a sustainable agriculture in the U.S. and throughout the temperate zone.

REFERENCES

Adams, J. A., and Pattinson, J. M. (1985). "Nitrate leaching losses under a legume-based crop rotation in Central Canterbury, New Zealand," *NZ. J. Agr. Res.* **28**:101–7.

Allison, F. E. (1968). "Soil aggregation—some facts and fallacies as seen by a microbiologist,"*Soil Sci.* **106**:136–43.

Baldock, J. O., R. L. Higgs, W. H. Paulson, J. A. Jackobs, and W. D. Shrader. (1981). "Legume and mineral N effects on crop yields in several crop sequences in the upper Mississippi Valley," *Agron. J.* **73**:885–90.

Beck, R. J. (1982). "Forecast and review," *Oil & Gas J.* **80(4)**:127–44.

Beijerinck, M. W. (1888). "Die bakterien der papdionaceen knollchen," *Bot. Zblt.* **46**:726–804.

Berry, J. T., and N. L. Hargett. (1987). *Fertilizer Summary Data*, Bull. Y-197, Natl. Fert. Development Center, TVA, Muscle Shoals, AL. p. 11.

Brown, P. L. (1964). "Legumes and grasses in dryland cropping systems in the northern and central Great Plains," *U.S. Dept. Agr. Misc. Publ. 952.* 64 pp.

Bruce, R. R., S. R. Wilkinson, and G. W. Langdale. (1987). "Legume effects on soil erosion and productivity," In *The Role of Legumes in Conservation Tillage Systems*, (Ed. J. F. Power), pp. 127-38. Soil Conserv. Soc. Am., Ankeny, IA.

Burton, J. C. (1979). "Rhizobium species," In *Microbial Technology*, vol. I., 2nd ed. pp. 29–45. Academic Press, New York.

Chepil, W. S. (1955). "Factors that influence clod structure and erodibility of soil by wind: Organic matter at various stages of decomposition," *Soil Sci.* **80**: 413–21.

Domanico, J. L., P. Madden, and E. J. Partenheimer. (1987). "Income effects of limiting soil erosion under organic, conventional, and no-till systems in eastern Pennsylvania," *Am. J. Alt. Agric.* **1**:75–82.

Doran, J. W., D. G. Fraser, M. N. Culik, and W. Liebhardt. (1987a). "Influence of alternative and conventional agricultural management on soil microbial processes and nitrogen availability," *Am. J. Alt. Agric.* **2**:99–106.

Doran, J. W., L. N. Mielke, and J. F. Power. (1987b). "Tillage/residue management interactions with soil environment, organic matter, and nutrient cycling," *Intecol Bull.* **15**:33–39.

Elliott, L. F., R. I. Papendick, and D. F. Bezdicek. (1987). "Cropping practices using legumes with conservation tillage and soil benefits." In *The Role of Legumes in Conservation Tillage Systems,* (Ed. J. F. Power). pp. 81-89. Soil Conserv. Soc. Am., Ankeny, IA.

Elson, J. (1941). "A comparison of the effect of fertilizer and manure, organic matter, and carbon-nitrogen ratio on water-stable soil aggregates," *Soil Sci. Soc. Am. Proc.* **6**: 86–90.

Emerich, D. W., and H. J. Evans. (1984). "Enhancing biological dinitrogen fixation in crop plants," In *Nitrogen in Crop Production,* (Ed. R. D. Hauck). pp. 133-44. Am. Soc. Agron., Madison, WI.

Evans, H. J., and L. E. Barber. (1977). "Biological nitrogen fixation for food and fiber production," *Science* **197**:332–39.

Gantzer, C. J., G. A. Buyanovsky, E. E. Alberts, and P. A. Remley. (1987). "Effects of corn and soybean residue decomposition on soil strength and splash detachment," *Soil Sci. Soc. Am. J.* **51**:202–6.

Gibson, A. H. (1978). "The influence of environmental and managerial practices on the legume-rhizobium symbiosis," In *A Treatise on Dinitrogen Fixation, Sec. IV: Agronomy and Ecology,* (Eds. R. F. W. Hardy and A. H. Gibson). pp. 393-450. Wiley, New York.

Goring, C. A. I., and D. A. Laskowski. (1982). "The effects of pesticides on nitrogen transformations in soils," In *Nitrogen in Agricultural Soils,* Agron. No. 22 (Ed. F. J. Stevenson). pp. 689-720. Am. Soc. Agron., Madison, WI.

Haas, H. F., J. F. Power, and G. A. Reichman. (1976). "Effect of crops and fertilizer on soil nitrogen, carbon, and water content, and on succeeding wheat yields and quality," *ARS-NC-38,* ARS, USDA, North Central Region. 21 pp.

Hall, J. K., N. L. Hartwig, and L. D. Hoffman. (1984). "Cyanazine losses in runoff from no-tillage corn in 'living' and dead mulches vs. unmulched conventional tillage," *J. Environ. Qual.* **13**:105–10.

Hardy, R. W. F., and V. D. Havelka. (1975). "Nitrogen fixation research: A key to world food," *Science* **188**:633–43.

Hardy, R. W. F., J. G. Criswell, and V. D. Havelka. (1977). "Investigations of possible limitations of nitrogen fixation by legumes." In *Nitrogen Fixation* (Ed. W. E. Newton). pp. 451-67. Academic Press, New York.

Hargrove, W. L. (1986). "Winter legumes as a nitrogen source for no-till grain sorghum," *Argon. J.* **78**:70–74.

Hargrove, W. L., and W. W. Frye. (1987). "The need for legume cover crops in conservation tillage production." In *The Role of Legumes in Conservation Tillage Systems* (Ed. J. F. Power). pp. 1-5. Soil Conserv. Soc. Am., Ankeny, IA.

Havelka, V. D., M. G. Boyle, and R. W. F. Hardy. (1982). "Biological nitrogen fixation." In *Nitrogen in Agricultural Soils,* Agron No. 22 (Ed. F. J. Stevenson). pp. 365-422. Am. Soc. Agron., Madison, WI.

Heichel, G. H. (1987). "Legumes as a source of nitrogen in conservation tillage systems." In *The Role of Legumes in Conservation Tillage Systems* (Ed. J. F. Power). pp. 29–34. Soil Conserv. Soc. Am., Ankeny, IA.

Hellriegel, H. (1886). "Welche stickstoffquellen steken der pflange zu gebote?" *Z. Ver. Ruebenzucker Ind. Dtsch. Reichs.* **38**:863–77.

Hesterman, O. B., M. P. Russelle, C. C. Shaeffer, and G. H. Heichel. (1987). "Nitrogen utilization from fertilizer and legume residues in legume-corn rotations," *Agron. J.* **79**:726–31.

Hesterman, O. B., C. C. Shaeffer, D. K. Barnes, W. E. Lueschen, and J. H. Ford. (1986). "Alfalfa dry matter and N production and fertilizer N response in legume-corn rotations," *Agron. J.* **78**:19–23.

Hoyt, G. D., and W. L. Hargrove. (1986). "Legume cover crops for improving crop and soil management in the southern United States," *Hort. Sci.* **21**:397–402.

Johnson, J. W., L. F. Welch, and L. T. Kurtz. (1975). "Environmental implications of nitrogen fixation by soybeans," *J. Environ. Qual.* **4**:303–6.

Johnston, J. R., G. M. Browning, and M. B. Russell. (1943). "The effect of cropping practices on aggregation, organic matter content, and loss of soil and water in Marshall silt loam," *Soil Sci. Soc. Am. Proc.* **7**:105–13.

Knight, W. E. (1987). "Germplasm resources for legumes in conservation tillage.", *The Role of Legumes in Conservation Tillage Systems* (Ed. J. F. Power). pp. 13-19. Soil Conserv. Soc. Am., Ankeny, IA.

Koerner, P. T., and J. F. Power. (1987). "Hairy vetch winter cover for continuous corn in Nebraska," In *The Role of Legumes in Conservation Tillage Systems* (Ed. J. F. Power). pp. 57-59. Soil Conserv. Soc. Am., Ankeny, IA.

Kurtz, L. T., L. V. Boone, T. R. Peck, and R. G. Hoeft. (1984). "Crop rotations for efficient nitrogen use," In *Nitrogen in Crop Production* (Ed. R. D. Hauck). pp. 295-306. Am. Soc. Agron., Madison, WI.

Ladd, J. N., and M. Amato. (1986). "The fate of nitrogen from legume and fertilizer sources in soils successively cropped with wheat under field conditions," *Soil Biol. Biochem.* **18**:417–25.

Laflen, J. M., and W. C. Moldenhauer. (1979). "Soil and water losses from corn-soybean rotations," *Soil Sci. Soc. Am. J.* **43**:1213–15.

Lutz, J. F. (1954). "Influence of cover crops on certain physical properties of soils." In *Winter Cover Crops in North Carolina,* North Carolina Agron. Res. Rpt. 12. pp. 6–9.

Mackay, A. D., and E. J. Kladivko. (1985). "Earthworms and rate of breakdown of soybean and maize residues in soil," *Soil Biol. Biochem.* **17**:851–57.

MacRae, R. J., and G. R. Mehuys. (1985). "The effect of green manuring on the physical properties of temperate-area soils," *Adv. Soil Sci.* **3**:71–94.

Meisinger, J. J. (1984). "Evaluating plant-available nitrogen in soil-crop systems." In *Nitrogen in Crop Production* (Ed. R. D. Hauck). pp. 391-416. Am. Soc. Agron., Madison, WI.

Muller, M. M. (1987). "Leaching of subterranean clover-derived N from a loam soil," *Plant & Soil* **102**:185–91.

Neely, C. L., K. A. McVay, and W. L. Hargrove. (1987). "Nitrogen contribution of winter legumes to no-till corn and grain sorghum." In *The Role of Legumes in Conservation Tillage Systems* (Ed. J. F. Power). pp. 48-49. Soil Conserv. Soc. Am., Ankeny, IA.

Odell, R. T., W. M. Walker, L. V. Boone, and M. G. Oldham. (1982). *The Morrow Plots: A Century of Learning.* Ill. Agric. Exp. Stn. Bull. 775. 22 pp.

Olmstead, L. B. (1947). "The effect of longtime cropping systems and tillage practices upon soil aggregation at Hays, Kansas," *Soil Sci. Soc. Am. Proc.* **11**:89–92.

Phillips, D. A., and T. M. DeJong. (1984). "Dinitrogen fixation in leguminous crops," In *Nitrogen in Crop Production* (Ed. R. D. Hauck), pp. 121-32. Am. Soc. Agron., Madison, WI.

Power, J. F. (ed). (1987). *The Role of Legumes in Conservation Tillage Systems.* Soil Conserv. Soc. Am., Ankeny, IA.

Power, J. F., and J. W. Doran. (1984). "Nitrogen use in organic farming," In *Nitrogen in Crop Production* (Ed. R. D. Hauck), pp. 585-98. Am. Soc. Agron., Madison, WI.

Power, J. F., and R. F. Follett. (1987). "Monoculture," *Scientific American* 256:79–86.

Power, J. F., J. W. Doran, and W. W. Wilhelm. (1986). "Uptake of nitrogen from soil, fertilizer, and crop residues by no-till corn and soybean," *Soil Sci. Soc. Am. J.* **50**:137–42.

Puckridge, D. W., and R. J. French. (1983). "The annual legume pasture in cereal-ley farming systems of southern Australia: A review," *Agric., Ecosystems, and Environ.* **9**:229–67.

Ramig, R. E. (1987). "Conservation tillage systems for green pea production in the Pacific Northwest." In *The Role of Legumes in Conservation Tillage Systems* (Ed. J. F. Power). pp. 93-94. Soil Conserv. Soc. Am., Ankeny, IA.

Randall, G. W. (1981). "Rotation nitrogen study, Waseca, 1980," Soil Series 109, *A Report on Field Research in Soils,* Agricultural Experiment Station Misc. Pub. 2 (revised)—1981, Department of Soil Science, University of Minnesota, St. Paul. pp. 115–16.

Ries, S. K., V. Wert, C. C. Sweeley, and R. A. Leavitt. (1977). "Triacontonal: A new, naturally occurring plant growth regulator," *Science* **195**:1339–41.

Russelle, M. D., O. B. Hesterman, C. C. Shaeffer, and G. H. Heichel. (1987). "Estimating nitrogen and rotation effects in legume-corn rotations." In *The Role of Legumes in Conservation Tillage Systems* (Ed. J. F. Power). pp. 41–42. Soil Conserv. Soc. of Am., Ankeny, IA.

Sears, O. H. (1953). *Legumes as Nitrogen Fixers.* Agron. Facts, SR8. University of Illinois, Urbana.

Shurley, W. D. (1987). "Economics of legume cover crops in corn production," In *The Role of Legumes in Conservation Tillage Systems* (Ed. J. F. Power). pp. 152-53. Soil Conserv. Soc. Am., Ankeny, IA.

Silver, W. S., and R. W. F. Hardy. (1976). "Newer developments in biological dinitrogen fixation of possible relevance to forage production," In *Biological N Fixation in Forage-Livestock Systems,* Spec. Publ. No. 28 (Ed. C. S. Hoveland). pp. 1-36. Am. Soc. Agron., Madison, WI.

Strickling, E. (1950). "The effect of soybeans on volume weight and water stability of soil aggregates, soil organic matter content, and crop yield," *Soil Sci. Soc. Am. Proc.* **15**:30–34.

Tisdall, J. M., and J. M. Oades. (1982). "Organic matter and water stable aggregates in soils," *J. Soil Sci.* **33**:141–63.

Utomo, M., R. L. Blevins, and W. W. Frye. (1987). "Effect of legume cover crops and tillage on soil water, temperature, and organic matter," In *The Role of*

Legumes in Conservation Tillage Systems (Ed. J. F. Power). pp. 5–6. Soil Conserv. Soc. Am., Ankeny, IA.

Wilson, H. A., and G. M. Browning. (1946). "Soil aggregation, yields, runoff, and erosion as affected by cropping systems," *Soil Sci. Soc. Am. Proc.* **10**:51–57.

Zachariassen, J. A., and J. F. Power. (1987). "Soil temperature and the growth, nitrogen uptake, dinitrogen fixation, and water use by legumes," In *The Role of Legumes in Conservation Tillage Systems* (Ed. J. F. Power). pp. 24-26. Soil Conserv. Soc. Am., Ankeny, IA.

7 MANAGEMENT AND SOIL BIOLOGY

JOHN W. DORAN
USDA-ARS,
University of Nebraska,
Lincoln, Nebraska

MATTHEW R. WERNER
Agroecology Program,
University of California,
Santa Cruz, California

HISTORICAL MANAGEMENT OF PRODUCTIVITY

Present-day agriculture has evolved as we have organized nature to meet our food and fiber needs and to support the urbanized structure of society. In our struggle to dominate a seemingly hostile natural environment, we have often failed to recognize the consequences of management upon balance and cycling of energy and matter in soil. Early settlers in America reaped immediate benefit from clearing and cultivating virgin prairie and forest soils, thus releasing stored soil nutrients to produce food crops. Within the lifetime of these early farmers, soil organic-matter levels, fertility, and productivity were reduced by half. Maintenance of crop productivity with continuous cropping required moving to virgin soils or returning harvested nutrients to cultivated soils. Initial supplementation of soil nutrient reserves was done with nitrogen-fixing legumes, animal manures, and soil-building green-manure crops.

Economic, social, and political forces have transformed agricultural ecosystems from what some have viewed as "domesticated" ecosystems, in relative harmony with the environment, into increasingly "fabricated" ecosystems that resemble urban-industrial systems with high energy and material demands and waste production (Odum, 1984). Fossil fuel and synthetic chemical inputs to agriculture have increased since World War II with the industrialization of agriculture. During the last 40 years, the increased use of synthetic N fertilizers for crop production in American agriculture has paralleled a corresponding decline in the use of legume cover crops and animal manures (Power and Papendick, 1985).

Increased monoculture production of cash grain crops and greater reliance on importation of chemical fertilizers and pesticides to promote crop growth have increased grain yields and labor efficiency. However, these practices have also been associated with increased soil erosion and organic-matter loss and the contamination of surface and ground water in the United States and elsewhere (Gliessman, 1984; Hallberg, 1987; Reganold et al., 1987). A 1980 U.S. Department of Agriculture study on organic farming reported increasing public concern about the adverse effects of our domestic agricultural production system, particularly with regard to intensive and continuous production of cash grains and the extensive and sometimes excessive use of agricultural chemicals (USDA, 1980). Motivations for shifting from intensive input management to reduced input farming include concern for protecting soil, human, and animal health from the potential hazards of pesticides; concern for protection of the environment and soil resources; and a need to lower production costs. Uncertain supply of energy sources, increased production costs, and low commodity prices within the past decade have led some to redefine crop productivity to include consideration of monetary and environmental input costs (Tangley, 1986). Cropping systems which promote efficient and sustainable production while minimizing external chemical inputs and degradation of soil and water resources represent alternatives to conventional production practices and, as such, deserve serious evaluation (Rodale, 1984).

MANAGEMENT AND SOIL PRODUCTIVITY

The productivity and stability of soil as a medium for plant growth depend greatly on the balance between living and nonliving components. Energy from the sun and nutrients essential for growth are stored as chemical energy in plant tissue. These resources are recovered for use through decomposition activities of micro- and macroorganisms in soil. To maintain productivity, soluble nutrients removed from soil by plant growth and harvest must be replaced, either as fertilizers or through biological decomposition of organic plant and animal matter in soil. The soil organic matter formed during decomposition serves as a continuous nutrient supply and stabilizes biological activity by reducing fluctuations in the soil physical environment. Microbial activity accounts for the majority (over 90%) of decomposition occurring in soils (Coleman and Sasson, 1978). Soil fauna, however, play an important role in regulating decomposition through preconditioning organic debris, feeding on microorganisms and altering the soil physical environment through movement and burrowing (Crossley, 1977; Werner and Dindal, 1987). Thus, the intricate balance of decomposition and nutrient cycling depends mainly on the varied activities of soil microorganisms and fauna. The effectiveness of these unseen citizens of the soil is largely determined by the regulation of food sources and the soil physical and chemical environment controlled by agricultural management (Juma and McGill, 1986). Where synthetic chemical use is reduced or eliminated, residue inputs and soil microbial and faunal activities become major determinants of nutrient cycling and plant growth.

In alternative or organic agricultural management systems, the recycling of nutrients and proper balance between organic matter, soil organisms, and plant diversity are necessary components of a productive soil. A basic tenet of organic farming is that an ecologically balanced soil environment results in healthy, vigorous plants and a stable soil environment (USDA, 1980). The use of synthetic chemical fertilizers and pesticides is thought to be harmful to microorganisms and other soil life forms and is therefore avoided. In research experiments, however, the use of chemical fertilizers or pesticides at recommended application rates has not proved to be harmful to overall microbial activity in soil as related to nutrient cycling (Wainwright, 1978; USDA, 1980; Goring and Laskowski, 1982). Decreased soil productivity, attributed to monoculture cash crops employing intensive use of chemical fertilizers and pesticides, more likely results directly from increased erosion, reduced plant cover, increased disease incidence, and greater export of nutrients in the harvested crop.

Alternate agricultural cropping systems are being sought to reduce on-farm inputs and offset rising production costs, decrease environmental and health hazards associated with the use of agricultural chemicals, and maintain soil fertility and productivity levels. Management systems using animal residues and legume and green-manure crops in rotation, as alternatives to synthetic chemicals, may require larger farm acreages to achieve cash grain production levels comparable to those with chemically intensive management (Council for Agricultural Science and Technology, 1980). Organic farming practices conserve C and N in the soil-plant system, and slower mineralization may result in mild N stress during periods of rapid plant growth (Power and Doran, 1984). Change from conventional to organic management systems may require a transition period during which the soil ecosystem adjusts to a new balance of biotic and abiotic components before comparable yields are achieved (Culik, 1983; Liebhardt et al., 1989). Limitations to crop yields during transition from conventional to alternative management include (i) reduced N availability, (ii) increased weed infestations, and (iii) incomplete user adjustment to new management techniques (USDA 1980; Radke and Liebhardt, 1988).

In this chapter we discuss the major influences of agricultural management practices on soil biological activity as it relates to nutrient cycling. We emphasize changes in C and N inputs, the soil physical environment, and synthetic chemical use that affect soil microbial and faunal activity. Examples of these relationships are largely confined to comparisons between conventional, chemically intensive cash grain production systems and alternative management systems employing animal manures, legume cover crops, and crop rotations for supply of nutrients and control of crop pests. Although discussion and examples are predominantly confined to management of field corn (*Zea mays* L.) and soybean [*Glycine max* (L.) Merrill], the principles presented are applicable to other agroecosystems as well. Major points for consideration include how our choice of agricultural management influences the biological balance of agricultural ecosystems and whether we should dominate the system or utilize soil biological resources to produce cash grain crops.

MANAGEMENT EFFECTS ON SOIL MICROFLORA AND FAUNA

Sustained agricultural productivity may depend on our ability to select management practices that enhance soil biological function in the fixation of atmospheric N and recycling of nutrients. The essence of conventional agriculture is to augment these functions with controlled physical manipulations and synthetic chemical inputs. Sustainable forms of agriculture aim to reduce external inputs and to utilize management practices that enhance soil ecological processes important to internal recycling of nutrient resources. All agricultural practices exert selection pressures on soil biota. Biotic populations may increase or decrease in abundance and activity in response to a specific practice. It is these responses that determine the structure and function of the soil community.

The soil is home to a diverse array of organisms. Most soil microorganisms are saprophytic and obtain nutriment directly from preformed organic substrates. Others are symbionts, predators, parasites, or autotrophs which derive energy from inorganic chemicals or sunlight. The five major groups of soil microorganisms include bacteria, fungi (including yeasts), actinomycetes, algae, and protozoans (Atlas and Bartha, 1987). The soil fauna is dominated by invertebrate animals, including nematodes, mites, Collembola and other insects, and earthworms. Some nematodes are plant root parasites, but most soil-dwelling species participate in organic-matter decomposition (Freckman and Caswell, 1985). Nonparasitic nematodes are especially abundant in agroecosystems when organic-matter levels are high (Van Gundy and Freckman, 1977; Yeates, 1979). Soil microarthropods (primarily mites and Collembola) influence microorganisms, nematodes, and other microinvertebrates through direct and selective feeding, organic residue incorporation, and inoculation of fresh residues (Werner and Dindal, 1987). Earthworms and other macroinvertebrates in the soil influence decomposition processes and have significant, positive effects on soil structure development (Lee, 1985).

Agricultural management influences the activity of soil microorganisms and soil fauna predominantly through changes in the soil environment, which limits biological activity. The most common limiting factors include soil temperature, water and aeration, availability and positioning of food sources, and an adequate physical and chemical habitat for growth and activity. The relative importance of a given management system on soil environment and thus biological activity, however, varies with soil, climate, and cropping history as do the relative predominance and importance of the aforementioned limiting factors. Consequently, generalizations on the effects of agricultural management on soil biological activity, although useful in evaluating the short- and long-term effects on biological productivity, can be misleading if not qualified for different soils, climates, and vegetation that exist within temperate regions.

To facilitate our discussion, we divide management effects on soil biological activity into three categories: tillage and residue management, use of synthetic chemicals, and crop rotations and animal manure. These categories are not mutually exclusive and in some instances overlap considerably.

Tillage and Residue Management

Physical disturbance of the soil caused by tillage and residue management is a crucial factor in determining biotic activity and species diversity in agroecosystems. Tillage usually disturbs at least 15 to 25 cm of surface soil and replaces stratified surface soil horizons with a tilled zone more homogeneous with respect to physical characteristics and residue distribution. The loss of a stratified soil microhabitat causes a decrease in the diversity of species that can inhabit agroecosystems. Various methods of tillage cause different degrees of disturbance, and regular disturbances by tillage prevent long-term successional development of soil ecosystems. Tillage effects are often related to the size of soil organisms. Single-cell microorganisms may respond positively to temporary increases in aeration and substrate availability created by tillage, while macroinvertebrates may be selected against as a result of the disruption of microhabitats and direct physical damage.

Reduced tillage, which maintains crop residues near the soil surface, is very effective in reducing wind and water erosion of soil. It also generally results in cooler, wetter, initially more compact, and less aerobic soil environment than that with conventional tillage. Conventional tillage, particularly with the moldboard plow, inverts and mixes the soil and distributes crop residues and organic matter to a greater depth in soil. By contrast, organic nutrient reserves with reduced tillage are stratified, with concentrations of organic matter and microbial populations being greatest near the soil surface. Stratification of crop residues, organic matter, and soil organisms often slows cycling of N as compared with conventional tillage with the moldboard plow (House et al., 1984; Doran, 1987). Increased microbial immobilization of soluble N in the surface of reduced tillage soils may require modified fertility or tillage management practices for optimal growth and yield of grain crops (Fox and Bandel, 1986; Doran, 1987).

The wetter, more compact soil environment often associated with reduced tillage management can also change the relative predominance of aerobic and anaerobic microbial populations in soil and, under certain conditions, result in greater losses of plant-available N through microbial denitrification. Greater soil water content and reduced air-filled porosity, shortly after rainfall or irrigation, may enhance denitrification and gaseous loss of NO_3-N in no-tillage as compared with plowed soils (Aulakh et al., 1982; Rice and Smith, 1982). The potential for less aerobic conditions with reduced tillage is site-specific and depends on climate, soil porosity, drainage characteristics, and the quantity of crop residues maintained on the soil surface (Doran and Smith, 1987). Significant reductions of soil aeration and resultant changes in biological activity with no tillage may be limited to poorly drained soils (Blevins et al., 1984).

Populations of most faunal groups are initially reduced by tillage (Sheals, 1956; Wallwork, 1976; Andren and Lagerlof, 1980; Loring et al., 1981). Tillage favors organisms with short life cycles, rapid dispersal, and small body size (Hendrix et al., 1986). Species with these characteristics may actually benefit from mechanical disturbance (Fjellberg, 1985). Populations of *Isotomodes productus* (Collembola), a species associated with drier conditions, increased slightly after cultivation (Sheals, 1956). The most successful groups in plowed soils may be those with a dormant life stage resistant to environmental stress (Wallwork, 1976).

Tillage has dramatic effects on the mite community in soil, and Oribatida especially are not tolerant of such disturbances (Sheals, 1955). Larger Mesostigmata, which tend to occupy surface litter habitats, are fairly rare in cultivated soils. Smaller species, especially Rhodacaridae, which are morphologically adapted for maneuvering through narrow soil pores, are more abundant (Sheals, 1956; Moore et al., 1984). Microarthropod numbers tend to be higher in no-tillage than in conventionally plowed soils (House and Parmalee, 1985; Hendrix et al., 1986), due in part to the increase of microhabitat diversity associated with residue buildup.

Earthworms are not favored by tillage. In general, the greater the intensity and frequency of tillage, the lower the population density of earthworms (Barnes and Ellis, 1979; Gerard and Hay, 1979; Edwards, 1980; Mackay and Kladivko, 1985). Lowered soil moisture caused by cultivation tends to influence earthworms negatively (Zicsi, 1967). Earthworms reported from agricultural soils in the temperate United States are almost always members of the introduced European family Lumbricidae, which includes species adapted to disturbance, low organic-matter content, and a lack of surface litter. The range and function of native species (Megascolecidae) is not well known (Fender and McKey-Fender, 1990).

Residue has an important effect on organic substrate availability and soil microclimatic characteristics. In a Pennsylvania study, corn residues chopped and left as a mulch supported higher populations of surface feeding earthworms, regardless of whether plots were conventionally or organically managed (Werner, 1990). Organic plots where all above-ground plant matter was harvested had smaller earthworm populations. Soils unprotected by a surface mulch will freeze much faster than mulched soils, and earthworm mortality increases in the absence of a gradual period of adjustment to decreasing temperatures (Slater and Hopp, 1947).

Synthetic Chemical Use

In the past three decades, increasing amounts of commercial fertilizers and an enormous array of biocide formulations have been applied to agricultural cropland in attempts to control unwanted organisms and increase crop yield. Commercial fertilizer use in the United States expanded steadily through 1981, when a record 21 million metric tons were consumed, over half of which was fertilizer nitrogen (USDA, 1988). Currently, more than 340,000 metric tons of biocides are used annually in U.S. agriculture [Eichers and Serletis, 1982 (in Pimentel and Levitan, 1986)]. Some of these chemicals are well known for their negative effects on nontarget organisms, including humans. For instance, organochlorine biocides accumulate in earthworm tissues and may reach levels toxic to birds that feed on earthworms (Lee, 1985). Alachlor and atrazine, the two most widely used herbicides in American agriculture today (Gianessi, 1986), are under scrutiny for possible toxic effects on human health.

The varied effects of management on physical and chemical soil environmental factors complicate direct evaluation of the effects of synthetic fertilizers and pesticides on soil biological activity and long-term productivity. Practices using plant or animal residues as sources of nutrients often increase soil organic-matter levels and associated soil biological activity. Where inorganic fertilizer use results in increased plant productivity and greater associated inputs of carbon, soil organic-matter and microbial activity levels are often greater than those of unamended soil (Martyniuk and Wagner, 1978). Long-term studies in Missouri indicate that crop yields and efficiency of N recovery in crops and soil organic matter are greater where a combination of animal manure and inorganic fertilizer are used than where either nutrient source is used alone or in large amounts (Smith, 1942).

Short-term (6 months) biological effects associated with the use of fertilizers or pesticides are generally related to disturbance of the chemical and biological balance of soil. The microbial processes of N fixation, nitrification, or mineralization of N and P can be temporarily repressed by excessively high levels of inorganic N or P, localized changes in soil pH near fertilizer sources, or changes in plant cover associated with herbicide use (Wainwright, 1978; Hauck, 1981; Goring and Laskowski, 1982; Biederbeck et al., 1987).

The chemical and physical form of commercial fertilizers, how they are applied to soil, and intensity of fertilization are dominant factors determining effects on soil microbial processes. The microsite at the center of a dissolved fertilizer granule may contain 1,000 to 10,000 times greater concentrations of inorganic solutes than those in unamended soil solutions (Hauck, 1981). Also, microsite changes in soil pH from alkalinity or acidity resulting from hydrolysis or nitrification of fertilizers can, depending on initial soil pH, markedly influence microbial activity and resultant soil fertility.

The application of anhydrous ammonia initially kills many soil microorganisms in the zone of application, but bacterial and actinomycete populations in acid and neutral soils recover within one to two weeks to levels severalfold higher than those of untreated soil (Eno and Blue, 1954). Populations of soil fungi may take as long as seven weeks to recover to population levels comparable to those of untreated soil. Although application of anhydrous ammonia normally affects a volume of soil equivalent to less than 10% of the plow zone, repeated applications may reduce surface soil organic-matter levels by organic C mobilization through alkaline hydrolysis and downward leaching; this effect may be increased by the presence of phosphate fertilizers (Myers and Thien, 1988). Farmers commonly report that the long-term use of synthetic fertilizers, especially anhydrous ammonia, can lead to soil compaction and poor tilth. At present, however there is little experimental evidence to support this supposition (Stone et al., 1982).

Indiscriminate use of commercial fertilizers can greatly reduce microbial activity and soil fertility. Pokorna-Kozova (1984) demonstrated that systematic treatment of uncropped soils with extremely high levels of N, P, and K fertilizers (480–980 kg N/ha) for long periods dramatically reduces microbial populations and cellulolytic activities, mineralization of added organic C and N, and free-living populations of N_2-fixing *Azotobacter*. The major mechanisms for impairment of microbial activity in this study were soil acidification (pH = 4.0) and salt damage resulting from overfertilization. This is an extreme case, however, which probably would not be encountered under conditions more typical of good agronomic management.

TABLE 7.1 Influence of Fertilizer Sources on Surface Soil Organic C Content and Biological Activities for a Wheat-Clover-Grass-Potato-Beet Rotation (Adapted from Brinton, 1979).

Fertility Treatment	Total N Applied	Soil Organic C	Relative Activity* Earth-worms	Soil Respiration	Soil Dehydrogenase
	kg N/ha	*(%)*			
Control	0	2.3	1.0	1.0	1.0
Raw manure	93	2.6	5.0	1.3	1.8
1/2 Manure + 1/2 NPK	61	2.1	2.4	1.1	1.4
NPK (low)	28	2.5	1.7	1.0	1.1
NPK (high)	111	2.7	0.7	1.0	1.2

* Activity relative to control (no treatment). N added as calcium nitrate or ammonium phosphate, P as superphosphate, K as potassium sulfate.

The influence of organic versus inorganic fertilizers on soil biological activity was evaluated in an 18-year study for a wheat-clover-grass-potato-beet rotation at Jarna, Sweden (Brinton, 1979). As shown in Table 7.1, the organic C levels of soils treated with manure or inorganic fertilizer were similar, but higher than the untreated soil. Soil respiration and dehydrogenase enzyme activities, both measures of soil biological activity, were stimulated by application of manure or synthetic fertilizers. Earthworm activity, as estimated by the predominance of wormholes in soil, was increased by the application of manure or low rates of synthetic fertilizers but then reduced by high fertilizer rates, presumably the result of higher salt concentrations in the soil solution.

Heavy application of inorganic fertilizers may cause immediate reductions in earthworm abundance (Edwards, 1980). Application of NH_4NO_3 (150 kg N/ha) caused declines in number of enchytraeids, nematodes, and Collembola in a forest soil, indicating a direct toxic effect (Huhta et al, 1983). Application of urea or superphosphate in another study had a similar negative effect on mites, Collembola, and fungi (Carlyle and Than, 1987). The application of inorganic fertilizers to experimental plots at the Rothamsted Experiment Station in England for 118 years caused decreases in earthworms and microarthropods (Edwards and Lofty, 1974). This may point to the long-term negative effect on soil fauna of moderate applications of inorganic fertilizers, an effect not seen in more typical short-term ecological studies.

Biocide effects on soil biota have been reviewed for invertebrates (Edwards and Thompson, 1973), earthworms (Lee, 1985), algae (McCann and Cullimore, 1979), and other microorganisms (Martin, 1963; Parr, 1974; Wainwright, 1978). The general consensus following more than 30 years of research is that pesticides, with the exception of fumigants and fungicides, have very few long-term deleterious effects on soil microbial processes and major biochemical cycles when applied at recommended field rates (Parr, 1974; Tu, 1978; Wainwright, 1978; Goring and Laskowski, 1982). In an excellent review of interactions between pesticides and microorganisms, Bollen (1961) concluded that the use of herbicides or insecticides at field application rates (2–3 ppm) had little effect on either soil respiration or the decomposition of added or native organic matter. Insecticides greatly reduce populations of earthworms and soil animals, but have very little deleterious effect on microorganisms and their activities in maintaining soil fertility (Martin, 1963; Tu, 1978).

Herbicide use can suppress activity of some soil microorganisms, particularly bacteria responsible for ammonium oxidation (nitrification) and nitrogen fixation. Field application of several herbicides can significantly reduce nitrification in soil for up to three or four months (Bollen, 1961; Chandra, 1964). Repeated applications, however, have less effect as soil microorganisms capable of decomposing the added herbicides become more predominant. Deleterious effects of many herbicides on symbiotic N_2 fixation have been attributed to reduced nodulation of plant roots and not a direct effect on the *Rhizobium* symbiont (Bollen, 1961). Recent comparisons between nonchemical and chemically intensive management systems in the subhumid Midwest and humid eastern United States confirm that cultural management affects soil biology more than chemical use. Type of crops grown in rotation or application of animal manure were the major factors regulating soil microbial populations and activities; little or no direct effect was observed from use of synthetic fertilizers, herbicides, or insecticides (Doran et al., 1987; Fraser et al., 1988). Increases in microbial and faunal activity in organic management systems, as compared with conventional systems using fertilizers and pesticides, result mainly from higher inputs of decomposable organic matter where animal manure or legumes are used to enhance soil fertility (Table 7.2). Greater soil microbial and faunal biomass and activity with organic management are also associated with buildup of potentially mineralizable N reserves in soil, enhancement of soil physical conditions, and better absorbance of added water.

Direct toxic effects of some herbicides on soil invertebrates have been demonstrated in lab assays. Such effects on soil biota, however, have not been convincingly demonstrated in the field (Edwards, 1980; Subagja and Snider, 1981; Mola et al., 1987). Yet reports of reduced invertebrates with herbicide use are common. Direct toxicological effects may play a role, but most authors conclude that reductions in biotic abundance or activity are mainly due to indirect effects caused by a loss in primary productivity, lowered organic matter input, and the loss of vegetative and litter cover (Edwards and Thompson, 1973; Greaves, 1979; Atkinson and Herbert, 1979; Edwards and Brown, 1982). This effect can be ameliorated by adding an organic mulch to the surface of herbicide-treated soils (Greaves, 1979).

TABLE 7.2 Biological and Physical Characteristics of Surface Soil for Conventional and Organic Fertility Management. Data Collected in July 1985 from Soils Cropped to Corn at Rodale Conversion Study in Southeastern Pennsylvania. Organic Treatments in Fifth Year after Conversion from Conventional Management.

Soil Measurement	Organic Using Crop Rotations		Conventional
	Manure	Hairy Vetch	Fertilizer & Pesticides
Microbial biomass (kg C/ha-7.5cm)	302	306	266
Potentially mineralizable N (kg N/ha-7.5cm)	1,110	1,220	1,020
Soil respiration (μg CO_2/g soil/24h)	217	243	176
Water infiltration (L/m^2/min)	38	38	24
Earthworms			
numbers (#/m^2)	11	4	5
biomass (g/m^2)	12	7	6
Microarthropods*			
saprovores (#/m^2)	51,000	33,000	30,000
predators (#/m^2)	2,000	2,500	2,300

* Saprovores include mites (Oribatada, Prostigmata, Astigmata) and Collembola; Predatory mites (Mesostigmata).

By way of illustration, microbial populations and activities in the surface of no-tillage soils treated with herbicides are often significantly higher than those in conventionally tilled soils (Doran, 1980). In this case, biological benefit from conservation of surface moisture and organic matter with reduced tillage exceeded detrimental effects of chemical use.

Methods by which pesticides are applied may have unique effects on certain groups of soil animals. Surface-feeding earthworms such as *Lumbricus terrestris* (L.) are most susceptible to surface applications of the herbicide benomyl and are less affected by incorporation of the biocide into the soil. *Lumbricus terrestris* forms permanent burrows and does not come into contact with subsurface soil beyond its burrow. However, subsurface species such as *Aporrectodea turgida* (Eisen), which continuously extend their burrows as they feed within soil, are most susceptible when benomyl is incorporated (Edwards and Brown, 1982). Surface dwelling Collembola may be more effected by herbicides than species that live in mineral soil layers (Loring et al., 1981; Moore et al., 1984).

Soil fungicides and fumigants cause the most drastic effect on the soil microflora. These chemicals are intentionally applied to soils as antimicrobial agents and at much higher rates (30 to 40 ppm) than herbicides or insecticides. While their action is directed towards pathogenic fungi and plant parasitic nematodes, it is seldom limited to pathogens. Fungicide application greatly alters microbial balance in soil; fungal populations may decrease two- to three-fold and numbers of heterotrophic bacteria rise as a result of decreased competition and increased mineralization of killed fungal cells (Anderson et al., 1981; Duah-Yetumi and Johnson, 1986). The overall effect of fungicide use is one of partial soil sterilization, in which microorganisms beneficial to soil tilth and fertility may be adversely affected for extended periods of time (Parr, 1974).

Crop Rotations and Animal Manure

Shifts in microbial and faunal numbers and activity are often related to changes in carbon inputs to soil as a result of management-associated changes in crop type or residue addition. Long-term benefits from use of animal manure or crop rotations include better soil tilth, improved water infiltration, and increased soil organic matter content (Smith, 1942; Odell et al., 1982). Systems that increase below-ground inputs of C and N through inclusion of legumes and/or fibrous rooted crops in rotation often increase microbial populations and activity more than that observed for conventional systems using commercial fertilizers (Bolton et al., 1985; Doran et al., 1987; Fraser et al., 1988).

Increasing the temporal or spatial diversity of plants in an agroecosystem tends to prevent the buildup of crop-specific pests above economically damaging threshold levels (Risch et al., 1983). Spatial diversity can be created by growing crop mixtures (polycultures or intercrops) or by tolerating some weed growth. Temporal diversity is created by growing different crops in rotation (Altieri, 1987). While increased spatial diversity is likely to increase soil faunal diversity, temporal diversity brought about by a sequence of crops may decrease soil animal diversity even more than continuous monoculture (Edwards and Lofty, 1969).

Each plant species influences soil ecosystems through the type and amount of organic matter released by roots, the type and amount of litter contributed to the soil, growth phenology, effects on soil microclimate, and characteristic mycorrhizal associations. While decomposer invertebrates are less discriminating than insect pest species in choosing food sources, the forementioned plant effects on the soil habitat greatly influence the decomposer community. Although choice of crop species probably has less effect on soil invertebrates than tillage or soil type (Wallwork, 1976), rotation does add to the level of disturbance in soil ecosystems (Kevan, 1962). Stinner et al. (1986) found that crop rotation in agroecosystems has a greater negative effect than insecticide use on microarthropods.

Monoculture tends to affect soil ecosystems more dramatically than other cropping systems (MacRae and Mehuys, 1985). Hopp and Hopkins

(1946) found the fewest earthworms in continuous corn culture, as compared to soybean or sod cover. Edwards (1980) found that earthworms tended to be more abundant in continuous cereal culture than in three or four crop rotations. He attributed these findings to the greater amount of organic matter added by the cereals.

Sustainable management systems generally include leguminous crops or green manures in rotation as a way to maintain soil fertility. Legumes as a nitrogen source in soils cropped with wheat support greater microbial biomass and activity, relative to inorganic nitrogen sources (Bolton et al., 1985).

The application of animal manures as nutrient sources generally increases the abundance and activity of soil biota. Microbial and protozoan activity is highest in organically fertilized agricultural soils (Tischler, 1955; Schnurer et al., 1985). Manure causes increases of saprovore, bacterivore, and predatory nematodes, while decreasing the abundance of plant parasites (Kevan, 1962; Marshall, 1973). Some predatory mites increase in abundance when manure is added to cropped soils, and may have a negative influence on nematode populations (Buhlmann, 1984). Collembola populations also tend to increase after manure application (Christiansen, 1964). The biological value of fresh manure was pointed out by Schnurer et al. (1985), who found that straw enhanced microbiological activity more than partially decomposed manure. Application of manure also increases the abundance and biomass of earthworms in cropped soils (Edwards and Lofty, 1974; Anderson, 1980).

NUTRIENT CYCLING

The success of agricultural cropping systems largely depends on an adequate supply of nutrients for optimum economic returns. The importance of nitrogen as the major element limiting growth of most cereal grains has been recognized for more than a century. Cost-effective sources of synthetic N fertilizer after World War II led to corn yields doubling between 1945 and 1980 (Power, 1987). Increased use of fertilizer and pesticides from nonrenewable energy sources largely replaced the use of renewable animal manures and legume-containing crop rotation for grain crop production. Increased dependence of agriculture on external sources of capital and energy, however, has contributed to environmental degradation and decreased profitability as a result of overproduction and lower commodity prices.

Alternative agricultural systems, which rely more on biological cycling of nutrients, may result in lower yields, but they often increase profitability by reducing capital inputs. The ultimate success of alternative production systems designed to create a sustainable agriculture will depend on our understanding of how soil biological activity can supply nutrients for crop production and maintain soil productivity.

Information on the functional role of soil animals in decomposition and nutrient cycling processes suggests that the increase of fungal- or bacterial-feeding soil invertebrates generally results in increased respiration, decomposition, and mineralization of N and P (Abrams and Mitchell, 1980; Coleman et al., 1983; Coleman, 1985). Where nutrient availability is limited, microbial-faunal interactions can increase mineralization and result in greater plant growth (Ingham et al., 1985). Research is needed to more thoroughly understand the role that faunal-microbial interactions play in regulating nutrient availability to plants in alternative management systems.

Conversion—Reducing Inputs

In low or reduced chemical input management systems, nutrient resources are internalized on the farm through use of legume-containing crop rotations and, where available, animal manures. Management-intensive cultural and biological practices replace synthetic chemical inputs and increase atmospheric N fixation, mineralization, and turnover of soil organic matter; these practices also control weeds, disease, and insects predominantly via biological control methods. Grain crop yields with reduced input management may be 10% to 50% lower, however, while the soil adjusts to new physical, chemical, and biological characteristics (USDA, 1980; Culik, 1983). Depending on climate and variations in tillage and residue management, levels of organic matter and associated organic C, N, and P contents in the surface 5 to 15 cm of low-input soils may be 10% to 40% greater than those of paired soils under conventional management (Bolton et al., 1985; Sahs and Lesoing, 1985; Doran et al., 1987). Increases in soil organic matter are also reflected by corresponding increases in soil microbial populations and activity. This is presumably the result of a more optimal soil physical environment and more continuous release of C and energy sources from animal and crop residues, as compared with conventional systems utilizing only one or two grain crops (Bolton et al., 1985; Fraser et al., 1988).

Availability of N, and perhaps other nutrients, during the conversion period from conventional to reduced-chemical management may limit the yield of crops such as corn. Culik (1983) reported that in southeastern Pennsylvania, corn yields in organic treatments during the second year of conversion from conventional management were 40% lower than those where fertilizer and herbicides were used. Decreased production resulted primarily from lack of adequate N and pressure from excessive weeds. In the same research study, Doran et al. (1987) reported soil nitrate levels at midseason that were highly correlated ($r^2 = 0.88$) with corn leaf N content at silking and were lowest with the organic management system.

Partitioning of Primary Productivity

Our most common measure of soil productivity is grain yield. Preoccupation with above-ground grain yield has diverted attention away from ecological benefits and an understanding of the overall effects and indirect benefits of various management systems. Ecological productivity of land is the net sum of above- and below-ground biological growth over an interval of time. As illustrated by the data presented in Table 7.3, the partitioning of above- and below-ground productivity between conventional and organic management systems in the early stages of conversion management can differ considerably. For this example, increases in dry matter and nitrogen of corn with conventional management were nearly counterbalanced by increases in weed growth and soil microbial biomass in the low-input organic management system. The slightly higher overall N yields of conventional management reflect the additional application of 123 kg/ha of fertilizer N, given an average recovery of 30% to 50%.

Increases in soil microbial biomass under alternative management are associated with increases in soil fungi and bacteria resulting primarily from additions of animal manure and crop food sources (Doran et al., 1987; Fraser et al., 1988). Type of crop grown can be important. Growth of red clover (*Trifolium pratense* L.) and oats (*Avena sativa* L.) increases soil microbial biomass levels more than corn or soybeans. This results from differences in length of time that the soil is cropped and the intensity of rooting patterns, as well as differences in the amounts of photosynthate C released from roots. Helal and Sauerbeck (1987) found that 43% of photosynthetically fixed C was transferred to corn (*Zea mays* L.) roots during the first 30 days of growth. They estimated that over 10% of this photosynthate was released to the soil and utilized by microorganisms. Martin (1987) found that 35% of wheat assimilate was translocated below ground between the two-leaf stage and harvest. This translocation varied with growth stage from a maximum of 50% at 5 weeks growth to a minimum of $< 5\%$ at 25 weeks. The type of crop grown in a management system and stage of plant growth may have a major influence on the size of the soil microbial biomass pool. Thus, during early transition from conventional management, more N and other elements may be partitioned into soil microbial biomass and may be less available for uptake by grain crops. The buildup of microbial and organic nutrient reserves, however, confers beneficial physical properties to soil and will likely become available for uptake by grain crops after a new equilibrium for soil organic-matter turnover is reached. Liebhardt et al. (1989) found that yields in a low-input organic system increased successively over a four-year period and by the fifth year were equivalent to those with conventional management.

Cropping rotation appears important to partitioning of above- and below-ground plant dry-matter and associated soil microbial biomass levels. In the central United States, Roder et al. (1988) found higher grain

TABLE 7.3 Above and Below-Ground (to 30-cm Soil Depth) Productivity of Corn, Weeds, and Microbial Biomass as Influenced by Farming Management Systems in the Second Year after Conversion from Conventional Management. (After Doran et al., 1987).

	Dry Matter Yields, kg/ha		Nitrogen Yields, kg N/ha*	
Biomass Component	Legume/Cash Grain Rotation	Conventional Cash Grain	Legume/Cash Grain Rotation	Conventional Cash Grain
Corn (grain, stover, roots)	12,910	22,550	127	242
Weeds (tops + roots)	6,280	210	51	2
Microbial Biomass (to 30 cm in soil)	2,620	1,980	157	119
Totals	21,810	24,740	335	363

* Nitrogen yields estimated assuming 1–1.2% N for corn stover, grain, and above-ground weeds; 0.7–0.8% N for roots; and 6% N for microbial biomass.

yields for sorghum (*Sorghum bicolor* [L.] Moench) and soybeans in rotation associated with a decreased production of root mass and lower soil microbial biomass levels as compared with either crop grown continuously. Photosynthate partitioning between above- and below-ground plant parts can also vary with level of inputs for crop production. Management-related increases in water and nutrient availability and weed and pest control alter the competitive nature of vegetative crop growth and generally require less diversion of photosynthate into root reserves. Crop varieties selected for high-input environments may not perform well under low-input management, where a greater proportion of plant photosynthate may be partitioned below ground to roots (Bramel-Cox, 1988).

Seasonal Availability of Nutrients

Nutrient supply to annual grain crops is most critical to growth and ultimate grain yield during the first 20 to 70 days after emergence when they take up over 60% of their requirements. For example, a 10 Mg/ha (180 bu/A) corn grain yield will require about 190 kg N/ha. About two-thirds will be required between the 25th and 75th day of growth. In conventional management, the farmer adds fertilizer N before or shortly after planting to supply N during this active growth phase. In alternative management systems, however, N supply to the crop depends largely on net gain in available soil N resulting from balance of mineralization and immobilization processes in soil. Levels of available N may at times be less than required for optimal growth and grain yield.

In reduced input alternative management systems, seasonal turnover of soil microbial biomass and other mineralizable soil organic N forms appears important to identifying management techniques that will supply adequate N for crop production (Patriquin, 1986). This approach is complicated because microbial growth and concurrent immobilization of N is enhanced by plant growth and by the same soil environmental conditions that favor plants. The varying effects of these relationships are illustrated in Figure 7.1. The data, taken from the Rodale Research Center Conversion Project, compare conventional and low-input management in the fifth year of conversion from conventional management. In the early growing season, before and near the time of corn planting, soil microbial biomass and potentially mineralizable N levels for low-input management with a hairy vetch cover crop averaged 62 and 250 kg N/ha greater, respectively, than those for conventional management. During the same period, soil nitrate-N levels for the low-input system with vetch were considerably less than the 36 kg N/ha level proposed by Magdoff et al. (1984) as limiting for corn yield in the northeastern United States. The large increase in soil nitrate-N (103 kg N/ha) after plow down of hairy vetch corresponded with declines in microbial biomass N and potentially mineralizable N of 52 and 74 kg N/ha, respectively. Midseason declines in available nitrate-N

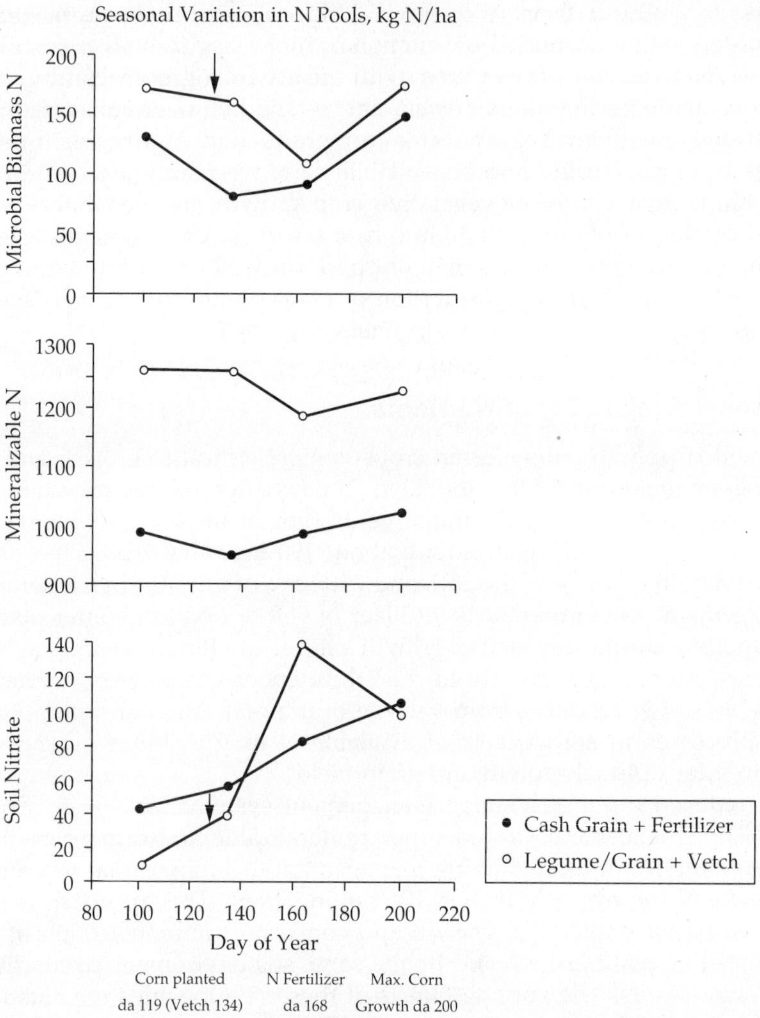

Figure 7.1 Surface soil (0–30 cm) fluctuations in soil microbial biomass N, potentially mineralizable N, and available nitrate-N contents as influenced by input management system during 1985 growing season. Arrow denotes date on which hairy vetch cover crop was plowed down in low-input rotation system (After Doran et al., 1987).

between day 160 (June 12) and 200 (July 19) resulted in part from uptake by the corn crop, which amounted to 30 to 70 kg N/ha. It is interesting to note that the increase in soil microbial biomass N during this period, apparently stimulated by corn growth, may also have contributed to the decline in available soil N. These data support the importance of understanding the synchrony between microbial and plant growth during the growing season and their association with immobilization/mineralization phenomena in low-input systems. Inclusion of legumes in a cropping system influences both the quantity and quality of soil microbial biomass and organic matter (Biederbeck et al., 1984). Subsequent mineralization of these plant-generated pools of soil organic N appears closely related to seasonal fluctuations in the soil physical environment (McGill et al., 1986). The economic production of grain crops in low-input management systems depends on our ability to synchronize supply of available N with periods of maximum need by the crop plant.

CONCLUSION

The biological balance of sustainable management systems, which depend on biological diversity for pest control and internal recycling of nutrients, differs from that of conventional management, where pest control and nutrients are supplied by external inputs of synthetic chemicals. Differences in levels of soil microbial and faunal populations and activity between systems appear largely controlled by varied inputs of organic carbon and other nutrients through addition of plant products or animal manure, and they are less influenced by the use of synthetic chemicals *per se*.

Lower grain yields during conversion from conventional to sustainable management with lower chemical inputs result in part from redistribution of C and N in nutrient sinks, such as increased weeds, soil microbial and faunal biomass, and perhaps a greater proportion of plant root mass. Crop yields in low-input management systems with increased time may well approach those with conventional management as a greater proportion of these soil pools of organic C and N build and mineralize more during the growing season.

Increased biological diversity and pool sizes with sustainable management will require understanding the balance between buildup of organic reserves in soil and the release of combined nutrients in forms available for grain crops. Understanding the synchrony between release of organic nitrogen (and other elements) and growth requirements for grain crops is important to the economic success of sustainable systems. Research is needed to devise tillage and residue management practices that optimize turnover of soil biological pools for nutrient release and increased availability to the crop during the growing season without depleting soil N reserves.

REFERENCES

Abrams, B. I., and M. J. Mitchell. (1980). "Role of nematode-bacterial interactions in heterotrophic systems with emphasis on sewage sludge decomposition," *Oikos* 35:404–10.

Altieri, M. A. (1987). *Agroecology, the Scientific Basis of Alternative Agriculture.* Westview Press, Boulder, CO.

Anderson, C. (1980). "The influence of farmyard manure and slurry on the earthworm population (Lumbricidae) in arable soil." In *Soil Biology as Related to Land Use Practices* (Ed. D. L. Dindal). pp. 325–35. EPA, Washington, DC.

Anderson, J. P. E., R. A. Armstrong, and S. N. Smith. (1981). "Methods to evaluate pesticide damage to the biomass of the soil microflora," *Soil Biol. Biochem.* **13**:149–53.

Andren, O., and J. Lagerlof. (1980). "The abundance of soil animals (Microarthropoda, Enchytraeidae, Nematoda) in a crop rotation dominated by a ley and in a rotation with varied crops." In *Soil Biology as Related to Land Use Practices* (Ed. D. L. Dindal). pp. 274–79. EPA, Washington, DC.

Atkinson, D., and R. F. Herbert. (1979). "Effects (of herbicides) on the soil with particular reference to orchard crops," *Ann. Appl. Biol.* **91**:125–29.

Atlas, R. M., and R. Bartha. (1987). *Microbial Ecology, Fundamentals and Applications,* 2nd ed. Benjamin/Cummings, Menlo Park, CA.

Aulakh, M. S., D. A. Rennie, and E. A. Paul. (1982). "Gaseous nitrogen losses from cropped and summer-fallowed soils," *Can. J. Soil Sci.* **62**:187–96.

Barnes, B. T., and F. B. Ellis. (1979). "Effects of different methods of cultivation and direct drilling and disposal of straw residues on populations of earthworms," *J. Soil Sci.* **30**:669–79.

Biederbeck, V. O., C. A. Campbell, and A. E. Smith. (1987). "Effects of long-term 2,4-D field applications on soil biochemical processes," *J. Environ. Qual.* **16**: 257–62.

Biederbeck, V. O., C. A. Campbell, and R. P. Zenter. (1984). "Effect of crop rotation and fertilization on some biological properties of a loam in southwestern Saskatchewan," *Can. J. Soil Sci.* **64**:335–67.

Blevins, R. L., M. S. Smith, and G. W. Thomas. (1984). "Changes in soil properties under no-tillage." In *No-tillage Agriculture* (Eds. R. E. Phillips and S. H. Phillips). pp. 190–230. Van Nostrand Reinhold, New York.

Bollen, W. B., (1961). "Interaction between pesticides and soil microorganisms," *Ann. Rev. Microbiol.* **15**:69–92.

Bolton, H. Jr., L. F. Elliott, R. I. Papendick, and D. F. Bezdicek. (1985). "Soil microbial biomass and selected enzyme activities: Effect of fertilization and cropping practices," *Soil Biol. Biochem.* **17(3)**:297–302.

Buhlmann, A. (1984). "Influence of agricultural practices on the population of *Alliphis halleri* (G.& R. Canestrini, 1881) (Acari: Gamasina)," *Acarology* **6(2)**: 901–9.

Bramel-Cox, P. (1988). "Resource efficient hybrids and varieties." In *Sustainable Agriculture in the Midwest* (Eds. C. A. Francis and J. W. King). pp. 21–23. North Central Regional Conference, March 22, 1988. Institute of Agriculture and Natural Resources, Lincoln, NE.

Brinton, W. F. Jr. (1979). *Effects of Organic and Inorganic Fertilizers on Soils and Crops.* (Results of a long-term field experiment in Sweden, translated from Swedish and German of the original work by B. D. Pettersson and E. V. Wistinghaussen.) Misc. Pub. no. 1., Woods End Agric. Inst., Temple, ME.

Carlyle, J. C. and U. B. Than. (1987). "Influence of urea and superphosphate fertilization on microarthropod numbers and fungal activity during a short-term incubation of material from a *Pinus radiata* D. Don forest floor," *Plant & Soil* **103**:143–46.

Chandra, P. (1964). "Herbicidal effects on certain soil microbial activities in some brown soils of Saskatchewan," *Weed Res.* **4**:54–63.

Christiansen, K. (1964). "Bionomics of Collembola," *Ann. Rev. Entomol.* **9**:147–78.

Coleman, D.C. (1985). "Through a ped darkly: An ecological assessment of root-soil-microbial-faunal interactions." In *Ecological Interactions in Soil: Plants, Microbes, and Animals* (Eds. A. H. Fitter, D. Atkinson, D. J. Read, and M. B. Usher). pp. 1-21. British Ecol. Soc. Spec. Pub. 4, Blackwell, Oxford.

Coleman, D. C. and A. Sasson. (1978). "Decomposer subsystem," In *Grasslands, Systems Analysis and Man* (Eds. A. J. Breymeyer and G. M. Van Dyne). pp. 609–55. IBP 19, Cambridge Univ. Press, Cambridge.

Coleman, D. C., C. P. P. Reid and C. V. Cole. (1983). "Biological strategies of nutrient cycling in soil systems," *Adv. Ecol. Res.* **13**:1-55.

Crossley, D. A. Jr. (1977). "The roles of terrestrial saprophagous arthropods in forest soils: Current status of concepts." In *The Role of Arthropods in Forest Ecosystems* (Ed. W. J. Mattson). pp. 49–56. Springer Verlag, New York.

Council for Agricultural Science and Technology. (1980). *Organic and Conventional Farming Compared,* Report No. 84. C.A.S.T., Ames, IA. p. 32.

Culik, M. N. (1983). "The conversion experiment: Reducing farming costs," *J. Soil Water Conserv.* **38**:333–35.

Doran, J. W. (1980). "Soil microbial and biochemical changes associated with reduced tillage," *Soil Sci. Soc. Am. J.* **44**:765–71.

Doran, J. W. (1987). "Microbial biomass and mineralizable nitrogen distributions in no-tillage and plowed soils," *Biol. Fertil. Soils* **5**:68–75.

Doran, J. W., D. G. Fraser, M. N. Culik, and W. C. Liebhardt. (1987). "Influence of alternative and conventional agricultural management on soil microbial processes and nitrogen availability," *Am. J. Alt. Agric.* **2**:99–106.

Doran, J. W., and M. S. Smith. (1987). "Organic matter management and utilization of soil and fertilizer nutrients." In *Soil Fertility and Organic Matter as Critical Components of Production Systems* (Eds. R. F. Follett, J. W. B. Stewart, and C. V. Cole). p. 53–72. Spec. Pub. No. 19, ASA, Madison, WI.

Duah-Yetumi, S., and R. B. Johnson. (1986). "Changes in soil microflora in response to repeated application of some pesticides," *Soil Biol. Biochem.* **18**: 629–35.

Edwards, C. A. (1980). "Interactions between agricultural practice and earthworms." In *Soil Biology as Related to Land Use Practices* (Ed. D.L. Dindal). pp. 3–12. Proc. VII Intl. Soil Zool. Colloq., EPA, Washington, DC.

Edwards, C. A., and J. R. Lofty. (1974). "The invertebrate fauna of the Park Grass plots, I. Soil fauna." *Rothamsted Report,* Part 2. pp. 133–54. Rothamsted Experiment Station, Harpenden, England.

Edwards, C. A. and J. R. Lofty. (1969). "The influence of agricultural practice on soil microarthropod populations." In *The Soil Ecosystem* (Ed. J. G. Sheals). pp. 237–47. Systematics Assoc. Pub. 8.

Edwards, C. A. and A. R. Thompson. (1973). "Pesticides and the soil fauna." *Residue Rev.* **45**:1–79.

Edwards, P. J. and S. M. Brown. (1982). "Use of grassland plots to study the effects of pesticides on earthworms," *Pedobiologia* **24**:145–50.

Eichers, T. R. and W. S. Serletis. (1982). "Farm pesticide supply-demand trends, 1982," *Agric. Econ. Rep.* **485**:1-23.

Eno C. F. and W. G. Blue. (1954). "The effect of anhydrous ammonia on nitrification and the microbiological population in sandy soils." *Soil Sci. Soc. Amer. Proc.* 18:178–81.

Fender, W. M., and D. McKey-Fender. (1990). "Oligochaeta: Megascolecidae and other earthworms in western North America." In *Soil Biology Guide* (Ed. D. L. Dindal). Wiley, New York (in press).

Fjellberg, A. (1985). "Recent advances and future needs in the study of Collembola biology and systematics," *Quaest. Entomol.* **21(4)**:559–70.

Fox, R. H. and V. A. Bandel. (1986). "Nitrogen utilization with no-tillage." In *No-tillage and Surface-tillage Agriculture: The Tillage Revolution* (Eds. M. A. Sprague and G. B. Triplett). pp. 117–48. Wiley, New York.

Fraser, D. G., J. W. Doran, W. W. Sahs, and G. W. Lesoing. (1988). "Soil microbial populations and activities under conventional and organic management," *J. Environ. Qual.* (in press).

Freckman, D. and E. P. Caswell. (1985). "The ecology of nematodes in agroecosystems," *Ann. R. Phyto.* **23**:275–96.

Gerard, B. M. and R. K. M. Hay. (1979). "The effect on earthworms of ploughing, tined cultivation, direct drilling, and nitrogen in a barley monoculture system." *J. Agric. Sci. (Camb.)* **93**:147–55.

Gianessi, L. P. (1986). "A national pesticide usage data base." A report to the office of standards and regulations, USEPA Renewable Resources Division, Resources for the Future, Washington, DC, p. 14.

Gliessman, S. R. (1984). "An agroecological approach to sustainable agriculture." In *Meeting the Expectations of the Land* (Eds. W. Jackson, W. Berry, and B. Colman). pp. 160–71. North Point Press, San Francisco.

Goring, C. A. I. and D. A. Laskowski. (1982). "The effects of pesticides on nitrogen transformations in soils." In *Nitrogen in Agricultural Soils* (Ed. F.J. Stevenson). pp. 689–720. Agron. Monograph 22, Am. Soc. Agron., Madison, WI.

Greaves, M. P. (1979). "Long-term effects of herbicides on soil microorganisms," *Ann. Appl. Biol.* **91**:129–32.

Hallberg, G. R. (1987). "Agricultural chemicals in ground water: Extent and implications," *Am. J. Alt. Agric.* **2**:3–15.

Hauck, R. (1981). "Nitrogen fertilizer effects on nitrogen cycle processes." In "Terrestrial nitrogen cycles" (Eds. F. E. Clark and T. Rosswall). *Ecol. Bull.* **33**:551–62. Swedish Natural Science Research Council, Stockholm, Sweden.

Helal, H. M. and D.Sauerbeck. (1987). "Direct and indirect influences of plant roots on organic matter and phosphorus turnover in soil," *Intecol Bull.* **15**:49–58. The International Association for Ecology, Athens, Georgia.

Hendrix, P. F., R. W. Parmalee, D. A. Crossley, Jr., D. C. Coleman, E. P. Odum, and P. M. Groffman. (1986). "Detritus food webs in conventional and no-tillage agroecosystems," *BioSci.* **36(6)**:374–80.

Hopp, H. and H. J. Hopkins. (1946). "The effect of cropping systems on the winter population of earthworms." *J. Soil Water Conserv.* **1**:85–88.

House, G. J. and R. W. Parmalee. (1985). "Comparison of soil arthropods and earthworms from conventional and no-tillage agroecosystems." *Soil Tillage Res.* **5**:351–60.

House, G. J., B. J. Stinner, D. A. Crossley Jr., E. P. Odum, and G. W. Langdale. (1984). "Nitrogen cycling in conventional and no-tillage agroecosystems in the Southern Piedmont," *J. Soil Water Conserv.* **39**:194–200.

Huhta, V., R. Hyvonen, A. Koskennieme, and P. Vilkamaa. (1983). "Role of pH in the effect of fertilization on Nematoda, Oligochaeta, and Microarthropoda." In *New Trends in Soil Biology* (Eds. P. Lebrun, H. M. Andre, A. De Medts, C. Gregoire-Wibo, and G. Wauthy). pp. 61–73. Dieu-Brichart, Louvain-la-Neuve, Belgium.

Ingham, R. E., J. A. Trofymow, E. R. Ingham, and D. C. Coleman. (1985). "Interactions of bacteria, fungi, and their nematode grazers: Effects on nutrient cycling and plant growth," *Ecol. Monogr.* **55(1)**:119–40.

Juma, N. G. and W. B. McGill. (1986). "Decomposition and nutrient cycling in agroecosystems." In *Microfloral and Faunal Interactions in Natural and Agro-Ecosystems* (Eds. M. J. Mitchell and J. P. Nakas). pp. 74–136. Martinus Nijhoff Junk, The Netherlands.

Kevan, D. K. MCE. (1962). *Soil Animals*. Philosophical Library, New York. p. 237.

Lee, K. E. (1985). *Earthworms, Their Ecology and Relationships with Soils and Land Use.* Academic Press, New York.

Liebhardt, W. C., R. W. Andrews, M. N. Culik, R. R. Harwood, R. R. Janke, J. K. Radke, and S. R. Swartz. (1989). "Crop production during conversion from conventional to low-input methods," *Agron. J.* **81**:150-159.

Loring, S. J., R. J. Snider, and L. S. Robertson. (1981). "The effects of three tillage practices on Collembola and Acarina populations," *Pedobiologia* **22**:172–84.

Mackay, A. D. and E. J. Kladivko. (1985). "Earthworms and rate of breakdown of soybean and maize residues in soil," *Soil Biol. Biochem.* **17(6)**:851–57.

MacRae, R. J. and G. R. Mehuys. (1985). "The effect of green manuring on the physical properties of temperate-area soils." *Adv. Soil Sci.* **3**:71–94.

Magdoff, F. R., D. Ross, and J. Amadon. (1984). "A soil test for nitrogen availability to corn," *Soil Sci. Soc. Amer. J.* **48**:1301–4.

Marshall, V. G. (1973). *Effects of manures and fertilizers on soil fauna: A review.* Commonwealth Bureau of Soils, Spec. Pub. 3. p. 79. Harpenden, England.

Martin, J. K. (1987). "Carbon flow through the rhizosphere of cereal crops: A review," *Inte col. Bull.* **15**:17–23. The International Association For Ecology, Athens, Georgia.

Martin, J. P. (1963). "Influence of pesticide residues on soil microbiological and chemical properties," *Residue Rev.* **4**:96–129.

Martyniuk S., and G. M. Wagner. (1978). "Quantitative and qualitative examination of soil microflora associated with different management systems," *Soil Sci.* **125**:343–50.

McCann, A. E., and D. R. Cullimore. (1979). "Influence of pesticides on the soil algal flora," *Residue Rev.* **72**:1–31.

McGill, W. B., K. R. Cannon, J. A. Robertson, and F. D. Cook. (1986). "Dynamics of soil microbial biomass and water-soluble organic C in Breton L after 50 years of cropping to two rotations," *Can. J. Soil Sci.* **66**:1–19.

Mola, L., M. A. Sabatini, B. Fratello, and R. Bertolani. (1987). "Effects of atrazine on two species of Collembola (Onychiuridae) in laboratory tests." *Pedobiologia* **30(3)**:145–49.

Moore, J. C., R. J. Snider, and L. S. Robertson. (1984). "Effects of different management practices on Collembola and Acarina in corn production systems. I. The effects of no–tillage and atrazine," *Pedobiologia* **26**:143–52.

Myers, R. G., and S. J. Thien. (1988). "Organic matter solubility and soil reaction in an ammonium and phosphorus application zone," *Soil Sci. Soc. Amer. J.* **52**: 516–22.

Odell, R. T., W. M. Walker, L. V. Boone, and M. G. Oldham. (1982). "The Morrow Plots: A century of learning." *Univ. Illinois Agric. Exp. Stn. Bull. 775,* Urbana.

Odum, E. P. (1984). "Properties of Agroecosystems." In *Agricultural Ecosystems, Unifying Concepts* (Eds. R. Lowrance, B. R. Stinner, and G. J. House). pp. 5–11. Wiley, New York.

Parr, J. F. (1974). "Effects of pesticides on microorganisms in soil and water." In *Pesticides in Soil and Water* (Ed. W. D. Guenzi). pp. 315–40. Soil Sci. Soc. Amer., Madison, WI.

Patriquin, D. G. (1986). "Biological husbandry and the nitrogen problem," *BAH* **3**:167–89.

Pimentel, D., and L. Levitan. (1986). "Pesticides: Amounts applied and amounts reaching pests," *BioSci.* **36(2)**:86–91.

Pokorna-Kozova, J. (1984). "Effects of long-term fertilization on the dynamics of changes of soil organic matter," *Zbl. Mikrobiol.* **139**:497–504.

Power, J. F. (1987). "Agricultural soil and crop practices." In *1988 McGraw-Hill Yearbook of Science and Technology,* pp. 12–16. McGraw-Hill, New York.

Power, J. F., and J. W. Doran. (1984). "Nitrogen use in organic farming." In *Nitrogen in Crop Production* (Ed. R. D. Hauck). pp. 585–98. Am. Soc. Agron., Madison, WI.

Power, J. F., and R. I. Papendick. (1985). "Organic sources of nutrients." In *Fertilizer Technology and Use* 3rd ed. (Ed. O. P. Engelstad). pp. 503–20. Soil Sci. Soc. Am., Madison, WI.

Reganold, J. P., L. F. Elliott, and Y. L. Unger. (1987). "Long-term effects of organic and conventional farming on soil erosion," *Nature* **330**:370–72.

Rice, C. W., and M. S. Smith. (1982). "Denitrification in no-till and plowed soils," *Soil Sci. Soc. Am. J.* **46**:1168–72.

Risch, S. J., D. Andow, and M. A. Altieri. (1983). "Agroecosystem diversity and pest control: Data, tentative conclusions, and new research directions," *Environ. Entomol.* **12**:625–29.

Rodale, R. (1984). "Alternative agriculture," *J. Soil Water Conserv.* **39**:294–95.

Roder, W., S. C. Mason, M. D. Clegg, J. W. Doran and K. R. Kniep. (1988). "Plant and microbial responses to sorghum-soybean cropping systems and fertility management," *Soil Sci. Soc. Amer. J.* **52(5)**:1337–42.

Sahs, W. W., and G. Lesoing. (1985). "Crop rotations and manure versus agricultural chemicals in dryland grain production," *J. Soil Water Conserv.* **40**:511–16.

Schnurer, J., M. Clarholm, and T. Rosswall. (1985). "Microbial biomass and activity in an agricultural soil with different organic matter contents," *Soil Biol. Biochem.* **17(5)**:611–18.

Sheals, J. G. (1955). "The effects of DDT and BHC on soil Collembola and Acarina." In *Soil Zoology* (Ed. D. K. MCE. Kevan). pp. 241–52. Academic Press, New York.

Sheals, J.G. (1956). "Soil population studies. I. The effect of cultivation and treatment with insecticides," *Bull. Entomol. Res.* **47**:803–22.

Slater, C. S., and H. Hopp. (1947). "Relation of fall protection to earthworm populations and soil physical conditions," *Soil Sci. Soc. Amer. Proc.* **12**: 508–11.

Smith, G. E. (1942). "Sanborn field: Fifty years of field experiments with crop rotations, manure, and fertilizers." *Missouri Agric. Exp. Stn. Bull. 458.* p. 61.

Stinner, B. R., H. R. Krueger, and D. A. McCartney. (1986). "Insecticide and tillage effects on pest and non-pest arthropods in corn agroecosystems," *Agric., Ecosystems, and Environ.* **15**:11–21.

Stone, L. R., R. Ellis, Jr., and D. A. Whitney. (1982). *The Effects of Nitrogen Fertilizer on Soil.* Cooperative Extension Ser. C-625. Kansas State Univ., Manhattan.

Subagja, J., and R. J. Snider. (1981). "The side effects of the herbicides atrazine and paraquat upon *Folsomia candida* and *Tullbergia granulata* (Insecta: Collembola)," *Pedobiologia* **22(3)**:111–52.

Tangley, L. (1986). "Crop productivity revisited," *BioSci.* **36**:142–47.

Tischler, W. (1955). "Effect of agricultural practices on the soil fauna." In *Soil Zoology* (Ed. D. K. MCE. Kevan). pp. 215–28. Academic Press, New York.

Tu, C. M. (1978). "Effects of insecticides on populations of microflora, nitrification, and respiration in soil," *Comm. Soil Sci. Plant Anal.* **9**:629–36.

USDA (1980). *Report and Recommendations on Organic Farming.* U.S. Government Printing Office, Washington, DC. pp. 94.

USDA (1988). *Fact Book of Agriculture.* Miscellaneous Publication No. 1063, Office of Governmental and Public Affairs, Washington, DC.

Van Gundy, S. D. and D. W. Freckman. (1977). "Phytoparasitic nematodes in below-ground agroecosystems." In (U. Lohm and T. Persson Eds.) *Soil Organisms as Components of Ecosystems,* Proc. VI Intl. Coll. Soil Zool. *Ecol. Bull.* **25**: 320–29.

Wainwright, M. (1978). "A review of the effects of pesticides on microbial activity in soils," *J. Soil Sci.* **29**:287–98.

Wallwork, J. A. (1976). *The Distribution and Diversity of Soil Fauna.* Academic Press, London.

Werner, M. R. (1990). "Earthworm community dynamics and conversion to organic agricultural practices." (in preparation).

Werner, M. R. and D. L. Dindal. (1987). "Nutritional ecology of soil arthropods." In *Nutritional Ecology of Insects, Mites and Spiders* (Eds. F. Slansky, Jr. and J. G. Rodriguez). pp. 815–36. Wiley, New York.

Yeates, G. W. (1979). "Soil nematodes in terrestrial ecosystems," *J. Nematol.* **11(3)**:213–29.

Zicsi, A. (1967). "About the effects of residue and soil tillage on activity of earthworms," *Pedobiologia* **9**:141–45. (English summary)

8 PASTURE MANAGEMENT

BILL MURPHY
Department of Plant and Soil Science,
University of Vermont,
Burlington, Vermont

Farmers, agricultural researchers, and extension personnel are beginning to realize that pasture is a forage crop and, as such, will respond positively to proper management. Not only that, but pasture is one of the forage crops most capable of intercepting and storing large amounts of solar energy, and consequently supporting high levels of livestock production at low cost. This realization seems simple and obvious to many of us now, but without it, pastures could not be managed well.

In the United States the pasture resource has been mismanaged, wasted, and ignored. This attitude probably developed as a result of previously having too much land available, too few animals to graze it well, a large grain surplus since World War II, and no pressing economic need to use land more efficiently. New Zealand, in contrast, contains about the same number of dairy cows as the United States, plus 70 million sheep, and all are fed on pasture alone within a land area the size of Colorado (Stephen Herbert, personal communication). New Zealand farmers obviously use their pasture resource more efficiently than we do. Due to the shrinking agricultural land base and farmers' financial problems, however, we are being forced to take a much closer look at all available resources.

Pasture is a tremendous resource. For example, in the northeastern United States alone there are 4 million ha of permanent pastureland that have been perceived to be practically worthless until now. This has been called marginal land, just because most of it can't be tilled and planted to row crops, due to soil and site limitations. Another 1.6 million ha of pastures rotated with other crops in the Northeast are used at a level far below their potential because of defective grazing management (Northeast Research Program Steering Committee, 1976). Other regions of the United States and elsewhere similarly contain vast areas of pastureland producing far below their potential because of poor management.

SWARD DYNAMICS

To manage pastures well, the biological basis for management must be understood. Then grazing management can be flexible, based on observation of plants, soil, and animals, rather than rigidly following a set schedule of calendar dates.

A great deal of work on the essential details of pasture sward dynamics under grazing has been done. However, most of that research involved either perennial ryegrass/white clover swards in New Zealand, or pure perennial ryegrass swards receiving 150 to 450 kg nitrogen fertilizer per ha annually and growing under relatively uniform cool, moist conditions in Ireland, Scotland, or England. It remains to be seen how relevant those research results are for the complex legume-based swards commonly grown under highly variable conditions in North America. Although much is known about the growth of temperate grasses, such as orchardgrass, Kentucky bluegrass, bromegrass, and timothy under cutting management in North America, relatively little information exists about their growth responses under grazing management that is much more controlled and intensive than that used in the past. Until research provides the needed information for optimum management of swards grown under our conditions, we can, with care, use existing knowledge of sward dynamics to begin managing pastures more efficiently. As experience accumulates and research results become available, our grazing management will improve.

Plant Tissue Flow in Grazed Pasture

Pasture is a dynamic community, with a continuous flow of new plant tissue forming and old tissue disappearing through the processes of aging, death, and decay, or consumption by grazing animals. This turnover of tissue can be very rapid. For example, in perennial ryegrass a new leaf appears on each tiller about every 11 days, but, since only three live leaves are maintained per tiller, the average life of each leaf is just 33 days.

Pastures are composed of populations of individual plant parts (tillers, stolons, crowns, stems) that vary in proportion depending on the species. Each plant part has an associated mass. Generally, as pasture mass (the total amount of above-ground live and dead herbage present per unit area at a certain time) increases, the size of each plant part increases, and the population of the parts decreases. Rotationally grazed swards have lower population densities and larger plant parts than continuously grazed swards at comparable stocking rates. Swards grazed mainly by cattle have lower population densities and larger plant parts than those grazed mainly by sheep. Any change in the equilibrium between number and sizes of plant parts potentially changes forage production.

Climate, soil fertility, and pasture species determine the rate of new herbage formation. Grazing management can influence this rate by maintaining as green and leafy a sward as possible, to facilitate photosynthesis at all levels of pasture mass. Sparse, stemmy, yellow postgrazing residual with little green leaf area remaining takes longer to reach a maximum rate of new herbage formation than a dense, leafy, green residual.

After grazing, or when plant growth begins in the spring, the rate of new herbage formation increases quickly as pasture mass, leaf area index, and light interception increase. The rate of new herbage formation reaches a maximum level at a pasture mass of 1,200 to 2,500 kg dry matter (DM)/ha, and remains at that level even if pasture mass continues to increase (Figure 8.1).

Net forage production is the amount of live and dead plant tissue that accumulates during a given period and is eaten by animals or is machine-harvested. It results from the balance between the processes of new herbage formation and disappearance. Net forage production, there-

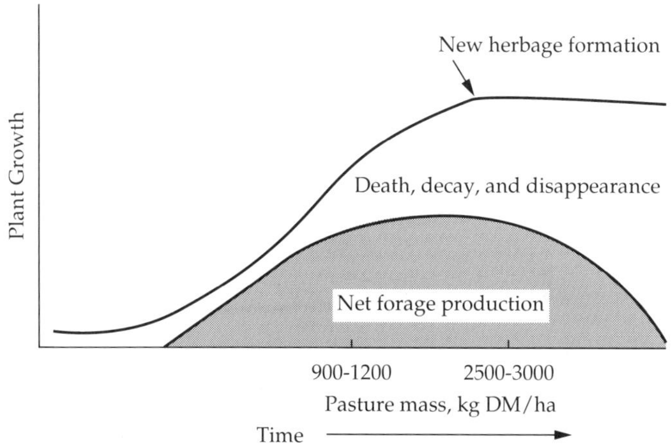

Figure 8.1 Influence of pasture mass on rates of new herbage formation, production, and death, decay, and disappearance of plants (adapted from Bircham and Korte, 1984).

fore, only represents part of new herbage formation. During regrowth, net forage production increases rapidly until a pasture mass of 900 to 1,200 kg DM/ha is reached. Then it remains at a fairly constant high level until a pasture mass of 2,500 to 3,000 kg DM/ha accumulates, when death, decay, and disappearance increase due to more shading of lower leaves and poorer utilization by grazing animals.

The interrelationship among new herbage formation, net forage production, and disappearance through death and decay influences, and is influenced by, grazing management. When the level of pasture mass is low, especially during periods of slow plant growth as in early spring or during drought, new herbage formation tends to limit net forage production. During the rest of the growing season when plant growth is more rapid, net forage production is limited by forage utilization, rather than by the rate of new herbage formation. It follows that the amount of forage allowed to die and decay mainly determines net forage production and utilization, and vice versa. Therefore, net forage production usually increases at higher stocking densities, because anything that improves utilization increases net forage production (Bircham and Korte, 1984; Blaser et al., 1986; Korte et al., 1987). [Note: Stocking density refers to the number of animals grazing a pasture or paddock (pasture subdivision) at a given moment. Stocking rate is the number of animals supported by a unit of land, usually the total pasture area or farm during part or all of the year. For example, a 20-ha farm carrying 300 ewes has an annual stocking rate of 15 ewes/ha. When the ewes are concentrated in a 1-ha paddock, stocking density is 300 ewes/ha.]

During regrowth, rates of herbage formation and death increase as progressively larger leaves are involved in the turnover of plant tissue. Changes in the rate of tissue death lag behind changes in the rate of new leaf formation, however, and this may result in the increased net forage production observed under rotational grazing management, as compared to continuous grazing (Parsons and Johnson, 1986).

Pasture Mass

Pre- and postgrazing pasture masses are the two most important parameters in determining animal production on pasture (Voisin, 1959; Bircham and Korte, 1984; Clark, 1984; Appleton, 1986). The value of using pasture mass in predicting pasture and animal performance is somewhat limited, however, because of variation in plant species composition, leaf to stem ratio, and green to dead ratio, but it is the most useful guide available for managing livestock in a given pasture.

Pasture grazed by cattle generally is more sparsely tillered and clumpy, with more erect tillers and leaves, than pasture grazed by sheep. Differences in light interception cause cattle pastures to need more postgrazing leaf area and mass than denser sheep pasture for maximum plant regrowth.

Pasture mass interrelates with forage quality in affecting productivity of grazing livestock. High pregrazing pasture masses over 2,400 to 3,000 kg DM/ha result from infrequent grazing or continuous grazing with low stocking densities. High pregrazing pasture mass can quickly result in decreased forage quality, patches of rank, low-quality ungrazed forage, a layer of dead plant material at the base of the sward, more upright-growing plants, patches with no live tillers, and a shift to undesirable species. For maximum pasture production, occasional close grazing down to a postgrazing mass of 900 to 1,100 kg DM/ha is needed to maintain high forage quality. At other times, pastures should fluctuate around an intermediate average mass of about 1,500 kg DM/ha for maximum plant and animal productivity (Korte et al., 1987).

Pasture Mass Estimates

Pasture masses can be estimated by (1) clipping, drying, and weighing forage from measured areas, (2) visually evaluating (eye-balling), (3) measuring sward height with a sward stick or ruler, (4) using a bulk density rising plate, (5) using an electronic capacitance meter, or (6) a combination of these methods (Johnstone-Wallace and Kennedy, 1944; Voisin, 1959; Frame, 1981; Fletcher, 1982; Vickery and Nicol, 1982; Bircham and Korte, 1984; Clark, 1984; Lowman et al., 1984; Appleton, 1986; Maxwell and Treacher, 1986; Smith et al., 1986; Korte et al., 1987; Rayburn, 1988; Griggs and Stringer, 1988).

Pasture mass usually is estimated by visual or actual measurements of sward height, which provides a guideline for managing pastures to influence herbage production and animal performance. Sward height affects pasture growth and senescence, and, consequently, influences net forage production and utilization. Sward height also affects forage intake and performance of grazing livestock. Maximum forage utilization by grazing animals per unit area occurs when perennial ryegrass/white clover swards are maintained at 4 to 6 cm tall for sheep (Maxwell and Treacher, 1986), and 8 to 10 cm tall for cattle. For other cool-season grasses combined with white clover, best pregrazing sward heights appear to be 8 to 10 cm tall for sheep (Voisin, 1959), and 10 to 15 cm tall for cattle (Johnstone-Wallace and Kennedy, 1944; Blaser et al., 1986). Optimum postgrazing sward heights are 2.5 cm for low-growing grasses (for example, perennial ryegrass, Kentucky bluegrass) and 5.0 cm for tall-growing grasses (for example, orchardgrass, timothy, bromegrass) (Blaser et al., 1986).

Using sward height to guide grazing management actually is not completely satisfactory, however, because it doesn't take into account variation in plant species and density. For uniform swards, such as nitrogen-fertilized perennial ryegrass growing under uniform conditions, sward height correlates closely with plant density. Sward height measurements may be misleading, however, for estimating pasture masses of complex, legume-based swards growing under variable conditions.

Since plant density is harder to estimate than plant height, visually estimating pasture mass requires experience and some actual physical measurements of the forage to calibrate visual estimates. Visual estimates may be checked against actual mass by first estimating the pasture mass in kg of forage DM/ha within small measured areas. Then the forage in those areas should be cut at the soil surface, dried at about 55°C for two days, and weighed, and the weight converted to kg DM/ha.

More accurate ways of estimating pasture mass, especially in complex swards, would be to take about 50 readings from throughout a paddock with a bulk density rising plate or an electronic capacitance meter, both of which take plant density and height into account. As with sward height and visual estimates, these instruments also must be calibrated with forage sampled from measured areas, but once calibration curves are developed, pasture mass in a paddock can be estimated quickly. Then the correct time to put animals in or remove them from paddocks can be determined regularly, based on measurements of pasture mass. Also the daily and seasonal net forage production and utilization of each paddock and the entire pasture area can be estimated for use in planning and balancing rations.

PASTURE SPECIES

People usually think that pasture species composition limits productivity, and generally they want to renovate pastures to introduce new species as a first step in improving pasture productivity. But where pastures are not being used near their maximum potential, introducing new species will have minimal effect on productivity. Differences in total pasture productivity among plant species rarely exceed 5% to 10% (Korte et al., 1987), and differences among cultivars are even less. These differences are very small compared with the effects of management and the environment. Changing the grazing management is by far the best way of improving pasture productivity.

Although pasture plant species differ little in total production, differences in seasonal distribution of production can be important. Seasonal forage distribution problems can be lessened by introducing new pasture species through overseeding. For example, perennial ryegrass could improve early spring and autumn growth of pastures, and Matua prairie grass (a bromegrass selected in New Zealand) might increase midsummer production. New pasture cultivars may also increase animal productivity through improved pasture forage quality (Korte et al., 1987).

GRAZING MANAGEMENT METHODS

Although methods of managing pasture for higher solar energy interception and increased plant and animal productivity have been known for many years, this knowledge is only beginning to be applied in the United States. As part of this learning process, we're realizing that animals are the tools for managing pasture and marketing its forage, not ends in themselves. Our primary concern in pasture management must be to take care of the plants and soil. The animals' needs will be met by the improved soil conditions and plant productivity. It follows that we need to manage pasture in ways that result in the highest quality, and most dependable and uniform supply of forage as possible at least cost. Then we need to select livestock that do well on what we are able to produce cheaply, rather than attempt to feed animals to a preselected genetic potential regardless of cost.

The problem then is how to manage the pasture resource to achieve that high quality, dependable and uniform supply of forage over the longest possible grazing season. Pasture management in the broadest sense has long- and short-term effects on pasture production. Our practices of grazing management, fertilization, and irrigation influence soil chemical and physical properties, species composition of pastures, and potential plant and animal productivity (Korte et al., 1987).

Confusion exists about the terms being used for the grazing management that we're trying to apply. Basically there are two methods of grazing management: continuous and rotational.

Continuous Grazing

Under continuous grazing, pasture plants are exposed to livestock continually during the growing season. Continuous grazing of pastures containing a single plant species (for example, perennial ryegrass) works fairly well in regions with relatively uniform growing conditions, such as in England. However, where growing conditions vary and/or pasture swards are composed of several species having different growth rates, it's especially difficult to avoid overgrazing (grazing of plant regrowth before energy reserves have been replenished) or undergrazing many plants that are continually exposed to livestock.

Since stock carrying capacity usually is limited by the pasture's least productive period during the growing season, numbers of stock carried under continuous grazing tend to be conservative. When excess forage is available, animals selectively graze, and repeatedly graze more palatable plants, leaving the rest to mature and multiply. During times of excess forage, individual animal productivity can be high, but production per unit area usually is low because under- or ungrazed plants compete with severely grazed plants for sunlight, water, and nutrients. Also, perennial grasses and legumes tend to disappear when overgrazed, allowing weeds and brush to encroach in ever-increasing amounts. Undergrazed patches become rank and even less palatable, and clovers are shaded out of the patches. As legume content in the sward decreases, soil nitrogen levels drop, and total production of the pasture eventually falls.

Selective grazing occurs when stocking density is too low to use most of the forage that has accumulated. Under continuous grazing, selection can be reduced by adding more animals to the pasture, or by restricting the area being grazed and machine-harvesting excess forage. When the amount of forage available decreases, however, stocking density must be lowered, by removing animals or by providing more grazing area, to prevent overgrazing. In practice, adjustments of stocking density to the amount of forage available are not made frequently enough or at all during the season, mainly because of the problem of what to do with animals that are removed from the pasture. The end result is that continuously grazed pastures in the United States generally are weedy and unproductive during much of the grazing season.

Rotational Grazing

The term *rotational grazing* is an especially poor one in the United States, because it has many different meanings, most of which are associated with previous defective grazing management and consequent failure to use pastures well. In the past, rotational grazing usually involved rotating animals among two to six paddocks on a rigid schedule that ignored plant growth rate and the amount of forage present in a pasture. As with continuous grazing, this kind of rotational grazing generally resulted in overgrazing of many plants, undergrazing of others, and overtrampling of soils. Consequently, pastures deteriorated and produced no more forage and livestock product than if they had been grazed continuously.

A different kind of rotational grazing is used in countries where pastures are essential to low-input, high levels of livestock production per unit area, such as in Europe and New Zealand. This is the rotational grazing defined by the late Andre Voisin (Voisin, 1959, 1960) of Normandy, France during the 1950s, which became the foundation of New Zealand's highly profitable and productive pasture-based, low-input agricultural economy (Smetham, 1973).

When Voisin defined the importance of time in rotational grazing management, it became possible to minimize overgrazing and undergrazing. He showed that overgrazing was unrelated to the number of animals present in a pasture, but was highly related to the time period (how long and when) during which plants were exposed to the animals. If animals remained in any one area for too long and grazed regrowth, or if they returned to an area before previously grazed plants had recovered, they overgrazed plants (Savory, 1988). Voisin called his flexible rotational grazing management method "Rational Grazing," because pasture forage is rationed out according to the needs of the animals (just as feed is rationed out in confinement feeding), while protecting the plants from overgrazing and achieving a high level of forage utilization.

Unfortunately, instead of using Voisin's term for his management method, several different names are being used for it: Intensive Rotational Grazing, Intensive Grazing Management, Short Duration Grazing, Savory Grazing, Controlled Grazing Management, and Voisin Grazing Management. This multiplicity of names for proper grazing management causes a great deal of confusion, since each implies a different method, but they are essentially all the same. Voisin Grazing Management will be used here for the method of rotational grazing defined by Voisin. Whatever the grazing management method is called, it must provide for the two essential points of recovery or rest periods that vary between grazings according to plant growth rate and pasture mass, and occupation periods of paddocks to prevent grazing of regrowth during any one rotation.

VOISIN GRAZING MANAGEMENT

Andre Voisin was a biologist and chemist who taught at the National Veterinary School of France and at the Institute of Tropical Veterinary Medicine in Paris. He also was a laureate member of France's Academy of Agriculture, and held an honorary doctorate from the University of Bonn, Germany. (Another French citizen honored in this way was Louis Pasteur.) Voisin also farmed in Normandy, France and he remained essentially a farmer in his outlook. That is, he was able to observe and understand natural interrelationships that often are missed by people not as perceptive and in tune with nature as farmers are.

Combining his scientific training and experience with diligent observation of livestock on pasture, Voisin defined a method of grazing management that interferes as little as possible with the pasture environment, while gently guiding it to benefit the farmer and protecting it from damage by grazing animals. It is a simple method that in essence just gives pasture plants a chance to photosynthesize and replenish energy reserves after each grazing. The Voisin method controls what and when livestock eat by dividing pastures into small areas (paddocks) and rotating animals through them, according to plant growth rate and pasture mass (Murphy, 1987).

The main objective of Voisin Grazing Management is to keep pasture plants within the steep part of their growth curve, so that the rate of new herbage formation always remains high (Figure 8.1). If they're kept within the steep part of the curve, regrowth occurs rapidly after plants are grazed. If postgrazing mass reaches less than 900 kg DM/ha, plants are then in the low part of the curve, and regrowth occurs slowly until adequate leaf surface develops. For this reason, in deciding when to move animals in and out of paddocks, pre- and postgrazing pasture masses in each paddock must be estimated.

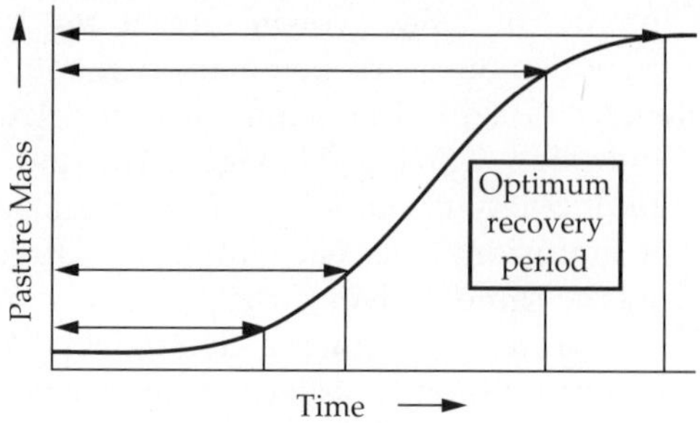

Figure 8.2 Effect of length of recovery period on accumulation of pasture mass.

Recovery Periods

The length of recovery periods considerably influences the amount of forage that accumulates between grazings (Figure 8.2). If the recovery period is cut to half of the optimum period, forage accumulation is reduced about two-thirds. If the recovery period is shortened even more, forage accumulation may drop to only 10% of that produced during the optimum recovery period. This short recovery period corresponds to what happens under continuous grazing: plants are grazed off every time they grow tall enough to be grasped in the animals' mouths. Conversely, if recovery periods are longer than the optimum, forage accumulation increases, but the increase is due mainly to more fiber, which lowers the forage feeding value.

Besides the variation of the plant growth curve, plant growth rate also differs within the season. One of the main rules of Voisin's method is that recovery periods between grazings must vary according to changes in plant growth rates, which reflect changes in growing conditions. In general, this means that recovery periods must get longer as plant growth rate slows and the season progresses.

For example, in the northeastern United States, plant growth rate in May through June is much faster than it is in August through September, and July is a transition time between the two. Therefore, recovery periods must be longer in August–September than in May–June (Figure 8.3). Of course, plant growth rates vary within regions and with prevailing climatic conditions in any season.

Basic recovery period guidelines are helpful for planning and beginning to use Voisin Grazing Management. As experience is gained, adjustments can be made to better suit local conditions and pasture plant communities. For example, in the Champlain Valley of Vermont, we have found that the following recovery periods work well: 12 to 15 days in late April to early May, 18 days by May 31, 24 days by July 1, 30 days by August 1, 36 days by September 1, and 42 days by October 1. These recovery periods are, of course, based on observation of plant regrowth and pasture mass.

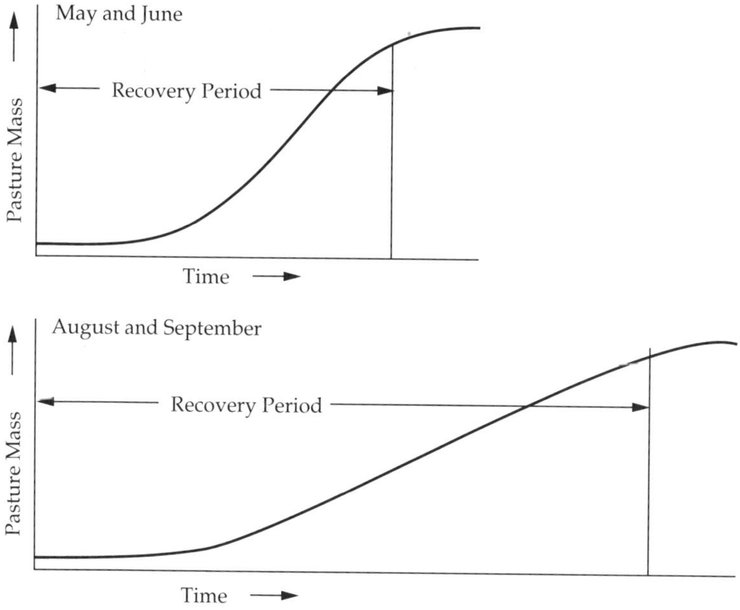

Figure 8.3 Effect of season on length of recovery period needed between grazings.

Paddocks are allowed to accumulate pregrazing pasture masses of 1,800 to 2,400 kg DM/ha in each rotation before animals are turned in. Animals are removed from paddocks when they are grazed down to 1,000 to 1,200 kg DM/ha. The days of recovery are only a guideline to facilitate planning and management. Movement of animals among paddocks must always be based on observation of plant regrowth and pasture mass. They are the best indicators of the condition of the plants.

Recovery periods needed in other climatic zones or for other pasture plant communities (for example, grass-alfalfa) can be determined by taking into account what is known about the plants' carbohydrate reserve cycles, experimenting with different periods, and observing the effects of different pre- and postgrazing pasture masses on plant growth and recovery rates. Under conditions that stress plants, such as drought or cold, longer recovery periods are needed. Warm moist conditions favor plant growth, and recovery periods may be shorter.

If the length of time required to graze a paddock to the required height decreases, the recovery period for the next paddock in the cycle will be shortened. If recovery periods begin to shorten by even 12 hours in any rotation, it is a warning signal that plant growth has slowed for some reason. For example, around June 20 recovery periods, which had been 22 days, begin to shorten by 12 hours in each paddock. Growth of plants in paddocks that the animals will go into next must be checked very carefully, and it must be decided whether or not to slow the rotation. If the plants haven't regrown enough in the next paddock, the rotation must be slowed to accommodate the slower plant growth. This can be accomplished by (1) increasing the pasture area and number of paddocks available for grazing, (2) removing all animals from the pasture and feeding them elsewhere, or (3) feeding the animals hay or other forage in a paddock or paddocks until recovery periods are adequate again.

If plants in the next paddocks to be grazed don't recover enough, less forage is available and, consequently, the movement of animals accelerates through the rotation at a time when it should be slowing down. Voisin referred to this faster movement of animals as "untoward acceleration." Quite suddenly the plants become exhausted, stop growing, and there is no more forage to graze.

An important question that must be answered is: How can the length of recovery period be adjusted during the growing season? There are three practical ways for doing this:

1. Usually about half of the pasture area must be set aside from grazing in the spring (or other periods of rapid plant growth) and machine-harvested, because too much forage is produced in May–June. Therefore, only about half of the total pasture area is grazed in May, June, and part of July. Because using Voisin Grazing Management doubles or triples plant productivity, forage must be machine-harvested from areas where an excess of forage probably has not been harvested before. If pastures are mainly on rough land, relatively level areas should be set aside where machinery can be used to harvest excess forage. The farmer must be prepared for the increase in forage production that will occur, otherwise the pasture won't be grazed properly and its full potential won't be realized. After the excess forage has been harvested, the areas should be rested to allow optimum regrowth, divided into paddocks, and included in the rotation. This increases the area available for grazing and automatically lengthens the recovery periods to what is needed for the remainder of the season.
2. Graze twice as many animals on the total pasture area in May, June, and first part of July as during the rest of the season. The decrease in animal numbers carried during the second half of the season lengthens recovery periods to what is needed. This really is the only way to graze a pasture well if it is all rough land that can't be harvested with machinery anywhere. About half of the animals will either have to be fed elsewhere after July 15 or sold.
3. A compromise way of keeping the pasture forage under control during times of excessive growth, on land where machine harvesting isn't possible, is to not graze down as low as usual, leaving more postgrazing pasture mass. The patchy grazing that probably will occur can be minimized by grazing two kinds of animals (for exam-

ple, sheep and cattle) on the same land, either simultaneously or one behind the other (Nolan and Connolly, 1977).

Periods of Stay and Occupation

The length of time that a group of animals is in a paddock per rotation is called the *period of stay*. If multiple groups of animals are used, the total time that all groups of animals occupy a paddock in any one rotation is called the *period of occupation* or *grazing period*. If only one group of animals is used, the period of stay equals the period of occupation. If two or more groups graze, their total periods of stay equal the period of occupation.

Severely grazed plants can regrow tall enough to be grazed again in the same rotation if the occupation period is too long. So the period(s) of stay must be relatively short to prevent grazing of regrowth. In the northeastern and north central United States, plant regrowth may be tall enough to be grazed again after 6 days in May–June and 12 days in August–September. Although most plants take about 12 days in August–September to regrow to grazing height, some continue to regrow tall enough to be grazed after only about 6 days. Therefore, periods of occupation must never be longer than 6 days to prevent grazing of regrowth in the same rotation, and really should be 2 days or less for best results.

Periods of stay for any one group of animals shouldn't be longer than 3 days, giving total occupation periods of 6 days for two groups of animals. This is because the longer animals are in a paddock, the less palatable the remaining forage becomes, and the more time and energy they spend searching for desirable feed. Periods of stay of 2 days or less if animals are grazed as one group, and 1 day or less for each of two groups, giving total occupation periods of 2 days or less, are better than longer periods of stay and occupation.

In practice then, the shorter the periods of stay and occupation, the better the conditions are for optimum plant and animal production. Milking, growing, and fattening animals shouldn't be in a paddock for longer than two days per rotation any time in the season, to keep them on a consistently high level of nutrition. Milking cows, goats, and sheep produce the most milk if they are given a fresh paddock after every milking. Not only is the forage of higher quality and grazed more uniformly than with less frequent moves to fresh paddocks, but milking animals let their milk down better, anticipating that they're going to a fresh paddock as soon as they're done milking. Growing and fattening animals, such as lambs and beef animals, also gain weight most rapidly if given a fresh paddock every 12 or 24 hours.

Paddocks must be small enough so that all forage in each paddock is grazed completely and uniformly within each occupation period. Occupation periods may need to change because plant growing conditions vary during the season, and the amount of forage available changes.

Lengthening occupation periods indicate that excess forage is available, and paddocks may need to be subdivided or removed from the rotation and machine-harvested, so that the pasture continues to be well grazed. Shortening occupation periods indicate that plant growth rate has slowed and insufficient pasture mass is accumulating.

When animals enter a paddock, plants should be about 15 cm tall for cattle and 8 to 10 cm tall for sheep or goats. These pregrazing plant height differences are necessary because plants are not as dense in pastures grazed by cattle as they are in pastures grazed by sheep or goats, as mentioned above. They also reflect the need to provide forage with physical characteristics that facilitate grazing and intake. Cattle can graze 15-cm tall forage efficiently, but sheep and goats require shorter forage to graze it well.

When animals leave a paddock, ideally all or most plants should have been grazed down to about 2.5 to 5.0 cm from the soil surface. Depending on plant species and local environment, postgrazing plant heights may need to be taller. But usually the taller the remaining plants are when animals are removed from a paddock, the more selective and uneven the grazing has been in the early stages of pasture improvement. It may be very difficult to have animals graze closer than 5.0 cm from the soil surface, because of stubble and matted plant parts from previous years of poor forage utilization. Getting animals to graze closely in areas that are machine-harvested will always be difficult, possibly because the dry, fibrous stubble left by the mowers irritates the animals' mouths.

Animals with high nutritional needs probably shouldn't graze forage down to 2.5 cm from the soil surface, because their production may be lowered. It may be best to follow high-producing animals with animals having lower nutritional needs to graze paddocks down closely.

Just as the pasture must be protected from being overgrazed by using short occupation periods, care must be taken so that the pasture isn't undergrazed as well. Any plants that are not grazed in one rotation probably won't be grazed again that season unless they are clipped. Time and money are saved, and utilization and net forage production increase if the animals graze as uniformly as possible down to a pasture mass of about 1,200 kg DM/ha in each rotation.

Dividing the Animals

The simplest way to graze is to include all animals in one group. Although only one source of water is needed when animals graze as one group, ideally drinking water should be provided in all paddocks so that animals, and their manure, remain in the paddocks being grazed. Also, if drinking water is always readily available in paddocks, animals don't waste energy walking to and from the source of water.

A more efficient way is to divide animals into groups according to their production levels and nutritional needs at different physiological states.

This allows close matching of pasture feeding value with animal needs. A dairy herd can be divided into two groups of (1) milking cows and (2) dry cows and heifers, or (1) milking does and (2) dry does and kids. A sheep flock being managed to produce lambs and wool can be divided into two groups of (1) weaned lambs and (2) ewes. If the sheep flock is also being milked, it can be divided into three groups of (1) milking ewes, (2) weaned lambs, and (3) dry ewes. Other kinds of livestock also can be divided into groups to improve efficiency.

The groups of animals can be handled in two ways:

1. Each group can be grazed in separate cells. (A grazing cell is an area of land planned for grazing management purposes normally as one unit subdivided into paddocks to ensure adequate timing of grazing and recovery periods [Savory, 1988]). Animals with the highest nutritional requirements (milking cows, goats, and sheep, growing lambs and beef cattle) are grazed on the best pasture available. Other animals (dry cows, does, and ewes) are grazed in a separate cell on lower-quality pasture that's being improved. An advantage of this method is that only one source of drinking water is needed in each cell, although providing water in all paddocks is preferable. The disadvantage is that more paddocks must be built, requiring more fencing materials.
2. Both groups graze within the same cell. Animals having the highest nutritional needs are turned into a paddock first so they can eat the best forage quickly. After the first group is removed from a paddock, the second group follows to clean up the remaining forage, which has a lower feeding value than the forage that was grazed first. Paddocks must be small enough so that the combined periods of stay of the two groups are less than six days (two days or less is best). When two groups of animals graze within the same cell, all paddocks must have a source of drinking water because at least one of the groups must stay in its paddock to keep the groups separate.

Number of Paddocks

When planning for managing grazing, it is important to know how many paddocks are needed to ensure adequate recovery periods. This number depends on the recovery periods, period(s) of stay or occupation of the animals in each paddock in each rotation, and the number of animal groups grazing. Since shorter periods of occupation favor higher plant and animal yields, the more paddocks used, the more productive the pasture tends to be. In deciding how many paddocks to build, the farmer should consider topography and soil fertility of the pastureland, pasture plant botanical composition and its potential yielding ability, maximum recovery periods likely to be needed in the area, livestock, fencing costs, and the farmer's financial situation.

First, the recovery period required during the time of slowest plant growth in the region's grazing season must be estimated. It may range from less than 36 to more than 100 days, depending on growing conditions. For example, while 36 to 42 days recovery are adequate for late summer and fall grazing in the northeastern United States, a 90-day recovery period may be needed in drier areas of other regions.

Second, one must estimate lengths of occupation periods and decide how many groups of animals will be used.

Third, the number of paddocks required is calculated as follows:

$$\frac{\text{recovery period}}{\text{occupation period}} + \text{number of animal groups} = \text{paddocks needed}$$

For example (Table 8.1), if 36-day recovery periods will be required, and one group of animals will graze for 12-hour occupation periods, the number of paddocks needed will be

$$36/0.5 + 1 = 73$$

With one group and one-day occupation periods, the number of paddocks needed will be

$$36/1 + 1 = 37$$

If one group grazes with two-day occupation periods, then

$$36/2 + 1 = 19$$

paddocks needed. If animals are divided into two groups and graze within the same cell, an additional paddock must be added:

$$36/0.5 + 2 = 74; \; 36/1 + 2 = 38; \text{ or } 36/2 + 2 = 20$$

TABLE 8.1 Example of the Number of Paddocks Needed for a 36-Day Recovery Period between Grazings.

Period of Stay for 1 Group, Days	Total Number of Paddocks Needed for: 1 Group	2 Groups
1/2	73	74
1	31	38
2	19	20
3	13	14

These are the total numbers of paddocks needed when pasture plants grow the slowest. During times of fastest growth, such as in May–June in the northeastern and north central United States, only about half as many paddocks will be needed, since the remainder of the pasture area will be set aside for machine harvesting.

The more paddocks that can be formed, the better. Voisin Grazing Management should never be attempted with fewer than ten paddocks, because the results will be very disappointing, and it may be concluded that the method doesn't work. The more paddocks used, the more flexibility exists for dealing with changing amounts of forage available during the season. For example, if only six paddocks are used, and for some reason the forage production in one of them drops off, the decline would affect 1/6 of the pasture area. In contrast, with 20 paddocks, a problem with one of them only affects 1/20 of the pasture area.

The most important issue is that enough paddocks be made available to provide the required adequate recovery periods, with the periods of stay and occupation that are to be used. The plants must be grazed at the optimum stage of growth for the plants and animals. The management plan should be thought of as a starting point. Farmers should not try to force the plants and animals to fit a rigid schedule. These are guidelines to help develop a management routine of planning, monitoring, controlling, and replanning if necessary (Savory, 1988).

Paddock Size

After deciding how many paddocks to use, the total pasture area is divided by the number of paddocks to calculate the average area of each paddock. Paddocks need not be equal in area, but they should produce more or less equal amounts of forage to facilitate moving animals on a fairly regular schedule, for the farmer's convenience. For example, pasture areas with fertile soils and good moisture conditions may produce twice as much forage as areas with poor, dry soils. Paddocks in the highly productive areas should then be about half the size of those in the poor areas.

With electric fencing it's always possible to relocate fences or subdivide more, if necessary. For example, as the pasture becomes more productive and the animals can't eat everything within the occupation period allotted (for example, 12 hours), just divide the paddocks in half, in thirds, or whatever it takes to reduce the amount of available forage enough. As stocking density increases, there is more competition among animals for feed and less selective grazing. Under heavy stocking density, even high-producing animals will graze uniformly and close to the ground, and they will perform well if pasture forage quality is high.

Paddock sizes must be adjusted according to the intensity of management desired. It's difficult to state the exact paddock size to use, because that depends on the productivity of the pasture and the numbers, kinds, and sizes of livestock grazing. But paddocks usually should be 1 ha or smaller for grass/white clover and similar plant communities. The flexibility of portable fencing enables easy machine harvest of excess forage from small paddocks, since subdivision fences can be removed for machine harvesting. Because conditions and situations vary so much, different paddock sizes need to be tried to discover what works best on individual farms.

The following examples may be helpful:

1. I have seen 1-ha paddocks in Vermont carry 70 milking Holsteins for three days in May–June, and still contain about 1,000 kg DM/ha that should have been grazed. In that case, I recommended cutting the paddocks in half. Obviously, if that same farmer wanted to provide a fresh paddock to the cows after each milking (every 12 hours), paddocks would have to be only 1/6 as large (six 12-hour periods in three days). That's about 0.17 ha in each paddock for 70 Holstein cows to graze 12 hours—on that soil, on that farm.
2. On well-developed pastures in northern Vermont, 0.10-ha (or less) paddocks carry about 100 growing lambs for 24 hours.
3. Many New Zealand farmers routinely, successfully, and profitably graze 200 milking Friesians in 0.5-ha paddocks for 12 hours. Their pastures are very well developed, with dense plant populations, and they show the potential of the well-managed pasture resource.

Paddock size is important, but not nearly as important as being absolutely certain that the required amounts of recovery time are provided between grazings. Just remember that having more paddocks is better than fewer paddocks, smaller paddocks are better than larger, shorter occupation periods are better than longer, and adequate recovery is essential.

PADDOCK LAYOUT AND FENCING

Right from the start, try to design an ideal paddock layout or cell that requires minimal work to use and is as intensive as possible. A less intensive level of grazing can be used for awhile, if the setting up costs and/or labor needs are too high. But it will always be possible to build toward the ideal layout, if it is designed that way in the beginning.

Several questions need to be answered before designing the paddock layout. How intensive will the grazing eventually be? How much can be spent in the first year to start using the method? Which areas will be set aside in spring for machine harvesting? Which animals will need to return to a central handling facility, and how convenient must it be? Will drinking water be provided to all paddocks? How will animals get through or around wet areas?

Keep the layout flexible for changes in production level and adding other kinds of livestock. Fences cost less, are easier and faster to build, and last longer when built in straight lines.

Topography and Handling Facilities

The shape of paddocks and the location of lanes and dividing fences depend on the topography of the pastureland and its location relative to the barn or other handling facilities. These logistics are especially important with dairy animals that must be milked twice a day. Use an aerial photograph of the farm if possible, or make a sketch of it, to design the paddock layout. Changes are easier to make if paddocks and lanes are drawn on paper, before any fences are built.

For a given enclosed area, square paddocks require less fencing material than long, narrow paddocks. If the pastureland is level-to-rolling, however, it can be divided into large, permanent or temporary rectangles. Inside the rectangles use portable fencing to subdivide into smaller paddocks.

Rough, Hilly Land. If the land is rough and hilly, as most permanent pastures tend to be, slope aspect and location of hill crests should be considered when designing the paddock layout. For example, land with north-facing slopes should be in paddocks separate from land with south-facing slopes, if the areas involved are large enough to make a difference and to form separate paddocks. South-facing soils warm up faster in spring, get hotter and dryer in summer, and stay warmer longer in fall than north-facing soils. So in early spring, south-facing slopes may be ready to graze two to three weeks before north-facing slopes. In midsummer, plants on south-facing slopes may grow very slowly because of the hot, dry conditions, while plants on north-facing slopes continue to grow rapidly. Plants on south-facing slopes can be grazed longer in fall than plants on north-facing slopes. Proper management of such areas is possible only if they are in separate paddocks.

Hill crests also should be contained in separate paddocks to promote uniform grazing and decrease nutrient transfer to favorite livestock resting sites (camps). Paddock divisions along hill crests encourage animals to graze sunny and shaded aspects of the hills, and prevent them from camping on sunny slopes after grazing the shaded slopes.

Level-to-Rolling Land. If the pastureland is level-to-rolling, large rectangular areas can be fenced with permanent perimeter high-tensile fencing or temporary portable fencing. Within the rectangles, subdivisions can be made with portable fencing that is moved ahead of and behind the animals. This kind of pasture division reduces the cost of fencing materials and facilitates machine harvesting of excess forage.

For example:

1. For cattle, build permanent one- or two-strand high-tensile fences around rectangular areas up to about 200 m wide and any desired length. Subdivide the rectangles across their widths by using a polywire (a plastic twine that can be electrified) in front of the animals, supported either on light portable fiberglass rods, plastic step-in posts, or rolling fence wheels (for example, Gallagher tumblewheels). Tumblewheels can be used to subdivide rectangles up to about 200 m wide, but 100-m widths are easier to work with, especially when the fence is moved every 12 hours to form paddocks for milking cows.

Follow the animals with a polywire L-shaped back fence, also supported on portable posts or tumblewheels. The back fence prevents animals from grazing regrowth, and the L shape forms a lane for them to go to the barn or to water, if a water tank is not moved along with them. The lane formed by the back fence can be alternated each rotation between the two sides of a rectangle, so that one side doesn't become too worn and compacted or muddy. This capability of changing the lane location is especially important during wet spring and fall conditions.

Move the front polywire to give the animals only enough forage for the desired occupation period. The advantage of using tumblewheels is that one person can very quickly give the animals a triangular strip of forage by moving every other end of the polywire at each setting. Reels at each end of the tumblewheels make moves easier, because the polywire can be let out or wound up as needed. Another way is to use a reel with a friction brake at one end of the tumblewheels, to allow pulling out the polywire as needed.

A front polywire supported on portable posts also works well, but requires more time and walking to move it. Portable posts that have a lip for stepping them in enable faster moving of the fences. A good way to move a front polywire on posts is to move one end first, and then reposition the posts without taking the polywire off the posts. Then just take up the slack with a reel at one end.

The back fence doesn't have to be moved every time the front wire is moved, but it should be moved frequently enough to prevent grazing of regrowth. The more frequently the back fence is moved, the better. This reduces treading damage, especially when soils are wet.

2. For sheep and goats, build rectangles that are multiples of 48 m wide and any desired length. The 48-m increments are needed because the electric net fencing for small livestock comes in 50-m rolls and some shrinkage occurs after a few years use.

Suppose, for example, that a five-strand, permanent high-tensile fence is built around a 192 × 500 m area. Then it can be subdivided lengthwise into four smaller rectangles, measuring 48 × 500 m, using portable posts and three stands of polywire. (Step-in posts are available that have five clips at correct positions. Multiple reel assemblies and single reel, multiple polywire combinations are also available for quickly and easily positioning three or more strands of polywire.) Then net fencing can be used to subdivide the 48-m wide rectangles.

Keep records of when and where portable fencing was placed, and relocate the fencing at the same positions every rotation. This is to be certain that all areas of the pasture receive adequate recovery periods. If the fencing isn't relocated at the same positions in every rotation, some areas may get less recovery time than they need and start the pasture into untoward acceleration. When using portable fencing, tie a short piece of twine or survey tape to the perimeter fence at each spot where the front wire is connected.

Lanes

Lanes should be as short and as direct as possible, to allow animals to reach all of the paddocks with the least amount of walking. Make lanes only as wide as needed, because forage in lanes usually becomes soiled and wasted. Lanes should be just wide enough to allow the animals and necessary machinery to get through. Animals will conform to the width of the lane and walk single file if they aren't rushed. For example, 4-m wide lanes work well for 35 cows or 350 sheep, but aren't wide enough for large machinery. Lanes 6 to 8 m wide are needed for greater numbers of animals, as well as for tractors with mowers, rakes, balers, and wagons.

Locate lanes on the highest and driest ground possible. Use culverts and gravel fill if necessary to make certain that animals can walk through the lanes without getting covered with mud. This is especially important in the earliest part of the season, when grazing must begin so that the forage doesn't become too mature, and during rainy periods. Fabrics are now available to be placed over muddy spots before covering with gravel. They keep the mud from mixing with the gravel, and form a stable surface over wet areas in lanes and around drinking tanks. For best results, the fabric and gravel fill should be put in when the areas are dry. In lane areas that are not as wet, spreading a thin layer of 8-cm crushed stone helps to support animals and machinery.

An ideal lane and gate arrangement (Figure 8.4) has gates at both paddock corners on the lane (note V-shaped openings), and in the paddock division fences. This design facilitates movement of animals in any direction. If the ideal setup can't be built, be sure to place lane gates in the corner of paddocks nearest the barn and/or water. Otherwise, some animals may not realize that they need to walk away from the barn to go through the gate into the lane before going to the barn. If this happens, instead of all cows coming to milk when called, someone will have to go to the paddock and chase out the few cows that get trapped in the paddock corner nearest the barn. This is especially a problem in the beginning stages of using Voisin Grazing Management and with new heifers.

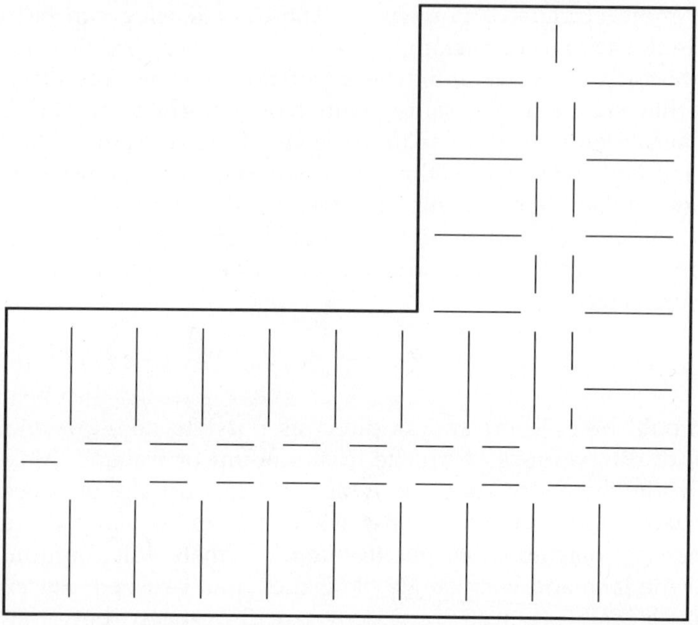

Figure 8.4 An example of a paddock and lane layout. Gates are at all paddock corners and in division fences to facilitate movement of livestock in all directions. (Courtesy of Gallagher Springtight Tunbridge, Vermont.)

If animals don't have to leave the pasture for drinking water or to be milked, gates can be put in paddock division fences, and the animals can be moved to fresh paddocks without entering the lane. The lane will then be used only once per rotation for returning animals to the first paddock to start the next rotation, and can be grazed as a paddock itself, since the forage in it will not have been soiled.

Day/Night Pasture Cells

If some pastureland for dairy cows is located 1.5 km or more from the barn, that area can be used as day pasture and the closer land as night pasture, so the cows can come in quickly for morning milking. (Cows can be called in ahead of time for the afternoon milking.) If day/night pastures are used, be certain to use adequate recovery periods within the two cells. Short (12-hour) occupation periods can be used, but more machine harvesting will be necessary for each cell, because rotation lengths are doubled with this dual cell system.

Drinking Water

Ideally, water should be provided in all paddocks, so animals can remain in the paddock they are grazing and drop their manure there. Otherwise, animals congregate around distant watering points, even if they aren't thirsty, and their manure is dropped and accumulates in lanes and around the watering points. This wastes nutrients, pollutes the environment around the watering points, possibly contaminating ground or stream water, and results in nutrient deficiencies within the paddocks. Also, if water is always available in all paddocks, animals use less energy walking to drink; the energy saved goes into improved animal productivity.

If drinking water can't be provided in all paddocks right from the beginning, at least plan for it in the paddock layout design, and start at a less-than-ideal level. When one group of animals grazes a cell, only one source of drinking water is needed, but they must be able to reach the water at all times. Use lanes to allow the animals to reach the water. When two groups of animals graze the same cell, water must be provided in each paddock, since one of the groups must be kept in a paddock to prevent the two groups from mixing together.

Water can be provided to all paddocks by using 2-cm black plastic tubing. In cool climates the tubing can be laid on the soil surface in the fence line of lanes. Plants growing over the tubing shade it and prevent the water in it from becoming too hot. In the fall, the tubing can be drained by opening both ends, picking it up at one end, and walking to the other end while holding the tubing up and moving it hand over hand. Then it can be left on the soil surface during winter without being damaged. In hot climates, the tubing should be buried to prevent the water from becoming too hot for the animals to drink.

Water can be supplied to paddocks from the tubing in at least two ways:

1. Make connections to full-flow float valves on oblong 200- to 400-1 tanks for cattle and horses, and place one tank under every other paddock division fence. In this way, two paddocks are served with each tank. Eight- to 40-1 tubs or buckets can be used in the same way for sheep, goats, and other small livestock.

2. Place garden hose connectors in the tubing at intervals that will allow you to use a garden hose to connect to a float valve for one tub or tank that you move along with the animals. (Lightweight plastic or fiberglass tanks are convenient to empty and move.) The connector intervals and length of garden hose should be such that you can service two or three paddocks from each connection point. This is especially useful for providing water to animals grazing within portable fencing in large subdivided rectangles. It's best to move an 8- to 40-1 tub along with small livestock every time they're moved within the net fences. With large livestock grazing within front and back fences, tanks also should be moved when the back fence is moved or less frequently by leaving the tank in a wide spot formed in the back fence lane.

A benefit of moving a water tank or tub through the paddocks with animals is that smaller tanks with full-flow float valves can be used, since the animals don't all need or want to drink at the same time. If they are forced to wait or walk to water somewhere else, however, larger tanks are needed.

Shade

The question of whether to provide shade for grazing animals isn't as simple to resolve as it might appear to be. If a pasture has too many trees, production can be improved by thinning them to less than 30% of the area with as uniform a distribution as possible. This ideal distribution provides adequate shade, and doesn't interfere significantly with solar energy interception by the pasture plants, nor does it result in damaging transfer and concentration of nutrients from manure around the trees.

The usual situation, however, is having two or three trees or groups of trees at various locations in the pasture. Animals tend to graze away from the trees, but camp around them. This transfers nutrients to the immediate area around the trees from the remainder of the pasture, and in the long term, pasture productivity decreases.

Establishing trees throughout a pasture to provide uniform shade is neither easy nor inexpensive. Besides the cost of buying and planting the trees, each sapling must be protected from being eaten by the grazing livestock. If something resembling a savanna can't be achieved or maintained, it's probably best from an economic and production standpoint to have no shade at all (Smith et al., 1986).

Some dairy farmers let their cows out to graze in the morning, and keep them in the barn during the heat of the afternoon. This requires some feeding of the animals in the barn, which substitutes for pasture forage they otherwise would have eaten, as well as cleaning the barn and spreading manure, all of which take time, labor, money, and energy. It also results in nutrient transfer from the pasture to lanes and barnyard, and wastes the animals' energy in walking to the barn.

FENCING

To use Voisin Grazing Management, livestock must be controlled. This means that animals must be placed where the manager wants them, for as long as they need to be there. Proper grazing management therefore requires dependable fencing. Fortunately, having good fencing is no longer a problem.

Energizers

With the development of low-impedance electric fence energizers in New Zealand, controlling livestock became much easier than it had been with ordinary fence chargers and nonelectric physical-barrier fences. Energizers are effective because they produce an extremely short, high-energy DC electrical pulse of about 6,000 volts that charges fence wires once per second or randomly every one to seven seconds. The short pulse brings the fence line up to high voltage so quickly that the energy isn't easily drained off, even if the fence wires are covered with rain-soaked plants or some of the wire has fallen on the soil. Also, the short pulse generates very little heat, so there's little danger of an energized fence causing a fire.

Energizers outperform all ordinary fence chargers in effective power by 2.5 to 20 times, depending on the size of the energizer. Although they dependably deliver a lot of power, energizers use very little electricity to produce it. For example, an energizer that will charge over 5 km of five-strand fence or 155 km of one-strand fence consumes only 8 watts of electricity.

Because energizers don't short out easily, animals (livestock or predators) that touch the fence always receive a strong electrical shock that makes them respect the fence and avoid it. This dependable shocking power is why energized fences can be psychological or mental barriers, rather than physical barriers. Mental-barrier fences can be greatly simplified, and they cost much less to build and/or maintain than physical-barrier fences, such as woven wire, multiple-strand barbwire, or non-electric smooth wire fences. For example, a five-strand, spring-loaded high-tensile fence for sheep, including the energizer, costs less than half as much per running meter than woven wire fencing.

Ordinary Chargers

Energizers make using Voisin Grazing Management easier, but they're not required, except for sheep. It's possible to control cattle, horses, goats, and pigs with ordinary chargers. But because chargers are so easily shorted out (their long pulse results in short circuits occurring more easily than with energizers) by even one animal or a wet weed touching the fence, they are not as dependable as energizers. A charger generally is one of the weakest links in a fencing layout, and should be replaced with an energizer if there are any doubts about the charger's ability to control livestock. Animals may not respect a fence powered by an ordinary charger, and can easily knock down all paddock division wires if the charger shorts out for even a little while.

A charger's performance can be improved by (1) making certain that it is well grounded, (2) keeping all fence lines clear of tall grass and weeds, and (3) making certain that all of the insulators are in good condition and that wires don't touch posts, trees, or brush.

Sheep can be kept in with ordinary fence chargers, but not easily, because their wool insulates them from the electrical pulse produced by ordinary chargers. Keeping sheep in with an ordinary charger requires a very strong charge, short fence, few or no plants touching the wires, and continual retraining of the animals. Ordinary chargers can't be used with the five-strand perimeter fencing needed to keep predators away from sheep or goats, because the lower wires must be close to the soil where many plants can touch them. Net fences for small livestock also touch plants and soil, which would easily short out an ordinary charger.

Training Animals

To avoid discouraging problems, animals should be trained to respect an electric fence before turning them out to pasture. To do this, build a strong multiple-strand electric fence around a barnyard or similar small area. Place some grain on the soil at various spots just inside of the electric wires. Connect the energizer or charger to the electric training fence. Confine the animals to the area for several hours (always have water available to them) until at least some of them have been shocked and they all respect the fence.

Physical-Barrier Fencing

If good physical-barrier fencing already exists, by all means use it, if it coincides with the plan for optimum paddock layout. It's not necessary to spend a lot of money building new fences to use Voisin Grazing Management. Quite the contrary: spend as little as possible getting started. Because electric fencing costs less than physical-barrier fencing, it's usually best to replace physical-barrier fences as they wear out with long-lasting, low-maintenance high-tension electric fences.

Physical-barrier fencing works perfectly well in many situations, especially as perimeter fencing for large livestock. Just make certain that the fencing is tight and in good condition. When used for paddock divisions, however, barbwire and nonelectric smooth wire don't work very well, because animals gradually loosen it by reaching underneath and through it to eat forage in adjacent paddocks. It's best to use portable or low-tension electric fencing to subdivide areas within good physical-barrier perimeter fencing.

The useful life of physical-barrier fencing can be extended by attaching offset brackets with insulators to the fences, and adding an electrified smooth wire to them. This prevents animals from pushing against and loosening the fences. The offset electrified wire can also be used to carry power to portable or permanent subdivision fences. Don't attempt to run an electrified strand along a nonelectric fence that is in poor condition, because it will cause more problems than it's worth by shorting out the electric fence.

Electric Fencing Materials

We are fortunate that New Zealanders didn't give up on pasturing livestock as others did, but continued to develop fencing materials and systems that would enable them to do a better job of grazing management. Because of this and recent developments in the United States, New Zealand energizer and fencing technology has been combined with Yankee ingenuity to make available extremely effective electric fencing. These fencing materials and systems are relatively inexpensive, easy to build and maintain, and long lasting. Excellent detailed instructions on building high-tension electric fences are available from dealers of fencing materials (Swayze, 1984; Gallagher, 1985abc; Premier, 1987).

Local university extension services and soil conservation services should have information about dealers that sell energizers and fencing materials and build high-tension electric fencing systems. But if they don't have the information, this list of companies should help to find a nearby dealer:

Gallagher/Spring-Tight Power Fence, RFD 1, Box 158, Tunbridge, VT 05077. Phone: 800/832-9482 (802/889-3737 in Vermont).

Kiwi Fence Systems, RD 2, Box 51A, Waynesburg, PA 15370. Phone: 412/627-8158.

Perfect Pastures, RR 1, Highland, WI 53543. Phone: 608/929-7654.

Premier Fence Systems, P.O. Box 89, Washington, IA 52353. Phone: 319/653-6631.

Snell/Gallagher Power Fencing Systems, Box 708900, San Antonio, TX 78270. Phone: 800/531-5908 (512/494-5211 in Texas).

Twin Mountain Supply, P.O. Box 2240, San Angelo, TX 76902. Phone: 800/331-0044 (800/527-0990 in Texas).

West Virginia Fence Corporation, U.S. Rt 219, Lindside, WV 24951. Phone: 304/753-4387.

RESULTS

Building 10 to 74 paddocks might seem too expensive, and the management required by Voisin's method may seem to take too much time and effort. However, the experience of farmers using Voisin Grazing Management correctly has been very favorable. Any costs that were involved in getting started have been paid back easily from the increased profitability of the farms. This is because feeding animals on pasture can cost as much as six times less than it does to feed them in confinement (Semple, 1970). Due to recent developments in fencing materials, the cost and building of paddocks is surprisingly low and easy to do. Labor requirements actually are significantly less with Voisin Grazing Management of livestock, compared to feeding them in confinement or supplementing them while continuously grazing (Burns and Jones, 1987).

Examples abound of farmers successfully and profitably using Voisin Grazing Management. The following are some case studies that have been published:

A dairy farm family in Maine gained about $3.70 in net profit for each $1.00 spent on setting up to use the Voisin method, or $101 net profit per cow in the first year. They stated that the greatest benefit from using Voisin Grazing Management was an improvement in the quantity and quality (butterfat) of the milk they produced. Milk production increased 327 kg per cow during the grazing season, and butterfat content increased by 0.2%. Like others using the method, this family realized that Voisin Grazing Management requires much less time, effort, and expense than feeding animals in confinement. Besides the economic benefits and savings in labor requirements, they felt that their cows were in better condition when grazing under the Voisin method. This could increase profitability even more by reducing veterinary costs and hoof problems and by improving conception rates (Burns and Jones, 1988).

On another Maine dairy farm, changing to Voisin Grazing Management from continuous grazing reduced the need for purchased grain concentrates by 48% and increased the milk production-to-grain ratio 83%. Although the annual herd average of milk production decreased slightly to 7,763 kg under the Voisin method, compared to 8,018 kg under continuous grazing, net profitability increased significantly because of the reduced need for expensive supplementation. Percent butterfat in the milk was not affected by changing to Voisin Grazing Management. Besides reducing the need for purchased inputs and thereby increasing net profitability, the farm family feels that Voisin Grazing Management also has important ecological benefits. Not only is their land, which is an important resource and investment, being intensively and efficiently used, but at the same time soil quality and diversity of plant growth have been improved. Consequently, they think that the farm is more sustainable now than it was in the past (Gage and Smith, 1988).

Another study of four Maine dairy farms showed net profit gains of $9 to $83 per cow, or $361 to $6,657 per farm, from using Voisin Grazing Management, compared to continuous grazing. The additional net profit resulted from (1) feeding less hay and grain during the grazing season, (2) harvesting extra hay on parts of pastures, and (3) higher milk production (Soil Conservation Service, 1986).

Changing to Voisin Grazing Management from continuous grazing greatly improved production and quality of pastures on two Maine sheep farms. The number of Animal Unit Days (amount of feed needed to support a 450-kg animal for one day) of grazing increased 255% on one farm and 910% on the other farm (Burns and Jones, 1987).

In 1987, the Soil Conservation Service estimated that 200 farms were using Voisin Grazing Management in New York. On nine farms (seven dairy, one beef, and one sheep farm) that were studied in detail, fencing and other setting up costs were more than recovered during the first year. Net profits increased on all farms in the study with the use of Voisin Grazing Management. Milk production increased on three of the dairy

TABLE 8.2 Average Analyses (Dry-Weight Basis) of 497 Forage Samples Taken from Permanent Pastures Grazed under Voisin Grazing Management on Six Vermont Dairy Farms from May 1 to October 1 in 1984. Dry Forage Yields Averaged 8.3 t/ha.

Analyses*						
DM %	CP %	AP %	ADF %	TDN %	ME Mcal/kg	NEL Mcal/kg
23	22	21	28	69	2.51	1.59

* DM = dry matter, CP = crude protein, AP = available protein, ADF = acid detergent fiber, TDN = total digestible nutrients, ME = metabolizable energy, NEL = net energy lactation.

farms and remained the same on the others. Net profit increases averaged $124 per dairy cow, $84 per beef animal unit, and $76 per sheep animal unit. Improved profitability resulted from (1) obtaining more nutritive value from pastures in terms of dry-matter intake and levels of protein and net energy, (2) reduced purchases of feed grains and hay, (3) expansion of the milking herd, (4) lower costs associated with forage harvesting, (5) lower manure handling costs, (6) extending the grazing season, (7) supporting more animals on pasture areas, and (8) improved herd health (Teague, 1987).

Three dairy farms monitored in Vermont during 1984 gained an average of 45 days extra grazing by switching from continuous grazing to Voisin Grazing Management. Their net profit increased an average of $67 per cow during the May 1 to October 1 grazing season (see Table 8.2 for pasture forage analyses of samples that included these farms) (Rice, 1984; Murphy et al., 1986; Murphy, 1987).

Besides increasing net forage production, applying Voisin Grazing Management results in high forage quality throughout the grazing season. As shown in Table 8.2, forage was sampled in 1984 on six northern Vermont dairy farms each time cows were about to enter a paddock. Analyses of the samples showed that well-managed pastures provide uniformly high-quality forage during the grazing season (Murphy et al., 1986).

CONCLUSION

The topic of pasture management is too large to be covered in one chapter, but this much will enable farmers to begin managing the pasture resource better and to learn as they go along. People involved with grazing livestock should read and attend workshops to learn as much as possible about pasture ecology, nutrient cycling, effects of grazing animals on pasture, and effects of pasture on grazing animals. With this knowledge, grazing management can become based more on understanding than on instructions, and farmers will become able to respond flexibly and effectively to the changes that are always occurring in pastures and livestock.

The evidence is overwhelming: Voisin Grazing Management works! Don't accept pastures for what they are now. Think what they could be if managed with the same amount of attention given to other crops. Incorporating pastures managed according to Voisin's method into farm feeding programs can reduce production costs, increase farm profitability, and help to make agriculture more sustainable.

REFERENCES

Appleton, M. (1986). "Advances in sheep grazing systems." In *Grazing* (Ed. J. Frame). British Grassland Society, Berkshire, England.

Bircham, J. S., and C. J. Korte. (1984). "Principles of herbage production." In *Pasture: The Export Earner* (Ed. A. M. Fordyce). New Zealand Institute Agricultural Science, Wellington, New Zealand.

Blaser, R. E., R. C. Hammes, Jr., J. P. Fontenot, H. T. Bryant, C. E. Polan, D. D. Wolf, F. S. McClaugherty, R. G. Kline, and J. S. Moore. (1986). *Forage-Animal Management Systems,* Virginia Polytechnic Institute and State Univ., Blacksburg, VA.

Burns, P. J., and C. R. Jones. (1987). "Voisin Rational Grazing—Sheep." Soil Conservation Service, USDA Office Bldg., Univ. Maine, Orono. Mimeo. 44 pp.

Burns, P. J., and C. R. Jones. (1988). "Economic Effects of Adoption of Rational Grazing." Soil Conservation Service, USDA Office Bldg., Univ. Maine, Orono. Brochure. 14 pp.

Clark, R. (1984). "Avoid that "T" grade," *NZ J. Agric.*, December: 3–6.

Fletcher, N. H. (1982). "Simplified Theoretical Analysis of the Pasture Meter Sensing Probe." Australian Animal Research Laboratory Technical Paper. Commonwealth Scientific and Industrial Research Organization, East Melbourne, Australia.

Frame, J. (1981). "Herbage mass." In *Sward Measurement Handbook* (Eds. J. Hodgson, R. D. Baker, Alison Davies, A. S. Laidlaw, and J. D. Leaver). British Grassland Society, Berkshire, England.

Gage, S. B., and S. N. Smith. (1988). *The Moore Dairy Farm: A Case Study in Reduced Input Farming*. Value Mode Agriculture Case Study Series, Tufts School of Nutrition, Medford, MA. Mimeo. 21 pp.

Gallagher Electronics Ltd. (1985a). "Gallagher Insultimber Power Fencing Manual." Hamilton, New Zealand. Brochure. 24 pp.

Gallagher Electronics Ltd. (1985b). "Gallagher Temporary Power Fencing Systems." Hamilton, New Zealand. Brochure. 6 pp.

Gallagher Electronics Ltd. (1985c). "Tumblewheels." Hamilton, New Zealand. Brochure. 2 pp.

Griggs, T. C., and W. C. Stringer. (1988). "Prediction of alfalfa herbage mass using sward height, ground cover, and disk technique," *Agron. J.* **80**:204–8.

Johnstone-Wallace, D. B., and K. Kennedy. (1944). "Grazing management practices and their relationship to the behavior and grazing habits of cattle," *J. Agric. Sci.* **34**:190–97.

Korte, C. J., A. C. P. Chu, and T. R. O. Field. (1987). "Pasture production." In *Livestock Feeding on Pasture* (Ed. G. K. Barrell). New Zealand Society of Animal Production, Hamilton, New Zealand.

Lowman, B. G., G. Swift, and S. A. Grant. (1984). "Grass Height—A Guide to Grassland Management." East Scotland College of Agriculture, Midlothian, Scotland. Technical Note.

Maxwell, T. J., and T. T. Treacher. (1986). "Decision rules for grassland management." In *Efficient Sheep Production from Grass* (Ed. G.E. Pollott). British Grassland Society, Institute of Grassland and Animal Production, Berkshire, England.

Murphy, W. M., J. R. Rice, and D. T. Dugdale. (1986). "Dairy farm feeding and income effects of using Voisin grazing management of permanent pastures," *Am. J. Alt. Agric.* **1(4)**:147–52.

Murphy, B. (1987). *Greener Pastures on Your Side of the Fence: Better Farming with Voisin Grazing Management.* Arriba, Colchester, VT. 215 pp.

Nolan, T., and J. Connolly. (1977). "Mixed stocking by sheep and steers: A review," *Herb. Abstr.* **47(11)**:367–74.

Northeast Research Program Steering Committee. (1976). "Forage Crops." College of Agriculture, Univ. Vermont, Burlington. Brochure. 40 pp.

Parsons, A. J., and I. R. Johnson. (1986). "The physiology of grass growth under grazing." In *Grazing* (Ed. J. Frame). British Grassland Society, Animal and Grassland Research Institute, Berkshire, England.

Premier Fence Systems. (1987). *The New Fencing Systems Made Simple: A Do-It-Yourself Guide to Buying and Building Better Fences.* Washington, IA. Brochure. 39 pp.

Rayburn, E. (1988). "The Seneca Trail Pasture Plate for Estimating Forage Yield." Seneca Trail Research and Development, Franklinville, NY. Mimeo.

Rice, J. (1984). "Pasture Management." Agriculture Fact Sheet. Univ. Vermont Extension Service, Burlington.

Savory, A. (1988). *Holistic Resource Management.* Island Press, Washington, DC. 564 pp.

Semple, A. T. (1970). *Grassland Improvement.* Leonard Hill Books, London. 400 pp.

Smetham, M. L. (1973). "Grazing management." In *Pastures and Pasture Plants* (Ed. R. H. M. Lauger). A. H. & A. W. Reed, Wellington, New Zealand.

Smith, B., P. S. Leung, and G. Love. (1986). *Intensive Grazing Management: Forage, Animals, Men, Profits.* Graziers Hui, Kamuela, HI. 350 pp.

Soil Conservation Service. (1986). "Short Duration Grazing Project." Soil and Water Conservation District Annual Report, Franklin County, ME. Mimeo. 1 p.

Swayze, H. S. (1984). *Gallagher Spring-Tight Power Fence Construction, Materials, and Costs.* Tunbridge, VT. Brochure. 8 pp.

Teague, P. D. (1987). "Pasture Economics." New York Soil Conservation Service. Mimeo. 1 p.

Vickery, P. J., and G. R. Nicol. (1982). "An Improved Electronic Capacitance Meter for Estimating Pasture Yield: Construction Details and Performance Test." Australian Animal Research Laboratory Technical Paper No. 9, Commonwealth Scientific and Industrial Research Organization, East Melbourne, Australia.

Voisin, A. (1959). *Grass Productivity.* Philosophical Library, New York. 353 pp.

Voisin, A. (1960). *Better Grassland Sward.* Crosby Lockwood, London. 341 pp.

9 CASE STUDY: A RESOURCE-EFFICIENT FARM WITH LIVESTOCK

DERRICK EXNER
Iowa State University,
Ames, Iowa

RICHARD THOMPSON, SHARON THOMPSON
Boone, Iowa

This chapter describes in some detail a case study of one farm in central Iowa. This farm has received wide publicity through on-farm field days and reports in the popular agricultural press. Richard and Sharon have put into practice many of the principles described in previous chapters. In some cases their practical management decisions predated the technical explanations of how and why certain practices worked for them. The Thompsons are currently on the frontier of on-farm research; they use large plots and replicated experiment designs to test potential new components of the systems on their farm (Rzewnicki et al., 1989). A number of practices and the explanations of why they work are not referenced in the literature as one would find in a technical book chapter; the credibility of these explanations is found in the experience of the chapter's authors and the practical success of the Thompson farm. Acres and bushels are used as local measures of land area and production (Editor's note).

BACKGROUND ON THE THOMPSON FARM

The farm of Richard and Sharon Thompson in Boone County, Iowa, is an admixture of the traditional and the experimental. Richard Thompson explains that his problem is not with technology per se, but with the fact that it often has been developed in ways that do not help the farmer. In 1953, Thompson himself earned a M.S. degree in animal production from what is now Iowa State University, and he "bought the whole package" of farming practices recommended at that time.

Eventually, a variety of circumstances—including routine illness in the livestock, pressures of aggressive farming, and spiritual crises—led the Thompsons to question many previously accepted norms. For a 16-year period, from 1967 to 1983, the farm was essentially a closed system in that no fertilizer or pesticides were purchased. During this time, the Thompsons experimented with composting and used crop rotations to maintain soil fertility. As they continued to search for answers, the Thompsons began to use moderate supplements of synthetic fertilizers. At the same time, they have continued efforts to "plug the leaks" in the nutrient cycles of the farm.

The initial searching and trial-and-error approach have evolved into an extensive on-farm research program, with statistically correct experiments covering much of the farm. Each year, 700 to 800 people visit the Thompson operation to observe the input-efficient methods used. They usually find Richard and Sharon Thompson as eager to learn as to teach. The Thompsons know that they do not have "the last word" on farming, and they prove it by constantly revising their practices. If they have a single guiding agricultural principle, it is a balance, provided partly through the diversity of Mother Nature. Diversity is cultivated, for example, in plant populations in the field and in microbial populations in the gut of the livestock. The biological and technological balances achieved by the Thompson operation provide a useful example for agriculture.

GENERAL DESCRIPTION OF THE THOMPSON OPERATION

The farm covers 300 acres, about average size for the state, on gently rolling glacial till soils in central Iowa. These soils are loams and clay loams, but subsoil clay accumulations impede internal drainage. Agriculture in this part of Iowa relies on tile drainage. At the same time, there is usually not sufficient precipitation during the cropping season to supply the crops' needs, so soil moisture reserves are critical.

Five outlying fields support a five-year rotation of corn-soybeans-corn-oats/hay-hay. Four smaller fields near the homestead are devoted to a six-year rotation of corn-soybeans-oats/meadow-pasture-pasture-pasture. Soybean yields are 45–55 bushels/acre (county average 40 bushels). Corn yields are 130–150 bushels/acre (county average 124 bushels). The corn is sufficient to feed the Thompsons' livestock for about six months of the year. Oat yields are generally 80–100 bushels/acre (county average 67 bushels). County averages are for the years 1979–1985, compiled and issued annually by Iowa Agricultural Statistics (formerly Iowa Crop and Livestock Reporting Service), Iowa Department of Agriculture and Land Stewardship.

The Thompsons keep 50 beef cows and two bulls for breeding stock. The hog operation retains 80 sows and finishes 1,200 to 1,300 pigs per year. The crops—including soybeans, which are extruded by a neighbor—are fed to the livestock. All the hogs and most of the cattle are sold through normal market channels. Six to eight head of cattle per year are sold to individuals.

CATTLE OPERATION

The cattle are "black baldies," an Angus-Hereford cross from the Sandhills area of Nebraska. Previously, the Thompsons raised exotic crosses. Those cattle had very high dressed weights, but they required so much corn in their diet that they were not profitable. The black baldies have been selected in an environment that provides mostly grass, and their protein requirements are modest.

These cattle have successfully dropped calves in below-zero weather. However, the Thompsons attempt to arrange their calving for the autumn. Spring is the traditional calving season in the area, but at that time the cattle are still in the lot. In their previous system, calving in the relatively crowded and contaminated pen sometimes led to scours (diarrhea caused by intestinal infection), so the Thompsons switched seasons. Now calves are born out in the pasture in September and October. Being six months off the conventional breeding cycle also imparts some marketing advantage, Richard Thompson believes.

Farmers around the Midwest are employing a number of strategies to produce leaner beef. The Thompsons have recently begun to leave male calves uncastrated in order to reduce fat accumulation by the animals. Hormonal growth stimulants have never been used on the Thompson farm.

During the growing season, the cattle live in the pastures. In winter, they are confined to a shelter and cattle yard from which manure is removed and stockpiled. Parasite problems are most severe in the cold months. The Thompsons use diatomaceous earth (D.E.), which they believe reduces the need for insecticides. The D.E. is applied externally twice a year and is also added to the rations. D.E., a very fine silicate material, acts as a physical irritant to insect larvae (Ross, 1981). Its effectiveness as a protectant for stored grain is well documented, but accounts of its use for internal pests have been limited so far to the popular press (for example, Winter, 1982). It is important to use D.E. of the correct particle size; too coarse a material ("swimming pool grind") is ineffective.

HOG OPERATION

The Thompsons raise hogs without routine use of antibiotics. The customary practice in Iowa is to include subtherapeutic levels of antibiotics in hog feed. If a crisis occurs, the Thompsons will use injectable antibiotics on a sick animal, but they have found that there are usually aspects of the environment that can be manipulated to solve disease problems.

A modified Cargill™ system of insulated prefabricated units with open fronts provides the hogs with sunshine, fresh air, and isolation—at less than half the cost of a confinement system. Pens are attached to the south sides of these units. It is important that a facility like this have a good windbreak planting to the north and west. Farrowing units have both liquid propane heaters and heat lamps. In the winter, two truck mudflaps are hung across the lower part of the nursery units' doors. This keeps out the wind but allows sufficient air circulation to prevent a buildup of humidity. Nipple waterers supply clean water on demand year-round at a fraction of the energy required to keep other watering devices from freezing.

Originally, the hogs were of the tall, narrow body type that has been popularized by livestock shows. The Thompsons now raise medium-framed animals with more lung capacity, which they believe to be better adapted to the outdoor environment. These hogs are crossbreds from Farmers Hybrid Companies, Inc., as are their boars. Replacement gilts are kept to replenish the sow herd.

Pig pens are scraped with a front-loader tractor every two weeks, but they have never been sterilized. Agricultural limestone ($CaCO_3$) is spread on the floors. Believing there is an association of disease problems with "sour" conditions, the Thompsons add lime to raise the pH of the environment above the level favored by potential pathogens.

Antibiotics create a biotic "vacuum" in which resistant strains may proliferate unhindered. The Thompsons' approach is to promote biotic diversity in the gut of their hogs. They regularly add to the feed *Bacillus subtillus,* whey cultured with *Lactobacillus* species, and a number of other cultures collectively known as "probiotics." There is evidence that certain of these microbes colonize the epithelial lining of the digestive tract of pigs, promoting motility and perhaps physically protecting the gut from enteric organisms that are potentially pathogenic (Tannock, 1984; Varley, 1987).

Baby pigs are prepared for life in the hog lot before they are born. Every week, manure is moved from the farrowing units back to the gestation pen. Immunity in the pregnant sow thus stays current with the changing microbial population in the hog lot. This immunity serves her offspring in their first vulnerable weeks of life. As a rule, passive immunity acquired by piglets through the colostrum of their mother's milk protects them for approximately 12 weeks (G. W. Beran, Iowa State University, 1988, personal communication). The pigs receive iron shots but no inoculations. Diatomaceous earth is used in the gestation unit for control of external parasites.

Diet is also designed to avoid gastrointestinal problems. Richard Thompson has said that he wouldn't even try to raise hogs without oats because of their value in preventing scours. The gestation ration includes oats and ground ear corn for bulk. The starter ration contains steamed rolled oats and no source of free sugars, such as molasses. Having concluded that too "hot" a feed mix was causing the young pigs to scour, the Thompsons were able to reduce the total protein content of the starter ration to 16.5% by adding manufactured lysine to improve the quality of the feed protein.

In confinement livestock systems, disease can spread quickly. Reasons include depressed immune response due to stress (Boehncke, 1985), the proximity of animals to one another, and the fact that urine and feces can move from stall to stall (beneath the floor grates or in cleaning operations). Physical isolation of units helps slow the transmission of disease. Careful management of these units helps to keep them healthy. Whereas hogs in a confinement operation live on concrete floors, each sleeping hutch in the Thompson operation has a 12-inch bedding board at the entrance to retain the straw and corn stalks that are pushed in every two weeks with the tractor. The bedding helps keep pigs inside dry and comfortable. The pigs chew on and rut around in the bedding, and Richard Thompson believes this helps keep them from chewing on each other. The pigs gradually work the bedding out of the hutches to the pens, where it serves to absorb excrement.

To ensure that the hutches stay clean, the pigs are chased out into the pens by 6:30 a.m. every day. This keeps them in the habit of dunging outside. Also, hutches are managed to maintain sufficient population density that the pigs tend not to foul the hutch. As the pigs grow, partitions are removed to increase the available space in the hutch, or the number of pigs in the unit is reduced. Every two weeks the pig pens are cleaned out with a front-loader tractor. The mixed bedding and manure is placed in a dump wagon and taken to the manure bunker.

Although this system requires punctuality and attention, it allows the handling of bedding and manure by tractor rather than by hand. The Thompsons say they have tried to "lower the stress on both the farmer and the pig."

MANURE HANDLING

An open bunker measuring 48 feet wide by 176 feet long by 12 feet deep was constructed to stockpile manure. Its design is similar to those used to store silage, except that its floor is sloped to retain the liquid fraction. The goal is to keep the manure anaerobic and cool, with minimum exposed surface area, until it is applied in the spring. For some years, the Thompsons composted their livestock manure to stabilize the nutrients and kill weed seeds. Eventually it became clear that quantities of potassium were leaching from the compost windrows. Nitrogen was also being lost, probably due both to leaching and volatilization.

The liquid that collects in the manure bunker is pumped out and used (at 100 gallons/acre) as a component of the starter fertilizer that is applied at corn planting. The analysis of this liquid in the spring of 1987 was 18 lbs. nitrogen, 2 lbs. potash, 33 lbs. phosphate per 1,000 gallons. This material, tested in the greenhouse of a private college, had no deleterious effect on germination, even when it was applied directly to the seed at 1,000 gallons/acre (R. Vos, Dordt College, Orange City, Iowa, 1987, personal communication).

The dump wagon deposits mixed manure and bedding over the edge of the bunker. This bunker also receives the municipal sludge of the nearby city of Boone (population 12,000). The sludge contains approximately 3.5% to 4.0% N, 1% phosphate, and 1.3% to 1.6% potash on a dry weight basis. Annually, around 209 dry tons, at 80% moisture, are delivered at no charge by city trucks.

The Boone sludge contains chromium from a tannery. The concentration is monitored by the city and averages around 2,000 parts per million. Sludge is mixed with other residues and manures on the farm before application. EPA guidelines suggest a safe annual concentration of 1,000 ppm for material applied to the land. The sludge, applied at an average annual rate of 0.8 dry tons/acre, contributes roughly 3 lbs/acre-year of chromium.

Notwithstanding the potential chromium problem, the Thompsons believe that municipal sludge is an important resource and that it should be utilized rather than allowed to become a pollutant. They feel similarly about the nutrients in their own manure. The storage bunker was a large investment ($25,000), but the Thompsons have found it an efficient and convenient way to handle the manure and sludge that they apply to their fields.

MANURE APPLICATION, PLANTING, AND WEED CONTROL

At planting time, barricade boards are removed from the upper end of the manure bunker. A bucket-loader tractor is then used to fill the Thompsons' 11-ton capacity manure spreader. The planter incorporates the manure/

sludge shortly after it has been deposited on the field by the spreader, thus limiting loss of nitrogen by volatilization.

The Thompsons have practiced ridge tillage since 1966. In ridge tillage, row crops are grown on parallel ridges as high as 8 to 9 inches. In the spring, these ridges warm up relatively quickly, providing a favorable environment for seed germination. The ridges are semipermanent, so wheel traffic can be kept off the planting rows, minimizing soil compaction there. Planting is done directly into the ridge left from the previous year, with no primary tillage to prepare a seedbed. After planting, the ground is bare for a short time, but the soil between the rows is rough and full of residue. The Thompsons' four-row Fleischer "Buffalo"™ planter is set for 36-inch rows and when new costs about $2,000/row.

Ridge tillage has been modified by the Thompsons to allow them to raise crops without herbicides. The sweep that runs ahead of the planter shoe is normally used to just skim the ridge, trimming off stalks and other residue. The Thompsons set this sweep lower, so that it removes the upper 2 inches of the ridge and throws the soil into the inter-row zone. This conveniently covers the manure and the winter cover crop. It also allows the row crop to be planted into soil that contains fewer weed seeds and that has not experienced the environmental cues—chiefly light, warmth, and oxygen—that cause weed seeds to germinate.

Richard Thompson has welded plates over the planter trash bars and extended the sweep back to these plates to prevent any removed soil (and weed seeds) from falling back into the row. He favors the planter shoe over disk openers to put the seed into the ground, because moist soil can accumulate between the openers and their depth gauge wheels. He also prefers a single, narrow press wheel and no "scratchers" behind the planter shoe. A narrow press wheel presses only directly over the row, firming the crop seedbed without creating a wide band of surface soil that is a good weed seedbed. Scratchers stir the soil, which, Thompson observes, stimulates weeds to sprout. Finally, the cover disks on the planter leave a small pyramid of loose soil over the seed row. If soil conditions lead to crusting, this cone is more easily penetrated by the crop seedlings than is a flat soil surface.

The Thompsons have found that weed control in the crop row is the critical factor for success. Richard Thompson plants corn and soybeans deeply enough to ensure that there will be an opportunity to hoe before the crops emerge. The 30-foot rotary hoe uses both a three-point hitch and rubber gauge wheels to give precise depth control, and it allows Thompson to hoe about 25 acres of corn per hour. Weeds are most vulnerable to hoeing before they have emerged from the soil, so the timing of this operation is important. After the crop emerges (beans need to be showing their first true leaves), the field is rotary-hoed a second time.

The Thompsons want vigorously growing seedlings in order to quickly establish a small shade canopy in the row—again, to remove weed germination cues. To this end, they plant 12 soybean seeds per row foot.

This was the same recommendation given by Iowa State University in the 1950s, before herbicides became popular. Tall corn hybrids are planted, at a rate to achieve 24,000 to 26,000 plants/acre in the field.

Rotary hoeing is effective in the rows, but the inter-row areas, being somewhat lower, turn green with weeds until the first cultivation. Corn and soybeans are now typically cultivated twice, though before they became more comfortable with their system, the Thompsons used to cultivate a third time. By the time of the second cultivation, in late June, the crops are of sufficient size that soil can be thrown back into the row, rebuilding the ridge. Many of the weed seeds reintroduced to the row at this time will attempt to grow and will die off from the shading and dry conditions. The Thompsons have found that, in terms of weed control for the following crop, this is the best time to build ridges.

The four-row, ridge-till Fleischer cultivator costs around $1,200/row when new. A Cultivision™ mirror, mounted low, near the front of the tractor, allows the driver to check the precise position of the cultivator without turning around in the seat. On the first cultivation, disks are mounted to clean the ridge to within 2.5 inches of the row. Adjustable shields ride over the rows at any desired height to protect the crop from the churning soil.

The Thompsons achieve sufficient weed control by these methods that they spend little time "walking" their soybeans to manually remove stray weeds. A replicated experiment using a corn-soybean rotation has, since 1984, compared weed populations in ridge-till without herbicides and ridge-till with a herbicide. The herbicide Dual (metalachlor) is used. There has so far been no consistent yield difference between the treatments.

TABLE 9.1 Costs of Producing Corn and Soybeans on an Acre and Bushel Basis by the Thompson and a Conventional Method (1989 Figures).

	Corn		Soybeans	
Item	Thompson	Conventional	Thompson	Conventional
		$ per Acre		
Machinery	84.51	85.86	65.50	53.44
Seed	22.10	15.60	15.60	13.00
Fertilizer, lime	10.50	50.80	5.10	27.90
Herbicides	0.00	18.20	0.00	16.40
Miscellaneous	16.25	10.25	10.50	10.50
Labor ($6.00/hr.)	15.39	12.46	11.45	12.43
Land	100.00	100.00	100.00	100.00
Total cost per acre	248.75	293.17	208.15	233.67
Cost per bushel:				
150 bu. corn	$1.66	$1.95		
50 bu. soybeans			$4.16	$4.67

However, the difference in velvetleaf (*Avutilon theophrasti*) populations has steadily grown, with greater numbers in the herbicide strips.

Richard Thompson says that farmers "select" their weeds with the herbicides they use. Before herbicides, the problem weeds were the large "horse weeds," such as hemp and ragweed. The herbicide 2, 4-D controlled these well, but then grasses such as foxtail (*Setaria* species) moved in from the fencerows. Now grass herbicides are popular, and velvetleaf is prominent in many soybean fields in Iowa.

CROP PRODUCTION COSTS

An older version of the following crop production tables appears in a case study written by Derrick Exner, Richard Thompson, and John Pesek for the NRC Board on Agriculture report *Alternative Agricculture* (1989). Tables 9.1 and 9.2 itemize expenses based on costs of operations given by "Estimated Costs of Crop Production in Iowa—1989," I.S.U. Extension publication FM-1712. These figures do not include interest on production loans because the Thompson operation does not utilize such

TABLE 9.2 Distribution of Costs of Producing Corn and Soybeans by the Thompsons (1989 Figures).

Operations	Corn (150 bu./Acre Yield Goal)	Soybeans (50 bu./Acre Yield Goal)
Offset disk (× 1/2)[a]	3.00	—
Ridge-till planting	10.65	10.65
Seed	22.10	15.60
Manure spreading	18.36	18.36
Purchased fertilizer		
30 lbs. Nitrogen ($.18/lb.)	5.40	—
30 lbs K_2O ($.17/lb.)	5.10	5.10
Rotary hoe (2 × $2.05)	4.10	4.10
Cultivation (2 × $3.45)	6.90	6.90
Corn picker	26.75	—
Combine	—	21.65
Grain transport and handling	11.30	3.84
Grain drying	0.00	—
Miscellaneous	16.25[b]	10.50
Chop stalks (× 1/2)[a]	3.45	—
Labor ($6/hour)	15.39	11.45
Land charge	100.00	100.00
Total cost per acre	248.75	208.15

[a] Cost shared with another crop or incurred only half of the time.
[b] Includes cost of custom shelling half the corn crop.

loans. Labor costs reflect only time spent in the field. A cash rent of $100/acre is used for comparison, although the Thompsons own the land they farm. Neither system is charged for costs of any final spot-treatment of weeds, whether mechanical or chemical. It is assumed that the sizes of farm and field equipment are similar for the Thompson system and a "conventional" farming operation.

For corn production, the conventional operations cited involve one-half trip across the field to spread dry fertilizer (this cost is shared with the subsequent soybean crop), one trip with a tandem disk, one trip to broadcast herbicide, one pass with a field cultivator, one trip to apply anhydrous ammonia, planting, one rotary hoeing, one cultivation, a harvest trip with the combine, and transport and handling costs. This is the series of operations used as the example of typical corn production in FM-1712. The Thompsons' rotational system leads to a corn operation that involves half of a disking (disking is shared with the previous or following crop), spreading manure, planting, hoeing and cultivating twice, harvesting the corn on the ear with a picker, transport and handling of the grain, and half of a pass to chop corn stalks (necessary for only one of the two corn crops in the rotation).

The conventional soybean operation, based on FM-1712, requires half a field trip to spread fertilizer, one field pass to chisel plow old corn stalks, one trip with a tandem disk, one pass to spray herbicide, one trip with a field cultivator, one trip to plant, one rotary hoeing, a harvest pass with the combine, and transport and handling. The Thompson soybean operation requires a pass with the manure spreader, planting, two rotary hoeings, two cultivations, a harvest trip across the field, and transport and handling.

SOIL FERTILITY AND COVER CROPS

The Thompsons' approach to soil fertility is to conserve what nutrients they can and purchase only what additional fertilizer is needed by the crop. Because they fertilize conservatively, they must closely monitor the nutrient status of crops, soil, and amendments.

Soil testing has shown that while there is abundant available nitrogen in June, soil nitrogen is limited in early May, when the corn is planted. The "biological" system, as the Thompsons call it, requires warmth before microbial action can release the nitrogen contained in manures and cover crops. Consequently, 10 to 20 lbs. of nitrogen (dry urea) is inserted near the corn seeds at planting. The manure bunker is helping to increase the nitrogen content of the manure/sludge, which may allow this starter urea nitrogen to be reduced in the future.

During the 16 years that no fertilizer from off the farm was used, soil tests for phosphorus gradually increased from 17 to 100 lbs P_2O_5/acre (Bray

#1 method). Soil potassium test levels remained in the low to medium range, but they were high enough that additional fertilizer was of questionable value (using Iowa State University soil test calibration figures). About the same time that the Thompsons began adding some additional nitrogen to the corn, potassium deficiency symptoms appeared, despite the fact that soil test results still indicated adequate potassium.

Experiences such as this have convinced the Thompsons of the value of tissue tests. Soil tests can be confounded by positional availability of nutrients, due, for instance, to the status of surface soil moisture or to placement of fertilizer in a band rather than broadcasting. Tissue analysis, the Thompsons feel, better reflects the nutrients that the crop actually takes up. In actuality, both soil and tissue tests are used on the farm.

Richard Thompson has concluded that banding promotes much more efficient utilization of fertilizer potassium, at least with a minimum tillage system and on these soils. This is not an unusual finding in the Midwest (Rehm, 1986; Randall and Hoeft, 1988). The Thompsons have initiated a replicated comparison of potassium chloride and potassium sulfate, applied in a band at planting. Potassium chloride costs about one-third as much as potassium sulfate for the same amount of nutrient. A number of farmers who follow alternative agricultural practices are concerned that materials containing the chloride ion may harm the soil biota. This field study should provide credible answers to the question.

The Thompsons place considerable value on the biological activity of their soil. Their farming system, which requires efficient nutrient cycling, depends on ample populations of organisms adapted to facilitate these processes. Part of the difficulty when farms convert from high-input, continuous row cropping to more input-efficient, rotated systems has been attributed to the relative scarcity of soil organisms capable of driving these cycles (Harwood, 1982; Doran et al., 1987).

Soil structure, permeability, and infiltration capacity are other properties related to biological activity in the soil. In particular, the role of earthworms (*Lumbricus* and other biological agents) in soil water movement is only now being recognized (Bevin and Germann, 1982; Priebe and Blackmer, 1989). The Thompsons favor earthworms for these attributes, and they also view the worms as a useful indicator organism by which to gauge general soil biological activity. They use a golf course cup cutter to sample fields (to a depth of 7 inches) for worms and worm eggs in late April. Richard Thompson has observed that earthworms require shelter and a food source, and that their populations seem particularly reduced by fall tillage.

Winter cover crops provide the shelter and substrate required by worms as well as other forms of soil life. They also protect the soil from wind and water erosion and improve soil structure. Cover crops can improve soil fertility by fixing nitrogen and by accumulating soil nutrients, holding them in available form near the soil surface. In addition, cover crops reduce weed pressure. Through direct competition and allelopathic antagonism (Barnes and Putnam, 1983; Worsham, 1984), cover crops can be used to establish what Thompson refers to as the "pecking order" of species dominance in the field.

The Thompsons avoid, when possible, leaving ground bare over the winter. Until a few years ago, they seeded after corn and soybean harvest a mixture of one bushel of oats and 15 lbs. of rye per acre. The oats winterkilled after achieving some ground cover, and the planter was capable of incorporating the small amount of rye the next spring. Of all cover crops used on the farm, rye has the greatest potential for depleting soil moisture and absorbing soil nitrogen needed by the following crop; timely and complete incorporation is important.

More recently, the Thompsons have employed an aerial application service to fly the cover crop seed onto corn and soybeans in early September. Soybeans drop their leaves a few weeks later, providing protection to the germinating cover crops. The Thompsons now seed 1 to 1.5 bushels of oats per acre together with 20 to 30 lbs. of hairy vetch (*Vicia villosa* Roth). The seed of this fairly winterhardy legume is purchased in northeast Nebraska for 40¢ to 50¢/lb.

Oats provide a quick ground cover and probably give some nurse crop benefit to the vetch. In the following spring, the oats are dead and the hairy vetch produces up to a foot of growth before being incorporated by the planter. Because hairy vetch is a legume, it can convert atmospheric nitrogen to plant-usable forms. Incorporated as green manure, vetch can thus supply a portion of the nitrogen required by the following crop. The Thompsons inoculate vetch seed with group C *Rhizobium* to ensure the presence of the bacterial symbiont that gives hairy vetch its nitrogen-fixing capacity.

Aerial seeding costs the Thompsons around $4.75/acre plus seed, which adds considerably to the expense of the practice. For several years, Richard Thompson seeded vetch from the cultivator at the last cultivation of corn. Results varied from excellent to very poor. Thompson has decided that higher populations of corn simply present too much light competition to vetch in the summertime. A secondhand, high-clearance tractor was recently purchased for late summer interseeding into corn.

DIVERSIFIED VS. CASH-GRAIN OPERATIONS

On the Thompson Farm

Richard Thompson has said that, whatever the market prices for cattle and hogs, livestock enterprises "just fit" on their farm. This is certainly the case in terms of nutrient cycling and rotations. It is also generally true that the diversified farm—that is, one raising a number of crops and livestock—

has economic risk management options not available to operations that produce only one or two commodities (Sonka and Patrick, 1984).

At the same time, the Thompsons recognize that not all farms today can—or wish to—have livestock. The "cash-grain" farmer, who raises grain only for sale, also needs options that will save input expenses while maintaining the soil and environmental quality. Two demonstrations on the Thompson farm illustrate alternatives for the cash-grain farmer.

One demonstration utilizes a three-year rotation of corn-soybeans-oats/annual alfalfa. The alfalfa is grown as a green manure for the following corn. The other demonstration is a two-year, corn-soybean rotation. In both rotations, hairy vetch/oats is aerially seeded in September to protect the soil and to accumulate/fix nitrogen. Tissue testing of the crops is used to help judge the adequacy of the fertility program. Early June soil nitrate levels are measured, using the method of Magdoff et al. (1984), to gauge nitrogen side-dress requirements, although at this writing calibration data for Iowa is incomplete (Blackmer et al., 1989; Pottker et al., 1987). The goal, in these demonstrations, is to show that cash-grain operations can profitably reduce purchased inputs and conserve resources.

Two Models

In order to appreciate the problems and constraints associated with integrated operations, and to perhaps understand why these operations are rare in certain parts of the country, consider two hypothetical two-operator farms. The first operation is a cash-grain farm that uses the two-year corn-soybean rotation. The profit maximization strategy here is to farm as many acres as possible, given weather and the capacities of equipment.

The second farm is a diversified operation that employs a six-year rotation of corn-soybeans-corn-oats/meadow-hay/pasture-pasture. An associated cow-calf enterprise is scaled to the hay and pasture available. The calves are raised to finished market weight. The corn not consumed by cattle, as well as all the oats raised, are used in a farrow-to-finish hog enterprise. This farm customarily markets cattle, hogs, and soybeans.

These models were constructed using Iowa State University estimates for 1988 and other published data for such parameters as average days suitable for fieldwork through the year, field capacity and cost of farm equipment, cost and feed requirements for livestock production. The cropping year is assumed to begin with week number 12 (the week of March 19). The last cropping "week," number 46, actually represents the 20 days from November 12 to December 1. Workweek formulas have been adjusted for this longer period. The assumed maximum workday is 12 hours. This, along with the days suitable for field work and the hours per acre required in a given week, determines the size of each farm. The wage rate is $6/hour.

For convenience, average per-acre figures for fixed and variable costs of field equipment were used where available. A more precise analysis of production costs would use the purchase and list price of each piece of equipment, as well as its field capacity and hours of annual use, to generate these costs. Effective field capacity is the measure of overall equipment efficiency *in the field*. Estimates of effective field capacity are available for different kinds and sizes of equipment. In calculating the economics of a whole enterprise or farming system, time spent in repairs and maintenance *out* of the field and time spent in transit *to* the field must be accounted for in some way. A rule of thumb is to add 50% to the effective field capacity time required by equipment to cover an acre or process a bushel. The 50% factor was used for the diversified farm. Because of the considerably greater size of the cash-grain operation, effective field times were increased by a factor of 75% for this farm.

Both farms are assumed to be on prime agricultural land. Crop yields are corn, 145 bushels/acre; soybeans, 46 bushels; oats, 80 bushels; and hay, 4.5 tons/acre. Commodity prices are taken to be soybeans, $5.85/bushel; beef cattle, $68/cwt; finished pigs, $44/cwt. Corn price is discussed below. The cost of land is taken as $100 per acre, a typical rental rate in central Iowa.

Both farms employ the ridge tillage system. By avoiding primary tillage—plowing, disking, field cultivation—and herbicide application prior to planting, the cash-grain operators avoid the spring labor bottleneck that has been a restraint on planted acres and on farm size. The cash-grain farm uses a 24-row planter and banded herbicide in the row at planting. Two 12-row cultivators are used to cultivate crops twice. Two 8-row combines are needed for harvest. Corn is dried to 15% moisture in high-temperature drying equipment.

The diversified farm uses 6-row equipment except for a 30-foot rotary hoe, which is used in lieu of herbicide. Like the Thompsons, these operators cultivate twice, but 25% of a third cultivation is included in production costs and time budgeting to cover the contingency of an extra cultivation. The corn is air-dried in cribs.

The figures show labor distribution through the cropping season in terms of the length of the workweek for an operator. This includes time spent in the field, time getting to the field, repair time, and time with livestock. It does not reflect time spent in management, that is, "desk time."

As Figure 9.1 indicates, the cash-grain operators farm in the spring and the fall. Harvest time is their busiest period. In a year of average weather, these operators put in 12 hours every day that is suitable for field work from week 39 through week 45. Weather problems or equipment breakdowns could necessitate still longer workdays. This farm is 1,821 acres in size. It offers each of the two operators 967 hours of employment per year, an average workweek of 18.6 hours.

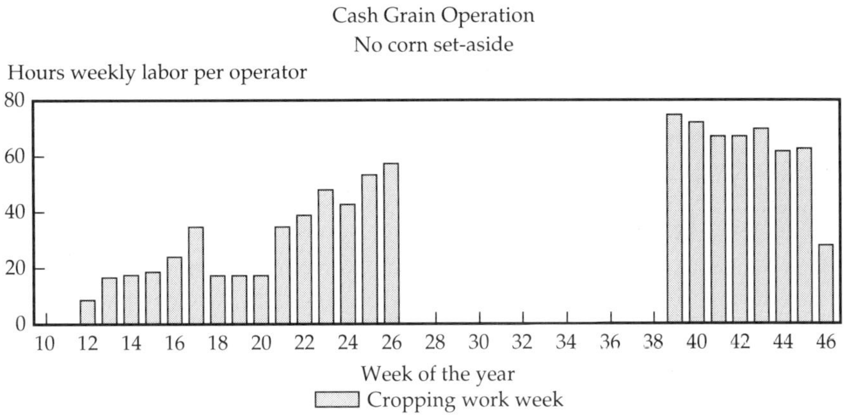

Figure 9.1 Workweek length by week on the cash-grain farm.

Weekly labor on the diversified farm is shown in Figure 9.2. The variety of crops distributes the workload over the season, and the livestock operations provide a relatively constant level of employment before, during, and after the crop year. The farm is 382 acres in size. It retains 62 stock cows and finishes 1,215 pigs per year. This operation furnishes 1,935 hours of work per operator for each of the two operators, for an average workweek of 37.2 hours.

In a sense, the cash-grain farmers clearly are underemployed. This farm does not offer enough simple wage labor to provide a living, should crop prices fall to break-even levels. Similar conclusions on the labor efficiency of cash-grain versus integrated farms are drawn by Guy et al. (1988),

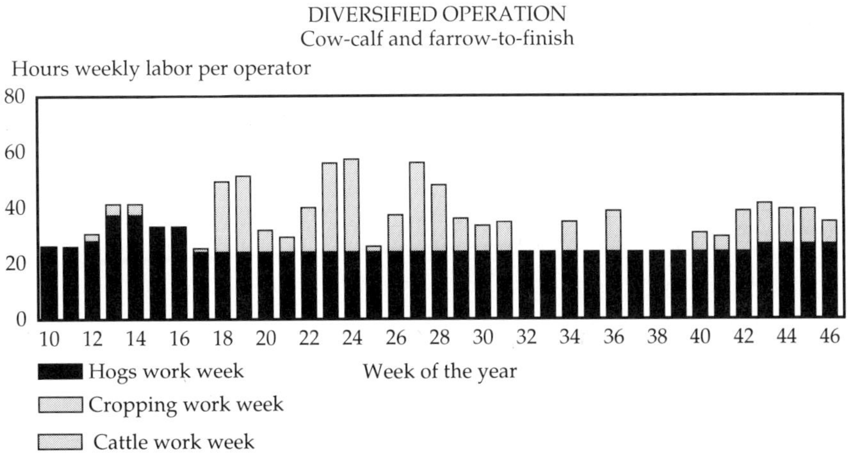

Figure 9.2 Workweek length through the cropping year on a diversified farm.

whose methodology was helpful in the development of the models discussed here.

Because it has fewer crops and greater acreage, the cash-grain farm requires fewer and more costly pieces of equipment. The total investment, based on purchase prices for new equipment, is about $609,000. The diversified farm represents a capital investment, based on new equipment, of $440,000. This includes more and smaller items of field equipment, as well as livestock and their facilities.

It is a moot point that the operators of the model cash-grain farm are underemployed if they are making a consistent profit, and that appears to be the case. Their costs of production, using the average fixed and variable expenses, are $1.88 per bushel of corn and $4.56 per bushel of soybeans. Corn prices have recently been as low as $1.20 per bushel, but the government has assured growers who participate in its feedgrain program the equivalent of a corn price around $3. Participation typically consists of the diversion to cover crops of 10% to 20% of the grower's customary corn acreage.

Without factoring in any particular level of participation in an acreage set-aside, and assuming the cash-grain farm receives a price of $2.50 per bushel on 100% of its corn acres, the return to land for the corn enterprise is $90.61/acre. Soybeans return $59.36/acre over and above land and labor costs. Each of the two cash-grain operators, by these figures, earns an impressive $74,075 per year, of which $5,803 is wage labor and $68,271 is profit on investment. If this projected income is unrealistically high, it may be that economies of scale translate imperfectly from paper to practice. What is clear is that the cash-grain operators in this model have a strong incentive to participate in the government program. In return for the assurance of income, however, each operator "forgoes" some additional hours of employment.

As configured here, the diversified farm markets its corn indirectly through livestock. It is reasonable that the farm does not participate in the corn acreage set-aside program, since the grain is needed for the cattle and hog enterprises. However, these operations must compete in livestock markets that are affected by corn prices which often are substantially below the cost of production ($1.70 per bushel on this farm).

Each operator on the diversified farm earns the very modest annual income of $17,661, of which $11,610 is hourly wages and $6,052 is profit. In terms of returns to land (over and above labor and the $100/acre rental), the soybean enterprise returns $80.91 per acre grown, the farrow-to-finish operation returns $39.25 per acre (based on the land required to grow the feed), and the cattle return just $2.56 per acre.

These figures are certainly approximations, and resourceful farmers find ways to cut expenses and increase production beyond the average values used here. Still, the results suggest something of a dilemma. The cow-calf enterprise justifies inclusion of the forage crop in the rotation, and this cover crop is integral to the nutrient cycling and the soil and water conservation strategies of the farm. The N-P-K fertilizer value alone of the green manure and cattle manure (after spreading cost, which is charged to corn production) accounts for roughly 13% of the profit margin in the soybean enterprise and 33% of the net profit in the hog operation. The cattle operation itself, however, is economically marginal.

The immediate reason for this is the annual cost of land. In its livestock enterprise budgets, the Iowa State University Extension Service sets the cost of pasture land at $20 to $35 per acre. This reflects the norm that the most marginal land is used for pasture, somewhat better land may be used to grow hay or silage, and the best land is reserved for row cropping. If a farm on fertile soils in central Iowa uses the six-year rotation discussed above, its operators may be making a sacrifice economically. If they hold clear title to the land, the roughly $15/acre they pay in taxes may give them an acceptable operating margin. If they rent, or if the land is not completely paid for, their landlord or bank may object to rotating out of row crops.

One can see the utility in the Thompsons' practice of leaving a field in hay for one year only. Their five-year rotation achieves crop rotation benefits, but minimizes the time that the land is devoted to the cattle enterprise.

Finally, this exercise in accounting has not placed a dollar value on environmental benefits of diversified farming, and it has not included macroeconomic effects. It is worth noting that a farm that grows corn only two years out of six contributes to both the production and conservation goals of the present federal feedgrain program, though it benefits relatively little from such a program because of its limited corn base. Michael Duffy (1988) has made this point in an examination of crop rotation data from north-central Iowa. Corn-intensive rotations were least profitable without the government program, but under the program they became the most profitable rotations. Duffy observed that the program designed to support corn prices has, in fact, encouraged the production of corn.

REFERENCES

Barnes, J. P. , and A. R. Putnam. (1983). "Rye residues contribute weed suppression in no-tillage cropping systems," *J. Chem. Ecol.* **9(8)**:1045–57.

Bevin, K., and P. Germann. (1982). "Macropores and water flow in soils," *Water Resour. Res.* **18(5)**:1311–25.

Blackmer, A. M., D. Pottker, M. E. Cerrato, and J. Webb. (1989). "Correlations between soil nitrate concentrations in late spring and corn yields in Iowa," *J. Prod. Agric.* **2(2)**:103–9.

Board on Agriculture, National Research Council. (1989). "Crop-livestock farming in Iowa: The Thompson farm. " In *Alternative Agriculture.* National Academy Press, Washington, DC.

Boehncke, E. (1985). "The role of animals in a biological farming system." In *Sustainable agriculture and integrated farming systems: 1984 conference proceedings,* (Eds. T. C. Edens, C. Fridgen, and S. L. Battenfield). Michigan State Univ. Press, East Lansing.

Doran, J. W. , D. G. Fraser, M. N. Culik, and W. G. Liebhardt. (1987). "Influence of alternative and conventional agricultural management on soil microbial processes and nitrogen availability," *Am. J. Alt. Agric.* **2(3)**:99–106.

Duffy, M., and C. Chase. (1989). "Impacts of the 1985 Food Security Act on crop rotations and fertilizer use," *Am. J. Alt. Agric.* (in press).

Guy, M., M. Lund, and M. Duffy. (1988). "Labor constraints on farm size and alternative farming systems." Unpublished paper.

Harwood, R. R. (1982). "Application of organic principles to small farms." In *Research for small farms: Proceedings of the special symposium* (Eds. H. W. Kerr, Jr. and L. Knutson). November 15–18, 1981, Beltsville, Maryland, USDA/ARS Misc. Pub. #1422.

Magdoff, F. R., D. Ross, and J. Amadon. (1984). "A soil test for nitrogen availability to corn," *Soil Sci. Soc. Am. J.* **48**:1301-4.

Pottker, D., A. M. Blackmer, and J. Webb. (1987). "Amounts and distribution of nitrate in Iowa soils in corn under various cropping systems," *Agron. Abstr.* 213.

Priebe, D. L., and A. M. Blackmer. (1989). "Preferential movement of ^{18}O-labeled water and ^{15}N-labeled urea through macropores in a Nicollet soil," *J. Environ. Qual.* **18**:66–72.

Randall, G. W., and R. G. Hoeft. (1988). "Placement methods for improved efficiency of P and K fertilizers: A review," *J. Prod. Agric.* **1(1)**:70–79.

Rehm, G. (1986). "P and K placement: Something old, something new," *Prod. 1986 Fertilizer, Aglime, and Pest Management Conference.* **25**:119–29. Madison, WI.

Ross, T. E. (1981). "Diatomaceous earth as a possible alternative to chemical insecticides," *Agric. Environ.* **6(1)**:43–51.

Rzewnicki, P. E., R. Thompson, G. W. Lesoing, R. W. Elmore, C. A. Francis, A. M. Parkhurst, and R. S. Moomaw. (1989). "On-farm experiment designs and implications for locating research sites," *Am. J. Alt. Agric.* **3(4)**:168–173.

Sonka, S. T., and G. F. Patrick. (1984). "Risk management and decision making in agricultural firms." In P. J. Barry, Ed., *Risk Management in Agriculture,* (Ed. P. J. Barry). Iowa State Univ. Press, Ames.

Tannock, G. W. (1984). "Control of gastrointestinal parasites by normal flora." In *Current perspectives in microbial ecology*. (Eds. M. J. Klug and C. A. Reddy). American Society for Microbiology, Washington, DC.

Varley, M. A. (1987). "The administration of a probiotic agent to early-weaned piglets." *Anim. Prod.* **44(3)**:464–65.

Winter, G., (1982). "Worming without chemicals," *New Farm* **4(1)**:44–47.

Worsham, A. D. (1984). "Crop residues kill weeds—allelopathy at work with wheat and rye," *Crops & Soils* **37(2)**:18–19.

10 CONVERTING TO SUSTAINABLE FARMING SYSTEMS

REBECCA W. ANDREWS, STEVEN E. PETERS, and RHONDA R. JANKE
Rodale Research Center,
Kutztown, Pennsylvania

WARREN W. SAHS
University of Nebraska,
Lincoln, Nebraska

Sustainable, productive, and environmentally benign farming systems benefit both farmers and society as a whole. A great challenge that must be addressed, however, is how a farmer can realistically move away from energy- and chemical-intensive farming practices and toward more sustainable, biologically based systems, and remain economically viable throughout this transition.

Any change from a traditional practice involves a certain amount of adjustment. Converting to a sustainable agricultural operation is particularly complex because it involves changing not only a technique, but a whole way of viewing the farm. Sustainability emphasizes *optimization* at the agroecosystem level rather than *maximization* of a single crop or component of the system. Changing to a sustainable system may necessitate redesigning the whole farm, including agronomic, ecological and socioeconomic factors, all of which interact differently in each individual farm. As one researcher aptly put it, "We're not talking about agriculture in a can anymore."

The process of conversion begins with the farmer's change in attitude and his/her motivation for altering current practices. Farmers switch to alternative agricultural systems for a variety of reasons. In 1987, *New Farm* magazine conducted a survey of its readers, asking them why they wanted to reduce or quit using chemicals. The respondents were allowed to check any number of the choices given. Results showed that 85% of farmers wanted to adopt low-input methods to cut their production costs. Seventy-five percent noted environmental concerns, and 73% were concerned about family health. Philosophical and quality of life issues such as independence and land stewardship were mentioned by 4% (Summary report, 1987). In a survey asking organic farmers what the advantages of organic farming were, 60% listed personal and family health, and nearly 50% mentioned livestock health. Benefits to the soil and the environment were also listed as important advantages. Only 2% of organic farmers questioned in this survey listed premium prices of organic produce as an important advantage (Lockeretz and Madden, 1987).

While the end results of a sustainable or organic farm can be environmentally and economically favorable, the means to achieve them can be difficult. Perhaps the most important prerequisites for executing a successful transition are a clearly articulated set of goals and a strong commitment to carrying them out. The next step is to develop and execute a plan of action. During (and beyond) the transition period, careful observations, record keeping, and experimentation by trial and error are essential activities. This enables the farmer to make intelligent decisions about fine-tuning equipment, labor needs, and cultural practices. In addition to information gathering and documentation, however, a farmer's creativity and intuition are needed to fit the pieces together into a functioning, integrated system. The final ingredients for surviving the transition period are patience and perseverance.

A farmer who has decided to develop a sustainable farming operation faces biological, managerial, informational, socioeconomic, and political barriers to successful adoption.

Biologically based problems can result from a change in fertility management or pest control. The elimination of chemical fertilizers from a cropping system may cause a reduction in crop yield until a new soil/plant equilibrium has been established. Repeated use of fungicides and anhydrous ammonia, for example, can suppress nitrifying bacteria, which are essential for N nutrition in a biologically based system (Alexander, 1977). Because insecticides can destroy important natural predators, elimination of these chemicals may result in pest infestations. Managerial and information based barriers include the lack of site-specific information that facilitates conversion. There are relatively few farmers who have gone through a transition, and if they have, it has been poorly documented, although there are a few exceptions (Patriquin et al., 1986). Since each farm is unique, the best crop or fertility regimen to start a conversion on a particular farm must often be developed through trial and error. Even with a substantial information base, however, the farmer must acquire the skills to manage new fertility and pest control programs, new equipment, new crops or rotations, and even different marketing strategies.

The socioeconomic and political barriers involve issues extending well beyond the farm gate. Government programs, for example, are still biased toward continuous monocultures of corn (Duffy et al., 1989), making it difficult for farmers to adopt alternative, diversified enterprises. Lending institutions generally perceive agricultural chemicals as the best tools for protecting their investment and hence more ecological approaches to fertility and pest management are discouraged. The greatest economic barrier to developing a sustainable system may be the narrow margin of profit in which most farmers operate, resulting in the sacrifice of long-term investments in soil and farm health for quick-fix solutions that are primarily concerned with this year's "bottom line."

This chapter focuses on the biological barriers faced when converting to sustainable systems and the appropriate change in cultural practices to overcome these obstacles. Field crop and livestock operations are emphasized, but the general principles are relevant to all agricultural enterprises. The economic consequences of conversion are also considered from results of research experiments.

In a recent survey conducted by Rodale Institute, 34% of farmers reported decreased yields when they switched to low-input methods (Summary report, 1987). Nearly half had no change in yields and 12% reported a yield increase. Forty-eight percent of those with decreased yields cited nutrient deficiency and 72% cited weeds as reasons for this decrease. Weeds, insects, and disease were cited as major problems in several other surveys as well (U.S. Department of Agriculture, 1980; Lockeretz and Madden, 1987; Baker and Smith, 1987). Therefore, the major areas of concern discussed in this chapter are soil improvement and pest management.

SOIL IMPROVEMENT

Nitrogen Dynamics

A major concern is providing adequate levels of nutrients to crops (and animals) to ensure a net farm profit during the transition toward a sustainable system. Nitrogen is particularly critical because it is required by plants in relatively large amounts and is subject to loss via leaching, volatilization, and denitrification.

A sustainable farm relies principally upon organic N sources, especially legumes and animal manures. Synthetic, soluble N sources are discouraged because they provide mainly short-term fertility, they require nonrenewable petrochemicals to manufacture them, and they can have negative effects on soil physical and biological properties (Arden-Clarke and Hodges, 1988).

Depending upon organic N sources means that, although much of the N will not be immediately available to the plant, long-term fertility will be enhanced. In an experiment comparing mineral fertilizers and legume residues labelled with ^{15}N in a wheat (*Triticum aestivum* L.) cropping system, Ladd and Amato (1986) found that, although wheat yields were unaffected by N source, 47% of the fertilizer N applied was taken up by the first wheat crop (tops only), versus only 17% of the legume-N. The other side of the story, however, was that at the time that the second wheat crop was sown (one year after planting the first crop), 67% of the legume-N initially applied was present in the soil, as compared to 33% of the fertilizer-N. The larger quantity of legume-N incorporated into the soil organic matter should continue to supply a small but significant amount of N to succeeding crops (Ladd and Amato, 1986). This study was conducted in the semi-arid region of south Australia. In a humid climate such as the eastern United States, leaching losses of N derived from synthetic fertilizers would probably be much greater.

Similar to plant residues, animal manures can supply a substantial amount of N that is slowly available to a crop. "The organic fraction of manure has many of the properties of the ideal N fertilizer—it is not subject to leaching or denitrification losses, it is not toxic to plants, and it mineralizes N at a rate dependent on the same climatic conditions as plant growth" (Bouldin et al., 1984).

The long-term positive effects of animal manure were demonstrated graphically in an experiment at Rothamsted, England, on a continuous barley (*Hordeum vulgare* L.) crop (Salter and Schollenberger, 1939). After 60 years of continuous cropping, a plot that received manure for only the first 20 years was still yielding twice as much as a plot that had never received manure. This long-term effect was the result of the manure's contribution to the organic matter of the soil, providing a substantial source of slowly mineralizable N.

Figure 10.1 illustrates C and N cycling through the various fractions of soil organic matter (Paul and Juma, 1981). Generally, the "old organic matter" pool is the largest. Nitrogen mineralization from this source is very slow, making it a negligible source of N for crop growth. However, it can be an important source of the ion exchange capacity of a soil, as well as being important for soil structure, both contributing significantly to the overall productivity of a soil (J.M. Duxbury, personal communication).

Although the labile pool is usually the smallest fraction of soil organic matter, it is the most biologically active portion and, therefore, the most important contributor of soil-derived, plant-available nutrients. In Fig. 10.1, the labile pool is broken down into an "active fraction," which includes microbial biomass, by-products of microbial activity, and other

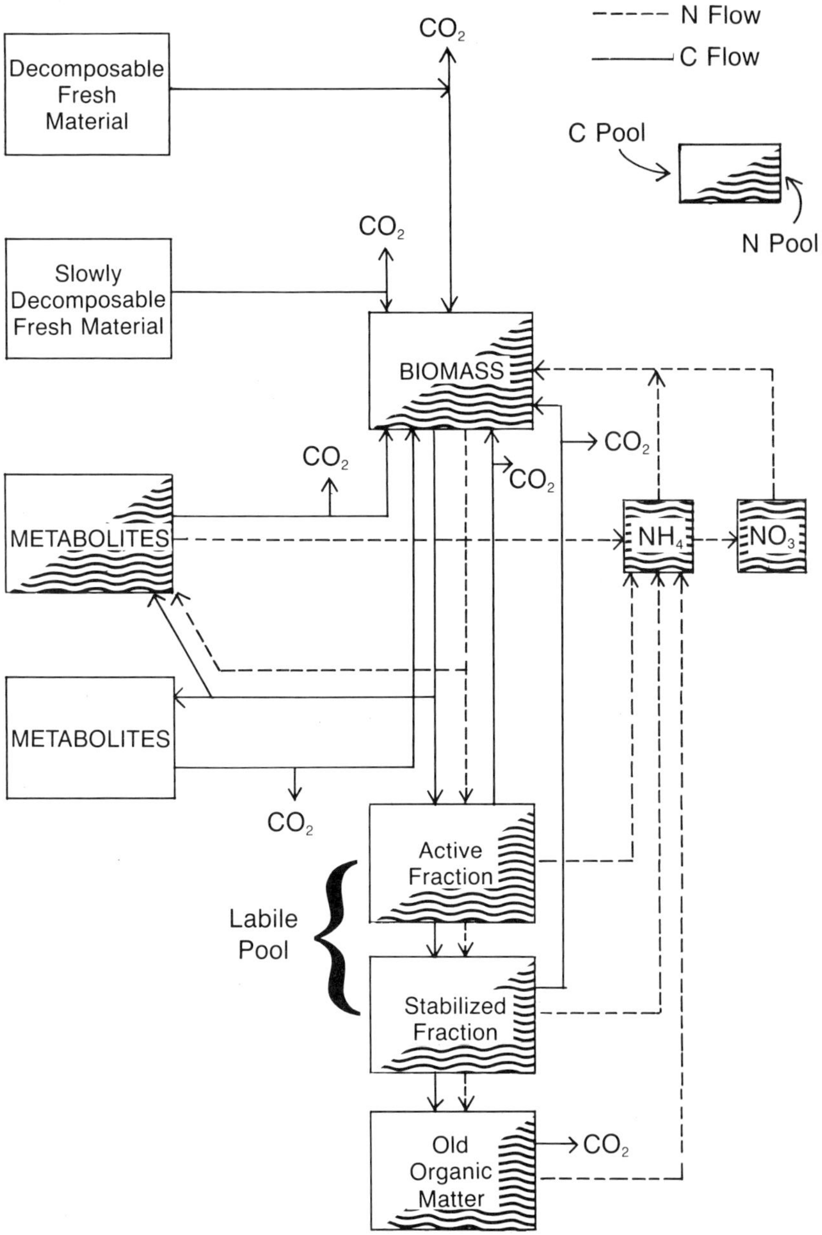

Figure 10.1 Flow chart of transfer of carbon and nitrogen in the soil (Paul and Juma, 1981).

highly available components, as well as a "stabilized fraction," which includes partially decomposed residues and is more gradually available (Paul and Juma, 1981). Although this component decomposes slower than the active fraction, it can be a significant source of available N in the soil. In Germany, annual additions of residue two to three times the size of the labile pool were required to maintain an equilibrium level in the soil (Sauerbeck and Gonzalez, 1976).

Studies quantifying potentially mineralizable N provide evidence of the importance of residue inputs in maintaining the labile pool. Stanford and Smith (1972) quantified potentially mineralizable N (PMN) from various soil types and cropping regimes. Increasing organic matter input, either by adding animal manure or by including hay crops in a rotation, doubled PMN relative to cropping systems without manure or hay crops. In the same study, Mollisols that were high in total soil N (0.19 to 0.29%) and intensively cropped with little or no fertilizer applied contained very little PMN. These soils had a large pool of resistant organic matter, but a very small labile pool, due to the lack of fresh organic matter inputs.

Results from the Rodale conversion experiment illustrate the importance of periodic inputs of organic materials. In 1987, corn (*Zea mays* L.) in a rotation that included a legume green manure every third year had yields equivalent to corn in a nonlegume rotation that received 145 kg N ha^{-1}. That spring, only 43 kg ha^{-1}N from the legume above-ground biomass was plowed under. Besides contributing to the pool of soil N, the additions of legumes improved soil structure and enhanced biological activity.

While green manures and animal manures are effective at creating stable humus over the long term, they must also build up stores of labile organic-N to ensure sufficient N for the present crop. This requires careful management because the potential for leaching or volatilization losses of N from these organic residues can match or even exceed those in systems using chemical fertilizers. Leaching and/or gaseous losses can be high if manure (animal or green) is applied in the fall or winter in areas of high winter precipitation. Fall-applied manure has been estimated to be only 58% as effective as spring-applied manure in sustaining crop growth (Bouldin et al., 1984). Therefore, animal and green manures should be applied as close to the time of crop uptake as possible (Aldrich, 1984).

Crop rotations that include hay can reduce N leaching relative to continuous row cropping. In an extensive agroecosystem study in Sweden, alfalfa (*Medicago sativa* L.) hay fixed about 384 kg N ha^{-1} yr^{-1}, and about 95% of this was retained in the plant-soil system (Long and Hall, 1987). Leaching losses were very low in the hay systems, at 1 kg N ha^{-1} yr^{-1}. Leaching losses from barley cropping systems in the same experiment were an order of magnitude higher, at 18 and 10 kg ha^{-1} yr^{-1} for fertilized and unfertilized barley, respectively.

Crop rotations can reduce leaching losses by synchronizing periods of greater N uptake by crops with greater availability of N in the soil

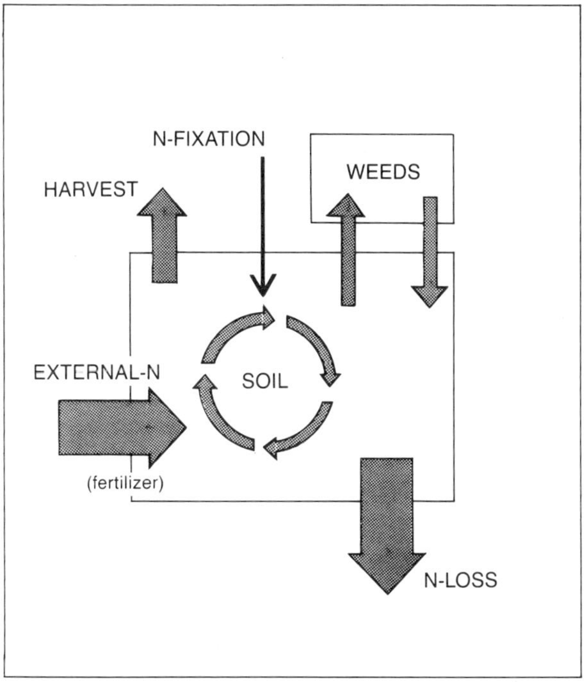

Figure 10.2a Nitrogen cycling changes (hypothesized) during the transition to organic forms of nitrogen inputs. (Before conversion. Nondiversified crop sequence.)

(Pierce and Rice, 1988). Legumes, including alfalfa and soybeans (*Glycine max* L. Merr), act as effective scavengers of NO_3-N in the soil following fertilized grain crops, thus reducing leaching losses (Schertz and Miller, 1972; Johnson et al., 1975). Crop residues and cover crops high in C, such as winter rye (*Secale cereale* L.) or annual ryegrass (*Lolium* sp.), can immobilize soil N, thereby preventing leaching loss, releasing N slowly through microbial degradation.

The diagrams in Figure 10.2 illustrate our hypothesis of N cycling dynamics during the process of conversion to a low-input system, based on our data and the work of others (Koepf et al., 1976; Patriquin, 1986; Radke et al., 1988). The conventional cropping system (Figure 10.2a), assumed here to be a cash-grain, row crop rotation, receives high inputs of fertilizer-N. Much of this N is lost from the plant-soil system, however, and relatively little is maintained in the soil.

During the conversion process (Figure 10.2b), N comes from a variety of sources, including animal manure additions and increased N-fixation from legume green manures and hay in place of some or all of the fertilizer-N. The decomposition of these residues will begin to increase the labile pool of soil organic matter. Nitrogen leaching loss may decrease, although

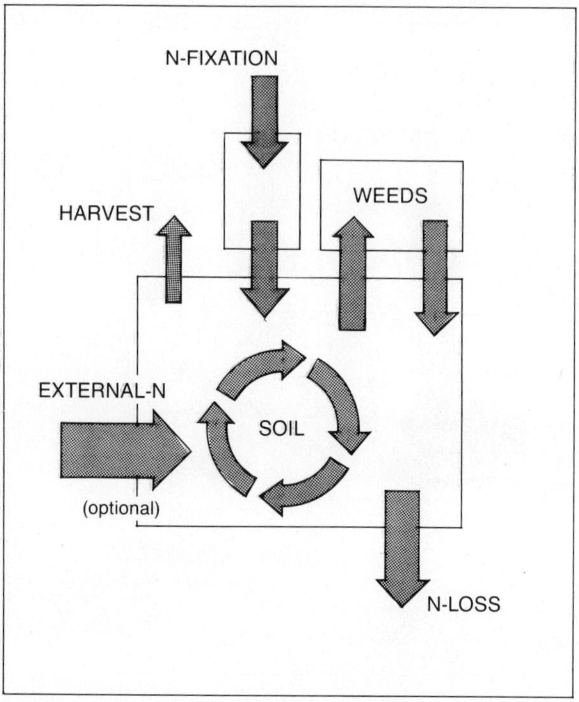

Figure 10.2b Nitrogen cycling changes (hypothesized) during the transition to organic forms of nitrogen inputs. (During conversion. Diverse crop rotation.)

this depends a great deal on how these amendments are managed. The amount of N that is cycled through weeds will probably increase, while the N taken off as harvest may or may not change.

As the conversion process continues (Figure 10.2c), the amount of N cycling through the labile pool continues to increase. At the same time, as the components in the system stabilize, crop yield increases and weed biomass decreases, although not necessarily to the preconversion levels.

Phosphorus and Potassium Dynamics

Due to the buildup of P and K in soil from years of fertilizer applications, the levels of these nutrients are often sufficient during the conversion to a low-input cropping system. Furthermore, some soils are naturally high in P and/or K. However, as these stores become depleted after several years of cropping without additions, the need eventually arises to add these minerals in some form. Unlike N, which can be added to the crop-soil system via symbiotic fixation, the soil supply of P and K can only be replenished by external sources. Since these nutrients are often present in forms that are slowly available to the crop, management techniques are of

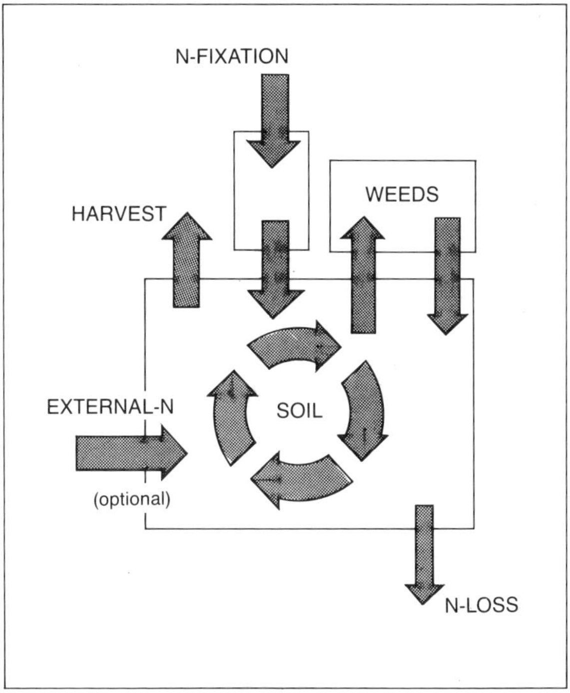

Figure 10.2c Nitrogen cycling changes (hypothesized) during the transition to organic forms of nitrogen inputs. (After conversion is completed. Diverse crop rotation.)

Note: Diagrams do not include all details of nitrogen cycling, such as rainwater inputs, volatilization.

utmost importance in increasing nutrient availability. Improving organic matter status and maintaining proper pH are important in increasing availability of both P and K.

The organic farming approach to fertilizer amendments is not simply avoidance of synthetic materials. Instead, it stresses the development of a "long-term soil management program with goals of resilient soil health, high levels of biological activity, and optimum nutrient balance. The program must also strive to achieve optimum levels of major, minor, and trace nutrients" (*Certification Handbook*, 1988).

The sources of P used to achieve these goals are still under discussion, although untreated colloidal and soft rock phosphate are the most common forms. Sulfuric acid-treated P (ordinary superphosphate) and phosphoric acid-treated P (triple superphosphate) are generally not permitted under organic certification guidelines.

Besides P and calcium, phosphate rock contains several essential minor and micronutrients, including copper, chromium, magnesium, barium, zinc, boron, and iodine. Although the acidified phosphates are much more water soluble, they are also devoid of many of these other nutrients (Minnich and Hunt, 1979). In addition, the manufacturing process of treated phosphate requires additional energy (Lockeretz, 1980) and hence is more expensive than untreated rock phosphate per unit of total P.

Due to its low degree of solubility, untreated rock phosphate usually does not supply sufficient P to an intensive rotation of short-season row crops (Tisdale and Nelson, 1975). Certain crops, however, such as lupins (*Lupinus* sp.), buckwheat (*Fagopyrum sagittatum*), clover (*Trifolium* sp.), sweet clover (*Melilotus* sp.) and brassicas (*Brassica* sp.), are very effective at utilizing P from untreated rock (Mengel and Kirby, 1982). Inclusion of these crops as green manures in the rotation can help provide adequate soil P to all crops over the long term. Applying the untreated rock phosphate to a compost pile, rather than spreading it directly on a field, may also increase P availability to plants (Mishra and Bangar, 1986).

Since most soil P reserves are present in soil organic matter, regular additions of animal and green manures are, just as for N, the best insurance for maintaining P fertility. During decomposition of the residues, P is taken up by soil microorganisms and consequently released in an available form (Stevenson, 1986).

Unlike N and P, K is not dependent on microbial biomass for its cycling through the soil. However, increasing organic matter may enhance availability and retention in the soil by raising the cation exchange capacity (Gershuny and Smillie, 1986). Liming may help protect K from leaching by the same mechanism, but in some soils, this may translate into lower availability to the crop (Brady, 1984). In addition to loss by leaching, excess soluble K is taken up by the crop as luxury consumption. However, most of this is probably returned to the soil as crop residues (or animal manure). As with P fertilization, organic and low-input farmers often utilize low-analysis sources, such as rock dust.

Organic certification guidelines vary, but generally organic sources of K are preferred, mined sources are allowed, and the use of manufactured or refined products is discouraged (*Certification Handbook*, 1988; MacCormack, 1989; *Organic Farm Certification*, 1989). The use of some forms of processed K is allowed during the transition period, but, in general, potassium chloride is not recommended for several reasons. These include excess solubility, high salt index, and high chlorine content, which may have inhibitory effects on soil microorganisms (*Organic Farm Certification*, 1989). The salt index of potassium chloride is 116 compared to 46 for potassium sulfate (Chambers, 1989), a form considered acceptable by several certification organizations.

Green manuring and crop rotation may utilize the ability of certain crops to pull nutrients from deeper layers of the soil, bringing into the plant-soil system nutrients that otherwise would be underutilized or lost from the system. Deep-rooted herbaceous plants are grown in certain pasture mixes in England, because of their ability to accumulate minerals from the lower depths of the soil (Foster, 1988). The use of legumes and other green manures or cover crops in rotation may, in this way, increase the proportion of total mineral nutrients in the soil profile available for crop growth. However, this mechanism would only be effective in soils in which the subsoil had substantial natural fertility. More research is needed to determine the effectiveness of using crop species in this way.

Soil testing is an important component of a conversion plan. However, fertilizer recommendations based on soil tests may vary widely among soil-testing laboratories. Subsamples of a homogeneous bulk soil sample were sent to several laboratories across the United States for analysis and fertilizer recommendations (Liebhardt, 1984). Some laboratories recommended fertilizer rates based on replacing the amount of nutrients removed by the crop, regardless of soil nutrient status. Others took soil nutrient status into account, but even when soils tested high in P and K, fertilizer was recommended to ensure availability to the crop. While some laboratories recommended fertilizer rates to build up reserves of P and K in the soil, others recommended no additional fertilizer if the soil tested high. Fertilizer recommendations should be evaluated carefully, and rates can often be reduced, whether the fertilizer used is "organic" or not.

PEST MANAGEMENT

An important characteristic of low-input cropping systems is the absence or reduced use of herbicides, insecticides, and fungicides. Withdrawing these chemicals from the system can result in pest problems, especially weed infestations, which are responsible for much of the yield reduction during the first few years of conversion. A farmer can employ strategies to reduce these losses, however.

Nonchemical pest management tactics include adjusting planting rate, time of planting, row and seed spacings, and altering choice of crop or variety. Mechanical cultivation is a mainstay for weed control in most low-input systems. Many farmers are adapting reduced tillage systems such as ridge-till to low-input cropping systems (See Chapter 9).

Crop rotation is an essential component of pest management strategies in low-input cropping systems. Numerous long-term rotation experiments have shown that yields decline under continuous row-cropping relative to crops grown in rotation, even in the presence of manure or fertilizers (Smith, 1942; Odell et al., 1974; Bolton et al., 1976). Differences in disease and insect presence, soil fertility, soil structure, or a combination of these factors may account for this "rotation effect" (Crookston, 1984).

Initiating a crop rotation can reduce insect damage by interrupting life cycles of pests. For example, corn rootworm (*Diabrotica* sp.) damage can be avoided almost completely by rotating corn each year with a legume such as soybeans, a crop which is not a host for the rootworm larvae (Krysan and Miller, 1986, citing Metcalf et al., 1962).

A study in Minnesota showed that a corn-soybean rotation reduced the need for herbicides as compared to continuous corn (Forcella and Lindstrom, 1988). Weed infestations also can be reduced by rotating between row crops and solid-seeded crops, and between warm-season and cool-season crops, thereby interrupting the life cycles of warm- and cool-season weeds (Harwood, 1985).

Crop rotation has also been shown to reduce crop disease. In studies in Washington, Goldstein (1986) found that fumigation of continuous wheat duplicated the effects of rotation on yields of wheat. The improved status of the crop due to either fumigation or rotation in turn reduced weed populations, as the crop competed more successfully for the available resources.

Crop rotation facilitates the conversion process in other ways as well. Rotating into crops that are less vulnerable to pest damage can reduce losses during the conversion. Incorporating hay and legume crops into a rotation can build up reserves of organic matter, N, and other nutrients in the soil, optimizing conditions for crop growth. Employing these strategies can help crops to compete successfully with weeds and resist other pests.

Cover crops and mulch crops can also help control weeds. In addition to providing some of the same benefits of crop rotation, such as interrupting pest cycles and building up reserves of soil nutrients, cover crops can suppress both late summer and early spring weeds. Techniques for including legumes in a cropping system include (1) overseeding legumes into soybeans that have begun leaf-yellowing in late summer, (2) drilling into small grains in midspring, (3) planting after small grain harvest in midsummer, and (4) overseeding into corn at final cultivation or after silking in late summer. Planting a crop directly into a dead mulch may control weeds well into the crop's growing season.

The quantity of weeds that a farmer will tolerate in a crop field can influence the success of the conversion. Studies at the Rodale Research Center have shown that up to 500 kg ha^{-1} weeds can be permitted in a corn field before yield is reduced (data not shown). Visual thresholds are often lower than actual economic thresholds, and economic returns can be improved by taking this into account when planning weed control strategies. Liebman and Janke (Chapter 4) discuss weed control in low-input systems in detail.

Careful monitoring of weed and insect populations, such as the scouting techniques used in Integrated Pest Management, can provide the farmer with valuable information to aid in decision making during the transition period and beyond. Awareness of the subtle effects of cropping systems on weeds and other pests is part of an integrated "bio-literate" approach to farming.

If suboptimal crop nutrient status is contributing to pest losses during the conversion, applications of small amounts of chemical fertilizer may improve the competitive ability of the crop. If weed or insect problems are especially severe, a farmer may decide to apply pesticides. However, because some pesticides may be harmful to beneficial organisms in the plant-soil system, it is preferable to use only "internal" pest controls through rotation and other cultural methods. If the crop is to be sold through organic marketing channels, the use of pesticides is generally limited to certain "biologically based" formulations, such as microbial pesticides and predator insects, for the control of insect pests. Also, insects and pathogens are available for biological control of weeds.

LOW-INPUT EXPERIMENTS

Few studies have been conducted on low-input cropping systems. However, interest in both the public and private sectors has prompted studies comparing low-input and "conventional" cropping systems, and, to a lesser extent, studies examining the process of conversion itself. The majority of these experiments have focused on individual components and processes, and have been invaluable for increasing our understanding of cropping systems, particularly low-input systems. However, studies looking at a system as a whole, while more unwieldy and difficult to analyze, are necessary to provide a perspective on how various components interact to form a viable low-input cropping system.

The following discussion focuses on two low-input cropping system studies which attempt to meet this need. The first investigates conversion to a low-input system on soil that had been conventionally cropped for several years. The second experiment compares several crop rotations, with varying levels of external inputs, with a continuous crop sequence, on land that had been in perennial hay for several years. Both experiments were still in operation as of the publication of this book.

The Rodale Research Center Conversion Experiment

This experiment was begun in 1981 at the Rodale Research Center in eastern Pennsylvania. The objectives of the experiment were to define yield-limiting factors that occur during the conversion process, to identify methods of minimizing yield reductions, and to identify physical, chemical, and biological changes that occur during conversion to low-input methods. A detailed description of the experiment can be found in Liebhardt et al. (1989).

The experiment is located on a 6.1 ha site adjacent to the Rodale Research Center on primarily silt loam soil (Typic Fragiudalf, fine-loamy, mixed mesic). For several years prior to 1981, the field had been farmed conventionally in a corn-wheat rotation. Chemical fertilizers and pesticides were applied at the discretion of the farmer. Phosphorus and K levels in the soil were high at the initiation of the experiment and organic matter was 2.4%, but levels of plant-available soil-N were probably low, based on both the cropping history of only cereals and the N stress evident in the first year of the experiment.

Three cropping systems are compared, each consisting of a five-year rotation (Table 10.1). The "low-input/livestock" system assumes a livestock component. Animal manure is used to provide nutrients, and forage and silage crops are harvested in addition to grains. The "low-input/cash-grain" system does not include an animal component. Plowed-down legumes are utilized as N sources, and a cash grain crop is harvested each year. The "conventional" system is a corn-soybean cash-grain rotation, with university-recommended rates of fertilizer and pesticides applied each year.

To assess the effect of the rotation starting point, each cropping system was initiated at three different phases of its rotation. Therefore, each of the three cropping systems includes three treatments in any one year, each one having begun its rotation at a different point, for a total of nine treatments. Field plots are 6.1 × 91.5 m, with eight replications of each treatment.

Soybean and small grain crops were not adversely affected by the conversion to a low-input system. There are no small grains in the conventional treatment for comparison with the low-input treatments, but wheat generally yielded only slightly lower than the average for Berks County, Pennsylvania, while oats yielded equal to or better than county averages (Liebhardt et al., 1989).

The low-input soybean yield was equal to or greater than conventional yield every year except 1987 (Table 10.2). Weed biomass in the low-input/livestock soybeans in 1987 was extremely high at 3.8 Mg ha^{-1}. A comparison of weeded to unweeded subplots showed that this amount of weed pressure caused a 9% reduction in yield. Reduced yields in the low-input/cash-grain treatment relative to the conventional soybeans in 1987 were caused by the presence of a wheat intercrop as well as weed competition, which was responsible for a 13% yield reduction ($P < 0.06$).

Corn grain yields in the conventional treatments were superior to yields in the two low-input systems for the first four years with one exception (Table 10.3). This occurred in 1983, a drought year, when the low-input/livestock corn yield was not significantly different from conventional yield. While the yields of all three treatments in 1983 were low, the conventional corn yields declined more than the low-input/cash-grain yields relative to 1982 (the low-input/livestock system cannot be compared because it did not include corn in 1982). This appears to be an effect of previous crop. The low-input/cash-grain corn preceded by a red clover green manure yielded 84% of the previous year's yield. The conventional corn following soybeans and the conventional corn following corn dropped to 74% and 62%, respectively, of the previous year's yield.

TABLE 10.1 Rodale Research Center Conversion Experiment Treatment Design.

Cropping System	Rotation Entry Pt.	Year: 1981	1982	1983	1984	1985	1986	1987
					Crops			
Low-input/ livestock (1)	1	Oats/ red clover	Red clover (2nd yr)	Corn	Soybean	Corn (silage)	Wheat/ legume mix (hay)	Hay
	2	Corn	Soybean	Corn (silage)	Wheat/ red clover	Red clover (2nd yr)	Corn	Soybean
	3	Corn (silage)	Wheat/ red clover	Red clover (2nd yr)	Corn	Soybean	Corn (silage)	Wheat/ legume mix
Low-input/ cash grain (2)	1	Oats/ red clover	Corn*	Oats/ red clover	Corn	Soybean	Oats/ legume mix	Corn
	2	Soybeans	Oats/ red clover	Corn*	Wheat/ hairy vetch	Corn	Barley/ soybean	Wheat/ legume mix
	3	Corn	Soybean	Oats/ red clover	Corn*	Oats/ red clover	Corn*	Wheat/ soybean
Conventional (3)	1	Corn	Corn	Soybean	Corn	Soybean	Corn	Corn
	2	Soybean	Corn	Corn	Soybean	Corn	Soybean	Corn
	3	Corn	Soybean	Corn	Corn	Soybean	Corn	Soybean

* Indicates a shorter-season variety of corn was used in this treatment, as compared to the other treatments in the same year.

TABLE 10.2 Soybean Seed Yield, Rodale Research Center Conversion Experiment[a].

Cropping System	1981	1982	1983	1984	1985	1986	1987
				Mg ha⁻¹			
Low-input/livestock	—	3.29a[b]	—	3.56a	3.42ab	—	2.96b
Low-input/cash grain	1.78a	3.31a	—	—	3.58a	3.22a	2.96b
Conventional	1.78a	2.87b	2.83	3.53a	3.31b[c] 3.56a[d]	2.82b	3.56a

[a] Seed yield adjusted to 13% moisture.
[b] Values for each year that are followed by the same letter are not significantly different (.05 level), as determined by Duncan's multiple range test.
[c] Rotation entry #1.
[d] Rotation entry #3.

Low-input and conventional corn grain yields from 1985 to 1987 did not differ significantly except for the lower yields in the low-input/cash-grain corn in 1986. This was mainly due to a short-season hybrid used in this treatment to allow sufficient time in the fall to establish a winter wheat crop.

The concentration of N in corn ear leaf tissue at silking is a good indicator of its N status and final grain yields. Corn leaf N concentration tended to increase in the low-input systems from 1981 to 1987, but the trend was not consistent (Table 10.4). The large difference in leaf N between the two low-input/livestock treatments in 1983 was reflected in a large difference in silage yield (11.8 vs. 8.0 Mg ha^{-1}). These differences were due in part

TABLE 10.3 Corn Grain Yield, Rodale Research Center Conversion Experiment.[a]

Cropping System	Previous Crop[b]	1981	1982	1983	1984	1985	1986	1987
					Mg ha⁻¹			
Low-input/ livestock[d]	Red clover	1.56b[c]	—	5.52ab	8.70b	—	10.78a	—
Low-input/ cash-grain	Legume	1.37b	5.47b[e]	4.60b[e]	7.41c 5.96d[e]	9.57a	8.62b[e]	9.03a
Conventional	Corn	2.37a	9.56a	5.92a	9.33ab	—	10.51a	8.18a
	Soybean	2.54a	8.59a	6.35a	9.69a	8.54a	10.42a	8.71a

[a] Grain yield adjusted to 15.5% moisture content.
[b] Crop preceding 1981 was wheat in all treatments.
[c] Values for each year that are followed by the same letter are not significantly different (.05 level) as determined by Duncan's multiple range test.
[d] Corn after soybeans in Low-input/livestock system was harvested as silage.
[e] Indicates a shorter-season variety of corn was used in this treatment as compared to the other treatments in the same year.

to the lower yielding treatment being preceded by corn (without manure) and soybeans, whereas the higher yielding treatment was preceded by two years of red clover. The occurrence of these differences, even though each treatment received manure in 1983, suggests that not enough manure N was readily available for crop use, so that previous cropping history was the main factor affecting yield and leaf N.

The highest leaf N concentration during the seven years was in the low-input/cash-grain corn in 1985 (31.7 g kg^{-1}). This crop was preceded by a hairy vetch (*Vicia villosa* Roth) green manure that contributed 180 kg N ha^{-1} (calculated from above-ground biomass only), 50 to 70 kg ha^{-1} more than had been contributed by red clover in previous years. The yield of this treatment was significantly greater than the conventional yield ($P<.10$).

Weed biomass in the low-input systems was consistently higher than in the conventional system during the first seven years (Table 10.5), although corn yields were not always correspondingly reduced. Weather and previous crop were the factors most affecting weed growth in the low-input corn.

In 1982, a wet spring prevented timely cultivation in the low-input/cash-grain corn, resulting in the highest weed biomass in all seven years. In contrast, dry conditions in 1983 allowed well-timed cultivations, and weed biomass was much lower than in the previous year.

The effect of previous crop on weed biomass can be seen in the low-input treatments when the same corn variety can be compared. In 1983 and 1986, the low-input/livestock corn following hay had significantly less weed biomass than did the low-input/livestock corn following soybeans. This was also true in 1985, when weed biomass in the low-input/cash-grain corn following hairy vetch was much lower than in the low-input/livestock corn following soybeans. Rotating away from a summer row crop for one year was probably the main reason for the reduction in weeds, but increased N availability to the crop following hay, indicated by ear leaf N in 1983 and 1985 (Table 10.4), may also have increased its ability to compete with the weeds.

The Rodale conversion experiment has demonstrated that when synthetic fertilizers and pesticides are removed from a cropping system, an initial transitional period occurs while the plant-soil system attains a new equilibrium. For example, corn yield in the low-input cropping systems did not reach the level of the conventional corn until the fifth year of the trial, but it has performed equally well or better since that time. Lack of available soil N and increased weed pressure appeared to be the primary yield-limiting factors in the low-input plots during the transition.

The low-input and conventional cropping systems in this experiment are different in two major ways. The first is the source and amount of fertilizer and pesticide used. The second, which may ultimately have a more profound effect, is the distinct crop rotations in each system. Conclusions made about each cropping system must consider both of these factors. Yearly climatic fluctuations must be taken into account as well.

TABLE 10.4 Corn Leaf Tissue Nitrogen Concentration, Rodale Research Center Conversion Experiment.

Cropping System	Previous Crop[a]	1981	1982	1983	1984	1985	1986	1987
					$g\ kg^{-1}$			
Low-input/ livestock	Hay	19.8c[b]	—	27.3a	25.8b	—	27.9a	—
	Soybean	24.1b	—	19.0d	—	25.0c	28.0a	—
Low-input/ cash grain	Green manure	20.3c	26.0a[c]	24.9b[c]	26.1b 25.8b[c]	31.7a	27.8a[c]	28.3a
Conventional	Corn	28.7a	27.4a	22.0c	29.8a	—	26.2b[d]	28.0a
	Soybean	29.5a	26.6a	23.9b	29.9a	29.2b	25.2b	28.7a

[a] Crop preceding 1981 was wheat in all treatments.
[b] Values for each year that are followed by the same letter are not significantly different (.05 level), as determined by Duncan's multiple range test.
[c] Indicates a shorter-season variety of corn was used in this treatment, as compared to the other treatments in the same year.
[d] Both CONV corn plots in 1986 followed a soybean crop.

TABLE 10.5 Weed Biomass in Corn Crop, Rodale Research Center Conversion Experiment, Sampled Mid-late August of Each Year.

Cropping System	Previous Crop[a]	1981	1982	1983	1984	1985	1986	1987
					Mg ha^{-1}			
Low-input/ livestock	Red clover	0.95a[bc]	—	0.44c	0.89a	—	0.25b	—
	Soybeans	0.89a[d]	—	0.63b	—	2.12a	0.75a	—
Low-input/ cash grain	Legume	0.83a	2.51a[e]	0.91a[e]	1.00a 0.69a[e]	0.27b	0.79a[e]	0.52a
Conventional	Corn	0.38b	0.09b	0.07d	0.24b	—	0.04b	0.12b
	Soybeans	0.29b	NA[f]	0.02d	0.26b	0.11b	0.18b	0.07b

[a] Crop preceding 1981 was wheat in all treatments.
[b] Values for each year that are followed by the same letter are not significantly different (.05 level), as determined by Duncan's multiple range test.
[c] Rotation entry #2, corn grain.
[d] Rotation entry #3, corn silage.
[e] Indicates a shorter-season variety of corn was used in this treatment, as compared to the other treatments in the same year.
[f] Data not available.

It is important to consider rotation starting point when converting to a low-input system. Lower corn yields in the low-input rotations for the first four years indicate that corn is a poor choice for starting the rotation. Soybeans require much less soil N, and yields were not affected by the lack of fertilizer N. Effective weed control depends upon timely cultivations, and this makes soybeans an intermediate choice for beginning the rotation. Winter wheat is also an intermediate choice. Although it can compete well with summer annual weeds, yield may be reduced if available soil N levels are low.

The ideal choice for starting the rotation is a leguminous hay or green manure crop. Established with a small grain nurse crop in the early spring, these crops effectively suppress weeds and contribute to improved soil structure and fertility, optimizing conditions for the crops that follow.

This experiment has also shown that additions of organic matter can have substantial effects on soil properties within a relatively short time. In 1985, corn plots in the low-input/cash-grain system that followed a plowdown of hairy vetch had greater water stable aggregates in the spring and more rapid water infiltration rates (Werner, 1988), greater microbial activity (Doran et al., 1987), and more earthworms, mites, collembola, and other soil invertebrates (Werner, 1988) than did plots in the conventional system. We believe that these physical and biological improvements have resulted in better crop root growth and increased efficiency of nutrient cycling.

The long-term benefits of continued additions of organic matter were demonstrated in a Washington State study that compared an organic farm to a conventionally managed farm (Reganold, 1988). The organically managed soil, which had received 40 years of green manures, had higher organic matter, greater cation exchange capacity, better structure, deeper topsoil (indicating less soil erosion), and a larger and more active microbial population than did the conventionally managed soil, which did not benefit from green manures.

An economic analysis was conducted to determine whether the starting crops in each system influenced net returns over one complete rotational cycle (five years) and whether the alternative systems produced significantly different returns relative to the conventional system (Duffy et al., 1989). A partial budgeting approach, which assumed that all crops were commercially marketed, was used to calculate net returns over variable costs. A standard linear regression analysis was used to test the starting crop question, and an analysis of variance was used to analyze the systems over a five-year period.

It was found that starting crop in the low-input rotations did influence returns to land and management (Table 10.6) (Duffy, M., personal communication). In the low-input/livestock system, starting the rotation with a legume hay and a nurse crop of oats (*Avena sativa* L.) resulted in a \$524 ha^{-1} average net return over five years, compared with \$358 ha^{-1}

TABLE 10.6 Return to Land and Management by System and Rotation in the Rodale Research Center Conversion Experiment, 1981–1985 (Duffy, M., Personal Communication).

Cropping System	Starting Crop (1981)	Average Annual Net Return $ ha^{-1}
Low-input/ livestock	Oat/legume	524
	Corn silage	489
	Corn grain	358
Low-input/ cash grain	Oat/legume	388
	Corn grain	277
	Soybeans	321
Conventional	Corn grain	398
	Soybeans	420
	Corn grain	341

when corn grain was the first crop. Beginning the rotation with corn silage resulted in an intermediate net return.

For the low-input/cash-grain system, starting the rotation with a legume with an oats nurse crop also resulted in the highest yearly average returns over five years. Beginning with either corn grain or soybeans achieved lower returns.

A comparison of each system over one complete rotational cycle (1981 to 1985) showed that the low-input/livestock system had significantly greater returns than the conventional system, as calculated using Duncan's multiple range test. However, there was no difference when Tukey's studentized range and Scheffe's means test were used. The average returns (combining rotational entry points) for both of these systems were significantly higher than the low-input/cash-grain system.

Nebraska Rotation Experiment

Researchers at the University of Nebraska initiated an experiment in 1976 to compare several cropping systems and input levels. Conventionally managed, continuous corn that received herbicide, insecticide, and synthetic fertilizer (CC-HFI) was compared to a corn-soybean-corn-oat/clover rotation with three levels of external inputs: herbicides and synthetic fertilizer (HF), fertilizer only (FO), and an organic treatment (ORG), which received no synthetic fertilizers or pesticides, but received nutrients from feedlot manure. The experimental site had been in alfalfa for five years prior to initiation of the experiment, and soil fertility was high. Further methodological details are provided in Sahs and Lesoing (1985).

Corn yields through 1987 are presented in Table 10.7. Until 1982, no clear trend was evident in comparing the ORG treatment to the other

TABLE 10.7 Nebraska Rotation Experiment Corn Yields.

	Treatment			
Year	Continuous Corn	Rotation Herb/Fert	Rotation Fert. Only	Rotation Organic
		$Mg\ ha^{-1}$		
1976	2.76a*	3.45a	3.51a	2.70a
1977	2.07a	2.07a	1.76a	0.88b
1978	8.03b	8.78a	8.28ab	8.34ab
1979	2.32ab	2.45a	2.32ab	1.69b
1980	3.83a	4.52a	4.58a	4.83a
1981	7.09ab	7.53a	7.21a	6.09b
1982	5.58b	6.65a	6.02b	5.83b
1983	1.69b	3.26a	2.82ab	2.89ab
1984	1.25b	3.89a	3.83a	3.95a
1985	6.46a	7.03a	6.84a	7.34a
1986	5.14b	7.15a	6.90a	6.40a
1987	3.70b	5.52a	5.52a	5.58a

* Values for each year (row) that are followed by the same letter are not significantly different (.05 level), as determined by Duncan's multiple range test.

three. The ORG corn yield was less than all others in three out of the first seven years (1977, 1979, 1981), when weed pressure was very high. Continuous corn (CC-HFI) yield was not significantly different from the other treatments during the first seven years, except in 1978, when it was significantly less than the HF corn.

Since 1982, the ORG treatment corn yields have not been significantly different from the other rotation (HF and FO) yields, except in 1982, when HF corn yielded significantly higher than all other treatments. This was a wet year, and weed control in this treatment was better than in the nonherbicide rotational systems. HF weed control was also better than in CC-HFI, which had smartweed (*Polygonum* sp.) and grass infestations.

The CC-HFI yields since 1982 have been lower than the other treatments, except in one year. Orthogonal contrasts showed that since 1982, CC-HFI corn yields were significantly lower than HF and FO yields, except in 1985, when the yield reduction was not statistically significant.

Corn yields in the various treatments appeared to be strongly influenced by weather conditions, particularly from 1980 to 1983. During this period, when corn experienced heat and drought stress, ORG corn yield was not significantly different from HF corn. In 1981, a year with adequate moisture and excellent growing conditions for corn, ORG yields were less than all other treatments, including CC-HFI. From 1984 on, however, weather had less impact on corn yields. ORG corn was generally comparable to, and CC-HFI was generally lower than, the other treatments, regardless of growing conditions.

Before 1980, any treatment interactions with weather were probably masked by the effects of the alfalfa that had been grown prior to the onset of the experiment. Soil moisture was lacking in all treatments at initiation of the experiment due to moisture depletion by the alfalfa. Dry soil conditions were intensified by reduced soil water recharge rates resulting from low rainfall conditions from 1975 through 1977. In addition, alfalfa's contribution to the organic-N pool in all treatments reduced N treatment effects during the early part of the experiment.

Soybean yields in the ORG treatment were significantly lower than the other two rotations in 1977, 1978, 1982, and 1983 (Table 10.8). Weed growth was high in these plots during those years. This is in contrast to results at Rodale Research Center, in which soybeans produced high yields during the conversion, even with high weed biomass. After 1983, soybean yields were similar among treatments.

Oat yields were not adversely affected by the practice of low-input methods (Table 10.9). In 1979 and 1985, ORG oat yields were significantly higher than the other treatments, and were equivalent in other years. No nutrients were applied through 1980. Beginning in 1981, however, oats in all treatments began receiving small amounts of N and P, according to recommendations based on soil tests taken the preceding fall. Oats in the ORG treatment received feedlot manure to supply the recommended N rate. Oat yields in this experiment are comparable to those from the Rodale experiment, where the conversion process did not greatly reduce small grain yields.

TABLE 10.8 Nebraska Rotation Experiment Soybean Yields.

	Treatment		
Year	Herbicide/ Fertilizer	Fertilizer Only	Organic
		Mg ha⁻¹	
1976	0.94a[a]	1.01a	0.87a
1977	2.96[b]	2.96	2.35
1978	2.02a	2.22a	1.61b
1979	1.75a	1.75a	1.61a
1980	2.76a	2.82a	2.69a
1981	3.49a	3.43a	3.29a
1982	3.29a	3.16b	2.82c
1983	2.82a	2.08a	1.68b
1984	1.14a	1.21a	1.08a
1985	2.62a	2.62a	2.62a
1986	3.02a	2.96a	2.89a
1987	2.62a	2.62a	2.49a

[a] Values for each year that are followed by the same letter are not significantly different (.05 level), as determined by Duncan's multiple range test.
[b] Statistics not available for 1977.

TABLE 10.9 Nebraska Rotation Experiment Oat Yields.

	Treatment		
Year	Herbicide/ Fertilizer	Fertilizer Only	Organic
		Mg ha^{-1}	
1976	1.94a[a]	2.01a	1.90a
1977	0.82a	0.75a	0.82a
1978	0.29a	0.22a	0.22a
1979	2.22b	2.33ab	2.69a
1980[b]	—	—	—
1981	1.58[c]	1.58	1.58
1982	2.29a	2.33a	2.54a
1983	2.22a	2.04a	1.90a
1984	2.44a	2.26a	2.40a
1985	3.19b	3.01b	3.73a
1986	3.62b	3.91ab	4.05a
1987	2.72a	2.87a	2.97a

[a] Values for each year that are followed by the same letter are not significantly different (.05 level), as determined by Duncan's multiple range test.
[b] Due to poor stands, oats were not harvested for grain in 1980.
[c] Statistics not available. This is an average yield across all treatments.

There were striking differences between the ORG treatment and the other three treatments in soil chemical characteristics by 1979, primarily resulting from the manure added to the ORG plots (Table 10.10). In 1979, the ORG soil was significantly higher than the other treatments in levels of P, K, organic matter, and pH. In 1986, the trend continued. Organic matter, P, and K were significantly higher in the ORG treatment, and pH was significantly lower in the CC-HFI treatment. Total soil-N was also highest in the ORG treatment in 1986, significantly higher than the CC-HFI treatment.

Soil chemical and microbial characteristics measured in 1981 and 1982 also showed significant differences between the ORG and the chemical treatments (Fraser et al., 1988). The use of pesticides did not appear to directly affect microbial activity, but additions of manure and the inclusion of hay crops in the rotations did have a significant effect. Total N and PMN were 22% to 40% higher, and soluble P was eight times higher, in the ORG treatment than in the nonmanured treatments. Soil microbial activity in manure-amended soils and soils planted to hay was higher than in other treatments.

In contrast to the RRC conversion experiment, no period of clearly reduced corn yields was observed at the initiation of a low-input crop rotation. This lack of yield reduction may be due largely to the effects of the preceding alfalfa crop. A stabilizing effect over time did occur,

TABLE 10.10 Selected Soil Chemical Properties (0 to 15cm depth) in 1979 and 1986, Nebraska Rotation Experiment.

Treatment	Organic Matter		Phosphorus		Potassium		pH		Total Nitrogen
	1979	1986	1979	1986	1979	1986	1979	1986	1986
	g kg^{-1}		*mg kg*$^{-1}$		*mg kg*$^{-1}$				*g kg*$^{-1}$
Continuous corn	30.5ab*	28.8b	10b	22b	291b	348b	6.33b	6.33b	1.36b
Herb./fert. rotation	32.1ab	31.2b	9b	18b	306b	375b	6.51b	6.88a	1.60ab
Fert. only rotation	30.0b	32.8b	7b	20b	287b	396b	6.63b	6.86a	1.59ab
Organic	36.0a	39.6a	49a	96a	478a	618a	6.93a	7.02a	1.81a

* Values that are followed by the same letter within year are not significantly different (.05 level), as determined by Duncan's multiple range test.

however, as the ORG corn yields are more consistently equivalent to the other rotation yields after the first six years. Stabilization of yield is due in part to N availability, because an ample supply of residual inorganic N from manure was available for good crop production. Increases in soil organic matter and improved soil physical characteristics (Fraser et al., 1988) also contributed to better crop growth.

This experiment was analyzed for net returns and risk (variability of returns) for the various systems over an eight-year period from 1978 to 1985 (Helmers et al., 1986). The first two years of the experiment were eliminated from the economic analysis to account for "start-up time" for the experiment, precluding a discussion of the economics of this period.

All of the rotation treatments performed better than continuous corn in terms of both net returns and risk. The different chemical treatments (HF, FO, or ORG) had little influence on net returns, since the lower input costs associated with the ORG treatment offset the reduced yields. If the cost of the manure fertilizer was charged at application cost only (a "lowest cost" scenario), net returns and variability of the ORG treatment were virtually the same as for the HF and FO treatments (average net return \$282 ha^{-1}). If manure is charged at the cost of an equivalent amount of chemical fertilizer (a "highest cost" scenario), the ORG treatment net return is significantly reduced, at \$230 ha^{-1}. This is still higher than the continuous corn treatment, however, which returned only \$183 ha^{-1}.

The Rodale and Nebraska studies demonstrate that organic cropping systems can be viable both biologically and economically over the long term. Several aspects of low-input cropping systems were not addressed in these experiments, however, and should be addressed in future farming systems studies. Many of these concerns were identified by Duffy et al. (1989).

The first major concern involves how changes in cultural practices and crop rotations would be integrated into a whole farm plan. Would a new crop mix, for example, fulfill the nutritional requirements of the farm's livestock? If not, is there access to a commercial market for these crops?

Second, labor must be evaluated not only in terms of availability, but also according to the degree of skill required and the need for timely field operations. In the Rodale study, for example, while overall economic returns in the low-input livestock treatment and conventional treatments were equivalent, the former required 33% more fieldwork time per hectare than the latter.

Third, risk must be more accurately quantified so that farmers can truly assess the benefits and shortcomings of the various types of cropping systems. Practitioners of organic agriculture, for example, must consider availability of labor and the limited number of days suitable for timely cultivations. A farmer using chemicals must consider the reduced effectiveness of herbicides under certain weather conditions and the health risks to the applicator.

Larger societal concerns that need further research are the long-term energy and environmental costs associated with a particular farming system. For example, some studies indicate that a farm using pesticides, synthetic fertilizers, and continuous row cropping practices requires more petrochemicals (Lockeretz et al., 1977) and poses a greater likelihood for soil erosion (Reganold, 1988) and water pollution than does an organic, diversified farm.

At the farm level, many of these concerns can be addressed with a thorough assessment of the farm's resources and the relative strengths and weaknesses of each resource (MacRae et al., 1990). This analysis should consider the soil, biotic, climatic, physical, and human aspects of the farm. A soil inventory not only involves type, capability class, and nutrient status, but biological activity and the presence of toxic residues. A biotic assessment includes (a) cropping history, (b) predominant weeds, insects (beneficial and pest), and diseases, (c) presence of livestock and manures, and (d) woodlands, meadows, and hedgerows within the farmstead. The climatic evaluation should consider any unique microclimate features present on the farm. The physical inventory includes (a) equipment, (b) housing and building facilities, (c) storage and processing capabilities, (d) borrowing and rental potential, and (e) water availability. The human resources include (a) family, (b) neighbors, (c) extension, (d) soil conservation service, (e) private consultants, (f) agricultural dealers and suppliers, and (g) government programs.

CONCLUSIONS

The barriers to reducing purchased inputs while creating a productive and sustainable farming system fall into three interrelated categories: biological, information and managerial, and socioeconomic and political. The biological barriers include improving soil fertility and minimizing outbreaks of pest insects, diseases, and weeds. These biological needs are fulfilled by making the necessary management decisions. The lack of farm-specific information, however, makes it difficult to make a smooth transition to alternative practices. Also, management decisions are primarily governed by the current economic and political realities that surround agriculture.

One of the cornerstones for a successful conversion is developing a crop rotation that not only improves soil conditions and prevents a buildup of pests, but also suits the needs of a specific farm. These needs include (1) growing appropriate crops for the soil/water/climate resources available, (2) having markets for the crops in rotation, (3) matching crops with available equipment, storage and processing facilities, (4) providing the necessary labor for the crops grown, (5) integrating the crops with livestock nutritional needs, and (6) identifying the effects that a new crop or new crop combination would have on the farm's seasonal cash flow (Kirschenmann, 1988).

Once these specific farm needs have been met, the farmer can concentrate on developing a soil building crop rotation that will provide long-term biological and financial stability. Including legumes in the rotation is a key component to achieving a sustainable system. Winter annual legume cover crops enable a farmer to continue raising the same acreage of cash crops. Perennial legume/grass mixtures are grown at the expense of some cash crop acreage, but they provide long-term improvement of soil physical and biological properties and can fulfill many nutritional needs of livestock.

The next step in creating a sustainable rotation is to choose the proper sequence of crops to minimize pest damage and avoid depletion of soil nutrients. Some general principles to follow include (1) following N-fixing legumes with high users of N, (2) alternating cool and warm season crops, (3) preceding a slow growing, noncompetitive crop with a weed-suppressing crop, (4) alternating deep- and shallow-rooted crops, (5) alternating high and low users of water if this resource is limited, (6) employing weed suppressing allelopathic crops such as oats or sorghum (*Sorghum* sp.) wherever possible, and (7) allowing a sufficient time interval before repeating a crop that is vulnerable to insect and disease pests (Vogtmann et al., 1986).

Most aspects of changing the farming system require good management skills and can be quite "information-intensive." Pest management with little or no use of pesticides is one example. Cultural and biological pest control requires information about a pest's life cycle and economic threshold levels in the crop. Familiarity with the many alternatives to chemical pest control is important to enable one to choose a set of tactics best suited for a particular cropping system.

Farmers can facilitate the conversion process by conducting on-farm research. An individual farmer can experiment with one field, or with strips of treatments in a field, in which a new technique or reduced rate of some input is compared to a standard practice. Participating in a network of farmers involved with on-farm research and demonstration can multiply the knowledge gained. Research results can be disseminated to other farmers at field days, workshops, and through newsletters and magazines, in addition to less formal networking with other farmers. Farmers are doing this now as individuals and as participants in various farmer organizations and networks throughout the United States. This grassroots type of information sharing may be an important part of overcoming many barriers to transition, and until recently has not been part of the traditional "researcher → extension agent → farmer" top-down model of research and extension.

Converting to a low-input/sustainable cropping system requires several "trade-offs" involving management or field time, money, materials, and risk. Though cultural control of pests requires fewer purchased inputs, some control measures may be time- and/or labor-intensive, particularly if scouting is involved. Mechanical cultivation for weed control can be time-consuming, especially on large acreages, and it may be necessary for the farmer to acquire additional equipment. Legume seeding may require the purchase of seed that is not yet widely available and, hence, expensive.

Economic risk is important when considering the trade-offs of conversion. Risk may be high in the early phase of transition. In the long run, however, risk should decline, as diversity of crops is increased, dependence on outside inputs is reduced, and yields stabilize as a result of improved soil conditions. In an economic analysis of various cropping systems by Helmers et al. (1986) that included data from the Nebraska experiment, rotations were found to exhibit lower variability of returns (a component of risk) than did continuous corn (data not shown).

A major challenge in low-input cropping systems research is to reduce the amount of risk the farmer must face during the conversion process. Any new practice being considered during the transition involves a degree of uncertainty, and the farmer must determine how much risk he or she is willing to tolerate and what the consequences of a crop failure or setback would be. Some tactics, such as crop rotation, can reduce risk by increasing the diversity of crops grown in a particular year. However, these strategies can also increase risk. For example, if a green manure needed to supply N to a subsequent crop fails, either another N source such as animal manure or mineral fertilizer will be necessary, or another crop will have to be substituted which will not need the additional N.

A realistic whole-farm conversion scenario that would reduce risk to the farmer is the gradual withdrawal of inputs and employment of new strategies, allowing the transition to take place over several years. For example, N rates for corn could be cut back 20 to 50 kg ha^{-1} yr^{-1}, until enough animal manure or green manure crops have been incorporated into the soil to build up the soil organic matter and the soil organic N. Spot spraying for problem perennial weeds could be continued. Postemergence grass and broadleaf weed herbicides could be used during the first season or two of adjusting to a new cultivator and as a backup during wet years when timely cultivation is not possible.

The conversion process could also be accomplished gradually by converting one field at a time. The experiment at the Rodale Research Center demonstrated that the optimal crop rotational entry point with which to begin the conversion process is a legume crop or a small grain/legume sequence. On a farm, once a diverse crop rotation that includes legumes is in place, the legume field can be the starting point for continuing the rotation under low-input conditions. Over a period of five or ten years, the whole farm can slowly be converted. This period of time also allows for learning to take place and a body of experience to be accumulated about what works best: which legumes to use, when to plant them, when to cultivate, how to set the cultivator depending on soil moisture conditions, and other details that are extremely important when managing, or rather orchestrating, an integrated farming system based on sound knowledge of ecological and agronomic principles.

Conversion to a sustainable agricultural system is not simply a substitutive process, such as replacing herbicide with mechanical cultivation for weed control, or replacing insecticide with a predator insect such as a wasp. Several techniques must be integrated into a strategy designed for a particular cropping system. Effective cultural control of weeds involves crop rotation and competitive crop cultivars combined with timely cultivation and/or mulch crops. Insecticide may be eliminated once a sound crop rotation is designed and combined with other strategies, including resistant or tolerant crop varieties, careful timing of planting and other operations, and possibly the introduction of a biocontrol organism or a cover crop that attracts beneficial insects.

An integrated approach also requires choosing cultural practices and plant species that are adapted to the specific climate and soils of a region. A farmer in Nebraska, for example, may find that sorghum is easier to work with than corn under low-input conditions, given the low rainfall of the region. While red clover (*Trifolium pratense* L.) may be suitable as a cover crop in Pennsylvania, crimson clover (*Trifolium incarnatum* L.) may be more appropriate for cropping systems further south. These differences can be used in cropping systems in innovative ways. For example, a legume that is not winter-hardy, such as crimson clover, could be used in a more northern cropping system where it will winter-kill and provide a dead mulch for planting into the following season (Janke et al., 1987).

Since a successful transition to a sustainable farming operation is *not* a substitutive process, some restructuring of the farm system is necessary to accommodate diverse crop rotations and integrated pest control. Biological restructuring of the plant-soil system, that is, changes in nutrient cycling due to crop diversification and organic matter accumulation, is also an integral part of the process. Labor and management restructuring may be necessary to accommodate changes in the crop and farming system. A farmer endeavoring to make this transition successfully requires not only excellent farming and managing skills, but also the support of a research and extension system willing to work with farmers in developing these alternatives.

REFERENCES

Aldrich, S. R. (1984). "Nitrogen management to minimize adverse effects on the environment." In *Nitrogen in Crop Production* (Ed. R. D. Hauck), pp. 663–73. American Society of Agronomy, Crop Science Society of America, and Soil Society of America, Madison, WI.

Alexander, M. (1977). *Introduction to Soil Microbiology*, 2nd ed. Wiley, New York.

Arden-Clarke, C., and R. D. Hodges. (1988). "The environmental effects of conventional and organic/biological farming systems. II. Soil ecology, soil fertility and nutrient cycles," *B.A.H.* **5**:223–87.

Baker, B. P., and D. B. Smith. (1987). "Self-identified research needs of New York organic farmers," *Am. J. Alt. Agric.* **II**:107–13.

Bolton, E. F., V. A. Dirks, and J. W. Aylesworth. (1976). "Some effects of alfalfa, fertilizer, and lime on corn yield in rotations on clay soil during a range of seasonal moisture conditions," *Can. J. Soil Sci.* **56**:21–25.

Bouldin, D. R., S. W. Klausner, and W. S. Reid. (1984). "Use of nitrogen from manure." In *Nitrogen in Crop Production* (Ed. R. D. Hauck), pp. 221–45. American Society of Agronomy, Crop Science Society of America, and Soil Society of America, Madison, WI.

Brady, N.C. (1984). *The Nature and Properties of Soils*, 9th ed., p. 357. Macmillan, New York.

Certification Handbook and Materials List. (1988). California Certified Organic Farmers, Santa Cruz, CA.

Chambers, R. D. (Ed.) (1989). *The Penn State Agronomy Guide 1989–1990.* Pennsylvania State Univ., University Park.

Crookston, R. K. (1984). "The rotation effect: What causes it to boost yields?" *Crops & Soils* **36**:12–14.

Doran, J. W., D. G. Fraser, M. N. Culik, and W. C. Liebhardt. (1987). "Influence of alternative and conventional agricultural management on soil microbial processes and nitrogen availability," *Am. J. Alt. Agri.* **2**:99–106.

Duffy, M., R. Ginder, and S. Nicholson. (1989). "An economic analysis of the Rodale conversion project: Overview." Dept. Agricultural Economics, Iowa State Univ. (unpublished).

Forcella, F., and M. J. Lindstrom. (1988). "Movement and germination of weed seeds in ridge-till crop production systems," *Weed Sci.* **36**:56–59.

Foster, L. H. (1988). "Herbs in pastures: Development and research in Britain, 1850–1984," *B.A.H.* **5**:97–133.

Fraser, D. G., J. W. Doran, W. W. Sahs, and G. W. Lesoing. (1988). "Soil microbial populations and activities under conventional and organic management," *J. Environ. Qual.* **17**:585–90.

Gershuny, G., and J. Smillie. (1986). *The Soul of Soil: A Guide to Ecological Soil Management*. Gaia Services, St. Johnsbury, VT.

Goldstein, W. A. (1986). *Alternative Crops, Rotations, and Management Systems for the Palouse*. Ph.D. dissertation, Washington State Univ., Pullman.

Harwood, R. R. (1985). "The integration efficiencies of cropping systems." In *Sustainable Agriculture and Integrated Farming Systems* (Eds. T. C. Edens, C. Fridgen, and S. L. Battenfield). pp. 64–75. Conference Proc. 1984, Michigan State Univ.

Helmers, G. A., M. R. Langemeier, and J. Atwood. (1986). "An economic analysis of alternative cropping systems for east-central Nebraska," *Am. J. Alt. Agric.* **1**:153–58.

Janke, R. R., R. Hofstetter, B. Volak, and J. K. Radke. (1987). "Legume interseeding cropping systems research at the Rodale Research Center." In *The Role of*

Legumes in Conservation Tillage Systems (Ed. J. F. Power). pp. 90–91. Soil Conservation Society of America, Ankeny, IA.

Johnson, J. W., L. F. Welch, and L. T. Kurtz. (1975). "Environmental implications of N fixation by soybeans," *J. Environ. Qual.* **4**:303–306.

Kirschenmann, F. (1988). *Switching to a Sustainable System*. Northern Plains Sustainable Agriculture Society, Windsor, ND.

Koepf, H. H., B. D. Pettersson, and W. Schaumann. (1976). *Biodynamic Agriculture: An Introduction*. Anthroposophic Press, Spring Valley, NY.

Krysan, J. L., and T. A. Miller (eds.). (1986). *Methods for the Study of Pest Diabrotica*. Springer-Verlag, New York.

Ladd, J. N., and M. Amato. (1986). "The fate of nitrogen from legume and fertilizer sources in soils successively cropped with wheat under field conditions," *Soil Biol. Biochem.* **18**:417–25.

Liebhardt, W. C., R. W. Andrews, M. N. Culik, R. R. Harwood, R. R. Janke, J. K. Radke, S. L. Rieger-Schwartz. (1989). "A comparison of crop production in conventional and low-input cropping systems during the initial conversion to low-input methods," *Agron. J.* **81**:150–59.

Liebhardt, W. C. (1984). "Variability of fertilizer recommendations by soil testing laboratories in the United States." In *Environmentally Sound Agriculture* (Ed. W. Lockeretz). Praeger, New York.

Lockeretz, W. (1980). Energy inputs for nitrogen, phosphorus, and potash fertilizers. In *Handbook of Energy Utilization in Agriculture* (Ed. D. Pimental), pp. 23–24. CRC Press, Boca Raton, FL.

Lockeretz, W., R. Klepper, B. Commoner, M. Gertler, S. Fast, D. O'Leary, and R. Blobaum. (1977). "Economic and energy comparison of crop production on organic and conventional corn belt farms." In *Agriculture and Energy* (Ed. W. Lockeretz). Academic Press, New York.

Lockeretz, W., and P. Madden. (1987). "Midwestern organic farming: A ten-year follow-up," *Am. J. Alt. Agric.* **II**:57–63.

Long, S. P., and D. O. Hall. "Nitrogen cycles in perspective," *Nature* **329**:584–85.

MacCormack, H. (1989). *Standards and Guidelines for Oregon Tilth Certified Organically Grown*. Oregon Tilth, Tualatin, OR.

MacRae, R. J., S. B. Hill, G. R. Mehuys, and J. Henning. (1990). "Farm-scale agronomic and economic conversion from conventional to sustainable agriculture," *Adv. Agron.* 43 (in press).

Mengel, K., and E. A. Kirby. (1982). *Principles of Plant Nutrition*, 3rd ed. International Potash Institute, Bern, Switzerland.

Metcalf, C. L., W. P. Flint, and R. L. Metcalf. (1962). *Destructive and Useful Insects*, 4th ed. McGraw-Hill, New York.

Minnich, J., and M. Hunt. (1979). *The Rodale Guide to Composting*. Rodale Press, Emmaus, PA.

Mishra, M. M., and K. C. Bangar. (1986). "Rock phosphate composting: Transformation of phosphorus forms and mechanisms of solubilization," *B.A.H.* **3**: 331–40.

Odell, R. F., W. M. Walker, L. V. Boone, and M. G. Oldham. (1974). *The Morrow Plots: A Century of Learning*. Univ. Of Ill. Agric. Exp. Sta. Bull. No. 775.

Organic Farm Certification Program. (1989). Natural Organic Farmers' Assn. New York, Ithaca, NY.

Patriquin, D. G. (1986). "Biological husbandry and the 'nitrogen problem,'" *B.A.H.* **3**: 167-89.

Patriquin, D. G., N. M. Hill, D. Baines, M. Bishop, and G. Allen. (1986). "Observations on a mixed farm during the transition to biological husbandry," *B.A.H.* **4**:69–154.

Paul, E. A., and N. G. Juma. (1981). "Mineralization and immobilization of soil nitrogen by microorganisms." In *Terrestrial Nitrogen Cycles* (Eds. F. E. Clark, and T. Rosswall). *Ecol. Bull.* **33**:179–95.

Pierce, F. J., and C. W. Rice. (1988). "Crop rotation and its impact on efficiency of water and nitrogen use." In *Cropping Strategies for Efficient Use of Water and Nitrogen* (Ed. W. L. Hargrove). American Society of Agronomy Spec. Publ. No. 51.

Radke, J. K., R. W. Andrews, R. R. Janke, and S. E. Peters. (1988). "Low-input cropping systems and efficiency of water and nitrogen use." In *Cropping Strategies for Efficient Use of Water and Nitrogen* (Ed. W. L. Hargrove). American Society of Agronomy Spec. Publ. No. 51.

Reganold, J. P. (1988). "Comparison of soil properties as influenced by organic and conventional farming systems," *Am. J. Alt. Agric.* 3:144–55

Sahs, W., and G. Lesoing. (1985). "Crop rotations and manure versus agricultural chemicals in dryland grain production," *J. Soil Water Conserv.* **40**:511–16.

Salter, R. M., and C. J. Schollenberger. (1939). *Farm Manure.* Bulletin No. 605, Ohio Agric. Exp. Sta.

Sauerbeck, D. R., and M. A. Gonzalez. (1976). "Field-decomposition of ^{14}C-labelled plant residues in different soils of Germany and Costa Rica." In *Soil Organic Matter Studies*, pp. 159–70. I.A.E.A., Vienna.

Schertz, D. L., and D. A. Miller. (1972). "Nitrate-N accumulation in the soil profile under alfalfa," *Agron. J.* **64**:660–64.

Smith, G. E. (1942). *Sanborn Field: Fifty Years of Field Experiments with Crop Rotation, Manure, and Fertilizers.* Bull. No. 458, Missouri Agric. Exp. Stn.

Stanford, G., and S. J. Smith. (1972). "Nitrogen mineralization potentials of soils," *Soil Sci. Soc. Amer. Proc.* **36**:465–72.

Stevenson, F. J. (1986). *Cycles of Soil.* pp. 231–84. Wiley-Interscience, New York.

"Summary report on farming operations, The Farmers' Own Network for Extension, updated 30 September 1987." (1987). *The New Farm* and the Regenerative Agriculture Assn., Emmaus, PA.

Tisdale, S. L., and W. L. Nelson. (1975). *Soil Fertility and Fertilizers,* 3rd ed. Macmillan, New York.

USDA. (1980). *Report and Recommendations on Organic Farming.* Washington, DC.

Vogtmann, H., E. Boencke, L. Woodward, and N. Lampkin. (1986). *Converting to Organic Farming* (Ed. N. Lampkin). Elm Farm Research Centre, Newbury, U.K.

Werner, M. R. (1988). "Impact of Conversion to Organic Agricultural Practices on Soil Invertebrate Ecosystems." Ph.D. dissertation, State University of New York, College of Environmental Science and Forestry, Syracuse.

11 THE ECONOMICS OF SUSTAINABLE LOW-INPUT FARMING SYSTEMS

J. PATRICK MADDEN
Glendale, California

Concerns about the sustainability of U.S. agriculture include a wide spectrum of trends and forces related to long-term productivity of the land, environmental hazards, rural communities, and human health risks (Madden, 1987a). Problems often attributed to farming technologies include the following:

- Risks of environmental damage due to chemicals.
- Depletion of nonrenewable resources such as underground water aquifers, petroleum, and phosphate deposits.
- Contamination of water supplies by agricultural chemicals.
- Possible health risks to farm workers and consumers associated with exposure to pesticides and their residues.
- Replacement of self-employed family farmers by large-scale absentee owners.
- Atrophy of social and commercial infrastructure necessary for rural communities.

Certainly not all American farmers use harmful farming practices. And many farmers would like to find profitable ways to reduce or eliminate their dependence on synthetic chemical pesticides and fertilizers, as well as other practices that are the source of growing social concern.

Sustainability in agriculture is much more than maintaining the status quo, and the economics of sustainable agriculture is concerned with much more than farm profits. Nonetheless, it is clear that if a farming method or system is not productive and efficient enough to be profitable, it cannot be considered sustainable. The purpose of this chapter is to review what is known about the profitability of alternative, low-input farming methods and systems generally considered more sustainable than conventional practices, and to suggest some challenges remaining in this important area.

The discussion begins by addressing conceptual issues regarding an alternative goal or norm for agricultural science and technology. Various economic perspectives are then outlined, as a prelude to a discussion of scientific studies on the profitability of low-input methods of farming. Rather than attempting an exhaustive treatment of the literature, this chapter emphasizes those areas of farming technology most dependent on chemical inputs, namely, methods of soil fertility management and pest control. These are components that must be understood before more sustainable farming systems can be widely and intelligently advanced. Studies pertaining to marketing and livestock production, while highly relevant to sustainable agriculture, are beyond the scope of this chapter. Some thoughts are offered regarding the likely impacts of widespread transition from conventional farming methods toward adoption of farming methods based on an alternative norm.

THE CONCEPTS OF SUSTAINABLE, LOW-INPUT, AND REGENERATIVE AGRICULTURE

What is the definition of sustainable agriculture? There exists no universally acceptable definitions. For the purposes of this chapter, a distinction is drawn between sustainable *agriculture* and sustainable *methods of farming*. It is best not to define sustainable agriculture in terms of a specific list of farming methods or practices; what farming methods are sustainable is a highly site-specific matter. Sustainable agriculture is a goal, a vision for the future that is humane, safe, and abundant. A sustainable agriculture provides an abundance of safe and nutritious food and fiber for present and future generations, at prices that are fair and reasonable, yet high enough to support a prosperous sector of family farms.

Sustainable agriculture is based on production methods and systems that require little or no input of synthetic chemical pesticides or fertilizers. A farmer, scientist, educator, or organization pursuing the goal of a more sustainable agriculture seeks profitable and socially acceptable ways for farmers to make a profitable transition toward less dependence on certain categories of synthetic chemical pesticides, highly soluble forms of fertilizers, and other substances that are known or suspected to be harmful to human health or environmental integrity. To the maximum extent feasible, this approach to farming relies on methods such as natural enemies or biological control agents, mechanical cultivation, and crop rotations to control pests. Renewable sources of soil nutrients such as manures and legumes (in rotation or overseeded into row crops) are largely or totally substituted for many chemical fertilizers, following a transitional phase.

This concept of sustainable agriculture is essentially compatible with the widely used definition found in the USDA 1980 report on organic agriculture, but with the added provision that, to be sustainable, farming methods must be "socially acceptable." Some of the attributes necessary to make farming methods more socially acceptable include the following:

- A decentralized land ownership structure of farms, featuring a prosperous sector of self-employed family farmers, and absence of monopoly power over land and the other factors of production.
- Freedom from dependence on huge public subsidies, artificial incentives, and burdensome regulations.
- Minimal threat of environmental damage, contamination of water supplies, or health risk to farm workers or consumers as a result of chemicals used on farms.

The comparison between conventional and sustainable farming methods is often a matter of degree rather than a black and white contrast. Compared with conventional farming practices, sustainable farming methods more effectively enhance the productivity and resilience of natural resources; they are safer to humans and less damaging to the environment. To be sustainable, farming methods must be profitable, but they must also be socially acceptable, just, and equitable.

When narrowly defined, sustainability is an inadequate norm for agriculture. Ruttan points out, for example, that in view of rapidly expanding world food needs, it is not enough just to sustain the current level of productivity; the goal of expanding production to avert widespread starvation is beyond mere sustainability (Ruttan, 1988). Rodale has expressed similar concerns, plus other considerations:

> Do we want to sustain the current quality of the agricultural resource base? Not at all. We must help to make the land better, the air cleaner, our selection of germ plasm richer, and the water we drink and use purer. Sustainability, therefore, speaks with a very weak voice to the problem side of the agricultural equation. My contention is that sustainability is only a part of the right question to ask. Far more important is an analysis of resource renewability . . . (Rodale, 1988).

The alternative norm proposed by Rodale is a "regenerative agriculture," which builds on nature's own inherent capacity to cope with pests, enhance soil fertility, and increase productivity. I personally concur with Rodale's preference for regeneration rather than sustainability as a norm for agriculture. *Regenerative* is a more constructive and proactive concept than *sustainable*, when that term is narrowly defined. In this chapter, the term is defined broadly to encompass the goals of regeneration. Some well-known examples of widely used sustainable or regenerative farming practices and systems include the following:

- Some of the integrated pest management systems, specifically those that emphasize protection and enhancement of beneficial species in place of regular use of pesticides.
- Biological control programs, which have resulted in massive reductions in the use of insecticides through release of natural enemies of the pest.
- Use of legumes such as alfalfa and clover (especially the improved varieties) to biologically fix nitrogen as a soil nutrient for other crops.

Many observers quarrel with the use of the term *low-input* in this context, under the belief that huge amounts of labor must be expended when chemical pesticides are reduced or eliminated. A caricature of low-input methods is a man enduring endless hours under a broiling hot sun, chopping weeds with a hoe. As used in this chapter, low-input/sustainable agriculture is not intended to imply a reduction in all inputs. Rather, it implies a substitution of improved management and scientific information, and sometimes more labor, in place of most or all of the inputs of synthetic chemical pesticides and fertilizers. Scientific research and development can make methods such as these more productive, less risky, and more profitable than they have been heretofore.

THE LOW-INPUT/SUSTAINABLE AGRICULTURE (LISA) PROGRAM

Congress created a new research and education program, "Agricultural Productivity Research" (Subtitle C of the Food Security Act of 1985, PL 99-198), and in December of 1987, appropriated $3.9 million to begin work under this program. The program is now known as "Low-Input/Sustainable Agriculture" (LISA). The central purpose of the program is to reduce environmental risks and human health hazards attributed to agricultural pesticides and fertilizers by improving the profitability of low-input alternatives. The primary thrust of the program is to encourage and provide financial support for a new generation of agricultural research and education that will increase the profits and reduce the risks of farming methods that use substantially lower than conventional levels of synthetic chemical pesticides and fertilizers.

As the LISA program has evolved, a set of guiding principles has been developed:

1. If a method of farming is not profitable, it cannot be sustainable.
2. Somewhat lower yields plus much lower costs equal higher profits.
3. While some low-input farming methods and systems may be adopted quickly and easily, others require several years of transition. The time required for the transition depends on existing residues of pesticides, soil organic matter, climate, and other factors. Changes essential to the transition include restoration of earthworms and other beneficial biota, emergence of a natural balance between pests and their natural enemies, improvements in soil tilth and nutrient balances, adjustments in cash flow caused by switching to legume-based crop rotations, and learning the labor and management skills required for successful operation. In many situations, adoption should be done slowly, through careful planning over several years.
4. Low-input/sustainable farming methods and systems are highly site-specific. The findings of each experiment and project must be defined in terms of weather, soil conditions, and level of labor and management skills required.
5. Farmers should be provided full, accurate, and readily usable information regarding the impacts that adoption of low-input/sustainable methods and systems are likely to have on the farm's cash flow and profits, labor and management requirements, financial risks, long-term productivity of the soil, impact on water quality, human health risks, and other ecological impacts.
6. The potential results of a farmer adopting a specific farming method usually cannot be anticipated except in the context of a whole-farm management plan. The plan must take into account the labor and capital requirements. It must also deal with the complex interactions among crops, livestock enterprises, soils, water, populations of pests and their natural enemies, and other environmental impacts.
7. Low-input/sustainable farming methods and systems can be made more practical and profitable, and less risky, through research and education.
8. Essential to the success of efforts to develop and promote the science and farming methods needed for a more sustainable agriculture is a multiorganization approach, including interdisciplinary team efforts, meaningful participation of operating farmers, and involvement of both public and private organizations. Reductionist

research can provide important components for sustainable agriculture, but it cannot stand alone.

9. Soil and water conservation agencies, both public and private, as well as farmers and extension agents should be encouraged to become full partners with experiment stations, universities, and other research organizations in the design and implementation of the LISA program.
10. The LISA program should be administered at the regional level, with the major decisions made by persons in the region who are aware of local climate, soils, crops, and other conditions, and with a minimum of administrative expense and bureaucratic hassle.

In each of the four regions (Northeastern, North Central, Southern, and Western) a management team was appointed for the purpose of developing selection criteria and evaluating the first set of proposals submitted for funding by the LISA program. The regional programs are being administered through the Universities of Georgia, Nebraska, Vermont, and California. A total of 371 proposals submitted by public and private organizations were evaluated in May and June 1988. Forty-nine projects (including some combinations of similar proposals) were funded, with grants ranging in size from $2,000 to $220,000 (Madden et al, 1988). Each region received $851,000, including $15,000 "seed money" to expedite the process. In the second year (FY 1989) the program received a 14% increase in federal funding, to $4.45 million (Madden, 1988a). A wide range of public and private organizations including farmers and private research and education organizations are participating in the projects funded under this program (Cramer, 1988).

PROFIT AND SUSTAINABILITY

The relationship between agricultural sustainability and economic conditions is a double-edged sword. The sustainability of agriculture is strongly influenced by powerful monetary and institutional forces that determine the profitability of conventional and alternative methods of farming. These forces include international commodity markets, availability and prices of inputs such as petroleum and phosphates, and provisions of governmental regulations and price support programs. In the long term, the sustainability of agriculture can also have a major impact on local, national, and world economic conditions, as related to food costs, depletion of essential resources, and possibly climatic changes such as the greenhouse effect or global warming trend.

Farmers decide whether or not to adopt various farming methods on the basis of their personal values and the information at hand. Clearly, farmers are concerned about environmental and human health impacts of farming methods. Their willingness to switch from chemical-intensive to sustainable methods may be strongly influenced, however, by the expectation of profits or losses in the current growing season, as well as in future decades and generations. Obviously a farmer must remain financially solvent in the short term to have a "sustainable" business.

An essential consideration in determining how a farm's profits and risks are affected by adoption of low-input farming methods is the stage of the transition. Depending on the starting condition of the farm (in terms of pesticide residues, populations of pests and the natural enemies of pests, cropping history, soil fertility), the farm's economic performance can change significantly during a several-year evolution from conventional to low-input practices (Dabbert and Madden, 1986). Beneficial insects, soil microorganisms, earthworms, and other biological resources may be suppressed by soil residues of certain synthetic chemical pesticides. Sometimes these populations can be reestablished quickly, or it may take as long as six years, according to some estimates (Koepf et al., 1976; USDA, 1980).

The full benefits of many low-input farming methods such as crop rotations can also take years to realize in terms of increasing yields and reduced dependence on fertilizers and pesticides. After the transition phase is completed, expenses for fertilizers and pesticides may be reduced. But in the first few years after initiating a crop rotation, farm income can decline in many situations.

There are no universally valid rules of thumb in this matter. The outcome depends on the resources and management history of each specific farm and the knowledge and managerial skills of the farmer. Farmers often make management mistakes, especially during the first few years after adopting any new method of production. This is particularly true of many low-input practices that require additional knowledge and management skills, which often take time and special information to acquire. Furthermore, as discussed below, the farm's losses due to adoption of a legume-based crop rotation are often magnified as a direct result of federal price support programs keyed to the farm's output of corn, wheat, or other politically salient commodities.

Farmers need more and better information if they are to make informed decisions regarding ways to make a transition (on all or some part of their farm) from synthetic chemical-intensive to low-input farming systems without running the risk of financial failure. Good management requires knowledge of probable outcomes and variability associated with low-input choices of farming system. This includes likely changes in farm income and off-farm income, changes in the uncertainty of yields and farm profits, and risks to human health and the environment.

Like most entrepreneurs, farmers typically make decisions as to the kinds of technology to use without fully understanding the financial, environmental, or other outcomes likely to result from those choices. They constantly deal with trade-offs between competing goals (current income vs. future income, practicality vs. aesthetics, production vs. conservation, work vs. leisure). As an aid to their decision making, farmers often seek advice from experts in the public sector (extension and research personnel at universities, for example) and in the private sector (sales representatives or hired consultants).

Many farmers use various low-input farming methods on all or part of their hectareage. Surveys and visits to these farms have provided several insights regarding the profitability and potential for widespread adoption of various sustainable farming methods. Successful adoption of low-input farming methods (IPM, organic, biodynamic, or regenerative) usually requires more skillful management than comparable size farms where major reliance is placed on use of synthetically formulated chemical pesticides and fertilizers. Several instances have been documented of farmers who are employing methods characterized as organic, regenerative, or similar terms, and earning profits comparable to (and sometimes greater than) those of similar farms using conventional farming methods.

Severe managerial difficulties are often (not universally) encountered by a farmer switching from chemical-intensive to low-input/sustainable methods. If a farmer chose (or was forced by regulatory or other pressure) to abruptly stop using all synthetic chemical pesticides and fertilizers, then yields and profits could decline sharply in the first few years of the transition. However, if the withdrawal of chemicals caused a significant drop in total market supply of major commodities, prices would tend to rise proportionately more than the reduction in output. Farmer profits could actually increase under these circumstances, though with the unfortunate effects of accelerating inflation, reducing farm commodity exports, and worsening our national balance of trade. Conversely, if a more gradual transition occurred, and with adequate investment in research and education to improve the productivity of the low-input alternatives, these adverse side-effects could be largely or totally avoided.

Most evaluations of the monetary impact of adopting low-input farming practices are done at the component level, rather than the whole-farm level of analysis. Specifically, these studies focus on the cost-and-return implications of adopting a specific farming method, as it affects an enterprise or its biological and physical components (an apple tree, a dairy cow, or an experimental plot of corn, for example). Economic analysis of an enterprise most often considers only a single farming practice or system, such as integrated pest management or crop rotation, while assuming other aspects of the farm business remain unchanged, and assuming that the relative market prices of the farm commodities and resource inputs do not change as a result of the farm's adoption of this low-input practice or system. Few studies have considered the impact on the whole farm, the farm economy, the environment, the nation, or other countries. Nonetheless, studies done at the enterprise or whole-farm level of analysis can provide valuable insights into the farmer's motivation to adopt certain farming practices and the likelihood that such farming methods may become widely adopted. Adoption is more likely to be profitable and sustainable if the studies provide valid and easily understandable information regarding the expected changes in income, risk, labor requirements, and other management considerations (Madden and Dobbs, 1989).

Given that a low-input farming method can only be sustainable if it is profitable, the results of several studies on the profitability of various low-input strategies are summarized in this chapter. Before discussing the findings of economic studies, a number of farmer surveys are examined regarding the practices and performance of farms using conventional and alternative sustainable farming methods.

FARM SURVEYS

Several surveys of farms using low-input methods and systems have been conducted. A comparative whole-farm study was done between 1974 and 1978 by the Center for the Study of the Biology of Natural Systems at Washington University in St. Louis (Lockeretz et al., 1981). This study, funded by the National Science Foundation, compared the incomes and other performance measures (including energy utilization) of commercial-scale organic and conventional farms in the Midwest. Two analyses were conducted. The 1974–76 segment included 14 matched pairs of organic and conventional farms. During 1977 and 1978, data were collected from 23 and 19 organic farms, respectively; these were compared with data from county statistical reports. During the first four years, when drought affected some parts of the study area, net income per hectare for the organic farms was approximately the same as for the comparison farms (or county reported data). In 1978, when growing conditions were generally favorable throughout the study area, the average net income per hectare on the organic farms was 13% below the the county averages. A similar pattern has been observed in more recent studies, including preliminary readings during the drought of 1988 in experiments funded by the LISA program; low-input experimental plots have higher yields than the conventional plots in several locations. A possible explanation of these differences is that many herbicides are activated only by soil moisture, and some herbicides may be toxic to certain crops in the current or subsequent seasons, especially during a drought.

A recent survey featured a second interview of Midwestern organic farmers, one decade after they were first interviewed in 1977 as part of the Washington University study (Lockeretz and Madden, 1987). Nearly two-thirds of the farmers reported their farms were still "entirely organic," as they had been a decade earlier. Another 12.5% had switched entirely to conventional practices. Only 8% reported they did not earn enough in 1986 to meet family living expenses; 18% reported a debt/asset ratio in excess of 0.4, the level considered "highly leveraged." In the survey, 66% of the farmers responded to the survey, and information was obtained from other sources concerning another 10% no longer farming.

The data obtained from studies such as these are problematic in that much of the information is based on the recall and perceptions of the farmers, rather than on direct measurements of yields, weed populations, income, and other attributes. Furthermore, the representativeness and other statistical properties of the samples are unknown. Ohio State University has taken the leadership in initiating a comprehensive longitudinal survey of some 900 farmers in Ohio (under the direction of Clive Edwards, Lynn Forster, and others). The second wave of survey data has been collected and, as of this writing, is being analyzed. Their survey is yielding an extremely rich data base regarding the production and marketing practices used, resources, and performance of farms, many of which are using sustainable farming practices. The organizers of this survey are now pursuing the possibility of expanding this survey to all or a significant number of states with funding provided by the federal LISA program.

Ongoing surveys by Economic Research Service are providing valuable information regarding use of various inputs (pesticides, fertilizers, manure) and practices such as crop rotations (Daberkow et al., 1988). The latest ERS survey of farmers in the ten major corn-producing states has documented a surprisingly high level of diversity of farming practices, dispelling many widely held beliefs about "conventional agriculture." One such belief is that most of the nation's corn is grown continuously year after year in the same fields. Many observers consider continuous production of corn to be the epitome of chemical-intensive agriculture, requiring high doses of insecticides, herbicides, and fertilizers for profitable production. This survey found that only 28% of the corn crop was grown on land that had been planted to corn in the two previous years. Crop rotations prevailed on 47% of the corn land. However, 39% of the corn crop was alternated with soybeans, a crop rotation known to cause severe erosion due to limited crop residue from soybeans (Francis et al., 1986) and to require extensive use of herbicides and insecticides. In contrast, an average of 8% of the corn crop in this ten-state area followed at least two years of alfalfa, a crop rotation widely favored by environmentalists and others concerned about agricultural sustainability. In Wisconsin, this type of rotation was used on 34% of the corn hectares.

In the ten states surveyed, 95% of the corn hectares were treated with herbicides and fertilizers. Manure was applied to 16% of the corn hectares; an average of 44.2 kg less nitrogen was applied to manured land, compared to other corn hectareage. And while an average of nearly 40% of the corn hectares received insecticide, continuous corn was much more likely to be treated—75% vs. 29% of noncontinuous corn. Crops in other areas, notably fruits and vegetables grown in areas with long growing seasons, are typically produced with heavy and frequent use of insecticides and other pesticides. U.S. farmers use a wide range of practices for pest control and soil fertility management. And while the conventional norm for agriculture in recent decades has been to rely heavily on these purchased inputs, it has become clear that the image of continuous, monoculture corn across the plains is a caricature. Clearly, not all farms are totally dependent on synthetic chemical pesticides and fertilizers, and the utilization of some of these materials has been declining in recent years.

For example, fertilizer application has declined from a high of 54 million tons in 1981 to 44 million tons in 1986, primarily due to decreasing hectareage of price-supported crops; the application rates per hectare have changed very little. The percentages of hectareage receiving fertilizer varies widely by crop—96% of all corn, 80% of cotton, 79% of wheat, and 33% of soybean hectareage in 1986, for example. Herbicides are very widely used on corn, 96% of the hectareage in the ten Cornbelt states, ranging from 81% in South Dakota to 99% in Iowa. A similar percentage of the soybean hectareage received herbicide treatment. The situation in wheat is highly variable, with winter wheat receiving the least herbicides (38% of the hectareage), spring wheat 86% and durum wheat 98%. Only 1% of the wheat grown in Missouri was treated with herbicide, compared with 94% in Oregon (Economic Research Service, 1987).

ALTERNATIVE SOIL MANAGEMENT STRATEGIES

The central methods of maintaining and enhancing soil fertility in low-input systems are the use of legumes in crop rotation or as overseeded cover crops, and the application of animal manure, compost, and other materials. Results of several of these strategies and their economic implications are discussed here.

One of the major economic advantages of including legumes in a crop rotation is their ability to biologically fix nitrogen. The amount of nitrogen provided by a legume crop depends on many factors. (See Chapter 6.) Different species and cultivars of legumes vary widely in the amount of atmospheric nitrogen they fix. A number of physical factors (soil acidity, temperature, existing amount of available nitrogen in the soil, drainage) as well as several managerial factors (such as number of cuttings of hay removed and whether foliage is turned under as green manure) determine the biological nitrogen fixation (BNF) performance of a given cultivar (Heichel, 1987; Zachariassen and Power, 1987). For example, Fox and Piekielek (1988) found that three different legumes (alfalfa, birdsfoot trefoil, and red clover) fixed 187, 169, and 147 kg of N per hectare during three years in central Pennsylvania.

The traditional rule of thumb that a crop of soybeans will contribute one pound of nitrogen per hectare for every bushel of yield has been overturned by recent findings. For example, Heichel (1987) reports that the amount of nitrogen fixation by soybeans ranged from a low of 19 kg/ha on soil already high in nitrogen and organic matter, to a high of 151 kg/ha in soils initially low in nitrogen and organic matter in Minnesota. Hanson et al. (1988) found that, in Missouri, a crop of soybeans provided an average of 94 kg of N/ha. In one of the two study locations, the most profitable level of nitrogen fertilization for a grain sorghum crop following soybeans is zero N, compared with 119 kg for continuous grain sorghum. Sorghum yields were consistently higher following soybeans in a two-year rotation than in continuous cropping.

Whereas most studies reporting on the monetary benefit of BNF have looked only at costs and returns on a per-acre basis, a study in southeastern Minnesota took a whole-farm approach in examining the nitrogen contribution and other benefits of legumes in a crop rotation with corn and/or soybeans (Kilkenny, 1984). The study used a linear programming model and assumed 1980–82 average prices of crops and nitrogen to project results for a 400-acre farm with a 60-cow dairy herd. The study concluded that if the monetary value of the nitrogen contributed were ignored and if nitrogen fertilizer is assumed to be free, the most profitable cropping system in terms of current net returns over cash operating costs is continuous corn on about two-thirds of the hectareage, with about one-third of the hectareage in a three-year corn-oats/alfalfa-alfalfa (C-O/A-A) rotation. As the assumed price of nitrogen increases, the profit-maximizing crop rotations feature increasing proportions of legumes. With an assumed nitrogen price of $0.052 per pound (the 1980–82 price), a corn-soybean rotation was found to be more profitable than continuous corn. Nitrogen fertilizer prices have declined slightly since 1982. However, the price of nitrogen would have to increase toward $0.313 per pound before the most profitable rotation would shift from corn-soybeans toward continuous soybeans on more of the hectareage in combination with the three-year C-O/A-A rotation.

This study was done prior to release of a new cultivar of alfalfa (Nitro), which biologically fixes about 105 kg N/ha in a two-year rotation with corn in Minnesota; this is approximately 59% more than the next best cultivar (Barnes et al, 1986). While the economic implications of this and similar innovations in legume genetics have yet to be fully realized, it is clear that acceleration of the BNF capability of a legume will increase its profitability and enhance the likelihood of its being included in crop rotations.

Another whole-farm study was conducted in two locations in South Dakota by Dobbs et al. (1988). At one of the locations, a corn-soybeans-spring wheat rotation using conventional (moldboard plow) and ridge tillage with herbicides and fertilizers was found to be about 9.3% more profitable than an alternative rotation of oats-alfalfa-soybeans-corn rotation with no herbicides or fertilizers. In this comparison, profit was measured as return to land, labor, and management. In the other study location, a soybeans-spring wheat-barley rotation with herbicides and fertilizers was found to be 2.5% less profitable than a nonchemical rotation of oats-sweet clover-soybeans-spring wheat. The alternative systems had much lower operating costs.

An experiment now in its thirteenth year conducted by Sahs, Francis, and others at the University of Nebraska (see Chapter 10) is comparing cropping systems, including an essentially "organic" rotation using manure for fertilizer and no herbicide or synthetic chemical fertilizers (Helmers et al., 1986). The crops grown include corn, soybeans, grain sorghum, and oats with sweetclover/red clover, grown in various rotations compared to continuous production of corn or other cash-grains. The results confirm the findings of studies done in the first half of this century (Heady, 1948; Heady and Jensen, 1951) using more primitive cultivars and no synthetic chemical pesticides: rotations have higher yields and net returns per hectare than continuous mono-cropping of corn, soybeans, or sorghum. The continuous cropping systems were found to require a higher expenditure for pesticides and to be subject to greater year-to-year variation in yields and profits per hectare compared with the various rotations.

Many farmers are reluctant to use legumes because the seed is expensive and a more profitable crop foregone while the legume is grown. Both of these disadvantages seem to have been overcome at least partially by an alternative rotation studied in the Palouse area of eastern Washington (Goldstein and Young, 1987). The legume tested was black medic, which reseeds itself like a common weed. In fact, it is considered a weed in some parts of the world, such as southern Greece. The low-input rotation is three years: peas plus medic the first year, followed by medic, then wheat. The only synthetic chemical applied during this rotation is an insecticide and a herbicide applied to the peas once every three years. The rotation controls almost all the weeds in wheat. When a few weeds appear, the farmer controls them by pulling a harrow over the field. The conventional rotation in this area is four years: wheat, barley, wheat, and pea. Herbicides are applied every year; insecticide is applied to the peas; and a fungicide is applied to the wheat. Commercial NPK fertilizer is applied to the wheat and barley in the conventional rotation; no fertilizer is applied to the low-input rotation.

The study found that crop yields under the conventional and low-input crop rotations were about the same during two trial years at three locations. The largest differences occurred during the drought of 1985, when average yields for the low-input plots were 83% above the conventional yields. In 1984, when rainfall was close to normal, the low-input wheat yields were 3% less than the conventional yields. Because almost no chemicals were used in the low-input rotation, its cost per hectare was less than half the cost of the conventional rotation—$56.28 vs. $129.40 per hectare.

The comparison of profits depends on what prices are used in the calculations. When the value of the crops is calculated based on 1986 market prices of wheat and barley (55% and 65% of government target prices, respectively), the low-input rotation was estimated to earn a net return of $27.94 per hectare, or about 30% more than the returns from the conventional rotation. But when profits are recalculated using the government target prices, the positions are reversed. The conventional rotation earned $65.59 compared with $51.63 from the low-input rotation. The cause of this aberration is that the low-input rotation produces about 50% less wheat than the conventional rotation—one year in three vs. two years in four. Consequently, the farmer receives much lower benefits from the federal price support payments. It is ironic that a federal program intended to reduce surplus wheat production actually rewards farmers for using a rotation that produces much more wheat than a low-input rotation that incidentally applies much lower amounts of pesticides and fertilizers.

Similar results were obtained by Duffy (1987) in his Iowa study. In addition to encouraging a much higher production of corn (which the government program is intended to limit), the program also encourages a massive increase in amounts of nitrogen fertilizer applied—268.8 kg N/ha for continuous corn vs. 44.8 for the low-input rotation. Again, it is ironic that the government price support program has the unintended effect of worsening both the corn surplus and the state's already severe groundwater pollution problem.

CONTROLLING PESTS WITH LOW-INPUT SYSTEMS

For many different reasons, an increasing number of farmers, scientists, and chemical manufacturers are turning their attention from synthetic chemical pesticides toward alternative, sustainable farming methods. The alternative pest control measures include the following:

- Enhancement of natural enemies (biological controls).
- Mating disruption by release of sex hormones (pheromones).
- Crop rotations to interrupt the reproductive cycle of pests and to deprive them of an essential food source.
- Control of weeds by mechanical cultivation, ridge tillage, and use of cover crops that emit allelopathic substances.
- Development of pest- and disease-resistant varieties of crops.

Many research studies have been done on these subjects. A few of the studies are discussed here to illustrate the economic implications of these approaches.

Many scientists, including most economists, seem to operate under the assumption that the ideal growing environment for crops is devoid of weeds and pest damage to the crop is zero. Other scientists have found strong evidence to the contrary in many locations and crops. For example, weeds often provide a food source for beneficial species, as well as serving as a ground cover that reduces soil temperatures and prevents erosion. (See Chapters 3 and 4.) Harris (1974) summarized and interpreted a number of studies showing that a small amount of insect damage actually stimulates crop growth. The biological reasons for this phenomenon are not well understood. For example, Harris cited a Russian study showing that the damage done by a single bollworm early in the season enhanced cotton yields by 23.8%. Another study in Czechoslovakia reported that insect damage up to 50% defoliation of potatoes from mid-June to the end of July increased tuber production by 13.2% to 26.2%. The economic implication of this startling result is that the benefits often attributed to total control of insect pests may be overstated. In many circumstances, the optimum pest population is not zero. This is more likely to occur in cases where pesticides are ineffective or are not permitted, and where a small population of pests provides an essential food source to maintain the population of an exotic or expensive parasite or predator serving as a biological control agent. This principle, which complicates the selection of most profitable pest control strategies, has not been incorporated into any economic studies I have found in the literature.

Antle (1988) has observed that economic evaluations of alternative pest control strategies should take into account the financial risks inherent in various methods. He found, for example, that IPM farmers who are rather risk-averse tend to apply more pesticides and use them more often than farmers who are less concerned about risks.

Other methodological and philosophical issues related to economic assessment of alternative sustainable farming methods include the need to take into account various off-farm monetary costs and benefits associated with use of synthetic chemical pesticides. Ideally, the economic assessment of various alternative pest control methods should take account of the full range of developmental, dissemination, and application costs, as well as benefits in terms of changes in monetary costs and returns to farmers. Market-level analysis should also be done to estimate the effects on prices of farm commodities (which may be suppressed if output is increased substantially), international balance of trade in farm commodities and purchased inputs such as petroleum (used to produce most pesticides), and the incomes of input supply firms (both those selling pesticides and those selling alternative inputs such as beneficials for use in biological control). The benefits of substituting biological controls for pesticides also include various externalities, such as reductions in costs borne by

- Municipal and private water sources for treating or replacing contaminated water supplies.

- Businesses and households whose incomes are affected by environmental damage by pesticides.
- Farm workers and others injured by chemicals.

All these factors are relevant considerations in the economics of sustainable pest control strategies. In practice, however, no study has attempted to measure or analyze all these factors. Several studies that have estimated some of the monetary implications of biological control programs and integrated pest management are discussed below.

Biological Control of Pests

In low-input/sustainable farming systems, the vast majority of weeds that sprout and various insects and other pests that hatch are killed by their natural enemies. Humans often upset the natural balance by growing crops that provide a favorable habitat for certain pests, or by using non/selective pesticides that kill many of the beneficial life forms, thus enabling pests that previously were not a problem to reproduce rapidly. Biological control strategies, an important aspect of sustainable methods of farming, are based on four principles:

1. Know the biology of the target pests and their natural enemies.
2. Make the habitat less favorable for the pests and more favorable to their natural enemies.
3. Where necessary, introduce natural enemies, either of native or exotic origin.
4. When all else fails and pesticides must be applied, use the smallest feasible dosage, and select materials that are as pest-specific as possible so as to protect populations of beneficial species.

An unobtrusive measure of the economic potential of biological controls is the fact that several major chemical manufacturers are now favoring the development of biological products rather than conventional chemical pesticides. One reason is the enormous cost of developing and gaining approval of a new chemical pesticide—between $50 and $60 million, according to officials in duPont, for example. Abbott Laboratories sells a fungus-based herbicide (Devine) that controls milkweed vine in citrus groves. Upjohn has developed another fungal product for control of northern joint vetch, a weed in rice and soybeans. Several different bacterial products are also widely used, including a probiotic medication to preempt digestive infections in baby pigs and calves, as well as several strains of *Bacillus thuringiensis* for control of insect pests and some disease pathogens (Cook and Baker, 1983).

Osteen et al. (1981) prepared an annotated bibliography on the economics of agricultural pest control, covering the literature from 1960 to

1980. This report includes many studies of biological pest control, as well as more conventional IPM programs. When an exotic natural enemy of a pest is found, reproduced in laboratories, and released into an area afflicted by that pest, this is known as classical biological control. This process requires careful testing to make sure the species being brought in is harmless to nontarget species and that it can survive where it is released. Certain weeds and many insects, mites, nematodes, and other pests are now controlled by classical biological control programs. Several diseases of crops and livestock are effectively and economically controlled by use of biological control agents (Cook and Baker, 1983).

A more passive biological control strategy is to enhance the habitat in ways that promote reproduction of naturally occurring beneficial species. This approach includes provision of ground cover, often including native weeds, that provide nectar or support a food source for the natural enemies. Some of these strategies can be implemented with rather simple and passive management; others require extensive scientific and managerial input.

Some studies report a benefit: cost (B:C) ratio, where estimated benefits in terms of market value of crop damage prevented in current and future years is discounted to a present value, and are compared with costs of developing and implementing the program. In one such example, Ervin et al. (1983) reported an economic analysis of a biological control program in California, in which natural enemies were introduced to control a tree fruit pest, comstock mealybug. The B:C ratio for this program ranged from $22 to $135 benefit for each dollar spent on the program, depending on the discount rate and crop loss assumptions used. The fact that the B:C ratio is much larger than 1:1 indicates the program was well worth the cost. The large variation in resultant B:C ratios illustrates the subjective nature of studies attempting to estimate the present value of a possible future benefit. Another approach to the analysis of the benefits and costs of biological control programs is to measure trends not just in monetary terms but also in nonmonetary outcomes, such as tonnage of pesticide utilization prevented, long-term reductions in groundwater contamination, and reductions in farmworker poisonings.

One of the most economically successful examples of classical biological control is the case of alfalfa weevil in the northeastern United States. This European pest was accidentally introduced in the 1950s. By the 1960s, it had spread throughout the northeastern region, and was causing tens of millions of dollars' worth of crop damage each year. Most farmers began spraying their hay fields at least twice a year to control the alfalfa weevil. Like all pests, the weevil tends to develop genetic resistance to pesticides over a period of several years. Through a federally funded research project, more than a dozen natural enemies of the alfalfa weevil were located, tested, and released in the region. The population of this pest has declined steadily as the natural enemies have spread and propagated more widely. Pesticide application has dropped sharply. Previously, virtually all the alfalfa hectareage was sprayed twice a year; as of 1981, less than 27% was sprayed even once. This biological control project cost about $1 million of federal funds over a 20-year period. Day (1981) estimated that farmers in the northeastern states would save about $8 million *per year* in reduced insecticide costs. Zavaleta and Ruesink (1980) estimated farmers in the region would save about $44 million per year in reduced insecticide costs and crop damage. That study estimated there would not be enough increase in production to suppress alfalfa prices appreciably. A nonmonetary dimension reported in the study is that 1,100 tons less insecticide would be applied, which would reduce environmental impacts and human health risks. Even when these nonmonetary benefits are ignored, however, the monetary benefits so far exceed the costs (44-fold annual return on the initial investment) that it would be superfluous to attempt to estimate a benefit:cost ratio based on the present value of future benefits. Many other examples of the economic effects of biological pest control are summarized in the annotated bibliography by Osteen et al. (1981).

Integrated Pest Management

Several integrated pest management (IPM) programs have been very successfully developed and widely adopted. Examples include cotton, alfalfa, soybeans, grapes, and apples (Frisbie and Adkisson, 1985; Allen et al., 1987). These programs include a wide array of strategies, including selection of resistant varieties of the crop; monitoring trends in the population of pests and their natural enemies and growth stage of the crop; and crop rotations to interrupt the reproductive cycles of various pests and disease pathogens. When I interviewed several state university leaders of IPM programs in 1983, I asked whether they attempted to enhance the populations of beneficial species. Some answered in the affirmative. One said definitely not: "Around here, IPM means responsible use of pesticides." In the more ecologically based IPM programs, pesticides are applied only as a last resort, with strong preference for pesticides that are pest-specific and as harmless as possible to nontarget organisms, especially the beneficial species. In many instances, farmers using IPM are able to greatly reduce and occasionally eliminate pesticide applications that would otherwise be routinely applied. In other instances, the IPM scouting for pest populations reveals an emerging pest problem that would have been missed otherwise; in these instances, the IPM program greatly increases yields and profits, but at the cost of higher levels of pesticide use.

Of crucial importance to an IPM program is determination of economic thresholds or action thresholds that indicate when the pest population that is approaching the level at which the farmer's profit would be diminished by withholding preventive action, because the expected value of the impending crop damage is likely to exceed the cost of treatment. Much research has been devoted to development of improved threshold procedures (Frisbie and Adkisson, 1985). Most economic thresholds determine the most profitable short-term pest control strategy, taking into account the prices of the crop or livestock at risk, the cost of the proposed action (usually application of a specific pesticide in precise timing), trends in populations of the target pest and its natural enemies, and anticipated damage under various action scenarios (including no action). Benefit:cost analyses at the farm level of aggregation take into account the increase in sales value, cost of pest scouting, and changes in pesticide application costs.

In addition to focusing on increasing current income, studies on development of economic thresholds must recognize specific biological features of the target pest and the long-term implications of current action. For example, it may be profitable in a one-year perspective to allow a certain number of weeds to grow in a field. From a longer term standpoint, there is concern over increases in the soil seed bank that may result from subthreshold populations of weeds in one year (Coble, 1985). Another important long-term implication of pesticide application is the trend in pest resistance. Thresholds that neglect pest resistance may lead to higher or more frequent applications of certain pesticides than would be in the long-term economic interest of the individual farmer, the industry, or the nation (Hueth and Regev, 1974).

POSSIBLE IMPACTS OF WIDESPREAD ADOPTION

How would American agriculture and rural communities be impacted if sustainable farming methods were to become much more widely adopted? Speculation abounds and opinions differ widely. While a number of studies have been done on the economic implications of the adoption of specific farming practices on individual farms or experimental plots, only one study has attempted to estimate the macroeconomic or market-level impact of widespread adoption of a certain type of low-input agriculture known as organic farming. A linear programming model was used to predict the potential effect of widespread adoption of what the authors defined as organic farming practices. The study concluded that total production of many commodities would decrease substantially. The export levels under an assumption of total adoption of organic methods was estimated to be 25% of the conventional level for wheat, 58% for feed grains, 37% for soybeans, and 66% for cotton. Because of the inelastic demand for farm products, farm income was expected to more than double, from $6 billion to $13 billion in 1975 dollars (Langley et al., 1983).

This study is severely faulted in its methods, assumptions, and data. For example, the crop yields assumed in what the authors called the organic option were based on historical 1944 yields, with some adjustment for improvements in cultivars, and assuming no fertilizer would be applied. This is not a valid representation of organic farming. Experiments both on farms and in experiment station plots have shown that organic methods can attain yields comparable to those of conventional practices (Helmers et al., 1986). The result of rotation, wherein hectareage of the main cash crops is reduced to accommodate legumes and small grains in the rotation, can significantly reduce output of corn and other major commodities (Dabbert and Madden, 1986). Yields of some farm commodities are not adversely affected by adoption of low-input methods. Many field crops such as wheat, corn, and soybeans can be produced in many locations with little or no use of synthetic chemical pesticides, and relying on legumes as the primary or sole source of soil nitrogen (Helmers et al., 1986; Duffy, 1987; Goldstein and Young, 1987). Surveys have shown that commercial-scale organic farms apply fertilizers—often using materials that are more expensive than conventional commercial fertilizers (Lockeretz and Madden, 1987).

Furthermore, organic or other low-input farming practices are likely to be adopted very slowly (if at all), causing only gradual shifts in prices and resource use. If farm commodity prices would begin to increase significantly, the resulting induced change in investments in research and technology would facilitate innovations that would tend to ameliorate the predicted long-term impact on production, prices, incomes, and exports (Ruttan, 1982; Ruttan and Pray, 1987). Among other deficiencies, the study also seems to have overstated the dependence of organic farms on livestock manure and underestimated the contribution of legume-based crop rotations to soil fertility. Because of these procedural flaws, the findings of this study are widely considered to be of no value. Consequently, there exists no reputable evidence regarding the macroeconomic impacts of possible widespread adoption of organic farming practices.

Several studies have estimated the aggregate economic effects of widespread adoption of integrated pest management (IPM). For example, a study of the potential economic impact of the adoption of cotton IPM in Texas estimated that, because of reduced pest damage to crops, production would be increased sufficiently to cause a 7.3% reduction in the farm-level price of cotton. Despite a decrease in pesticide costs and crop losses, the net result of the reduced price of cotton would be a $43.9 million decline in aggregate farm income of cotton growers. The researchers concluded, however, that the monetary benefit of lower prices to consumers would exceed the income lost by farmers, thereby yielding a net social gain (Taylor and Lacewell, 1977). The method used to estimate net social benefit, consumer surplus, is subject to controversy in the economics literature (Morey, 1984).

Changes in other monetary costs associated with off-farm externalities were beyond the scope of that study. Examples of these externalities include largely unmeasurable costs associated with human suffering, increased monetary cost of treatment (or replacement in severe cases) of contaminated water supplies, health costs due to farm worker exposure to pesticides, financial losses to commercial fishing, recreational businesses, and other industries affected by environmental degradation.

Economic theory suggests that if all farmers were forced by prohibitively high pesticide and fertilizer prices, regulatory action, or other institutional changes to adopt low-input farming methods, major changes in the structure of agriculture and rural communities would result. Regional patterns of production would be expected to shift away from locations heavily dependent on synthetic chemical pesticides (such as Florida vegetable producers) toward areas where cold winters and shorter growing seasons make it possible for natural enemies to more effectively control insect pests. For example, Lichtenberg et al. (1988) estimated that cancelling the insecticide parathion would increase marginal costs by quite different amounts in various locations, ranging from 0.4% in California to 4.5% in the Umatilla Valley of Oregon. Production cost per pound of prunes would increase by 0.07 cents in Michigan and 2.6 cents in Oregon. In general, the prices of the more pesticide-dependent commodities would likely increase, especially during the winter and early spring. As relative prices of various commodities would change, household consumption patterns would likely shift toward vegetables and fruits that could be produced efficiently with low-input methods. And, according to Ruttan's induced innovation theorem (1982), rising commodity prices would tend to encourage an acceleration of research and technology development and dissemination needed to make low-input methods more productive, thereby increasing supply and reducing the inflationary pressure on food prices.

Where increases in management and labor are required, and where these services are not available from reliable custom firms, widespread adoption of low-input/sustainable farming methods would be difficult if not impractical on super-large farms. But in some locations (such as Florida and California) a well-developed custom service industry has emerged to provide pest control scouting, composting, planting, and harvesting services. In these situations, the size of the farm is not as severely limited by the management and labor provided by the farmer, compared with locations lacking such services. Very large farms prosper in these states, in part because of the availability of such services. Many low-input farming practices, such as legumes in crop rotations, biological controls, cover crops, and use of pest-resistant cultivars, can be profitably and efficiently applied on a wide range of farm sizes, from very small to very large operations. However, low-input practices that are highly management-intensive are more feasible for moderate-scale owner-operated farms than on huge farms operated by hired labor and hired managers. These low-input methods are scale-negative, favoring family-scale farms.

Significant changes would also occur in employment and income patterns in rural areas, particularly among firms supplying synthetic chemical inputs. With an increase in the prices of many commodities, farm exports would decline and imports would increase. Consumers would expend a higher percentage of their income on food. On the other hand, certain costs might decline, including costs associated with the externalities of agricultural chemical use, such as detoxification of water supplies, medical costs, and losses of productivity (beyond the immeasurable value of unnecessary loss of human life) associated with acute and chronic exposure to pesticides (Stokes and Brace, 1988). Further analysis based on actual experience is needed to corroborate these results.

Producers of some commodities require (and sometimes receive) premium prices necessary to cover the added per-unit costs associated with increased use of labor (as in the case of hand weeding of vegetables) and lower yields of some crops (especially where insect damage reduces the percentage of marketable product). In most situations, however, farmers producing crops with few or no synthetic chemical pesticides are able to compete in regular markets. Case studies of such farms conducted by this author are presented in the June 1989 report by the National Research Council, *Alternative Agriculture*. One of these farms, Pavich Brothers, produces more than 1% of the U.S. fresh grape output. They use almost entirely nonchemical methods, earn a rather substantial profit, and are gradually expanding their share of the fresh grape market (Madden, 1987b). Farmers who use various integrated pest management programs often report lower costs and higher yields than nonusers (Allen et al., 1987).

Several studies done in Europe indicate organic farmers would be unable to meet their production costs without premium prices on several commodities. In the United States, the situation is mixed. A recent survey of Midwestern organic farmers showed that 22% sold the majority of their crops through an organic market of one kind or another in 1987. This percentage is up from 11% reported by the same sample of farms a decade earlier. Only 13% of those with livestock reported they sold most of their livestock or livestock products through organic marketing channels (Lockeretz and Madden, 1987). Nonetheless, the market for certified, organically grown produce is rather substantial and rapidly growing. Franco (1988) reports that in California the market value of organic produce was between $54 and $68 million in 1987, a 41% increase over the previous year. Anecdotal evidence from other states supports this conclusion. Some 14 states now have passed organic certification laws, and a movement is under way to come up with a national standard for certification of organically grown foods.

UNFINISHED AGENDA

The economic implications of alternative low-input/sustainable farming methods and systems are not well understood. Very little (and for the most part, poor quality) research has been done on the macroeconomic implications for market prices, international trade, and economic growth and stability. Only a small fraction of the studies on component technologies contributing to a more regenerative agriculture have been carefully analyzed to assess the implications of their findings for farm profits, risk, and environmental impacts in various locations. Nor have the findings of the vast majority of studies been translated and integrated into an information delivery system that will enable farmers to make more informed decisions, and reduce the financial perils of a transition to less dependence on synthetic chemical pesticides (Madden, 1988b).

Environmentalists, consumer advocates, and other citizens often express an urgent, at times impatient, desire for farmers to abandon the use of synthetic chemical pesticides and greatly reduce their use of chemical fertilizers. Both coercion and persuasion have been tried. Regulatory actions by state and federal agencies are sometimes employed to force socially desired changes in operating practices of businesses; to date, farmers in most states have been largely immune to these sanctions.

Persuasion is gentler, but often less effective, than coercion. It would be futile, as well as unethical, to attempt to persuade or coerce farmers to adopt farming methods that would very likely lead to their financial ruin. There has to be a better way. Farmers need better options. We must develop the science, the technology, and an effective information delivery system that will enable rank-and-file commercial-scale farmers to profitably make a transition from heavy dependence on synthetic chemical pesticides and fertilizers to low-input/sustainable methods of farming. The goal is a more regenerative agriculture, where family-scale farms (as well as larger and smaller sizes) can prosper, while using methods that are less hazardous to the environment and human health and more sustainable for generations to come.

Research is needed to achieve this goal. The new federal grants program, LISA, is an important step in the right direction, but its support by the new administration of USDA is questionable. Public policies (including price support programs) have a massive effect on farmer incentives to adopt low-input farming methods and systems. (See, for example, the 1987 studies by Duffy and by Goldstein and Young). Macroeconomic research could provide insights into the possible effects that widespread adoption of low-input farming methods would have upon prices of various agricultural inputs and commodities, as well as the likely effects on the environment, water quality, international trade, employment, economic development, incomes of various categories of farmers, and the overall structure of agriculture and rural society. However, the validity of such research is directly linked to the accuracy of the data assumptions used to represent various alternative production methods. Just as the massive death toll of the recent earthquake in Armenia has been attributed to shoddy construction practices, notably the use of hollow, unreinforced concrete beams in tall buildings, grand economic models based on naive assumptions can lead to disastrously misleading guidance to policymakers.

Present and future generations of this nation and all nations deserve better than we have done thus far. Profitable alternatives to synthetic chemical pesticides and herbicides are being developed—gradually, and for only a small segment of agriculture. Past failures and present difficulties enrich the challenge. Past successes and present ideals light the way to the future.

REFERENCES

Allen, W. A., R. F. Kazmeirczak, M. T. Lambur, G. W. Norton, E.G. Rajotte. (1987). *The National Evaluation of Extension's Integrated Pest Management (IPM) Programs*. Virginia Coop. Ext. Services, VCES Publ. 491–10, Blacksburg.

Antle, J. M. (1988). "Integrated pest management: It needs to recognize risks, too," *Choices*, third quarter, 1988. pp. 8–11.

Barnes, D., G. Heichel, and C. Sheaffer. (1986). "Nitro alfalfa may foster new cropping system," *News*. Minnesota Extension Service, St. Paul, MN, November 20, 1986.

Batra, S. W. (1981). "Biological control of weeds: Principles and prospects." In, *Biological Control in Crop Production* (Eds. Papavizas et al.). pp. 45–59. Allenheld, Osmun, Totowa, NJ.

Coble, H. D. (1985). "Development and implementation of economic thresholds for soybeans." In *Integrated Pest Management on Major Agricultural Systems* (Eds. R. E. Frisbie and P. L. Adkinsson). pp. 295–307.

Cook, J. R. and K. F. Baker. (1983). *The Nature and Practice of Biological Control of Plant Pathogens.* American Phytopathological Society, St. Paul, MN.

Cramer, C. (1988). "Low-input research in high gear," *New Farm* **10(7)**: 27–31. (1986).

Dabbert, S. and P. Madden. (1986). "The transition to organic agriculture: A multi-year model of a Pennsylvania farm," *Am. J. Alt. Agric.* **1(3)**: 99-107.

Daberkow, S., L. Hansen, and H. Vroomen. (1988). "Low-input practices." Economic Research Service, USDA: *Agricultural Outlook* AO-149, December 1988: 22–25.

Day, W. H. (1981). "Biological control of alfalfa weevil in the northeastern United States." In *Biological Control in Crop Production* (Eds. Papavizas et al.). pp. 361–74. Allenheld, Osmun, Totowa, NJ.

Dobbs, T. L., M. G. Leddy, and J. D. Smolik. (1988). "Factors influencing the economic potential for alternative farming systems: Case analyses in South Dakota," *Am. J. Alt. Agric.* **3(1)**: 26–34

Duffy, M. (1987). "Impacts of the 1985 Food Security Act." Iowa State Univ., Ames.

Economic Research Service. (1987). "Agricultural resources—Inputs situation and outlook report." USDA, ERS report AR-5, Jan. 1987.

Edwards, C. A., G. J. House, R. Lal, J. P. Madden, and R. H. Miller. (1989). *Sustainable Agriculture Systems.* Soil and Water Conservation Society, Ankeny, Iowa. (Proceedings of conference in Columbus, OH, Sept. 1988.)

Ervin, R. T., L. J. Moffitt, and D. E. Meyerdirk. (1983). "Comstock mealybug (Homoptera: Pseudococcidae): Cost analysis of a biological control program in California," J. Econ. Entom **76(3)**: 605–9.

Fox, R. H. and W. P. Peikeilek. (1988). "Fertilizer N equivalence of alfalfa, birdsfoot trefoil, and red clover for succeeding corn crops," *J. Prod. Agric.* **1(4)**: 313–17.

Francis, C., K. Wittler, A. Jones, K. Crookston, and S. Goodman. (1986). "Strip cropping corn and grain legumes: A review," *Am. J. Alt. Agric.* **1(4)**:159–64.

Franco, J. (1988). "Analysis of the California market for organically grown produce." Annual meetings of Western Economic Association in Los Angeles. Graduate School of Management, University of California, Davis.

Frisbie, R. E. and P. L. Adkisson. (1985). *Integrated Pest Management on Major Agricultural Systems.* Texas Agri. Expt. Station MP-1616, College Station.

Goldstein, W. A., and D. L. Young. (1987). "An economic comparison of a conventional and a low-input cropping system in the Palouse," *Am. J. Alt. Agric.* **2(2)**: 51–56.

Hanson, R. G., J. A. Stecker, and S. R. Maledy. (1988). "Effect of soybean rotation on the response of sorghum to fertilizer nitrogen," *J. Prod. Agric.* **1(4)**: 318–21.

Harris, P. (1974). "A possible explanation of plant yield increases following insect damage," *Agro-Ecosystems.* **1**: 219–25.

Heady, E. O. (1948). "The economics of rotations with farm and production policy applications," *J. Farm Econ.* **30(4)**: 645–64.

Heady, E. O., and H. R. Jensen. (1951). "The economics of crop rotations and land use." Iowa Agri. Expt. Sta. Bul. 383, August.

Heichel, G. H. (1987). "Legumes as a source of nitrogen in conservation tillage systems." In *1988 McGraw-Hill Yearbook of Science and Technology* (Ed. J.F. Power). pp. 29–35. McGraw-Hill , New York.

Helmers, G. A., M. R. Langemeier, and J. Atwood. (1986). "An economic analysis of alternative cropping systems for east-central Nebraska," *Am. J. Alt. Agric.* **1(4)**: 153–58.

Hueth, D., and U. Regev. (1974). "Optimal agricultural pest management with increasing pest resistance," *Am. J. Agri. Econ.* **56(3)**: 543–52.

Kilkenny, M. R. (1984). *An Economic Assessment of Biological Nitrogen Fixation in a Farming System of Southeast Minnesota.* Unpublished M.S. thesis, Univ. of Minnesota, St. Paul.

Koepf, H. H., B. D. Pettersson, and W. Schaumann. (1976). *Biodynamic Agriculture: An Introduction.* Anthroposophic Press, Spring Valley, NY.

Langley, J. A., E. O. Heady, and K. D. Olson. (1983). "The macro implications of a complete transformation of U.S. agricultural production to organic farming practices," *Agric. Ecosystems & Environ.* **10(4)**: 323–33.

Lichtenberg, E., D. D. Parker, and D. Zilberman. (1988). "Marginal analysis of welfare costs of environmental policies: The case of pesticide regulation," *Am. J. Agri. Econ.* **70(4)**: 867–74.

Lockeretz, W., and P. Madden. (1987). "Midwestern organic farming: A ten-year follow-up," *Am. J. Alt. Agric.*

Lockeretz, W., G. Shearer, and D. H. Kohl. (1981). "Organic farming in the corn belt," *Science* **211**:540–7.

Madden, P. (1987a). "Can sustainable agriculture be profitable?" *Environment* **41(4)**: 18–34.

——. (1988). Staff Paper 147, "Regenerative agriculture—Concepts and case studies of low-input, sustainable farming methods," Pennsylvania State University, University Park.

——. (1988a). "LISA 89 Guidelines" CSRS, U.S. Dept. of Agriculture, Washington DC.

——. (1988b). "Low-input/sustainable agriculture research and education—Challenges to the agricultural economics profession," *Am. J. Agric. Econ.* **70(s)**: 110-19.

Madden, P., and T. L. Dobbs. (1989). "The role of economics in achieving low-input/sustainable farming systems." In *Sustainable Agricultural Systems*, C. A. Edwards et al., Eds. Soil Water Cons. Soc., Ankeny, Iowa (in press).

Madden, P., D. Hubbard, P. O'Connell, and V. Jennings. (1988). "Low-input/sustainable agriculture research and education projects funded under the Agricultural Productivity Act." CDRS, USDA, Washington, DC. July 28, 1988.

Madden, P., and P. O'Connell. (1988). "Guidelines for preparing regional plans of work on low-input farming systems research and education." CDRS, USDA, Washington, DC. Feb. 16, 1988.

Morey, E. R. (1984). "Confuser surplus," *Am. Econ. Rev.* **74**: 163–73.

National Research Council. (1989). *Alternative Agriculture.* Academy Press, Washington, DC.

Osteen, C. D., E. B. Bradley, and L. J. Moffitt. (1981). *The Economics of Agricultural Pest Control.* USDA, Economics and Statistics Service, Bibliographies and Literature of Agriculture No. 14.

Papavizas, G. C., B. Y. Endo, D. L. Klingman, L. V. Knutson, R. D. Lumsden, and J. L. Vaughn. (1981). *Biological Control in Crop Production.* Beltsville Symposia in Agricultural Research, No. 5. Allanheld, Osmun, Totowa, NJ.

Rodale Institute. (1987). "On-farm research and demonstration sites." Rodale Press, Emmaus, PA.

Rodale, R. (1988). "Agricultural systems: The Importance of Sustainability," *National Forum, the Phi Kappa Phi Journal* **68(2)**: 2–6.

Ruttan, V. W. (1982). *Agricultural Research Policy.* Univ. of Minneapolis Press, Minneapolis.

——. (1988). "Sustainability is not enough." Symposium on Creating a Sustainable Agriculture, St. Paul, Minnesota, April 30, 1988.

Ruttan, V. W., and C. E. Pray. (1987). *Policy for Agricultural Research.* Westview Press, Boulder, CO.

Stokes, C. S., and K. D. Brace. (1988). "Agricultural chemical use and cancer mortality in selected rural counties," *J. Rural Studies.*

Taylor, C. R., and R. D. Lacewell. (1977). "Boll weevil control strategies: Regional benefits and costs," *South. J. Agric. Econ.* **9(1)**: 129–135.

USDA, Study Team on Organic Farming. (1980). *Report and Recommendations on Organic Farming.* Washington, DC.

Zachariassen, J. A., and J. F. Power. (1987). "Soil temperature and the growth, nitrogen uptake, dinitrogen fixation, and water use by legumes." In Power, 1987; pp. 24–26.

Zavaleta, L. R., and W. G. Ruesink. (1980). "Expected benefits from nonchemical methods of alfalfa weevil control," *Am. J. Agric. Econ.* **62(4)**: 801–5.

1988 McGraw-Hill Yearbook of Science and Technology (Ed. J. F. Power). pp. 24–26. McGraw-Hill, New York.

12 SUSTAINABILITY OF AGRICULTURE AND RURAL COMMUNITIES

CORNELIA BUTLER FLORA
Virginia Polytechnic Institute and State University
Blacksburg, Virginia

Community can be defined in social psychological terms of personal identification and perception of reality and in social organizational terms as an organization whose structure and function provides for basic interactions that give individuals their sense of identity and their basic institutional setting (Rubin, 1969). Following Haskinger and Pinkerton (1986), I will define a community as an area in which groups and individuals interact as they carry on daily activities and solve common problems.

According to Warren (1978), there are a number of functions that communities perform: providing opportunity for making a living, socializing community members, exercising social control, participating in group activities, and caring for those in need in crisis situations. Communities do this through both formal and informal groups, facilitated by both paid specialists and volunteer lay people.

This chapter examines how a shift to more sustainable, low-input agriculture would affect the ability of a community to solve its common problems and carry on the functions described above. The changes in management and economic activity triggered by a shift to more sustainable agricultural practices have implications for the communities that serve the farm families involved. In this chapter, the hetereogeneity of rural community types is presented, along with the macroeconomic forces impacting them. Trends in farming that have resulted from these macroeconomic shifts are outlined, as well as the impacts on community. Characteristics of communities that have responded positively to these changes are presented. The potential implications of adopting sustainable agricultural practices on communities are then discussed.

ECONOMIC AND CULTURAL BACKGROUND OF RURAL COMMUNITIES

While rural communities may have in common relatively small population size and relatively low population density, they are otherwise very diverse in the temperate zones around the world. In the United States, Bender et al. (1985) have classified, through systematic analysis of secondary data, seven distinct types of counties. These are (1) farming-dependent counties, (2) manufacturing-dependent counties, (3) mining-dependent counties, (4) specialized government counties, (5) persistent poverty counties, (6) federal land counties, and (7) retirement counties. In some communities several of these functions dominate, whereas in a few none of them is dominant. One of the community survival strategies that may occur with a shift to more sustainable agricultural practices is an acceleration of the current trend of farming-dependent counties to move into other categories. This would occur through diversification led by more mixed cropping and local processing and value-added activities.

The number of farming dependent counties declined 14% between 1979 and 1984 (Henry et al., 1986). This decline of dependence on agriculture should not in itself be viewed as a decline in the sustainability of the community, *if* the community were able to provide alternative mechanisms for making a living and alternative sources of social identity.

Macroeconomic Shifts and Shifting Factor Costs

Rural communities, particularly those in the interior region of the United States, have undergone extreme shocks in the last 15 years. These communities tended to be dependent on single, traditional economic sectors: agriculture, mining, or light manufacturing. These economic bases performed relatively well in the United States into the 1970s, until the world terms of trade shifted. Oil became more expensive, increasing the purchasing power of oil-exporting, developing nations, and demand increased for all commodities, increasing the price of agricultural products, timber, and minerals. This led to relatively good times in rural areas. Further, the cost of labor in the United States declined vis-à-vis other areas, because of the falling value of the U.S. dollar internationally. Exports increased as a result, particularly favoring areas in the United States with cheap labor, such as the rural South. During this period, high inflation accompanied high prices. Land purchase is often used as a hedge against inflation, and thus rural and urban land prices during the 1970s rose much faster than the land could sustain in terms of potential earnings and return to the investment.

Land became a speculative investment, not a productive one. Borrowing became logical, because money could always be repaid in dollars that were increasingly worth less because of inflation. Borrowing to buy land seemed particularly safe, since it was assumed that land could always be sold to repay the debt—and still leave a profit for the investor.

Capital was relatively cheap and land relatively expensive, encouraging the use of high levels of fertilizer and chemical inputs to maximize yields and provide for export markets. Some of the purchase of inputs was local, providing a multiplier effect in the community. However, in many areas the larger farmers tended to skip over the local community to purchase inputs, and thus benefited disproportionately from the shifting factor costs. Despite that, the 1970s brought a decline in the average age of farm operators. Young people, for the first time in nearly 50 years, saw agriculture as a potentially viable way to make a living.

Another sharp shift occurred in factor costs in the 1980s. Capital, which had once been cheap, suddenly became very expensive as inflation slowed and the U.S. deficit increased because of shifts in monetary and fiscal policy. Capital suddenly became the expensive factor of production. Land decreased in value, the dollar increased in value vis-á-vis other currencies, and export markets fall sharply. As a result, surpluses began to accumulate, farm incomes dropped, and debt became a major problem in many agriculturally dependent rural communities.

Input use, particularly by medium-sized farmers who purchased locally, fell dramatically, and farm machinery dealers and feed and seed businesses closed their doors. This was due in part to decreased sales and in part to overextended lines of credit. As a result, much of the damage to "small community mainstreet" caused by reduction in purchased inputs has already been done.

Increasing Separation of Producers from Consumers in Markets

Agriculture became more specialized during this period of time, and producers became increasingly separated from consumers and markets. Indeed, particularly in the 1980s, a very large part of agriculture survived on government payments, with almost no local market interaction. Field- and farm-level decisions often were based on "farming the farm program," rather than responding to domestic consumer demands. As a result, there was considerable resistance to the growing number of consumers seeking foods free of pesticide, residues, and leaner and hormone-free meat products.

The large government buffer between producers and markets tended to further isolate rural communities as well as farmers.

Increasing Internationalization of Agriculture and the U.S. Economy

There were dramatic increases in percentage of agricultural products exported as well as agricultural goods imported during the 1970s and 1980s. As an example, most of the wheat produced in the 1960s was consumed domestically; by the 1970s most was exported. This dependence on world markets increased the vulnerability of U.S. producers to shifts in the fortunes of both developing and developed countries. It also made our policies more dependent and more reactive to the policies of other countries. Rural communities increased their international dependency but not their international awareness. If they are not aware of international settings and trends, communities cannot respond to the opportunities and competition that are increasingly present.

Increasing Importance of Off-Farm Income in Rural Communities

A very high proportion of farmers nationwide now get most of their income off the farm. This reduces the risk from farming, but also may reduce the amount of time that can be spent on farm management. Further, while sustainable agriculture uses few chemical inputs, it is high in management inputs. Individuals and firms specializing in Integrated Pest Management, marketing, and other management-related services will increase with the advent of sustainable agriculture, providing highly skilled off-farm jobs and increasing community sustainability.

Overcommitment of Resources to Agriculture

As a result of cheap capital and inflated demand caused by subsidies, there has been an overcommitment of resources to agriculture in rural communities. This overcommitment early on was described as simply having too much labor or too many farmers. However, more recent analysis suggests that there is too much land and too much capital in agricultural production. Capital is represented in costs of land, machinery, and chemical inputs.

Implications for Rural Communities

There were four trends that had a marked impact on rural communities after 1970. They were increasing export dependence, increasing use of capital, increasing importance of off-farm income, and increasing dependence on federal intervention.

Export Dependence

The increasing dependence of rural communities on exports (Busch and Lacy, 1984), particularly of a single crop, has made many communities extremely vulnerable to weather trends or policy shifts. This, in turn, has decreased their ability to engage in mutual problem-solving behavior. Further, export dependence on a reduced number of crops tends to make the agricultural cycle extreme, causing intense work loads at certain periods of the year and very little work during other periods of the year. This, in turn, has created part-time community members with the capacity to leave, who go south for the winter or otherwise seasonally disengage themselves from that community and its problems. Further, export dependence has made it possible once again for "suitcase farming" (living outside the county or state where crops are raised and coming in only to plant and harvest) to occur, particularly in ecologically vulnerable areas in the Great Plains. Suitcase farmers, who generally farm in several different counties as a speculative rather than risk-reducing strategy, do not contribute to their local communities. The number of "eyes per acre" needed to respond to problems declines (Jackson et al., 1984).

Increasing Use of Capital

Capital from rural communities was heavily invested in agriculture through 1982. This, in turn, has increased the debt load and the financial problems in many rural communities for farmers, the retail trade, and financial institutions. The traditional assumption by rural investors that capital should be used for agriculture has meant that when agriculture has not been a profitable investment, many rural banks have been simply moving capital outside the community rather than looking for new opportunities for investment within the community. The extremely low loan/deposit ratio in many rural banks across the nation shows the degree to which the use of capital in agriculture has not been supplemented by local investment in other areas. With the exception of a few entrepreneurial rural communities, under traditional high chemical-input agricultural regimes, there has not been the motivation or imagination for alternative local investment.

Increasing Importance of Off-farm Income

Because of the increasing importance of off-farm income for rural communities, those communities that have diversified their economic bases have been able to survive. There is a growing recognition of the interdependence between agriculture and other economic activities within the community. Increasing off-farm income in some rural communities has led farmers to complain about competing employers who drive up labor cost. The alternative income opportunities also have increased the ability of many farmers to remain in farming and to have the economic flexibility to innovate in terms of sustainable agriculture practices. The increasing number of farm women in the labor force, in particular, provides that flexibility from off-farm income.

Increasing Dependence on Federal Payments

In many rural communities in the United States, over half of farm income comes directly from federal payments. On the one hand, for communities this often means temporary survival, but on the other, it signals a narrowing of economic options. The farmers begin to focus on only those crops that are supported by government subsidy payments. Moreover, set-aside land uses fewer agricultural inputs and contributes directly only to participating farmer income.

As a result of these four trends, farms have changed in the following ways.

Expansion in Size

Increasing export dependence, increasing use of capital, and increasing dependence on federal payments (encouraging production of crops that were eligible for subsidy payments) were highly related to the number of large farms in many agriculturally dependent communities, particularly in the Great Plains and the West (J. Flora and C. Flora, 1988). This has been offset to a degree nationwide by an increase of small, part-time farms, particularly in nonfarming dependent counties where reliance on off-farm incomes is high. The famous concern with the disappearing middle is hotly debated among rural sociologists and agricultural economists. But it is important to note that studies are showing that the decline in medium-sized farms is associated with a number of measures of decline in community economic and social viability (Reif, 1987; J. Flora and C. Flora, 1988).

Increased Mechanization

Mechanization reduces on-farm labor but increases specialized off-farm labor for repairs and servicing, which are now more complicated than most farmers can do themselves. Increasing mechanization, in terms of sustainability of agriculture, generally means an increase in compaction of the soil and an increasing destruction of soil-conserving structures or methods that are not suitable for large machines, such as terracing and contour plowing and windbreaks. For communities, it may mean the introduction of specialized mechanics who now must carry out the repairs that farmers once did for themselves. This increases the circulation of capital within the community. Sustainable agriculture will probably be highly mechanized, but with small-scale machinery that is more adaptable to diversification. Therefore, specialized mechanics will remain an integral part of the community.

Specialization

Specialization in a few crops means reduced labor demand (Dovring and Yanagida, 1979). It also means that communities become more specialized and thus more dependent on government and world trends. Communities' options are thus becoming more limited.

Intensification

As U.S. farming became more intensified, the chemical inputs to the land increased (Youngberg and Buttel, 1984). This is in contrast to some highly intensified agroforestry systems and intensified home gardens in Southeast Asia that are without chemicals. While this means a move toward reduced environmental sustainability, it would appear to be good for those who supply the chemical inputs. Among the other off-setting aspects for the community, declining water quality and potential for pesticide residue must be addressed.

OPTIONS FOR RURAL COMMUNITIES THROUGH SUSTAINABLE AGRICULTURE

There is great need for institutional change in rural communities. Both public and private institutions need to change to facilitate a move toward sustainable agriculture and to integrate that agriculture into the community. Cooperative extension will be even more important, because there is less profit as incentive for input dealers and salespeople introducing new technologies. A new, low-input technology based on internal resources will mainly be profitable for the farmer.

There will also be the need to create, on the local level, businesses related to risk-shifting strategies such as insurance, the futures market, and call-put options (CAST, 1988). Sustainable agriculture, which will be diversified and more market oriented, will require a greater division of labor within the community. Some enterprises would identify markets, others would respond to market identification. Sustainable agriculture, like conventional agriculture, will depend more and more on off-farm, community-based institutions to be competitive. These institutions include

1. Diversified sources of debt, equity, capital, and income.
2. Economies in marketing and production.
3. Asset portfolio manipulation.
4. Use of sophisticated technology, such as computers, risk management strategies, and paid consultants (CAST, 1988).

Characteristics of Viable Rural Communities

Viable rural communities are ones with active participating citizens, including farm and nonfarm populations. These citizens participate in collective problem solving. Sustainable rural communities will have

1. Diversified farming systems.
2. Better links to consumers/markets.
3. More participation and responsibility in community affairs.
4. Legitimation of innovation.
5. Mechanism to make capital available for nonagriculture development instead of the current situation of overinvestment in agriculture. Such capital saving measures include
   a. Diversification, by using more complex rotations and adding animal enterprises.
   b. Reduced pesticide inputs to zero levels or economic thresholds.
   c. Adaptation of soil- and money-saving tillage practices without herbicides.
   d. Use of nitrogen fixing legumes as cover crops and in rotation to help decrease the need for nitrogen fertilizer (CAST, 1988).

These practices are more profitable and more environmentally sound, leaving capital for other use and reducing the current excess capacity in agriculture.

Entrepreneurial Rural Communities

Farming-dependent communities that have displayed local initiative, according to our research (C. Flora and J. Flora, 1988), share the following attributes:

1. Acceptance of controversy as normal, indicated by a weekly newspaper willing to print controversy.
2. Long-term emphasis on academics (compared to sports) in the school.
3. Generation of enough surplus, often from slightly larger than average family farms, to allow for collective risk taking.
4. Willingness to invest that surplus to local private initiatives.
5. Willingness to tax themselves and to invest in the maintenance of rural infrastructure.
6. Ability to define community broadly, so that consolidation has meant larger boundaries for small communities, not a win-lose battle.

7. Ability to network vertically and horizontally to direct resources, particularly information, to the community.
8. A flexible, dispersed community leadership.

Acceptance of Controversy

In rural communities, especially those with a high density of acquaintanceship (Freudenberg, 1986), there is a great deal of what sociologists call role homogeneity. That means that there is a high degree of overlap among the different roles that community members perform. Your banker buys at your hardware store; he is also treasurer of the Rotary Club where you are president, coach of your daughter's Little League baseball team, a fellow member of the school board, a deacon in your church, and the parent of your son's date to the junior-senior prom. In such situations, with a high degree of interaction and interdependence, there is a tendency to repress controversy, to insist that "we are all just folks," instead of raising and discussing clear differences of opinion. As a result, when disagreements do surface, they have been nurtured so long that they burst into the open as full-fledged conflicts, often deeply splitting a community.

In such communities, the weekly newspaper tends to be long on ads and short on news. There is a great emphasis on the biggest zucchini of the season and the scores of the girls' volleyball game. But there is seldom a reporter at the school board meeting, and no hint of any bad news is allowed to appear in print. People generally love this kind of home town paper. But those towns tend to be dying towns. Only in the minority of communities, where the editor is willing to offend people and print disagreements and potential problems, are community members prepared to mobilize and act to control their own destinies. In contrast, people who are not informed tend to react, often with great emotion but little concrete action, when they realize the finality of the changes that have taken place. In those communities, citizens lack the basic information and the debate on the information that can foster informed public decision making (Tichenor et al., 1980).

Emphasis on Academics

Schools have traditionally provided the symbolic center of Great Plains rural communities. They have been the center of social life and an active indicator that the community is alive and functioning. However, that focus on community solidarity has many times placed undue emphasis on extracurricular activities. Academic excellence and the provision of more exotic courses like foreign languages and laboratory sciences have been neglected. Further, the emphasis on consensus and distaste for controversy often drives out teachers, principals, and school superintendents who raise new ideas or question the sacred nature of the way things have always been done. In these communities, school board elections are seldom contested (it would be bad manners to run against a neighbor). The only school issues debated are the hiring of the football coach or the theme for the homecoming dance.

Presence of Surplus

Although there is a widely quoted old saw that "necessity is the mother of invention," anthropologists persuasively argue that "surplus is the mother of invention." Only when there are enough resources to ensure the provision of basic necessities will individuals or communities innovate. Innovation means risk. Only when basic necessities are assured will people risk that which they value but do not depend on for survival. Where land is highly concentrated, such collective risk taking does not occur.

Willingness to Invest Private Capital Locally

We have found that communities having relative equality and a number of slightly larger family farms are more likely than similar communities with smaller farms to pool resources and invest (take risk) in community-based enterprises. This is in contrast to communities constantly sending investment capital out of the community, either directly or through local financial institutions. In one community of this type that we have studied, residents have galvanized local resources to build a home town carnival, a feedlot, a dairy, and a movie theater. Most recently, they mobilized funds to buy a factory and attempted to buy out the FDIC when a local bank went under. In none of these cases was there any single major investor. Multiple community members put up relatively equal amounts that were enough to form a solid investment, but not enough to force anyone to leave town broke if the venture did not pay off.

Willingness to Support Local Services Through Taxes

A low tax ideology generally predominates in rural areas (Vidich and Bensman, 1978). This is logical, given that rural people, particularly in agriculture, are "land poor." However, as a result, not only do communities delay construction and repair of needed infrastructure, but they depend on state and federal governments to provide the capital. That establishes a dependency in attitude and action that relies on the outside to provide—or not provide—the basic community needs and to set the agenda, based on national priorities, for what gets done locally. In contrast, communities that are willing to raise necessary local capital through local taxation develop a sense of empowerment that allows them the independence to recognize local needs and act collectively to meet them.

Broad Definition of Community

The rural community has been a major source of identity and participation for its residents. However, as populations decline and services of various sorts, like schools, are consolidated and shared, only those communities that have wide and relatively permeable boundaries have not been split. Such mechanisms as a countywide chamber of commerce and the organization of multicommunity events are helpful in bringing about a broad definition of community.

Vertical and Horizontal Linkages

Although entrepreneurial communities are not dependent on outside agencies to initiate action, they are active in seeking out resources from similar communities and from state and federal entities. They generally participate in regional planning groups, contact the cooperative extension agent and specialists, and apply for federal block grants. Further, they engage in lateral learning from other communities. When a community member visits another community and sees a process of interest, she or he is likely to go back home and organize a group to visit and learn how they can do the same thing. Further, they encourage other communities to learn from them.

Lemaire (1985) predicts that by the year 2000 the food industry will be made up of a few "mega" international food companies, plus a number of small entrepreneurial companies in niches between the production lines of the giant food companies. Communities with sustainable agriculture will have to provide institutions for forward contracting to the food companies in ways that assure a profit for the farmer or develop the local companies to identify the niche markets and fill them. Local, flexible, sustainable agriculture should further this networking.

THE CONTRIBUTIONS OF SUSTAINABLE AGRICULTURE TO VIABLE RURAL COMMUNITIES

Sustainable agriculture is much more complex than traditional agriculture. On some sustainable farms, with diversified farming systems and low chemical input technology, master farmers are able to keep in their heads an entire production system and how to make it work in a productive way. However, for many other farmers, the great diversity of tasks is both intellectually taxing and beyond their frame of reference in terms of calculating everything from markets to soil balance to optimum cultivation times. There will be a need in communities that are fostering sustainable agriculture to develop service industries that respond to the needs of low-input agriculture. Those needs include sources of information and alternatives similar to the input dealers and feed and seed promoters, who have in the past showed the benefits of using their products. The same potential and ability to mutually profit should be developed for the new strategy.

Management Intensity and Complexity

The very nature of sustainable agriculture will require broadly educated (in the formal and informal sense of education) and flexible individuals, who can contribute to community growth and development through management skills which are especially important in the new global economy. The ability to solve problems on-farm in a creative and responsive way rather than a prescriptive way could be transferable to other communities. Thus, economic development, like profitability, would begin to be defined as utilizing what is available locally, rather than applying the set formulas that supposedly work for everyone across a region.

There are already data to suggest that IPM (Integrated Pest Management) takes more management input, as does reduced dependence on preventative drug dosing of animals. Low-input, sustainable agriculture will require farmers able to use problem solving skills and approaches instead of simple application. Studies by William Owen of Ohio State University, cited in May–June 1988 issue of *The New Farm,* found that beef and hog producers using fewer drugs had costs that were 20 times lower than those of other farmers. Producers using a low chemical-input strategy had fewer disease losses, because they paid attention to management and environmental factors, such as ventilation, crowding, sanitation, nutrition, and good water. The ability to think creatively, demonstrated by these responsive management activities, may mean a turnover of farmers but a revitalization of community mentality from one of dependence and specificity to one of empowerment and ability to generalize.

Cost-Minimization

Low-input agriculture contributes to sustainability for the community by reducing farm expenditures, not by maximizing output. Reducing the cost of capital debt servicing and purchased inputs may mean a reorganization of the financial and retail trade in rural communities to more service-oriented, diversified kinds of investments. The fact that the agriculture in most rural communities is already above capacity and is an overuser of capital means that cost-minimizing strategies could lead to the use of capital for further community diversification, including using local resources to seek out niche markets worldwide.

Sustainable agriculture increases the viability of rural communities in a variety of ways.

Diversified Farming Systems

By increasing the number of crops grown and rotating crops, the economy of the entire community becomes more stable.

Better Links to Consumers/Markets

Sustainable agriculture is, in part, consumer oriented, particularly relating to consumers' concern for food quality. This, in turn, increases the price per unit available and the economic gain to community members as a whole.

More Participation and Responsibility in Community Affairs

The necessity of sustainable agriculture to be linked to larger community institutions will force the increased participation of farmers, decreasing their isolation and helping to build mutual problem-solving capacities in rural communities.

Legitimation of Innovation

Sustainable agriculture is innovative, requiring a shift from habitual patterns for most farmers. This increased emphasis on sustainable agriculture and the kind of location-specific knowledge necessary to make it work will increase legitimation of innovation within the community.

Making Capital Available for Nonagricultural Development

Agriculture has in the past overused capital because of the incentives built into agricultural policy; therefore, a shift of capital away from agriculture should free it for nonagricultural development. Such capital savings measures include the following:

1. Diversify using more complex rotations and adding animal enterprises, which, in turn, add to the diversity of community production. This should stimulate related industries within the community.
2. Reduce pesticide inputs to economic thresholds. This careful pest management practice may often involve the use of professional services in the community, which could again increase the division of labor within the community, at the same time improving water equality.
3. Adapt soil and money-saving tillage practices without herbicides. Once again, these practices imply more management and may require community-based sharing of knowledge on how to use them. This, in turn, increases community solidarity.
4. Use nitrogen-fixing legumes as cover crops in rotation to help decrease the need for nitrogen fertilizer (CAST, 1988). This practice is more profitable and more environmentally sound, leaving capital for other uses and reducing the current excess production capacity in agriculture that has led to low market prices.

Buttel et al. (1986) point out that "reduced input agriculture is not likely to lead to a renaissance of small or family farms." However, reduced chemical input agriculture could slow the trend to bifurcation of the farm economy. Cacek and Langner (1986), in examining studies that compared the actual economic performances of organic farms and conventional high chemical-input farms, found that actual net income was not that different. In comparing the net returns to farms that were matched with similar nonorganic farms, the organic farms had equal or greater net returns despite their disadvantaged tax situation (Lockeretz et al., 1978; Roberts et al., 1979). The disposable income of farm families would not necessarily fall if they were to use sustainable agricultural practices, particularly if the policy environment were to become more favorable (see Chapter 13). Further, more of the income could be recirculated in the community.

WHAT WILL HAPPEN TO BUSINESSES BASED ON HIGH-INPUT AGRICULTURE?

It is important to recognize that by the late 1980s, most businesses that depended solely on agriculture had faded from the local scene. A move to lower chemical inputs would be done by *solvent* farmers, relieving the remaining dealers from having to cope with bad debts that hurt them badly in the early 1980s. The problem with many agricultural businesses was not simply *volume*, but also unpaid bills.

Large farmers tend to buy from major wholesalers, while medium-sized farmers are increasingly becoming informal sector entrepreneurs, selling feed and seed at the same time they are farming. In fact, there has been a trade revolution in communities at the retail level, as the "Wal Marting and Seven Elevening" of rural America has taken place. In terms of the agricultural sector, the retail trade for agricultural inputs has moved from the feed and seed store to farmer entrepreneurs who, like the Avon lady, have the advantage of the informal sector. They do not have to pay overhead for buildings or help and are able to combine their labor as farmers with selling agricultural inputs.

A move to low chemical-input agriculture, which would increase farmers' profits by reducing production costs, should not have a major effect on input dealers, since they have increasingly become the farmers themselves.

Further, any movement toward changing agricultural practices will occur as part of a continuum. Farming systems studies indicate that farmers adopt technology in small increments and do not convert their entire operation at one time because of the risk involved and the uncertainty that such major changes cause. It must be understood that movement towards sustainability will occur in such low increments. There are those who would hope for an overnight solution of problems and situations. This will not likely occur.

Sustainable agriculture is risk reducing and diversifying. Thus, sustainable agriculture would increase the use of management mechanisms related to diversification, such as equipment leasing, land rental, business arrangements, forward contracting, futures in hedging, off-farm employment, marginal land retirement, social services, cooperative and group action, and related factors that reduce risk and improve sustainability for individual farmers. Further, many of these risk-reducing strategies contribute to community sustainability as well.

What Are the Options for Off-Farm Work?

With the decreasing use of capital in agriculture, more capital will be available for innovative creation of off-farm employment in rural communities. This diversification of the economy, in turn, will provide more off-farm jobs and, more importantly, *better* off-farm jobs than are currently available.

QUALITY OF COMMUNITY LIFE AND SUSTAINABLE AGRICULTURE

Sustainable agriculture will require dramatic policy shifts, including a move away from agricultural subsidies. Currently, "high support prices are capitalized into land values enriching current owners, but raising the cost of entry for new farms" (Resources for the Future, 1988). Low-input agriculture combined with low land prices could encourage the entry of innovators into farming who could help add dynamism to rural communities.

CONCLUSIONS

Sustainable agricultural practices are not antithetical to viable rural communities. Indeed, the movement of capital out of agriculture and into diversified enterprises within the community should be a major help in sustaining communities that are facing major crises throughout rural America. As a result, communities will have more options in performing the key functions described by Warren (1978).

The environmental quality of communities will improve, partly as a result of the current concern with water quality. The social environment should improve as well, because the kind of management style required by sustainable agriculture increases individual empowerment and belief that individuals can effect their own life chances.

The damage to the rural retail establishments, those that are both agriculturally related and non-agriculturally related, has already occurred in response to other macroeconomic trends. Indeed, one could argue that sustainable agriculture, by freeing capital and freeing people's lives from set patterns of dependence, should contribute substantially to the sustainability of rural communities in the temperate zone.

REFERENCES

Bender, L. D., B. L. Green, T. F. Hady, J. A. Kuehn, M. K. Nelson, L. B. Perkinson, and P. J. Ross. (1985). *The Diverse Social and Economic Structure of Nonmetropolitan America*, Agriculture and Rural Economics Division, Economic Research Service, USDA, Rural Development Research Report No. 49, Washington, DC.

Busch, L., and W. B. Lacy. (1984). "Agriculture policy: Issues for the 80s and beyond," *Agric. & Human Values* **1**:5–9.

Buttel, F. H., G. W. Gillespie, Jr., R. Janke, B. Caldwell, and M. Sarrantonio. (1986). "Reduced-input agricultural systems: Rationale and prospects," *Am. J. Alt. Agric.* **1**:58–64.

Cacek, T., and L. L. Langner. (1986). "The economic implications of organic farming," *Am. J. Alt. Agric.* **1**:25–29.

CAST (Council for Agricultural Science and Technology). (1988). *Long-Term Viability of U.S. Agriculture*, Report No. 114, CAST, Ames, IA.

Dovring, F. , and J. F. Yanagida. (1979). *Monoculture and Productivity: A Study of Private Profit and Social Product on Grain Farms and Livestock Farming in Illinois*, AE-447, Illinois Agricultural Experiment Station, Univ. Illinois, Urbana. AE-4477.

Flora, C. B., and J. L. Flora. (1988a). "Characteristics of entrepreneurial communities in a time of crisis," *Rural Develop. N.* **12(2)**:1–4.

Flora, J. L., and C. B. Flora. (1988b). "Public policy, farm size, and community well-being in wheat and livestock farming systems." In *Agricultural and Community Change in the U.S.* (Ed. L.E. Swanson). Westview; Boulder, CO.

Freudenberg, W. R. (1986). "The density of acquaintanceship: An overlooked variable in community research?" *Am. J. Soc.* **92**:27–63.

Haskinger, E. W., and J. R. Pinkerton. (1986). *The Human Community.* Macmillan, New York.

Henry, M., M. Drabenstott, and L. Gibson. (1986). "A changing rural America," *Econ. Rev. Fed. Res. Bank Kansas City* **71**:23–41.

Jackson, W., W. Berry, and B. Coleman. (1984). *Meeting the Expectations of the Land.* North Point, San Francisco.

Lemaire, W. H. (1985). "Food in the year 2000," *Food Eng.* **57**:90–117.

Lockeretz, W., G. Shearer, R. Klepper, and S. Sweeney. (1978). "Field crop production on organic farms in the Midwest," *J. Soil Water Conserv.* **33**:130–34.

Owen, W. (1988). As cited in *New Farm*, May–June:9.

Reif, L. L. (1987). "Farm structure, industry structure, and socioeconomic conditions in the United States," *Rural Sociol.* **52**:462–82.

Resources for the Future. (1988). *Mutual Disarmanent in World Agriculture: A Declaration on Agricultural Trade.* Washington, DC.

Roberts, K. J., P. F. Warnken, and K. C. Schneeberger. (1979). "The economics of organic crop production in the western corn belt," Agricultural Economics Paper No. 1979-6, Univ. Missouri, Columbia.

Rubin, I. (1969). "Function and structure of community: Conceptual and theoretical analysis," *Int. Rev. Com. Dev.*, **21-22**:111-119.

Tichenor, P. J., G. A. Donohue, and C. N. Oliere. (1980). *Community Conflict and the Press.* Sage; Beverly Hills, CA.

Vidich, A., and J. Bensman. (1968). *Small Town in Mass Society.* Princeton University Press, Princeton.

Warren, R. L. (1978). *The Community in America,* 3rd ed. Rand McNally, Chicago.

Youngberg, G., and F. H. Buttel. (1984). "Public policy and socio-political factors affecting the future of sustainable farming systems." In *Organic Farming: Current Technologies and Its Role in a Sustainable Agriculture,* pp. 167–87. Amer. Soc. Agronomy, Madison, WI.

13 POLICY ISSUES AND AGRICULTURAL SUSTAINABILITY

CORNELIA BUTLER FLORA
Virginia Polytechnic Institute and State University
Blacksburg, Virginia

Sustainable agriculture is based on decisions made by individual farmers. Yet those decisions are a function of a wide variety of other decisions made at other system levels. This chapter examines the interaction of individual farmer decisions with policy decisions, particularly state and national level policies that have tremendous implications for the range of choices farmers can make and the probability of those choices leading to the economic and biological sustainability of the individual farm.

Lowrance et al. (1986) have suggested a useful hierarchical approach to sustainable agriculture, which looks at nested system levels. The nesting moves from field systems (agronomic) to farm systems (micro economic) to watershed (ecological) to regional or national (macro economic) systems. The factors that most constrain the sustainability of agriculture must be defined for each of these systems. Something that may be sustainable on a farmer's field may not be sustainable for the farm as a whole because of the implications for other parts of the farming system or the fact that it is not profitable. Even if it is profitable for that farmer, the impact on the watershed, particularly on water quality, may lead to inability of the watershed to sustain itself. Practices that may be sustainable for the watershed may not be sustainable on a regional or national level because of either economic costs or pollution implications.

Policy interactions with other factors affect all system levels by providing constraints to what can be attempted at an agronomic level, to what is profitable at macroeconomic levels, to what is tolerated at the ecological level and even in the international context. The fact that almost all agricultural economies are international, combined with the artificiality of national boundaries, means that there must be multilateral agreement on the general rules of the agricultural game. If not, systematic policy on the national level to aid sustainable agriculture could disadvantage a nation's agriculture in the world market.

Current public policies in developed temperate zone countries channel an excess of resources into agriculture (Zietz and Valdes, 1988), resulting in the high use of chemical inputs (Reichelderfer and Phipps, 1988). The resulting overproduction stimulates agricultural trade wars, which attempt to dispose of the surplus. Although laws at a local, state, or national level are important for making low chemical-input agriculture an appropriate and profitable activity, unless the problem is addressed on a world scale, these are only stop gap measures.

HISTORY OF NATIONAL AGRICULTURAL POLICY IN TEMPERATE ZONE COUNTRIES

Governments have long believed that they have a role in determining what farmers do on individual fields, defined in terms of the national welfare. National governments also have been responsive to farmers as organized political pressure groups, as well as other groups within society, who have defined their welfare as related to agriculture. Agricultural policy can then be seen as the jockeying between various societal groups attempting to define their particular good as a generalized national good (Flora, 1986; Browne, 1988).

GOALS OF AGRICULTURAL POLICY

There are at least five conflicting goals that agricultural policy has attempted to achieve in temperate zone countries: (1) supply control, (2) conservation of natural resources, (3) price stabilization, (4) farm income maintenance, and (5) export enhancement.

Supply Control

While in the United States from at least the mid-1800s the specter of overproduction has been the major problem of supply, European common market countries and Japan have experienced food shortages in the 20th century. Thus, one of the early goals of governments in Japan and the European Community was to assure sufficient food. In the United States, on the other hand, supply control policies often were aimed at reducing the surplus with the concommitant goals of stabilizing price and, thus, assuring income for farmers (Benedict and Bauer, 1960). Supply control mechanisms included (1) acreage limitations, or set-aside acreage, generally on a voluntary basis for which the farmers or land owners receive a premium or subsidy for the crops produced on the rest of their acreage; (2) production controls or allotments, whereby a certain amount of a given product can be marketed at above market price, such as the tobacco program; and (3) exporting surplus, using a variety of programs aimed at moving the surplus overseas, including export subsidies disguised as foreign assistance or marketing loans.

Conservation

Such programs as the Soil Bank of the 1950s and the Conservation Reserve Program of the 1980s were justified in terms of soil conservation. However, the major motivation behind these programs was to reduce production, not to conserve land or water. As a result, as soon as the surplus was seen to have been reduced, the land was again brought into production. There was no systematic, long-term consideration of how best to conserve land and water resources.

Price Stabilization

Agricultural prices historically have been subject to extreme price fluctuations, creating unstable conditions for producers and consumers. Further price fluctuation has led to speculation and hoarding, disrupting state and individual planning. A variety of mechanisms have been implemented in the United States and the European Community (EC) to maintain price stability.

Beginning in 1973, the United States instituted a complicated system of floor prices (loan rates) and price support ceilings (target prices) involving a nonrecourse loan that allows for the surplus to be acquired by the federal government when market prices are too low and released by the federal government when the surpluses are reduced. The difference between the market price (received by the farmer) and the target price is made up by the federal government as a deficiency payment. The EC has an intervention price, similar to the U.S. loan rate. However, prices are supported by raising the price of imported products and by reducing the price of products for exports. Both systems resulted in government payments of about $3 per bushel on wheat in 1986–87 (Newman et al., 1987).

Such price stabilization programs were designed, in part, to give farmers a secure and stable price they could count on for calculating appropriate costs of production. The result of price stabilization was movement of land into production of the relatively few supported crops (Blanpied, 1984).

Income Maintenance

A major tenet of populist ideology in the United States is the independence of the farmer. As a result, direct income-maintenance programs were unacceptable. A variety of indirect mechanisms have been instituted by local, state, and federal governments to provide income for farmers when the market failed. For example, on a county level in many parts of the dustbowl in the early 1930s (before the institution of the major agricultural programs under Franklin Delano Roosevelt), counties paid farmers to plow their own ground for planting and to make roads along their section boundaries (Edwards, 1939). This provided income for work, not "welfare."

The perennial problem of low farm incomes amid boom-bust cycles promulgated a variety of income maintenance programs in Japan, the United States, and the European Community. Income maintenance in the United States after the 1930s was achieved through payment for products, rather than direct payment for work. Often, indirect income-maintenance programs, such as deficiency payments, contributed to other agricultural problems, such as supply control. In most countries in the temperate zone, strong political factions resist any change in price supports, as a threat to rural income maintenance and to national food security.

Prices for the EC are set by annual price reviews undertaken by ministers of agriculture. They require unanimity. The major concern has been the income of farmers rather than a balanced budget for the European common market. The cost has been extremely high. That policy has kept marginal land in agricultural production and has encouraged the high use of chemical inputs.

Japan has a similar policy of high import restrictions, high taxes on imports, and high subsidies to the politically powerful farm lobby.

Export Enhancement

For many developed countries, agricultural exports contribute substantially to their foreign exchange, both by stimulating exports and limiting inputs. The politically determined internal price is maintained for most agricultural products by imposing a variable levy on imports. This is an amount equal to the difference between the "threshold price" and the world market price. This policy prevents competition by foreign suppliers and is very troubling to the European trading partners. European Community funds are used, in turn, to "dump" the products on the world markets or divert them to inferior uses, such as animal feed (Bale and Koester, 1983). U.S. export enhancement programs have followed similar patterns. That presents an apparent conflict with much of development theory, which predicted that developed countries would export manufactured goods and import raw materials, including agricultural goods, from developing countries. In fact, the differential labor costs between developed and developing countries, particularly newly industrialized countries, has led to a greater dependence on agriculture for foreign exchange in the developed countries. Thus, in the 1980s, exports from the United States, Europe, and even Japan have been heavily subsidized under the euphemisms of "foreign assistance," "food for peace," and "marketing loans." These programs not only distort the world commodity market, they send misleading signals to producers, encouraging maximum chemical input per acre.

EMERGENCY MEASURES VS. LONG-TERM PROGRAMS

Agricultural policy generally is based on short-term response to current farm problems rather than systematic planning of national economic goals (Flora, 1986). The programs that began as emergency measures responding to the major problems of the Great Depression in the 1930s were presumably to be in place only a few years until agriculture got back on its feet. The assumption by many policymakers has been that there was a time when agriculture was truly profitable in the temperate zones. However, systematic historic analysis suggests that this has not been the case without government guarantee of price. The theory is that we just have to wait it out and pay for expensive government programs until the mythical normality returns. However, these programs have now become long-term. Farmers' complaints about the programs in the 1930s are echoed by farmers in 1988. One such criticism has been that the government delayed too long their announcement of the set-aside acreage and that most government subsidies tended to go to those that were the worst stewards of the soil (Edwards, 1939). (The unique early announcement in 1988 of the set-aside acres for wheat may have been part of a strategy to attempt to influence the Uruguay Round GATT (General Agreement on Tariffs and Trade) negotiations, which dealt with agricultural concerns.)

The very nature of recent policymaking in temperate zones seems to mitigate against any historic perspective in learning from past policy mistakes or from any attempt to extrapolate into the future the implications of short-term measures to deal with immediate problems in the agrarian sector.

Agricultural policies that have been instituted to meet the short-term goals of a variety of interest groups have led to a series of countervailing results. Programs aimed at reducing acreage (a mechanism of supply control) have resulted in increased use of inputs on the area farmed (influencing conservation in terms of water quality). Mechanisms aimed at income support have resulted in the generation of surplus. Other similar contradictions, apparent and real, have been discussed elsewhere (Flora, 1986; Langley et al., 1987).

The current contradictory agricultural policies of the United States, Japan, and the European Community have encouraged extremely high chemical-input agriculture, with implications for environmental sustainability. They also have resulted in (1) "extravagant expenditures to support farm prices," which have "alienated Community taxpayers;" (2) high food prices (in Europe and Japan but not the United States.), which worried the consumers; and (3) subsidized exports, which have angered trading partners. Further, farmers in these countries claim that the benefits have been inadequate and inequitably distributed (Bale and Koester, 1983).

In EC countries, the Common Agricultural Policy (CAP) supports domestic prices well above world market levels by a complex system of tariffs and variable import levies. The CAP derived from the attempt to gain food security following World War II and to integrate the divergent economies in a politically viable way. Thus, political expediency took precedence over either economic viability or agricultural balance (Bale and Koester, 1983). Certainly, environmental issues and agronomic limitations were not considered at all.

IMPLICATIONS OF THE 1985 FOOD SECURITY ACT FOR SUSTAINABLE AGRICULTURE

Implications for Crop Diversity

The current U.S. agricultural program offers subsidies for a limited number of commodities, which encourages monocropping and discourages crop rotation. Farmers are extremely hesitant to put land into non-subsidized crops for fear of losing their base acreage, which determines the area on which they will be able to produce, once their set-aside land has been taken out. Further, the guaranteed price per unit produced increases the incentive to use levels of chemicals and fertilizers that will maximize production of program commodities on the allotted acres.

Implications for Input Use

High prices increase chemical input use: between 1964 and 1985, farm use of pesticides increased almost 170%, or almost tripled. In 1971, herbicide was applied to 71% of U.S. farmland in row crops and 38% of U.S.

farmland in small grains. By 1982, 91% of row crops and 41% of small grains had herbicide applied (Healy et al. 1986). The farm program indirectly discourages farmers from adopting low chemical-input methods. As Fleming (1987) describes, ". . . several eligibility requirements and subsidy incentives in federal farm support programs discourage farmers from adopting strategies that would reduce their usage of agrichemicals." He points out that legumes and cover crops are not included in the farm program. Most studies have been unable to detect a statistically significant effect of farm programs on aggregate fertilizer and pesticide use. That lack of a relationship may in part be due to high prices (generally present when farm programs are allowed to lapse) providing an incentive in the short-term for high chemical inputs.

Farmers must plant at least half their base to crops included in support programs in order to maintain full support eligibility. This penalizes farmers who use crop rotations that run three years or more. Federal support programs are focused on four crops that account for at least 65% of all agrichemical use: corn, wheat, cotton, and soybeans. (Soybeans are not part of a price support program, but can be rotated with corn or cotton.) Price supports are paid on a "per-unit-produced" basis, and base area is determined by land area. This forces farmers to increase their use of agrichemicals and fertilizers, because the more the total production is (which is now limited by crop produced per hectare), the more support will be received. Fleming (1987) points out that the price supports for corn may induce farmers to more than double the amount of fertilizer that they would apply to crops in the absence of government subsidies. The result is an ineffective set-aside program with a great deal of slippage. For example, reducing corn planting by 25% only reduces corn production 7%, because farmers add more nitrogen fertilizer to the planted acres. If the role of set-aside is to reduce crop production, it does not do so in the way desired. Besides the increased use of inputs, slippage occurs because poorer land is put into acreage reduction programs and non-program lands are brought into production.

Implications for Soil and Water Quality

The increased total amount of chemicals used and their concentration on the non-set-aside acres may increase runoff and resulting water pollution (Fleming, 1987). Between 1970 and 1981, 55 million acres were brought into crop production. Much of that land put into crops had been previously idle under government programs or was converted from pasture, forest, or wetlands (Healy et al., 1986).

The Conservation Reserve Program (CRP) is an attempt to temporarily retire vulnerable land from production, but its ability to protect land is exceedingly limited. In many areas where soil loss is great but productivity is high, it is more profitable to plant continuous, subsidized crops rather than put land into the CRP.

The current farm program rewards farmers, through deficiency payments, not only for planting strictly program crops (leading to such anomalies as the current importing of oats to the United States), but for maximizing production on the non-set-aside land through the intensified use of chemical and fertilizer inputs. A result is the contamination of both ground and surface water from high levels of nitrate and pesticides (Hallberg, 1987). In this case, economic sustainability of the farming system is being improved (although the farm family may decide to purchase bottled water), while the sustainability of the ecosystem is being degraded by the lowering of water quality and the resulting health hazards. The national system further is being hurt by the high cost associated with subsidizing high yields per acre. This then has an echo effect internationally.

NONAGRICULTURAL POLICIES THAT AFFECT SUSTAINABILITY OF AGRICULTURE

Monetary Policy

Monetary policy is the degree to which governments allow the money supply to expand or contract. The erratic and overly expansionary monetary policies in the 1970s brought excess debt, inflation, and an unsustainable demand for agricultural products that resulted in an oversized agricultural capacity (CAST, 1988). Chemical inputs and mechanical inputs into agriculture increased. The monetary policy of the 1970s that encouraged spending and capital use in agriculture was reversed by the 1980s, resulting in high real interest rates and low inflation, which increased the cost of borrowing. The 1980s have shown a marked decrease in the purchase of expensive new machinery as a result. Some chemical inputs have been reduced as well.

Fiscal Policy

Fiscal policy is the set of legislation and administrative decisions that determine how much a government takes in from where and what it spends on what. Federal deficits in the 1980s brought high real interest and exchange rates, falling land prices, falling exports, and farm financial stress. Financial stress, in turn, leads to a very short-term planning horizon for farmers, encouraging the mining of land and water in attempts to maintain the farm. Further, as the CAST (1988) report on the viability of U.S. agriculture points out, high real interest rates (caused by a combination of monetary and fiscal policy) discourage long-term investment in conservation measures. The dramatic shifts in monetary and fiscal policy lead to ever shorter time horizons in judging the sustainability of the farm system and land productivity.

Trade Policy

Protectionism that keeps prices artificially high encourages the use of expensive inputs, including capital and chemicals. Protectionist trade policies in all the temperate zone countries have led to an over investment in agriculture, resulting in agricultural surplus. Further, a trade policy that subsidizes exports also artificially encourages a higher productivity through high use of chemical inputs. A trade policy that dumps products in the international market, such as the current marketing loans in the United States, encourages monoculture and discourages farmers from diversifying and seeking out international niche markets. Protectionist trade policies include our current strict curtailing of sugar imports from countries under the U.S. sugar quota, countries that receive our high, subsidized price. This has resulted in bringing fragile lands in the southern United States into sugar production, as well as artificially increasing the demand for corn, because corn sweeteners cannot otherwise compete with sugar at world market prices. This policy not only adversely affects wetlands in the South by bringing them into sugarcane production, but favors high chemical-input monoculture in the Midwest. The inclusion of agriculture under the GATT would be a major mechanism to allow the cutting back of agriculture subsidies on an international basis (Petit et al., 1988).

Tax Policies

Interest Deductions. Deduction of interest paid from farm tax liability encourages the use of capital. It helps most those producers with adequate resources and those who are indebted. This policy favors high chemical-input agriculture by giving input-intensive operators the tax breaks that are lost by those who do not use their interest deductions. However, it also favors longer term investment in land improvement and erosion reduction, which otherwise have little monetary incentive (van Es, 1982; Mueller et al., 1985).

Accelerated Depreciation. Accelerated depreciation, particularly in the late 1970s and early 1980s, increased purchase of new equipment. It was subsidized by tax codes that allowed interest deduction from taxable income. Accelerated depreciation allowance and investment tax credits favored larger farmers (Raup, 1978). The increase in the use of large machinery, which tends to be somewhat specialized and thus discourages crop diversity, also encourages larger field units and removal of conservation structures.

Investment Tax Credits. Investment tax credits serve as subsidies to farm expansion, as does accelerated depreciation. Investment tax credits in the 1970s encouraged the plowing up of virgin land and increased soil erosion

(Watts et al., 1983). This provision of the tax code had a negative effect on watershed sustainability. Those who could take advantage of investment tax credits tended to be wealthy individuals who had tax liabilities from other income sources. Often they were absentee owners who contributed little to the community and did not recycle the economic advantages gained locally. These policies increased large farmers' income at the expense of the environment. These policies, particularly capital gains allowances, favored high land turnover for speculation purposes, because in order to realize capital gains advantages, assets must be sold. As Watts et al. (1983) point out, "The capital gains features provide greater incentive to those at higher marginal tax rates who are not going to retain the cropland for production but who are going to take capital gains as soon as other tax advantages are dissipated." This, in turn, leads to decline in community sustainability.

Energy Policy

A policy of cheap energy encourages mechanization and the use of petroleum-based chemical inputs (Rushefsky, 1980). Energy policy includes subsidization of the oil and gas industry and promotion of alternatives to nonrenewable energy sources.

Food Price Policy

Both Europe and Japan have passed some of the cost of maintaining input-intensive agriculture to food consumers. The price of food in those countries is relatively high compared to the United States. The American taxpayer has absorbed that cost, however, because one of the roles of U.S. agriculture is viewed as providing low-cost food, defended as being of special benefit to low-income people (CAST, 1988). This policy assumes that, in general, there will be worldwide low purchasing power. The United States, the European Community, and Japan all use government funds to artificially lower the cost of agricultural exports to maintain or increase world market share. For the United States, "export enhancement" is supported as necessary to help the balance of trade and earn foreign exchange to service international debt.

POLICIES UNDER WHICH LOW-INPUT AGRICULTURE WOULD ALSO BE PROFITABLE

Targeted Programs

Currently, programs are aimed at growers based on the commodities that they produce. Targeted programs would more specifically identify the precise goals of agricultural policies and target funds to reach those goals. Such targeted policies would be aimed at small/poor farmers, if the goal were income maintenance, and would be aimed at large farmers, if the goal were supply control. Targeted programs would decrease the current confusion of direct and indirect goals which these policies seek to address.

Decoupling

By the late 1980s, the high costs and multiple objectives of agriculture programs have led to a desire to decouple the income maintenance objectives from the other objectives of the farm programs. The attempt is to separate the legitimate farm income objectives from the crops subsidies and to deal with them directly. The current method of dealing with problems of low farm income in the developed temperate zones is rationalized as saving the family farm through the use of subsidies, which are also incentives to maximize productivity using high levels of inputs.

Most of the decoupling programs proposed involve letting prices approach world market levels and cushioning the farmers' adjustment to immediate income drops by specific income maintenance programs. It is clear that decoupling, that is to say, removing subsidies from agriculture, would mean a movement of capital out of agriculture. It would initially lower farm incomes, as world market prices are now lower than the target prices farmers receive.

Decoupling has been faulted by some farmers' organizations as being a direct welfare program, and thus unacceptable ideologically. They also point out that decoupling has the potential of ending conservation and environmental incentives that are currently tied to subsidy programs (National Farmers Union, 1989).

If decoupling became an international policy—and it can only be an international policy if it is incorporated into existing treaties such as the GATT—this adjustment ultimately could lead to increased prices, reduced supply, and lower costs of production. An immediate way to lower production costs is to reduce inputs.

Marketing Quotas

Production controls through marketing quotas allow farmers to market a limited amount of subsidized crops on any amount of land desired. The U.S. tobacco program is an example of production limitations. Such a program would encourage low chemical and fertilizer input use to maximize profit per bushel rather than profit per hectare. Positive impacts noted by the USDA include declines in erosion and a decline in the delivering of non-point pollutants (ERS/USDA, 1987).

Production controls are not appealing to any particular country, because their institution means giving up market share internationally. The high internationalization of agricultural trade in the latter half of the 20th century has meant that production controls applied unilaterally are not popular policy instruments. Yet they have positive impacts on income maintenance.

Terminate Price Subsidies

Subsidies can be seen as leading to high chemical-input agriculture and distorting the balance of the larger economic and social systems in which agriculture is embedded. Ending subsidies would eliminate an incentive for excessive use of agrichemicals and extend the range of crops grown by allowing farmers to respond to market demand, relating supply of commodities to demand rather than to government set prices. Diversification, which could result when the incentive to overproduce subsidized crops is gone, would also reduce risk and thus the need to use agrichemicals, which is another way of overcoming risk.

"Subsidies transfer wealth, distort resource allocation, and affect economic activity. Export subsidies enhance demand, raise prices, and lead to greater production than otherwise would occur. This bids resources away from other uses, leading to higher production costs for nonsubsidized goods, lower output, and higher prices" (Wilson, 1988).

Elimination of Trade Barriers

The current costs of high chemical-input, high-yield systems of agricultural production, which may be efficient in use of land but inefficient in use of capital and natural resources, could be reduced by eliminating tariff and nontariff barriers to world trade. Tariff barriers limit the amount or highly taxed agricultural products that enter a country. Nontariff barriers, particularly sanitary and phytosanitary regulations, also serve to limit imports. One could argue that certain sanitary regulations, particularly those regarding pesticide residues, might be made part of the international trade agreement. But many psuedosanitary problems have been raised merely to protect domestic production systems as a successful, if temporary, nontariff trade barrier. Also to be eliminated would be the variable levies and minimum prices of imports that are currently in effect in Europe. The key to this would be bringing agriculture under the GATT, which would eliminate export subsidies.

Based on data from on-farm trials and modeling in the Palouse region of eastern Washington, Goldstein and Young (1987) conclude that "if world market competition and domestic policy pressures reduce crop price supports and grain prices in the future, low-input systems would become more profitable than conventional practices."

Soil and Water Conservation Programs

Soil and water conservation programs that have the best chance of being effective should (1) calculate the externalities of soil loss and water pollution to quantify the real long-term costs, (2) monitor the impact of pollution on farmers' fields and on society in general, and (3) reimburse farmers for nonrecoverable expenses in instituting soil and water conservation. Fleming (1987) suggests taxing agrichemicals so that their price reflects the real social and environmental costs of their use. Money raised then could be used for monitoring ground water quality and for research aimed at reducing the need for agrichemical use.

Herbicides account for 85% for all pesticide use (Fleming, 1987), and more than half of the herbicides applied to field crops are applied to corn. Corn also accounts for almost half of all fertilizer used in the United States. If current patterns continue, about 40% of the revenues from a general agriculture tax is likely to be paid by corn farmers. Since the Corn Belt suffers from widespread groundwater contamination, and since corn farmers also receive a large portion of government support payments, a tax on agrichemicals could help solve several problems at once (Fleming, 1987).

Monoculture corn or even corn after soybeans has proven to be a potentially unsustainable system in terms of impacts on soil and water quality (watershed), as well as in terms of the highly subsidized price currently paid by taxpayers worldwide (nation and world systems). Further, our current trade barriers/subsidies to sugar help make corn sweeteners economically competitive with sugar in the United States. If corn sweeteners were to compete against the world sugar price, the degree of corn use domestically would decline. Further, we are currently able to ship corn gluten (a corn wet milling by-product from the production of corn sweeteners) to Europe for animal feed, because it was developed after current tariff barriers were set. If the price of sugar were allowed to fall to world market prices, demand for corn would drop even further, and substantial changes in cropping patterns would have to take place in the Corn Belt. This could have positive implications for soil and water conservation.

In terms of the individual farm system, it is doubtful that most conservation technologies are sustainable economically. Indeed, a number of authors argue that the cost of most conservation technology exceeds the benefits on a short-term basis (Pampel and van Es, 1977; Lovejoy and Parent, 1982; van Es, 1982). Indeed, some authors looking at specific crops question the long-term profitability of conservation tillage, one of the most cost-saving of the conservation technologies (Mueller et al., 1985).

It has been argued that soil and water degradation would not substantially affect long-term U.S. productive capacity (Crosson, 1982, 1986, 1987; Pierce et al., 1984). The off-farm costs of erosion and runoff are substantial and well in excess of the on-farm costs (Buttel and Swanson, 1985;

Clark et al., 1985; Huzar and Piper, 1986; Strohbehn, 1986). This would suggest that a market mechanism to ensure soil conservation is made ineffective, because the people who must pay the costs of conservation are not principal beneficiaries of their own actions. Thus, voluntary programs of soil conservation are unlikely to be effective, except among those farmers of deep convictions and/or deep pockets. The 1987 staff report of the Soil Conservation Service Task Force stressed that conservation practices should be mandatory for participation in government farm programs and "that conservation compliance policies should also be effective . . ." (Soil Conservation Service, 1987). As the 1988 CAST report on the viability of U.S. agriculture points out, "Any successful soil conservation program will have to meet off-site as well as on-site erosion reduction goals." This could ultimately mean a move to mandatory rather than voluntary conservation programs (Buttel and Swanson, 1985). The CAST report (1988) further points out that "reduced input practices may only become relevant if farmers are forced, through regulation, to reduce the surface and subsurface transport of soil, chemicals and salt that adversely affect off-site resources."

There is a need to permanently retire vulnerable land or land irrigated with nonrenewable water supplies and shift to grass, trees, recreation, or other soil and water conserving uses. Mechanisms for soil retirement might include easements and purchases by land trusts, rather than the current, temporary land retirement systems (Ward, 1988). The conservation program should be supplemented by instituting cropland easements on highly erodible land and allowing the land to be used for grazing and haying, but not for annual row crop production.

Highly erodible land must be made ineligible for regular price support and diversion programs, if soil conservation measures are to be effective, as occurs in 1990 under the conservation compliance provision of Title 13 of the 1985 Food Security Act. It will be necessary further to increase the proportion of land in highly erodible counties eligible for the CRP. The argument that the CRP is detrimental to community viability has not been borne out by research (Mitchell, 1987). To make the CRP effective, the payments or easements must be indexed to cash rents to reduce the risk of inflation. A highly inflationary period would make it very profitable for farmers to plow up their conservation lands, pay back the government, and continue to lose soil at high rates. CAST suggests expanding the CRP or easements to control the use of nonrenewable groundwater supplies for irrigation on a countywide basis.

The CAST report (1988) also suggests expanding the "sodbuster" and "swampbuster" provisions of the 1985 farm bill to deny price supports to farmers whose production methods contribute to ground water contamination, as it already does to those who cultivate highly erodible land or who convert designated wetland to cropping. This would involve establishing limits on agrichemical usage and monitoring application, a highly complex and expensive endeavor.

Water quality regulation is a particular problem, but one that can be addressed on a state basis through laws such as that passed in Iowa in 1987 funding education, research, and regulation related to groundwater protection, particularly protection from agricultural chemicals. Abdallah and Hill (1987) outline some public policy options to attempt to avert future contamination of ground water. These include (1) taxes on practices or products that contribute to groundwater contamination, (2) subsidies to promote farm activities that prevent or reduce contamination to lower their relative costs, and (3) redefining property rights by developing regulations or prohibitions for certain farming activities or products that contaminate ground water. These are all policies that could be applied on the level of a watershed, state, or the nation and, in fact, might be best applied at more local levels, depending on degree of contamination and willingness to monitor and enforce.

There is currently little or no monitoring of pesticide use. Taxing of pesticides, which makes them more expensive, might result in reduced application. Stricter laws governing application can only work if mechanisms of monitoring and enforcement are put in place.

Mandatory regulation of agricultural chemical use is another way to reduce water pollution. Currently there is no monitoring of chemical use or runoff. There is a tendency for many farmers to apply preventative, rather than treatment, doses of agricultural chemicals. Regulation might include licensing and monitoring those allowed to apply nonbiodegradable agrichemicals.

Controls of Food Quality

Food quality programs, including grain quality controls, continued limiting of pesticide residue on foods and pharmaceutical residues in meat, and the support of such institutions as the Federal Grain Inspection Service of USDA, can all help reward farmers who use low chemical-input agricultural methods. Such programs should include investment in better monitoring and grading, with emphasis on continuous rather than discrete grading that would discourage adding low-quality materials to grain shipments. Today the product delivered often meets only the minimum standard of the grade purchased, a "value-added" practice of the international grain trade, which gives no return to the farmer. In fact, U.S. grain is often considered "inferior" on the international market because of dilution with low-quality material—to the detriment of our grain exports and returns to the producer.

Direct Poverty Programs

Policies that deal directly with the problem of poverty by increasing the incomes of poor people would ultimately serve to increase effective demand and raise farm prices to a sustainable level. This would contrast with today's implicit cheap food policies that assume that agricultural prices must be kept low to help the poor at home and abroad. Such modified policies would aid development in the United States and abroad, as well as direct payments to poor who cannot be in the labor force, even if jobs were available. This would suggest increased payments through Aid to Families with Dependent Children, as well as programs that encourage increased wage levels for the working poor.

Changing Criteria for Successful Agricultural Research

For sustainable agriculture to become part of the research dialogue and agenda, we need criteria that do not simply measure agricultural productivity and progress by yield per acre. We must consider a return to the other factors of production. For example, research that is reported in terms of bushels per dollar invested or bushel of crop produced per bushel of top soil lost would give us a different notion of a particular agricultural innovation. However, this reliance on a single production input would be just as misleading. Phipps (personal communication, 1988) suggests measures of productivity that account for purchased inputs (land, labor, machinery, and chemicals) *and* nonpurchased environmental impacts such as water pollution, soil erosion, and destruction of wildlife habitat.

At present, return to land (not a scarce resource in most temperate zones with the exception of Japan) distorts our perception of what are effective and efficient agricultural technologies. Even the CAST report on the viability of U.S. agriculture (1988) assumed that yield per acre planted had to improve. As long as productivity is defined as yield per acre, there is a logical bias toward high chemical-input agriculture. This bias has been a particular handicap for farmers, because increased yield per acre is almost always accompanied by decreased price per unit produced. As supply continues to outstrip demand and the overcapacity of the United States and other temperate zone countries is maintained, sustainability of U.S. agriculture is threatened on the farm and national system levels (Avery, 1985).

Research on Low Chemical Input Agricultural Techniques

Research is under way on low chemical-input agricultural techniques in both private and public institutions. In public institutions, the emphasis has been on integrated pest management (IPM) and more recently, low-input/sustainable agriculture (LISA), whereas more far-reaching "blue sky" work, for example, developing perennial grains, is taking place in private institutions, such as the Land Institute in Kansas and Rodale Research Center in Pennsylvania.

Research to further low-input agricultural techniques must include basic research on agroecosystems (Jackson and Bender, 1984; Lowrance et al., 1984), plant physiology, plant and insect pathology, and soil microbiology, among other areas. There is also a need for applied research, including analysis of farming systems through on-farm trials and systematic comparisons to conventional high-input agriculture.

CONCLUSIONS

The widespread adoption of low-input agricultural techniques for crop and livestock production will depend on policy changes at the macronational and global levels. We cannot depend on conversion of cropping systems by individual farmers to make a difference at the field or farm system level, unless those changes are made profitable and possible by changes at the higher system levels. It will be difficult to break the Gordian knot of policies that indirectly favor high-input agriculture unilaterally, because the policy bias toward such techniques is worldwide in the temperate zones. Collective action, probably through existing mechanisms such as the GATT, will be necessary for the "mutual agricultural disarmament" that is necessary for public sustainability of agriculture in the temperate zones.

REFERENCES

Abdallah, H. F. , and L. D. Hill. (1987). "Agriculture and ground water quality: A public policy perspective," *Am. J. Alt. Agric.* **2** (Winter):37–40.

Avery, D. (1985). "U.S. farm dilemma: The global bad news is wrong," *Science* **230**:408–12.

Bale, M. D. , and U. Koester. (1983). "Maginot line of European farm policies," *World Econ.* **6**:373–91.

Benedict, M. R. , and E. K. Bauer. (1960). *Farm Surpluses: U.S. Burden or World Asset?* University of California, Division of Agricultural Sciences, Berkeley.

Blanpied, N. A. (1984). *Farm Policy: The Politics of Soil, Surpluses and Subsidies.* Congressional Quarterly, Washington, DC.

Browne, W. P. (1988). *Agriculture, Private Interests, Public Policy, and American Agriculture.* Univ. Press of Kansas, Lawrence.

Buttel, F. H. , and L. E. Swanson. (1985). "Soil and water conservation: A farm structural and public policy context." In *Conserving Soil: Insights from Socioeconomic Research* (Eds. S. Lovejoy and T. Napier). pp. 26-39. Soil and Conservation Society of America, Ankeny, IA.

CAST (Council for Agricultural Science and Technology). (1988). *Long-Term Viability of U.S. Agriculture.* Report No. 114, CAST, Ames, IA.

Clark, E. H. , J. Haverkamp, and W. Chapman. (1985). *Eroding Soils: The Off-Farm Impacts.* Conservation Foundation, Washington, DC.

Crosson, P. R. (1982). "The long-term adequacy of agricultural land in the United States." In *The Cropland Crisis: Myth or Reality?* (Ed. P. R. Corran). pp. 1–22. John Hopkins University Press, Baltimore, MD.

Crosson, P. R. (1986). "Soil erosion and policy issues." In *Agriculture and the Environment* (Eds. T. Phipps, P. Crosson, and K. Price). Resources for the Future, Washington, DC.

Crosson, P. R. (1987). *The Long-Term Adequacy of Land and Water Resources in the United States,* Resources for the Future, Washington, DC.

Edwards, A. D. (1939). *Influences of Drought and Depression on a Rural Community: A Case Study in Haskell County, Kansas.* Social Reports No. 7, Farm Security Administration and USDA, Bureau of Agricultural Economics. Government Printing Office, Washington, DC. January.

ERS/USDA. (1987). *Mandatory Production Controls.* Agriculture Information Bulletin No. 520, Washington, DC.

Fleming, M. H. (1987). "Agricultural chemicals in ground water: Preventing contamination by removing barriers against low-input farm management," *Am. J. Alt. Agric.* **2**:124–30.

Flora, C. B. (1986). "Values and the agricultural crisis: Differential problems, solutions, and value constraints," *Agric. & Human Values* **3**:16–23.

Goldstein, W. A. , and D. L. Young. (1987). "An agronomic and economic comparison of conventional and a low-input cropping system in the Palouse," *Am. J. Alt. Agric.* **2**:51–56.

Hallberg, G. R. (1987). "Agricultural chemicals in ground water: Extent and implications," *Am. J. Alt. Agric.* **2**:3–15.

Healy, R., T. Waddell, and K. Cook. (1986). *Agriculture and the Environment in a Changing World Economy.* Conservation Foundation, Washington, DC.

Huzar, P. C. , and S. L. Piper. (1986). "Estimating the off-site costs of wind erosion in New Mexico," *J. Soil Water Conserv.* **41**:414–16.

Jackson, W., and M. Bender. (1984). "An alternative to till agriculture as a dominant means of food production." In *Food Security in the United States* (Eds. L. Busch and W. B. Lacy). pp. 27-45. Westview, Boulder, CO.

Langley, J., K. Reichelderfer, and J. Sharples. (1987). *The Policy Web Affecting Agriculture: Trade-Offs, Conflicts, and Paradoxes.* USDA, ERS, Agricultural Information Bulletin No. 524.

Lovejoy, S. B., and D. Parent. (1982). "Conservation behavior: A look at the explanatory power of the traditional adoption-diffusion model." Rural Sociological Society Meeting, San Francisco.

Lowrance, R., P. F. Hendrix, and E. P. Odum. (1986). "A hierarchical approach to sustainable agriculture," *Am. J. Alt. Agric.* **1**:169–73.

Lowrance, W. G., B. R. Stinner, and G. J. House (eds.). (1984). *Agricultural Ecosystem.* Wiley-Interscience, New York.

Mitchell, J. E. (ed.). (1987). *Impacts of the Conservation Reserve Program in the Great Plains: Symposium Proceedings.* USDA, Forest Service, General Technical Report RM-158, Fort Collins, CO.

Mueller, D. H., R. M. Klemme, and T. C. Daniel. (1985). "Short- and long-term cost comparisons of conventional and conservation tillage systems in corn production," *J. Soil Water Conserv.* **40**:466–70.

——. (1989). "Mutual disarmament in world agriculture: A declaration on agricultural trade," *Choices* **3(1)**:32–33.

National Farmers Union. (1989). *Implications of Decoupling.* Denver, CO.

Newman, M., T. Fulton, and L. Glaser. (1987). *A Comparison of Agriculture in the United States and the European Community,* ERS Foreign Agricultural Economic Report No. 233, Washington DC.

Pampel, F. and J. van Es. (1977). "Environmental quality and issues of adoption research," *Rural Sociol.* **42** (Spring):57–71.

Petit, M., G. E. Rossmiller, and M. A. Tutwiler. (1988). "International agricultural negotiations: The United States and the European Community square off." In *U.S. Agriculture in a Global Setting: An Agenda for the Future,* pp. 88-104. Resources for the Future, Washington, DC.

Pierce, L., R. Dowdy, W. Larson, and W. Graham. (1984). "Soil productivity in cornbelt," *J. Soil Water Conserv.* **39**:131–36.

Raup, P. M. (1978). "Some questions of value and scale in agriculture," *Am. J. Agric. Econ.* **60**:303–8.

Reichelderfer, K., and T. T. Phipps. (1988). "Agricultural policy and environmental quality." Briefing Book, Resources for the Future, National Center for Food and Agricultural Quality, Washington, DC.

Rushefsky, M. E. (1980). "Policy implications of alternative agriculture," *Pol. Stud. J.* **8**:772–84.

Soil Conservation Service, USDA. (1987). *Agricultural Trends and Resource Conservation: Implications and Issues.* Staff Report, Appraisal Program Development Division, Economics and Social Science Division, Washington, DC.

Strohbehn, R. (ed.) (1986). *An Economic Analysis of USDA Erosion Control Programs: A New Perspective.* Agriculture Economic Report No. 560, Economic Research Service, USDA, Washington, DC.

van Es, J. (1982). "The adoption/diffusion tradition applied to resource conservation: Inappropriate use of existing knowledge." Rural Sociological Society Meeting, San Francisco.

Ward, J. (1988). *Conservation Easements: Prospects for Sustainable Agriculture.* Natural Resource Defense Council, Washington, DC.

Watts, M. J., L. D. Bender, and J. B. Johnson. (1983). "Economic incentives for converting rangeland into cropland." Cooperative Extension Service Bulletin 1302, Montana State University, Bozeman. November.

Wilson, E. M. (1988). "Export subsidies on value-added products: Effects may differ from policy objectives," *Choices* **3(2)**:5–7.

Youngberg, G., and F. H. Buttel. (1984). "Public policy and socio-political factors affecting the future of sustainable farming systems." In *Organic Farming: Current Technologies and Its Role in a Sustainable Agriculture,* pp. 167–85. Amer. Soc. Agron., Madison, WI.

Zietz, J., and A. Valdes. (1988). *Agriculture and the GATT: An Analysis of Alternative Approaches to Reform.* IFPRI, Washington, DC.

14 AGRICULTURE WITH NATURE AS ANALOGY

WES JACKSON
The Land Institute,
Salina, Kansas

Nearly all researchers in sustainable agriculture work with species already in the crop inventory. They design experiments to explore alternative strategies for seedbed preparation, planting, maintenance, and eventual harvest. Most strategies involve diversity over time (rotations that usually include a legume) or space (polycultures). The motivation may be an interest in lowering production costs for farmers (which could help more farmers to stay on the land) or in reducing the impact on the environment or dependency on nonrenewable resources such as fossil fuels.

The time frame for most investigators is the near future. This is a practical approach designed to help the farmer and reduce environmental damage now. In some cases, the research is aimed at achieving a scientific understanding of certain traditional farming methods, partly for the purpose of adding legitimacy to the wisdom of tradition and perhaps with the hope that the added scientific respect and more finely-tuned knowledge will lead to an expansion of sustainable agriculture everywhere. Examples of such research can be found in the work of Altieri (1987), Francis (1986), Gliessman (1984), the Rodale Research Center, and elsewhere.

Workers at The Land Institute share the values of these researchers, but our research has a somewhat different motivation, approach, and time consideration. Rather than deal with problems *in* agriculture here and now, we address the "problem *of* agriculture," a problem that has been with both humanity and the earth since agriculture began some 8,000 to 10,000 years ago. This long backward look in time has helped us to consider corrective measures for a time into the future that may be longer than that of most agricultural researchers. Rather than the ten years or less that most workers in sustainable agriculture have in mind as they design experiments or promote certain practices, we think the major benefit of our work will be realized after perhaps 25 but more likely in the next 50 years or more. We think there may be some short-term benefits, but most will be in the 50 to 100 year time consideration (Jackson, 1985; Jackson, Berry, and Colman, 1984).

NATURE AS ANALOGY

The starting point in our search for solutions is nature. The term *nature* is, of course, elusive. Consequently, it is an elusive or slippery standard. Nevertheless, it seems to us at The Land Institute less slippery or elusive than any other standard when sustainability is our primary criterion. The nature we look to is the native prairie. And when we compare prairie with the ordinary grain crop, such as a corn or wheat field, important differences become apparent. From the sloping fields of humanity, valuable nutrients run toward the sea where, for all practical purposes, most of them are gone for good. The prairie, on the other hand, by sucking nutrients from parent rock material or subsoil, all the while bathing them with the chemicals produced by life, actually builds soil. Even though diversity may not yield stability overall, chemical diversity inherent in plant species diversity on the prairie confronts insects and pathogens so that the epidemic, so common to agricultural monocultures, is rare on the prairie. Since no creature has an all-consuming enzyme system, diversity does yield some protection. The prairie does not require, therefore, the introduction of chemicals with which humans have had no evolutionary experience. The prairie, like nearly all of nature, runs mostly on contemporary sunlight while our modern agricultural fields benefit from the stored sunlight of floras extinct for hundreds of millions of years. A casual examination of the ordinary differences between a prairie ecosystem and an agricultural field will cause us to see that the prairie features perennials in a polyculture, while agriculture features annuals in monoculture. Our work at The Land Institute, therefore, is devoted primarily to exploring the feasibility of an agriculture that features the growth of herbaceous perennials in a mixture for seed production as substitutes for annual monocultures grown on ground that can erode. There would be domestic prairies, cropping systems analogous to the vegetative structure of the prairie, not like the prairie exactly, not a template of the prairie (Jackson, 1985).

Dr. Peter Kulakow, our plant breeder at The Land Institute, has summarized our overall scheme for illustration purposes in Figure 14.1. In studying this figure, it is useful to keep in mind the four basic questions we address in our experiments at The Land:

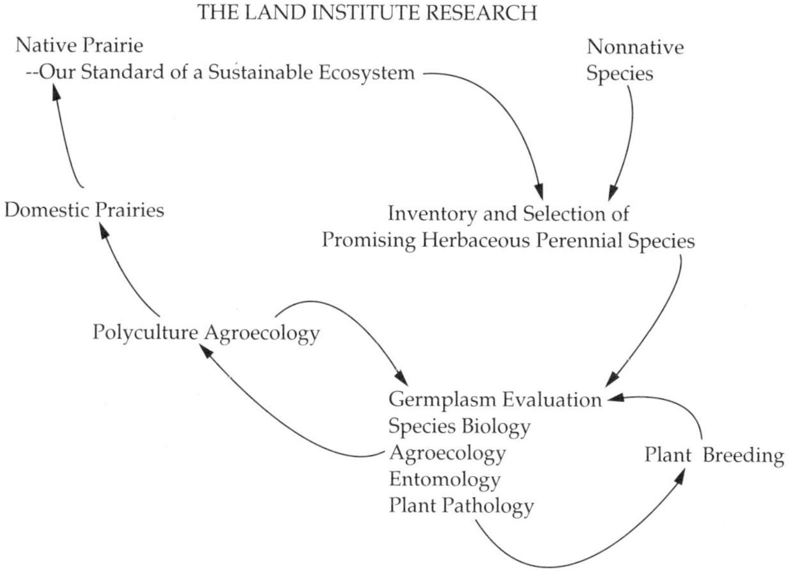

Figure 14.1 The Land Institute research.

1. Can herbaceous perennialism and high seed yield go together? Perennial plants must reallocate some resources to parts below-ground. It could be difficult, therefore, to breed perennials to outyield annuals which died after seed set. Perennial species differ, however, in relative and absolute amounts of resources devoted to seed. It seems worthwhile therefore, to search for species with high seed yield. Once species are discovered, breeding work could begin.

Before we begin to breed for stable high seed yields in a herbaceous perennial, however, we need to determine its genetic potential. Whether we start with a wild species that has been introduced or a wild native, the development of perennial seed-producing polycultures will require that we select varieties that perform well in polyculture. The potential improvement, therefore, depends on the range of existing genetic variability in the wild. To assess this variation requires adequate sampling across the geographic range of the species and then an evaluation of the collection within a common garden.

So far I have mentioned two categories of herbaceous perennials: the wild introduced and the wild native species. A third category is nonwinter-hardy, a domestic annual that we wish to turn into a winter-hardy perennial through hybridization with a close relative.

2. Can a polyculture of perennial seed producers outyield the same species grown in monoculture? Overyielding occurs when competition between species is less intense than between members of the same species. Polycultures typically yield more than monocultures due to differences in resource use and timing of demand.

3. Can a perennial polyculture provide most of its own fertility? Can biological nitrogen-fixation and phosphate accumulation compensate for harvested nutrients? Soil nutrient pools, nutrients in the seed, and capacity of crop plants to enrich the soil needs to be documented.

4. Can a perennial mixture successfully manage phytophagous insects, pathogens, and weeds? Breeding for resistant lines and studies on the effects of species diversity must converge if we are to protect a crop. Insect pests can be managed through a combination of predator attraction and preventing an insect from locating host plants. Mixtures of species, and of genotypes within species, may reduce the incidence and spread of disease. Weeds may be controlled either allelopathically or via continuous shading of the soil surface by the perennials (Jackson, 1985).

SEARCHING FOR HIGH SEED-YIELDING HERBACEOUS PERENNIALS

Procedures for Discovering Candidates

The most pressing biological question at The Land is whether perennialism and high seed yield can go together. Stated otherwise, how do the annual and long-term seed yields of herbaceous perennial plants compare with those of annual grain crops? To begin this investigation, we started an inventory:

1. We reviewed the literature of seed yield in winter-hardy herbaceous perennials.
2. We collected seed and plants in nature and developed an herbary consisting of herbaceous, perennial winter-hardy species, approximately 300 species in all, each grown in 5 rows.
3. We planted more than 4,100 accessions of over 100 species representing seven cool season grass genera collected worldwide. Our inventory effort continues, even though in recent years we have only worked with five species plus a hybrid of our making.

The relationship between perennialism and high yield involves a question that gives emphasis to sustained *production*. Prairies, after all, feature perennials, but they do not feature high seed yield. Ultimately we have to explore the optimum balance between sustainability and yield.

Research on the perennialism and high yield question will necessarily require several years, for it amounts to an investigation of long-term demographic patterns in perennial seed production, a field that is largely unexplored. Studies thus far at The Land have shown increases, decreases, or oscillations in seed yield over time. Details of these results will be presented here, but first some background on this part of our inventory phase.

Literature Review—Seed Yield. The literature on seed yield in herbaceous perennials is dominated by results gained from species used as forage. It is of biological importance to note that in essentially every case, before a species can be suitable for forage, the problem of seed production and harvest must be solved. This means that yield, shatter resistance, and determinate seed set must be sufficiently high that a harvestable seed crop is practical. Vegetative propagation for pasture establishment is impractical. Therefore, the seed yield results for herbaceous perennials from forage crops *is* relevant for our work. Selection for seed in herbaceous perennials for pastures has probably been under way for millenia.

Tables 14.1–14.5 summarize our literature survey on yield and Table 14.6 summarizes data on protein quality. These tables were generated by Marty Bender, former research associate at The Land Institute.

Observations from Nature and Herbary Establishment. At the same time we were searching for literature for yield records among perennials, our own field experiences and those of colleagues were important. Plants that seemed to be high seed producers were observed or came to mind. Seeds were gathered from nature. These apparent high yielders were planted at The Land so that we might more carefully observe the problems of establishment and seed yield under cultivated conditions. Soon thereafter, we concluded that any winter-hardy, herbaceous perennial flowering plant should be planted in a 5 meter row for observation. Approximately 300 species were established in the herbary with varying degrees of success.

We started the herbary at The Land Institute in 1979, mostly from seeds provided by the Plant Materials Center (USDA Soil Conservation Service) near Manhattan, Kansas. In 1987, the herbary had 265 species. Over the years, agricultural interns at The Land have recorded flowering dates, plant height at flowering, and date of seed collection, and they have taken photos for reference and miscellaneous notes. In 1987, plants were rated on a four-point scale by intern Randy Kempa as follows:

1. Plant not there or declining (spring ephemerals sometimes included).
2. Plot is growing well but some plants are missing.
3. All plants are growing vigorously.
4. Plants growing out of their plot and invading alleys and other plots either by seeding or clonal spread.

The 4,100 Accessions Involving Six Grass Genera. Based on the results of the literature survey, we picked six grass genera as candidates with a high likelihood of having other species with high seed potential. We

TABLE 14.1 Irrigated Grass Yields.

Species	Seed Yields (kg/ha)	Field Conditions	Reference
Buchloe dactyloides	1,537*	27 kg N, 27 kg P	Ahring, 1964
Festuca arundinacea	1,299	Fertilized, 16 ha	USDA, 1976
Dactylis glomerata	979	Great Britain	Whyte, 1959
Eragrostis curvula	979	Fertilized	Wenger et al, 1943
Sporobolus cryptandrus	890	— — —	USDA 1948
Elymus canadensis	890	Fertile soil	Cooper, 1957
Agropyron cristatum	886	Fertilized, 2nd yr.	Wheeler, 1957
Bromus inermis	838	Fertilized, 1st yr.	Wheeler, 1957
Lolium perenne	828	12,950 ha average, 1yr.	Wheeler, 1957
Lolium multiflorum	828	44,516 ha average, 1yr	Wheeler, 1957

* Includes the burs.

TABLE 14.2 Dryland Grass Yields.

Species	Seed Yields (kg/ha)	Field Conditions	Reference
Sporobolus cryptandrus	801	Native stand	Brown, 1943
Bromus inermis	445	Fertile soil	Cooper, 1957
Agropyron trichophorum	445	32 kg N	Thornberg, 1971
Agropyron intermedium	401	Fertile soil	Cooper, 1957
Festuca arundinacea	356	Fertile soil	Cooper, 1957
Agropyron trachycaulum	356	32 kg N	Thornberg, 1971
Stipa viridula	356	Fertile soil	Cooper, 1957
Agropyron elongatum	356	32 kg N	Thornberg, 1971
Agropyron spicatum	356	32 kg N	Thornberg, 1971
Elymus junceus	267	Fertile soil	Cooper, 1957

TABLE 14.3 Irrigated Native Grass Yields.

Species	Seed Yields (kg/ha)	Field Conditions	Reference
Buchloe dactyloides	1,537*	27 kg N, 27 kg P, 4th yr.	Ahring, 1964
Elymus canadensis	890	Fertile soil	Cooper, 1957
Sporobolus cryptandrus	890	— — —	Wheeler, 1957
Bromus marginatus	803	14 kg N, 1st year	Klages & Stark, 1949
Panicum virgatum	534	Fertile soil	Cooper, 1957
Eragostis trichoides	445	Fertile soil	Cooper, 1957
Agropyron smithii	445	Fertile soil	Cooper, 1957
Bouteloua curtipendula	445	Fertile soil	Cooper, 1957
Sorghastrum nutans	445	Fertile soil	Cooper, 1957
Bouteloua gracilis	445	Fertile soil	Cooper, 1957

* Includes the burs.

TABLE 14.4 Dryland Native Grass Yields.

Species	Seed Yields (kg/ha)	Field Conditions	Reference
Sporobolus cryptandrus	801	Native stand	Brown, 1943
Stipa viridula	356	Fertile soil	Cooper, 1957
Sorghastrum nutans	267	Fertile soil	Cooper, 1957
Bouteloua curtipendula	267	Fertile soile	Cooper, 1957
Eragrostis trichoides	223	Fertile soil	Cooper, 1957
Panicum virgatum	223	Fertile soil	Cooper, 1957
Elymus canadensis	178	Fertile soil	Cooper, 1957
Orzopsis hymenoides	178	32 kg N	Thornberg, 1971
Agropyron smithii	178	Fertile soil	Cooper, 1957
Andropogon gerardi	178	Fertile soil	Cooper, 1957

TABLE 14.5 Irrigated Legume Yields.

Species	Seed Yields (kg/ha)	Field Conditions	Reference
Desmanthus illinoensis	1,058	5-year average	USDA, 1978
Astragalus cicer	890	Fertilized	USDA, 1966
Onobrychis viciaefolia	890	32 kg N	Thornberg, 1971
Medicago sativa	445	32 kg N	Thornberg, 1971
Trifolium pratense	445	32 kg N	Wheeler, 1957
Lotus corniculatus	445	32 kg N	Thornberg, 1971
Trifolium hybridum	378	1,700 ha avg.	Wheeler, 1957
Vicia villosa	356	10,522 ha avg, 1 yr	Wheeler, 1957

TABLE 14.6 Seed Protein Yields.

Species	Seed Yield[a] (kg/ha)	Percent[b] Protein	Protein Yield[b] (kg/ha)
Annuals			
Corn	5,055 (100.8 bu)	9	455
Wheat	1,687 (31.6 bu)	12	203
Soybeans	1,575 (29.5 bu)	42	661
Perennials			
Dactylis glomerata	979	27	264
Elymus canadensis	890	27	240
Bromus inermis	838	19	159
Festuca arundinacea	1,326	22	286
Desmanthus illinoensis	1,058	38	360
Astragalus cicer	890	40	356
Onobrychis viciaefolia	890	40	356
Medicago sativa	445	37	165

[a] USDA, 1980

[b] Earle and Jones, 1962; Jones and Earle, 1966; Piper et al. 1988.

TABLE 14.7 Listing of Accessions of Species and Hybrids Involving Six Grass Genera Planted at The Land Institute from a Worldwide Collection.

Genus	Number of Species	Number of Hybrids	Number of Accessions
Agropyron	59	3	1207
Bromus	58		629
Elymus	26		289
Festuca	47		1255
Lolium	9	1	673
Sporabolus	13		73
Total	212		4126

Note: Data on each species have been entered on a computer data base for future reference when we begin to build our domestic prairies.

asked the USDA Plant Introduction Center in Pullman, Washington, if they could provide us with their entire collection of all accessions of all species within these genera from all over the world. They were generous in meeting our request, for they supplied us with seeds from over 200 species and four hybrids represented in more than 4,100 total accessions from over 60 countries. We planted them all, and 95% of all the accessions were established long enough for us to evaluate. Table 14.7 summarizes the number of species, hybrids, and accessions.

We have tried to avoid being too hasty in adding a taxon to our list of species with which we would work. Neither the data from the literature nor our own observations from a single year's yields results satisfied our criteria. For example, buffalo grass (*Buchloe dactyloides*) yields at The Land Institute did not come close to the claims in the literature. *Agropyron intermedium* had high yields in the first year, but the ability of the plants to persist as perennials was not strong. So from our inventory prompted by the literature and growing out the 4,000 accessions, we added only one species to our inventory of plants for improvement, *Leymus racemosus*.

Taxa Selected and Reasons for Adoption

Wild Senna (Cassia marilandica). This perennial herb sprouts erect stems every spring from a woody caudex. At maturity, these stems will vary from 0.5 to 2 m high. The yellow flowers are pollinated by insects. Flowering is from July to September. The many-flowered racemes emerge in the upper axils and in terminal panicles. The flat mature pods vary from 7 to 11 cm in length and 8 to 11 mm in width. The black ovate mature seeds measure 4 to 5 mm long and 2 to 2.5 mm wide. We have counted the chromosomes from the root tips of 10 plants in each of our 97 accessions and never found a departure from $2n=28$. This is a fairly deep-rooted legume that does not appear to form symbiotic associations with the nitrogen-fixing *Rhizobium* (Leonard, 1925). Several members of this genus are important medicinally (Duke, 1981) and as forage for livestock (Taparia et al., 1978).

The habitat description in two floras (Steyermark, 1963; Great Plains Flora Association, 1986) confirms our field experience concerning habitat and distribution. It is usually represented by only a few plants when found in nature, but sometimes is quite common along ravines or creek banks. It likes alluvial thickets, open woodlands, especially at the base of slopes and bluffs, and wet meadows.

The species ranges from Pennsylvania to Iowa and Kansas and from southeastern Nebraska to Florida and Texas.

In both the field and in the herbary, this species appeared to be both vigorous and high yielding. The arrangement of the pods made harvest easy. Even though flowering is spread over several weeks, the pods do hang on and are not inclined to dehisce upon ripening. From the start, we doubted that it had little, if any, economic utility. Plant parts of a close relative are used as a laxative, but for such purposes, a little can go a long way. We selected it because we thought it might be our best bet to help us answer our most basic of biological questions: whether it were possible for perennialism and high yield to go together.

Illinois Bundle Flower (Desmanthus illinoensis). This herbaceous perennial, like its legume relative, wild senna, also sends forth erect new stems each spring from a woody caudex. Plants may reach 2 m in height, although 1 m is more typical. When it blooms (June to August across the range) some 30 to 50 flowers will be borne in a bundle, yielding a tight-fisted cluster or bundle of brown pods. Individual pods will vary from 1.5 to 2.5 cm long and 4.5 to 6 mm wide. When mature, these flattened pods will usually contain brown seeds 3 to 5 mm long and nearly as wide. It fixes nitrogen (Kulakow et al., in press).

Its natural distribution and preferred habitat would be hard to know anymore, since it has been so widely used in range revegetation programs and as habitat for wildlife enhancement projects. The current range is from as far west as New Mexico and Colorado. South and eastward, it ranges from Texas to Alabama with the most southern locality in Florida. In the north, it can be found in Ohio, Indiana, Illinois, Minnesota, and North Dakota (Steyermark 1963; Great Plains Flora Association, 1986).

IBF appears to be a high seed-yielding species, and though our primary interest is in perennial polycultures as substitutes for annual grains, we are always aware of the livestock component. The forage of IBF is preferentially chosen by grazers (Latting, 1961). Seeds contain 38% protein (Piper et al., 1988). Besides its economic potential, it can be used to answer the question as to whether perennialism and high seed yield could go together. We hope to use this species as the primary nitrogen fixer in our polycultures.

Maximilian Sunflower (Helianthus maximilianii). Stems rise, usually in clusters, but often singly each spring from short but thick and spreading rootstocks. There are reports from the literature of plants achieving heights of 2.5 m. This species is highly responsive to habitat in both the achievement of height and flowering. Flowering heads can be numerously arranged in the upper leaf axils or on an elongated spike or racemelike inflorescence. One notable feature we have observed at The Land is that while last year's seeds are struggling to germinate and establish themselves as seedlings, the young shoots from the parental root stocks are experiencing rapid growth sufficient to establish a quick canopy in early spring.

The species ranges from Minnesota and Manitoba to Saskatchewan south to Missouri, Oklahoma, and Texas, Maine to North Carolina westward to the Rocky Mountains and from southern Canada to Texas. It has been cultivated in gardens with the cultivar displaying more flowers and broader leaves.

It is found in numerous habitats from damp and open prairies to dry conditions. It is found in sandy waste ground and on limestone glades, ledges, and bald knobs, as well as in loess hills and deep rich soil. (Great Plains Flora Association, 1986.)

Sunflowers are reported to be allelopathic (Rice, 1984). Whether they are or not, they are highly competitive. This characteristic, along with their relatively high yield, makes this native species most attractive. We have noticed Maximilian to be most effective in suppressing other species when they are in the seedling stage. Once another species is established, however, little effect is experienced. We imagine a mixture of species being established in a polyculture followed by Maximilian sunflower to provide weed control.

Eastern Gama Grass (Tripsacum dactyloides). This tall, stout perennial emerges each spring from a proaxis and achieves a height at flowering of 1 to 2 m. The terminal flowers of a primary shoot usually bear three spikelike branches. The lateral shoots bear only one. The lower parts of these branches are pistillate and the upper two-thirds or so are staminate. The arrangement of the staminate spikelet is important; at any one locus is a pair of double-flowered florets. The pistillate part of the spikelet usually carries a single flower per locus. Seeds contain 27% protein, roughly three times that of corn and twice that of wheat. Seeds may be 1.8 times higher in methionine than corn (Bates, Bender and Jackson, 1981). Some plants are apomictic. Two apparently distinct chromosome races predominate: a diploid of 36 and a tetraploid of 72. It is not surprising to find an intermediate $3n = 56$ in nature.

The species covers great diversity of habitats including sand, sandstone, limestone slopes at the edge of woods and thickets, along roadsides, railroads, prairies, and even in cultivated fields. The distribution in the northern hemisphere, according to Hitchcock (1935), is from Massachusetts to Michigan, Iowa, and Nebraska, south to Florida, Oklahoma, Texas, and the West Indies.

This widely distributed warm-season grass has an abundance of genetic variation even within the same ploidy levels. The basis for genetic improvement, therefore, is very great. The high protein content of the seed and high levels of methionine make this an attractive food for both humans and livestock.

A few years ago, Bob Dayton of the USDA Plant Materials Center near Manhattan, Kansas, discovered an important double recessive mutant displayed by a plant grown from seed he had collected along a roadside in Ottawa County, Kansas, approximately 20 miles north of The Land Institute. Dayton's mutant is responsible for converting the male part of the flower into female, yielding from 12 to 20 times the number of seed as a normal plant, though only, at this date, 2 or 3 times the yield in weight. This high seed number results from the condition mentioned above; the four ovaries per *locus* in what is ordinarily male results in four seeds per locus in the mutant.

This species was a candidate seed producer before the mutant was discovered and before we at The Land discovered that small quantities of nitrogen were fixed in the rhizosphere of its roots. Its attractiveness as a perennial seed producer continues to improve. Forage yield and quality have been well known by stockmen for a long time.

Leymus (Leymus racemosus). This rhizomatous perennial of the Triticeae (which also includes wheat, rye, and barley) has culms up to 12 mm thick and, in our plots at The Land Institute, stands over 1 m tall. The flowering spikes range from 15 to 35 cm in length and 10 to 20 mm in width. The spike is densely packed with three to eight spikelets at each node with each spikelet containing four to six florets. These large spikes are borne on a strong stalk.

Agriculture intern Brad Burritt (1986) has learned much of the natural history, use, and value of the grain. He reports that this cool-season native of Europe and central Asia generally grows well on dry, sandy soils in the native southeastern European countries of Bulgaria, Romania, Turkey, and parts of Russia (Melderis et al., 1980). Its annual growth cycle is similar to that of winter wheat. It has been introduced only a few times into the United States. Green shoots begin to emerge at The Land as early as late January. Our material at The Land Institute came from the USSR and possibly explains the late January to early February emergence of green shoots. Annual above-ground biomass production rivals native tallgrass prairie (Gernes and Piper, 1987). Seeds have been eaten as a supplemental grain by Russian peoples especially during drought years when annual small grains failed (Komarov, 1934).

Indigenous peoples and Viking settlers have used various species of the genus. Northern California natives ate *L. cinereus* grains. Eskimos and Aleuts ate *L. Mollis* seeds (Weiner, 1980; Klebesadel, 1985). Archaeological evidence in Iceland points directly to the likelihood that Viking settlers cultivated *L. arenarius* until this century (Griffin and Rowlett, 1981). Various Leymus species have been grazed by sheep, cattle, and horses.

When the grain of *L. arenarius* was analyzed by Griffin and Rowlett (1981), they found that the protein content rivaled red beans and salmon and that its fatty acid content surpassed amaranth, wheat, and high protein hybrid corn, rice, and oats. They also noted that the yield of *L. arenarius* was similar to domesticated cereals cultivated in subsistence agriculture conditions.

Burritt further notes that the genus has caught the attention of other scientists. Researchers in the Soviet Union from 1940 to 1975 attempted to hybridize wheat and Leymus species (Bodrov, 1960; Tsitsin, 1978) with only moderate success.

Nevertheless, because of its strong perennial nature and soil-holding capacity, because it is in the human inventory already, because of the nutritional value of the various species, and because scientists have found the genus attractive enough to attempt hybrids with wheat, we have found it worthy of further exploration for incorporation into perennial polyculture.

***Winter-Hardy Grain Sorghum—a Hybrid Derivative Involving Johnson Grass* (S. halepense) *and Commercial Sorghum* (S. bicolor).** *Sorghum bicolor*, an important grain and ensilage crop, particularly in dry regions of the United States, is a perennial that is not winter-hardy. Commercial breeders have, therefore, treated it as an annual in their breeding programs. *Sorghum halepense*, commonly known as Johnson grass, is an aggressive weed, particularly in warm areas of high rainfall and humidity. It is common particularly throughout the Southeast from Florida to Texas and northward to Kansas, Nebraska, and Iowa eastward to Illinois, Indiana, Ohio, New York, and Massachusetts. It is also found in the West Indies and Central and South America. It was introduced into the New World from Turkey around 1830 as a forage crop. This native of Europe and Asia normally stands less than head high, although some plants have been reported to reach 3 m. Its open, loosely spreading inflorescence is unlike the densely packed heads of its domestic relatives.

The creeping rhizomes of Johnson grass have some degree of winter-hardiness. I have seen the plant growing as far north as northern Iowa. We saw this physiological quality as a possible source of winter-hardiness to be combined with a tetraploid version of *S. bicolor* mentioned above. Because we are worried about the possible spread of Johnson grass plants, as well as any hybrid or hybrid derivative, all plants are grown in plastic pots. We expect, however, the hybrid derivative to be much less aggressive than its Johnson grass parent.

If a relatively nonaggressive hybrid derivative could be established as a winter-hardy perennial, it would amount to the winterizing of a grain crop already important in the crop inventory. Perennial sorghum could be included in polycultures and, as such, would represent the first domesticated crop to be integrated with recently wild species.

THE ECOLOGICAL INVENTORY

Early Work (1977–1983).

The ecological inventory at The Land in this era was very general. At one time, two researchers spent most of a spring and summer conducting research to determine whether an unplowed native prairie that had never had forced grazing practiced "companion planting." We wondered if two species were more likely to be associated with one another than with any third species. James Peterson and Phil Haves investigated this question at the Fent prairie, eight miles north of The Land. Their studies indicated no preferential growth patterns.

In those early years we also began to catalog the flora of The Land. Marty Bender and student Joy Hasker began this effort, which continues under the direction of our ecologist, Jon Piper.

Phytomass Survey of Two Native Prairies (1984–Present).

In a 1986 preliminary investigation, Jon Piper asked (1) how much aboveground plant life is supported each year by the prairie and (2) what are the proportions of grasses, legumes, and composites. These happen to be the plant families comprising most of our temperate agricultural species. Net production at his grassland sites (500–700 g/m^2) was similar to that of many midwestern crops. At their peaks, grasses composed 67% to 94% of plant matter; legumes and composites represented 16% and 11% of vegetation, respectively. Piper concluded that these encouraging results suggest that polycultures of perennial warm- and cool-season grasses and legumes are feasible (Piper, 1986).

In 1987, intern Mark Gernes joined Jon Piper in the continuation of the study. This time, they evaluated the species mix of the same three families and estimated the richness, diversity, and evenness of species in this part of the tallgrass prairie and examined the phenological differences among the major families. Their data source came from clipping "aboveground phytomass within 12 0.25 m^2 quadrats on each of three prairie sites during early May, early June, and mid-August. Average phytomass per site ranged from 71.8 to 151.1 g/m^2 in May and from 377.3 to 1077.2 g/m^2 in August. Species diversity varied among sites and dates from 2.3 to 9.6. Low diversities and high phytomass resulted from dominance by *Andropogon gerardii* on all sites. Species in the Gramineae, Euphorbiacea,

Malvaceae, Labiatae, and Compositae were present in all three sites, but only Gramineae and Compositae were sampled in all sites on all dates. There were no significant differences in mean flowering and fruiting time among Gramineae, Leguminosae, and Compositae. The data suggest that our domestic polyculture be dominated by a C_4 grass, but associated with a C_3 grass and a few forbs (Gernes & Piper, 1987).

Insects and Pathogens: Prairie, Herbary, Plots.

In 1985, intern Danielle Carré examined the insects and plant pathogens qualitatively in nine experimental plots at The Land. Every week from May through August and every other week from September through October, insects were collected with a sweep net or from individual plants. All diseased plants were sent to the Disease Diagnosis Laboratory at Kansas State University for pathogen identification. All sampled plots showed a diversity of both beneficial and harmful insects. Several foliar diseases were present, but few were serious. A fungus on Illinois bundle flower caused significant leaf loss. The majority of the wild senna accessions were infected with a virus (Carré, 1985).

Mark Slater continued this inventory in 1986, but with modifications. The prairie was sampled using sweep nets every third week from April to September and in the research plots from June to September. Adults were identified to the family level and the immatures to the order. While more families were observed in the prairie than in any single plot, the aggregate of all plots yielded more families than were present on the prairie. The ecological roles of both the adults and immatures of each family were assigned and compared across the sampled areas. Insect damage was low both in the prairie and in the research plots, with two notable exceptions. A curculionid stem cutter, *Rynchites sp.*, cut the flowering stems in the *Helianthus* species and a chrysomelid, *Anomoea laticlavia*, ate the pods of Illinois bundle flower (Slater, 1986).

In 1986, Mary Handley, our plant pathologist, had an exceptionally good year to examine leaf diseases. Plenty of cool weather and moisture throughout the growing season followed a mild winter. All six of our selected species were surveyed. Wild senna plantings experienced a near uniform virus infection. Yellow blotches on leaflets distorted their shape and reduced their size. Handley did not observe this on native stands (Handley, 1986).

Continuing our inventory of pests, we have noted that Maximilian sunflower displayed a powdery mildew on the lower leaves of a few plants in mid-July. The pathogen remained restricted to the lower part of the canopy and never reached significant proportions. Senescing leaf surfaces also had extensive growth of *Penicillium* and *Aspergillus* species.

Eastern gama grass experienced two or three diseases throughout the season. *Puccinia sorghi*, a leaf rust, was widespread and highly variable in severity. What initially were regarded as two kinds of leafspots turned out to be the same disease. Handley observed defined lesions with reddish margins and yellow-tan centers and irregular margins with a gray-tan center. Handley observed this disease in all eastern gama grass plots at The Land, as did researchers at the USDA station in Woodward, Oklahoma.

Handley reported four or five diseases on *Leymus*, including ergot, stem rust, head discoloration, and one or two leafspots. Although the incidence was high, none was severe. She concluded that head discoloration was partially related to maturation, but fungal growth *was* observed on the heads. The fungi observed were mostly saprophytes: *Penicillium, Aspergillus, Alternaria*, and some others.

The hybrid between *Sorghum bicolor* and *S. halepense* developed the leaf rust *Puccinia sorghi* late in the season, but at a low incidence.

By June 1986, Handley observed that the Illinois bundle flower plots with a 40-cm spacing between plants had been defoliated on the lower two-thirds of stems, whereas plots with a 7-cm spacing were only about half defoliated. Plants at the closer spacings were taller, so that even though total height of defoliation was greater at closer spacings, the ratio of defoliation to total height was lower. By early July, all plots were about 40% defoliated. The actual heights of defoliation had progressed equally in all plots, but those at lower densities had grown more than the higher density plots. Apparently, a density-dependent factor was influencing disease progress early in the growing season.

Early season defoliation may influence flowering and seed yield either directly or indirectly. The 1986 yield was 36% lower overall than the yield in 1985, although there were no significant differences among density treatments. Since there was only a brief period during which differences in defoliation between densities occurred, measurements of overall yield may not be sensitive enough to detect yield reduction due to disease (Handley, 1986).

Two major disease epidemics visited Eastern gama grass in 1986, making this species the most interesting crop, pathologically speaking. According to Handley, "the more obvious disease was leaf rust, *Puccinia Sorghi*, which occurred on 92% of the plants evaluated. Twenty-two percent of the plants were severely, or very severely, infected, with more than 50% of their total leaf area destroyed by fungal lesions. In addition to rust, many of the plants were infected with anthracnose, *Colletotrichum Graminicola*, which further reduced healthy leaf area.

"Large differences in severity of rust and anthracnose were observed both among individual plants and among accessions. Individual plants ranged from healthy (rating of 0) to very severely infected (rating of 7). Accessions with more than five plants had mean rust ratings ranging from 0.8 to 4.9, with an accession mean rust rating of 2.8. Accession means for anthracnose ratings were much lower, with a range of 0.8 to 5.4 and a mean of 2.5 (out of a possible total of 14). These differences are likely due, in part, to genetic differences among plants."

With *Leymus racemosus*, none of the diseases observed became severe during the growing season. The disease most significant to human or animal consumption was ergot. Although most heads had no ergot sclerotia, up to thirteen sclerotia were observed on one head. Stem rust levels were high on flowering tillers shortly before harvest, with 38% of tillers showing sporulation of *P. graminis*. The disease did not worsen the rest of the growing season. Rust did not develop to any great extent on nonflowering tillers. Whether this reflects resistance of nonflowering tillers to infection, or unfavorable environmental conditions is not certain. The *forma specialist* of *P. graminis* occurring on *Leymus* was not determined (Handley, personal communication, 1986).

Among 200 heads selected for yield, 90% showed some degree of leafspot; 40% moderate to severe. Only 12% showed moderate head discoloration. The high incidence of leafspot indicates a potentially serious problem. It will be important to clarify physiological leafspot and pathological leafspot, and determine the relative importance of each. Physiological leafspot is common on cool season grains during periods of high temperature, especially during seed fill (Handley, personal communication, 1986).

This baseline information for disease severity provides us ongoing information that can be correlated with plant growth, flowering, sex-ratio genotype, yield, winter-hardiness, and other important characters.

Upshot of the Pathogen Inventory. Mary Handley concludes that perennial plants have many of the same types of disease problems as our annual crops. Our observations were restricted to leaf diseases; the situation on roots and crowns is probably even more complex. Development of these species as crops will require us to understand the nature of their susceptibilities in order to develop ways to minimize disease impact. In a sustainable perennial polyculture there will be two main ways to minimize diseases: breeding for genetic resistance and employing cultural and management practices that reduce incidence or spread. Each host/pathogen interaction will require a different approach to control.

Leaf rust is a serious problem on *Tripsacum*. This pathogen also infects corn and sorghum, and is common in the Great Plains. In our accessions, there is already great variation in susceptibility to rust. Good resistance to this pathogen is present in corn, and has remained useful over a period of many years (Don Duvick, Pioneer Seed Co., personal communication). Thus, it is reasonable to expect that resistant *Tripsacum* can be developed that will remain resistant over many years. As with leaf rust, differences in leafspot severity among accessions were distinct. It will be important to test plants in replicated trials over several years and in several environments.

Handley's next step is to identify the pathogens to species and learn about their life cycles in order to better understand the role they play in the ecosystem. Survey work of both the six species being studied as crops and of native prairie plants will continue to provide insight into how diseases fit into the total picture.

The Soil Environment

This is a necessary category for future studies. Studies were initiated in 1988 to evaluate the water relationships, nutrients, and organic matter within three species in monoculture. The invertebrates, bacteria, and fungi need emphasis as well as the role of the soil herbivores.

LESSONS FROM OUR BIOLOGICAL STUDIES

Some of our early experiments had to be scrapped as the consequence of our ignorance about the requirements of these particular wild species. For example, requirements for germination and an adequate seedbed preparation had to be learned. We lacked the advantage that comes from working with agronomic species. Nevertheless, successful establishment of experiments has greatly improved over the years, partly because in 1986, Rob Peterson conducted an experiment to determine the best methods of propagating Illinois bundle flower, Maximilian sunflower, and Leymus. He stratified, scarified, and soaked seeds with household bleach solutions featuring different stratification periods, field and greenhouse conditions, and planting dates ranging from early spring to late fall. As we began to understand the requirements of each species from planting to harvest to storage and back to planting again, our experimental work accelerated. Several years of preliminary work were necessary before we could seriously address the major questions that follow.

Can Herbaceous Perennialism and High Seed Yield Go Together?

This straightforward question does not have a straightforward answer. If we were stuck only at the level of results gained from our inventory phase where there has been no breeding for high seed yield or chemical inputs, the answer to this question would be no. Of course, the yields of the ancestors of our herbaceous annual crops before agriculture yielded much less than today. An important but perhaps unanswerable question might be: how did the best of the herbaceous perennials compare with the best of the annuals in seed yield before agriculture?

These considerations are raised to illustrate that the answer to the perennialism and high yield question is not likely to be resolved in any satisfactory manner without years of breeding. Questions about yield potential invariably involve yield improvement through breeding as well as improved agronomic measures.

In our experiments with the five species we have used to answer this question, seed yield over time may increase with plant size, but it may also decrease with plant aging or soil nutrient depletion, or yield may simply oscillate predictably or irregularly. No one model has helped us predict the patterns we have observed in the five perennial monocultures maintained at The Land, where there was no fertilizer or irrigation beyond establishment. From our own data, wild senna yields have oscillated, but averaged 2092 ± 247 kg/ha ($n = 89$ rows) in the second year. Illinois bundle flower produced relatively high yields during the first three years, with a peak of 1967 ± 41 kg/ha ($n = 32$ rows). *Leymus racemosus* yielded 510 ± 24 kg/ha ($n = 34$ rows) and Maximilian sunflower 304 ± 15 kg/ha ($n = 32$ rows), but both steadily declined over time. Eastern gama grass yields increased fourfold to 884 ± 337 kg/ha ($n = 6$ rows) from the first to the second year (Piper & Towne, 1988).

These baseline yields under zero input conditions for monocultures are important. Among other comparisons, they will be used to compare with yields achieved for these species in polyculture. This should tell us whether complementarity among compatible species can compensate for declining yields under low-input conditions.

In the early stages of our research, the theory of r and K selection was rather widely held as a reality. This theory holds that r-selected plants allocate more energy and materials to reproduction than to vegetative parts. For K-selected plants, it is the other way around. Once established, a K-selected plant is more devoted to staying alive than to reproducing. Annuals would generally be r-selected and perennials K-selected. It is an intuitively satisfying notion, but in recent years a widespread questioning has developed of the validity of the theory.

Investigations of Wild Senna. Annuals grown to produce fruit and seed usually have a high yield and high sexual reproductive effort or SRE (Singh and Stoskopf, 1971; Fisher and Turner, 1978; Johnson and Major, 1979; Wardlaw, 1980). SRE is the effort that a plant exerts in producing sexual organs (Hickman and Pitelka, 1975; Abrahamson, 1979, 1982; Thompson and Stewart, 1981). SRE can be measured in various ways, such as measuring the carbohydrates or nutrients allocated to various plant parts. One simply dries the plant to some standardized percentage of moisture, then divides a plant up into its various parts and weighs the parts (Hickman, 1975; Thompson, 1981). SRE is an important measurement because it is positively correlated with high yield in our cereal crops, and plant breeders use it in their selection programs (Fisher and Turner, 1978). Another measurement is the harvest index, which involves a somewhat simpler approach. Here the seed and/or fruit weight is divided by the above-ground weight (seed weight/above ground weight).

TABLE 14.8 Mean and Standard Deviations of Fruit Weight/Total Biomass for Wild Senna and Other Perennials (Burris, 1984).

Species	Fruit Weight/ Total Biomass%	Reference
Cassia marilandica (wild senna)	27 ± 10.2*	
Hieracium floribundum	34 ± 6.6	Abrahamson, 1979
Hesperis matronalis	41 ± 8.3	
Houstonia caerulea	36 ± 11.5	
Hypericum perforatum	32 ± 5.1	
Rudbeckia hirta	39 ± 9.2	
Rudbeckia triloba	32 ± 1.8	
Sisyrinchium angustifolium	32 ± 7.5	
Arisaema atrorubens	39 ± 10.5	
Rumex crispus	46	Weaver and Cavers, 1980
Rumex obtusifolius	35	

* n = 19

David Burris, a 1983 intern, measured the SRE for wild senna (Burris, 1983), which of course cannot be used as the only indicator of seed yield. The very small *Houstonia caerulea* (see Table 14.8) is a plant no one would expect to compete with the seed yield of an annual grain. Nevertheless, it compares somewhat favorably. Burris' harvest index measurement (Table 14.9) shows it to be within or exceeding the ranges for five of the seven listed crops.

TABLE 14.9 Ranges and Means for Harvest Indices of Wild Senna and Some Annual Agronomic Crops (Burris, 1984).

Species	Harvest Index	References
Wild senna	.31*	
Corn	.32–.47	Deloughery and Crookston, 1979
	.39	
	.37	Alessi and Power, 1976
	.26	Fisher and Turner, 1978
Wheat	.29	
(high yielding)	.40	Johnson and Major, 1979
(average)	.34	
Rice (high yielding)	.47–.57	
(average)	.23–.37	
Barley (winter)	.45	Singh and Stoskopf, 1971
(spring)	.51	
Rye	.27	
Oats	.41	
	.44	McDaniel and Dunphy, 1978
Soybean	.30–.41	Johnson and Major, 1979

* Standard deviation = .09, n = 57.

Two important conclusions of the Burris study are that this herbaceous perennial has both a relatively high sexual reproductive effort and harvest index.

Other research with perennial grasses has shown that seed yields can vary from year to year (Cornelius, 1950); how much and with what predictability has to be known for each species.

In a 1984 study, intern Mike Berghoef and staff ecologist Judy Soule measured third-year yield in 29 accessions of wild senna (Berghoef and Soule, 1984). Although yield had declined drastically by the third year, one accession from Chapman, Kansas, averaged 856 kg/ha. The highest yield row in this accession gave 1417 kg/ha, an accession that had averaged only 337 kg/ha in the previous year and 618 kg/ha in the first year. This accession had the highest average over all.

In 1985, intern Vern Stiefel found total seed yield had continued to decline. But when plots were fertilized that year, three- and four-year-old accessions had higher yields than in their first year by 13% and 5%, respectively (Stiefel, 1985). Apparently it is not senesence that causes yield decline, but the need for fertility.

Correlations among yearly yields were examined over the four years. There was a strong positive correlation between yield and accession in the mostly unfertilized plots. Apparently genotype and yield are linked, although this relationship can be masked by fertilization. Stiefel concluded that "when fertilizer is applied, the predictability of how one accession will yield compared to another for a given year may be lost." We should select for yield in an environment that has not been homogenized with inputs.

In 1985, two interns, Danielle Carré and Michel Cavigelli, planted seeds of wild senna on Tobin silt loam at The Land at four different densities. They recorded the yields under different densities, seed yield patterns among individual plants over time, and determined how these patterns correlate with basal stem diameter and planting density. In rows 0.75 m apart, seeds were planted with in-row spacings of 5, 10, 20, and 40 cm. Area yields were significantly higher with 5 and 10 cm spacings than those with 20 and 40 cm spacings (Carré & Cavigelli, 1985).

In September 1985, Jon Piper randomly selected 90 wild senna plants within one plot and then measured basal stem diameter (mm), distance to the two nearest neighbors within the row (cm), and seedmass (g). He repeated these measurements on the same plants in 1986 and 1987. Average seed yield in the first year was 10.5 g per plant. In 1986, the mean yield rose to 28.9 g, but fell in 1987 to 5.8 g (Piper, 1987). What was responsible for the surprising unpredictability across the three years? Plant size accounted for only 30% to 46% of the variation in yield. Temperatures and precipitation were about the same, although there was a storm with winds close to 50 m sec^{-1} in mid-July which flattened many of the plants. Seed predation by larva, probably *Sennius abbreviatus* (Say) (Baskin and Baskin, 1977) was high. The mean distance to the nearest neighbor within the row was negatively correlated with yield in 1985, positively correlated with yield in 1986, but independent of yield in 1987. Although plot density remained constant over the three years, plant size and yield oscillated and population size declined somewhat. From these three years of observations, Piper concluded that "the *long-term* prospects for seed yield in *C. marilandica* are still uncertain. . . . Extrapolated mean yields ranging from about 1000 kg/ha in the first year to about 2000 kg/ha in the second are similar to the fertilized yields of *Triticum aestivum* and *Glycine max* in central Kansas." We realize that to seriously extrapolate from single plant yields to kg/ha is highly questionable.

Investigations of Illinois Bundle Flower. Illinois bundle flower (IBF), The Land's only nitrogen-fixing legume, is one of our favorite species because of its ultimate utility as a livestock food (see species description in inventory section). Consequently, we have done more research with this organism than with wild senna. Earlier I mentioned our efforts at propagation of all our species, including IBF. We have also explored (1) its breeding mechanism, (2) what yields have been achieved, (3) the optimum density for highest yield, (4) what morphological features correlate with yield, and (5) how much genetic variation is present. This last question must be answered in order to predict the possibility for yield increase and competitive ability.

1. Breeding Mechanism. In 1986, intern Patrick Bohlen determined that IBF is self-compatible by noting that the inflorescences that had been bagged set seed (Bohlen, 1986). Latting (1961) had suggested the species may be wind-pollinated. Bohlen considered this improbable because pollen production is small and IBF is relatively large and sticky at dehiscence. Such insects as Chrysomelidae, Halictidae, Apidae, Sphecidae, Bombidae, and Braconidae visited bundle flower in our plots that summer and are likely pollen vectors.

   Bohlen concluded that, to a large extent, this species is self-compatible and that our chances of obtaining seed from selfed plants and maintaining inbred lines was very good. We also have evidence of some outcrossing.
2. Yield Studies. In 1984, intern Martin Gursky planted and harvested seed from 30 accessions. A year later, Mary Bruns recorded seed yields from 45 rows of the 1984 planting and compared her results to those of Gursky. The first year yields recorded by Gursky ranged from 393 to 3,412 kg/ha with a mean of 1,344 across all accessions (Gursky, 1984).

   The second year range was from 539 to 2,416 kg/ha with a mean of 1,164 (Bruns, 1985). Although yields were generally lower in 1985, five of the top ten yielders in 1984 were among the top ten in 1985. The highest yielding row in 1984 ranked third in 1985, whereas the second highest 1984 yielder also ranked second in 1985. The lowest yielding row in 1984 was also lowest in 1985.

3. Yield and Density Studies. Interns Danielle Carré and Michel Cavigelli established another experiment in 1985 (Carré and Cavigelli, 1985). We needed to know what densities of IBF would give optimum yield or whether density was an important factor at all. The experiment was established in a randomized complete block design with eight replicates. Plots consisted of three 8 m rows (two border and one data) per density treatment, with 0.75 m between the rows. The nitrogen-fixing bacterium, *Rhizobium*, was used to inoculate seed sowed by hand. Eventually densities of 7, 14, 28, and 40 cm in-row spacings were achieved.

   In August and September, hand harvest from the middle 5 meters of the data rows brought yields ranging between 1,382 and 1,538 kg/ha. There was no significant difference in yield across the four different spacings.

   In the second year, there was no significant difference in yield among the four planting densities, but there was a 36% yield drop, a decline which we thought might be partially explained by a disease which had appeared just before flowering (Collins, 1986).

   No yield differences were observed among the four spacing treatments in the third year, and yield declined again, 18.1% from 1986 and 44.8% from 1985. The highest producing row this year was 1,167 kg/ha, compared to 1,942 kg/ha in 1986 and 2,013 kg/ha in 1985. Yields in all treatments declined significantly from 1986 ($p < .05$) (Liebovitz, 1987).

   To observe if the defoliation was transmitted through soil splash, Liebovitz randomly chose two replications and mulched with wheat straw in early May. Six replications were left unmulched as a control. Mulching reduced defoliation, but in the unmulched plots, the amount of defoliation had no effect on seed yield (Liebovitz, 1987). We have yet to explain the cause of this defoliation.

4. Morphological Yield Indicators. We have sought some quick and easy yield indicators on individual plants. In 1986, intern Dennis Rinehart measured individual stems of plants at two sites among our accessions. He measured 40 plants on a well-drained upland slope in a biculture experiment with wild senna, and measured 33 stems in a two-year-old plot of accessions in moist bottomland. He concluded that basal stem diameter is a "fairly reliable index of both aboveground biomass and yield" (Rinehart, 1986).

5. Germplasm Evaluation for Breeding Purposes. By 1987, enough groundwork had been laid to move to the next phase of exploration of the possibility of increased yield in Illinois bundle flower.

Genetic variability is necessary for any breeding program to be successful. We want Illinois bundle flower to produce seed, forage, and fix nitrogen all in a mixed cropping system. Initial studies under the direction of Peter Kulakow have served to generate and define useful descriptors for future breeding work (Mecko-Ray, 1987).

From intern Mecko-Ray's observations, "timing and duration of flowering were important in determining harvest and total seed yield. Accessions that bloomed late or had second flowering periods produced few bundles. We may select, therefore, for synchronous flowering that should result in only one harvest. On the other hand, a reduced flowering period could make seed yield more vulnerable to heat or drought. Clustering of inflorescences in determinate branches may be one method of increasing synchrony and thereby facilitating harvest. Seven accessions had 90% or more of the bundles collected at the peak harvest, indicating some synchrony of maturity. The three accessions with the most bundles collected had over 95% of the bundles gathered at peak harvest. An alternative method of cultivation is practiced in Texas, where two seed harvests (June and October) are made each season. This practice may require a longer growing season than we have in Salina, Kansas, however. If a long flowering period and a one-time harvest is chosen, then higher resistance to seed shattering will be necessary.

"Timing and duration of flowering may also be important for compatibility of Illinois bundle flower in a perennial polyculture. Depending on flowering patterns and relative nutrient and water demands of other species in a polyculture, accessions may be required that minimize overlaps in flowering.

"Branching pattern may be important for seed yield and compatibility in polycultures. For example, it is difficult to harvest prostrate forms with branches at ground level, whereas erect branches may allow better infiltration of light into the canopy and perhaps enable compatibility in a mixed cropping system."

Mecko-Ray goes on to report "a positive correlation between mean height and seed yield. Taller plants are more likely to have a higher seed yield than short ones. One might expect such a correlation, considering the generally indeterminate growth of IBF. Plant height may be important for compatibility in mixtures; a shorter plant would be less likely to shade surrounding plants. Smith and Francis (1986) describe several examples of how reducing the height of a dominant crop allows increased productivity of other crops. IBF is tall and bushy, and we'll have to see how this habit affects seed yield or forage value of surrounding plants.

"The wide range of seed yields in this 1987 study may be explained largely by environmental variation. For example, initial seedling size may have influenced seed yield. Variable soil conditions due to the gradual sloping nature of the site may also have affected growth and yield. Yet, several accessions had consistently high or low seed yields in both replications, indicating that there were genetic differences. This year's yields were low compared to the yields of a previous two-year study of IBF accessions (Bruns, 1985); this was probably due to the relatively poor soil of this year's nursery."

Veronica Mecko-Ray made "no observations on nodulation or nitrogen-fixation in this study. Future studies of IBF will include these factors and also consider if there is a trade-off between nitrogen-fixation and seed yield. Since the nitrogen-fixation is so critical to a perennial polyculture, maximizing seed yield in IBF may not be the predominant goal of breeding. Future studies should also observe uptake and utilization of essential nutrients, as legumes tend to have high phosphorus requirements (Williams, 1983)."

"Illinois bundle flower accessions need observation for several years to determine how variation patterns change as a stand ages. Because a perennial polyculture would be harvested for several years, we must determine the length of evaluation necessary to assess adequately the potential of Illinois bundle flower germplasm. Future evaluations should test germplasm with both monoculture and polyculture plantings to assess differential performance between planting arrangements (Williams, 1983)."

Sunflower Studies.

1. Common Genomes and Yield Among Helianthus Species. In 1984, we became interested in the diploid and hexaploid perennial relatives of the high-yielding diploid annual, *Helianthus annuus*. We had concluded that Maximilian was probably the highest yielding diploid perennial sunflower. Studies by Kostoff had led him to conclude that the hexaploid *Helianthus* species contain a genome of the high-yielding annual (c.f. Heiser and Smith, 1964). We wondered if that common genome might be correlated with the high yield characteristic of the diploid annual.

   Paul Adelman planted eight *Helianthus* species in rows ten feet long in random order in four adjacent replications. In this experiment, two of the three perennial diploids outyielded all four of the perennial hexaploids. The lowest yielding perennial diploid, *H. mollis*, outyielded the two lowest yielding hexaploid perennials. We concluded that the genome gave those hexaploids no yield advantage over the perennial diploids lacking the annual diploid genome (Adelman, 1984).

2. Sunflower Hybrids. In 1984, intern Janine Calsbeek set out to learn whether the hybrid between two perennial sunflower species, sawtooth sunflower (*Helianthus grosseserratus*) and Maximilian sunflower could outyield the parents. Both had been top yielders, and Long (1959) had reported on the extreme vigor of the F_1 hybrid.

   Germination problems made it impossible to obtain data on sawtooth yield, but Calsbeek did compare the yield of a race of Maximilian (three generations from the wild) with the F_1 hybrid, sawtooth times Maximilian. They were planted in a randomized block design at two densities with four replications. The Maximilian yield averaged 595 kg/ha. In spite of the extremely impressive vegetative growth of the hybrid, its seed yield was only 204 kg/ha (Calsbeek, 1984).

In 1985, intern Mary Bruns established plantings of a hybrid annual sunflower (Triumph 505C confectionery), an open-pollinated annual sunflower (mammoth) and Maximilian. The mean yield for the mammoth was 931 kg/ha, which included an estimate of seed lost to insects and birds. The yield of the hybrid was 2,166 kg/ha, including an estimate of seed lost to insects (bird damage was minimal). Mean yield for the perennial was 325 kg/ha, a very low yield mostly caused by an early heavy frost (Bruns, 1985). From here on, our concentration has been on Maximilian sunflower only.

3. Investigations of Seed Yield in Maximilian Sunflower. Density Trials: From 1983 through 1987 we set out to determine the optimum density for high yield in Maximilian. In 1983, we learned that the optimum spacing in a row lies between 3 and 30 cm. In 1984, intern Kirk Riley established four densities in rows about 6 m long and about 1 m apart. Four within-row spacings ranged between 7.5 and 18 cm. He randomized these densities throughout the four replications. Seed yield varied between 237 and 614 kg/ha and averaged 456 kg/ha (Riley, 1984). Variation in seed yield, however, was not significantly related to density. These results are consistent with observations in two other experiments conducted for different purposes at The Land. Walter Pickett found no differences in spacings of 3, 6, and 12 inches, respectively, nor did Janine Calsbeek with the Maximilian times sawtooth hybrids (Calsbeek, 1984). This lack of correlation between density and yield is supported by field observations where greater branching occurs at the lower densities, filling in the available space and the increase in apical dominance at the higher densities.

We learned, however, that these results of no difference in seed yield among various planting densities do not mean that we should spare seed when planting. The greater apical dominance at the higher densities places the seed-bearing areas in a more favorable position for harvest.

Riley's results suggested the need to gather more information about growing this potential crop. The highest yielding planting density the *first year* may not be optimum later on. If so, we may have to control density by cultural practices in *subsequent years* or plant at less than optimal first-year density.

Juli Kois, a 1985 intern, set out to confirm whether there were yield differences associated with four different spacings of plants (2, 6, 10, and 15 cm) within rows that were 75 cm apart. Using a randomized complete block design, she recorded plant height, branching pattern, and number of seed heads per plant, all yield and growth characteristics. Again, yield did not differ significantly across the density ranges but did differ among individual plants according to density. Plant height and number of seedheads were affected most, not branching pattern. Kois's yields were 40% lower than Riley's in 1984, but 1985 was the year of the unusually heavy frost, which came two weeks after flowering began (Kois, 1985).

In 1986, intern Mark Gernes continued Kois's experiment. He split her plots, thinned half of them by hand and by rototilling, and left the other half alone. Seed yield did not differ among the unthinned densities. The number of seed heads per stem increased. Plant height decreased and density increased. Branches per stem and stem height decreased from 1985 to 1986. The nonthinned treatment with the heighest density was three times more stem dense than the thinned control. However, the seed yield for the thinned control was two times greater. Rototilling early in the season reduced stem density but increased yield (Gernes, 1986).

This experiment was initiated by Kois in 1985, carried into the second year by Gernes in 1986, and continued in 1987 by intern Roger Liebovitz. Recall that in 1986, each of the eight blocks was divided. Half were thinned to the original spacings and the other half were allowed to experience normal growth as a control. On October 19, 1987, only the control half was harvested. Again, there were no significant differences in seed production among the four density treatments. Yield was a disappointing 33.8 kg/ha, a decline of 83.7% compared to 143.4 kg/ha in 1986 and 183.4 kg/ha in 1985 (Liebovitz, 1987). There are several possible reasons for this decline: nutrient depletion in the soils, crowding, senescence, or a combination.

In 1986, intern Dennis Rinehart examined the relationships among basal stem diameter, above-ground phytomass, and fruit yield in Maximilian. He examined 40 stems growing in a low, moist area. In every case, basal stem diameter predicted more of the variation in plant phytomass than in reproductive yield. Nevertheless, it seems that basal stem diameter is a fairly reliable and useful index for crude estimate of yield. These plants were grown under favorable conditions (Rinehart, 1986). To know what the response might be under resource limits will require further research.

Investigations Involving Eastern Gama Grass. We hope the pistillate form of EGG has the potential to greatly increase seed production in eastern gama grass. In 1986, intern Patrick Bohlen established over 4,000 plants from seed, representing 122 half-sib families (opr crosses) of this species (Bohlen, 1986). Each cross or family was segregating for the pistillate mutant.

Bohlen's objective in 1986 was to establish a population containing as many genotypes of the pistillate mutant as practical. From data taken we now have a record of the transplant survivorship, a better idea of the differences in gross plant morphology in both the normal and mutant, and a comparison of the incidence and level of rust and leafspot diseases in the normal and mutant. We have more than a feeling of the percentage flowering and flower type. All of these categories were recorded that summer for 77 of those 122 accessions established at The Land. Knowing that 93% of the transplants survived, 9% flowering this first year of establishment, and 23% of those that did were mutants is useful in-house data. Perhaps more importantly, we have a better idea of the kind of tasks facing those who would seek to develop perennial grasses. Accessions varied in their susceptibility to both rust and leaf spot diseases. Generally, the level of infection of leafspot was lower than the rust. On a scale of 0 to 7, the overall mean of leafspot was 2.5, which was lower than the overall mean for rust at 2.8.

From Bohlen's experiments we learned more about the culturing of this wild species. For example, survival is best when seedlings are planted in flats, overwintered outdoors, and then transplanted into a prepared seed bed. Spring-emerged seedlings can be planted directly into the field and thus avoid the intermediate step of transferring to peat pots. Direct seeding into a fall seedbed can save labor and materials, and thinning or transplanting to the necessary density can wait until spring. We now know with confidence that the most vigorous plants have the most flowers, and that usually vegetative shoots do not become reproductive until they are a year or two old (Dewald and Louthan, 1979). Even though 10% of ours flowered the first year, perhaps because growth conditions were ideal, there was still not enough flowering on one plant to harvest.

Mary Handley and Patti Boehner, intern in 1987, studied the variability in resistance to anthracnose and leaf rust (Handley & Boehmer, 1987). They selected 10 of these 122 half-sib families for planting in 1987. These were selected on the basis of mean leaf rust severity in 1986. They first divided the large field of over 4,000 plants into five blocks and then selected families with the lowest and highest mean rust severity from each block. Thirty plants were chosen from within each of the ten selected families for transplanting. They dug these 300 plants in early April and divided them into six clones each. They established one planting at The Land (three clones times 30 plants times 10 families = 900 plants) and the other at the Kerr Center for Sustainable Agriculture in east central Oklahoma near Poteau. The three replicates of each of the 300 plants were set out in a randomized complete block design at both localities. They were spaced 1 m apart with two border rows around the entire planting. The rototiller and intern hands did the weeding.

Handley and Boehner (1987) reported differences both between and within families, indicating the likelihood of a genetic basis of plant susceptibility. Within-clone variation was low. There is a negative correlation between rust and anthracnose incidence. Happily, the mutant does not seem to be closely associated with unthriftiness and disease susceptibility, suggesting considerable promise for the use of the mutant character to increase seed yield.

Leymus racemosus. We established a pure stand of *Leymus racemosus* at The Land in April 1985. This 16 by 20 m plot consisted of 33 rows, which first flowered in mid-May 1986.

Brad Burritt, an agricultural intern, conducted an extensive study of this plot in 1986 (Burritt, 1986). His seed yield estimates were based on heads he collected from 34 row segments measuring 5 X 0.5 m. Rows were selected where no gaps were obvious and two harvests were conducted in order to minimize shattering loss. After cleaning, seed were dried at 40°C overnight before weighing.

According to Burritt, extrapolated yield values for the 34 row segments ranged from 258 to 900 kg/ha with a mean of 510 kg/ha. There was variability among these row segments, which accounted for the wide range in yield results.

Threshing percentage was nearly constant among rows, while head number and total weight of heads per row were significantly correlated with yield.

Burritt determined the harvest index for two rows to be 6.7% and 7.7%. The latter also had the highest seed yield. Forage yields extrapolated from the total biomass yields of these two row samples were 8,500 and 11,000 kg/ha, respectively.

Ergot is a serious problem in *Leymus*. And since deep plowing and crop rotation are antithetical to a perennial crop on a yearly basis, other possibilities will need to be explored. Ergot-free seed is possible, but with the perennial, infestation from nearby stands of wild grasses is always possible. We can burn at the growing season's end, but in the long run, resistance varieties will be necessary. Varietal resistance to ergot is variable since many pathogenic races exist (Jones and Clifford, 1978). Burritt concludes that "since the fibrous glumes must be removed from the seed before it can be eaten by people anyway, an extra step in the milling process to remove the sclerotia poses no problem if other solutions fail."

Winter-Hardy Perennial Sorghum. Our sorghum research began in 1983 when a tetraploid domestic sorghum, *Sorghum bicolor* L. Moench, was crossed with Johnson grass, *S. halepense* L. Perso. The F_1 was planted in 1984, with 465 plants surviving the winter. In 1985, we planted the F_2 generation. These plants displayed considerable variation in such important characteristics as flowering time, height, shattering, and panicle compactness. In 1986 Peter Kulakow acquired agronomically more favorable lines of the tetraploid *S. bicolor* and collected 27 accessions of *S. halepense* from sites in Kansas, Oklahoma, and California. Several additional techniques were employed in crossing, and a few hundred hybrid seeds were produced. Kulakow now has a major effort underway to develop and explore the potential of this hybrid derivative.

Work on this artificial hybrid is of great biological importance, as it represents a major effort to take a domesticated species that, for all practical purposes, is an annual in temperate zones and cross it with a relative that has no economic value except negatively as an aggressive weed. The biology of a perennialized domestic crop is of great importance because it would be an example of beginning with a known crop rather than developing a wild species into a domesticated form. This fusion of the domesticated and wild has philosophical implications for the "nature as measure" paradigm.

As philosophically, biologically, and practically interesting as this process and its product might be, we intend to be careful about the spread of these hybrids and hybrid derivatives. The weedy potential of the perennial sorghum lines is not yet known, and the potential for the perennial sorghum to cross with the cultivated annual could cause weedy threats in the field. A perennial sorghum of our making could cross with the weedy populations, transferring cultivated genes to the weed populations. It is unlikely, but the competitive ability could be enhanced by mimicking the crop or enhancing overall vigor.

Can Perennial Polycultures Outyield Perennial Monocultures?

We began to address this question in our experiments in 1983 with a biculture of two legumes: Illinois bundle flower (IBF), a nitrogen fixer, and wild senna, a non-nitrogen fixer. Interns Mark Böhlke and Juli Neander, who established this experiment, planted four replications of five treatments: monocultures of IBF and wild senna at 12-inch within-row spacing, alternate row biculture at 12-inch spacing, and alternate plant biculture at two densities corresponding to 6 inch and 12-inch spacings. Each treatment consisted of nine 27-foot rows spaced 3 feet apart, with the inner 21 feet of the seven middle rows used for data. The upland soil on which this experiment was planted was low in nitrogen and organic matter. The IBF was not inoculated with *Rhizobium* bacteria at planting time. We did inoculate the established plants in 1984 by pouring the *Rhizobium* into the furrows alongside the plants.

Dana Price and Martin Gursky, agriculture interns in 1984, took charge of the second year of this experiment (Price and Gursky, 1984). Where there were gaps, the 1984 interns replaced wild senna with same-age rootstock and IBF with transplanted seedlings. In 1983, the first year of establishment, there was no overyielding. Monocultures of the two species collectively had the same yield as the bicultures; relative yield totals (RYT) for the three biculture treatments ranged from 0.9 to 1.1. The second year, 1984, was a different story. Wild senna yielded higher than IBF in monoculture even though *relative* yields of wild senna were consistently lower than those of IBF in biculture. For 10 of the 11 biculture plots, the relative yield exceeded one. The differences among the three biculture treatments were not significant. Considered together, the bicultures significantly outyielded the monocultures by 29% on the average. Overall yields from the bicultures were near 1,000 kg/ha.

1985 intern Lois Braun continued this experiment. Third-year results revealed an average relative yield for the three bicultures to be 1.94. In only one of the blocks was the relative yield less than 1.0, although in one block it exceeded 3.0. The relative yields of IBF and wild senna were almost equal. It is important to note that the absolute yields of the biculture were significantly higher than the monocultures in the second and third years, but biculture yields remained nearly constant (Braun, 1985).

From these results, we began to develop some ideas about planting density and planting combinations. Significant differences were found between the alternate-row and alternate-plant bicultures and between the two densities of alternate-plant bicultures. Also, the 30 cm spacing of the alternate-plant biculture overyielded more than the other two bicultures.

In 1985, we started a triculture from seed consisting of *Leymus racemosus*, IBF, and Maximilian sunflower. They were established in alternating rows with each species being a neighbor of the other. We wanted to observe the growth and survivorship of the interplanted species and identify problems that would suggest alternative designs. We had problems establishing Leymus that first year, which made any comparison of seed yields of the three species meaningless. But we saved the plot, and in 1986, Dr. Teresa Maurer, resident postdoctoral affiliate, conducted a preliminary study to evaluate the effects of cutting Maximilian back. After cutting she evaluated the overall growth of the triculture and seed head production of Maximilian. Seed head production and lodging decreased as the number of cuttings increased. Plants that had been cut twice continued to produce seed heads, suggesting that this plant has the potential for both forage and seed production. This is important for management purposes (Maurer, 1986).

Based on what we had learned from this triculture, we established a second triculture in 1987. This time we included two grasses: Leymus (cool-season) and eastern gama grass (EGG), a warm-season grass, and our nitrogen-fixing legume, IBF. Agriculture intern Bruce Kendall assumed primary responsibility for this small demonstration plot. Our ultimate purpose was to gain information on inter- versus intra-specific competition. Kendall monitored growth rates, flowering, and fruiting patterns and seed yield. The three species appeared to be compatible. IBF adjacent to the grasses produced more flowers and seeds than individual plants next to their own kind. Individuals next to EGG had a more synchronous harvest than those neighbored by Leymus. A majority of later-flowering individuals were bordered by their own kind (Kendall, 1987).

Can Our "Domestic Prairie" Ecosystem Sponsor Its Own Nitrogen Fertility?

The first two of our studies dealing with this question constitute baseline information necessary for future experiments.

In 1985, we had anecdotal speculation from a physiological ecologist that agricultural species tended to respond better to ammonium, whereas native prairie species responded better to a nitrate environment. We wondered whether our ecosystem should be considered agricultural, natural, or both. Intern Martin Gursky conducted a greenhouse experiment in which he compared seven herbaceous perennial species for response to different nitrogen sources (Gursky, 1985). Each species was divided into two groups: one receiving nitrogen in nitrate form, the other as ammonium. Curly dock (*Rumex crispus*) in the buckwheat family, Maximilian sunflower (a composite), three grasses (eastern gama grass, *Leymus*, and luna pubescent wheat grass), and *Elytrigia intermedium* var. *trichophorum* grew equally well under both conditions. Our two legumes, wild senna and IBF, grew poorly in both treatments, although with these legumes, the nitrate-fed plants performed slightly better overall.

In 1987, agriculture intern Doug Dittman identified and monitored the soils of different plots at The Land to observe and compare the changes within the soil over the growing season. Besides knowing our own place better, we hoped this information would help us predict which species would compete with, or complement, other species within a perennial polyculture. Dittman had a second purpose: to increase our understanding of soil nutrient and moisture use patterns within native prairie to establish standards against which to compare our agriculture plots. He monitored soil moisture from May through October and measured soil nutrients in early July, August, and September. His sites included two tallgrass prairie areas at The Land that had never been plowed, three monocultures in the bottomland of our species where our experiments were growing, and an adjacent milo field.

Most of the differences among these plots were attributable to differences in soil type and plant root structure. Most interesting was the observation that the prairie growth patterns more closely followed moisture in the B horizon than in the A. This was also the case in IBF. Shallow rootedness seemed to tie flowering and fruiting to the A horizon, as Dittman observed in both *Leymus* and EGG. Overall, Dittman's results suggest complementarity in the use of the soil resource among these species if grown together in polyculture (Dittman, 1987).

Can a Perennial Polyculture Adequately Control Pests?

So far we have only explored, through experiments, how the weed pests may be controlled. Several authors have suggested that allelopathic interactions have an important role to play in biological weed control (Putnam and Duke, 1974; Lovett, 1982; Gliessman, 1983). Our Maximilian sunflower stands have been nearly weed-free. Our interests intensified in 1983 when two-year-old plantings were bare in early spring while an adjacent area was covered with wild mustard. To get a more precise assessment of this species' potential, in 1984, intern Dana Price studied the relationship between density of Maximilian sunflower and growth of the weeds.

Employing a randomized block design with four replications, Price planted Maximilian seeds at six densities. Each density plot consisted of two sunflower rows 1 m apart, 18.5 m long, with the inner 5 m the data source. Her goal was to measure weeds between these rows. Since sunflowers germinate more slowly than some of the weeds, plots were weeded within rows once and mowed between rows twice during the first 40 days. This assured both weed and sunflower establishment. At the end of the growing season in this establishment year, Price analyzed the percentage cover using a 0.25 m quadrat to compare weediness among treatments. She also took data on the predominant weed species. Covariance analysis showed a significant effect of sunflower density this first year. The denser the within-row plantings, the fewer the weeds. The density regression was significant (Price, 1984).

A 1985 intern, Michel Cavigelli, assumed responsibility for this experiment, which involved mostly data collection. Cavigelli found that the species mix changed. Weed collections were made May 23, July 14, and September 23. Again, the weed biomass declined with an increase in sunflower density. The slope of the regression was similar on the latter dates. Sunflower density in the early part of the growing season has a less pronounced effect on weed biomass. Cavigelli has considered that because we had to weed during the establishment year, Maximilian seedlings may be unable to suppress weeds. These plants were planted, however, in late spring, at the time of peak emergence. Cavigelli has observed that Maximilian does become established without any weeding when broadcast in tilled ground (Cavigelli, 1985).

Intern Mark Gernes took over the third year of this experiment and once again recorded total weed biomass decline with a decrease in the within-row sunflower density. There were major changes in the weed flora during the three years of this study. The first year and second years, weeds were mostly annuals. By the third year, most of the weed species were perennial, but as a percentage of total weed weight, the contribution of perennials was very small (Gernes, 1986).

One of the most interesting pilot projects we have conducted had to do with the effect of Maximilian sunflower on subsequent plantings, a project carried out by 1985 intern Carol LaLiberte. We had a fortuitous circumstance awaiting her in 1985. We had an observation plot of Maximilian rows. Some had been there four years, some three, some two, and some one. They were arranged side by side and adjacent to an open area. La Liberte tore up the entire area and removed each Maximilian plant, roots and all. She kept track by stakes where each row had been. She then planted four different species on this well-worked ground in rows perpendicular to the former Maximilian arrangement. In other words, IBF, curly dock, wild rye, and wild senna were planted on ground where Maximilian had grown four, three, two, one and zero years. This gave us a chance to determine if there was any residual effect due to the former presence of Maximilian. We reasoned that if persistent allelopathy or nutrient depletion were important factors, we might expect poorer germination and ultimately lower yields correlated with the length of time that the sunflowers had grown in a particular area. Neither pattern was seen. The only significant differences were for percentage emergence in curly dock and yield in IBF, but neither had anything to do with how long the sunflower had been growing in an area (La Liberte, 1985).

RESEARCH AGENDA FOR THE FUTURE

Research Agenda for the Land Institute

We intend to plant our first large-scale polyculture at The Land in the spring of 1991. With this as our goal, especially during the last two years, our experiments have been designed to gather the essential information. As of 1989, 12 experiments were under way, all long-term efforts built on past results. The perennialism/high-yield question remains paramount. We continue in our effort to establish baseline seed yields over a five-year period. Current monocultures of Leymus, eastern gama grass, and Illinois bundle flower are in their fourth and fifth years. In previous years, yields in herbaceous perennials have declined or oscillated over time. We don't expect these plants to compete with the subsidized annual crops, and we intend to stick with our plan, since long-term patterns of seed yield in plots without inputs remains largely unexplored.

The Leymus plot was established in 1985. Because plants were vegetative during the first year, 1989 will represent the fourth year of yield data. The Illinois bundle flower plot was established as a density trial in 1985. During the last four years, there has been no effect of density on seed yield, although yields have declined steadily since 1985. This will mark the fifth year of data collection for that plot.

We continue to conduct our search for relevant patterns in the never-plowed native prairie at The Land, because the native prairie remains our standard for sustainability and we seek insights into the successful development of perennial polycultures. In 1986, under the direction of our ecologist, Dr. Jon Piper, we began documenting annual above-ground productivity and spatial and seasonal heterogeneity on three native prairie sites involving various soils. This, too, is a long-term project. So far productivity has differed in each year, and sites have responded deferentially to burning and drought. These are factors to which we must pay close attention. Cool-season grasses, a few legumes, and spring ephemerals are important early in the growing season. Warm-season grasses and composites predominate from late summer onward. Piper has discovered that, for our prairie, legumes are at their highest densities on stony, marginal soils and that, overall, species diversity is inversely related to productivity at these sites.

Piper will continue to direct a study begun in 1988 of monitoring changes in soil moisture and nutrients in plots of Leymus, eastern gama grass, and Illinois bundle flower. We need to understand what differences exist among these species in nutritional requirements, what portions of the soil are occupied by roots, and what the seasonal developmental patterns are for moisture and nutrient uptake. This is essential for determining optimum planting densities and optimum species mixes for a large-scale polyculture featuring these plants.

To measure moisture, we use ohm readings, as well as wires attached to gypsum blocks buried at depths of 10 cm, 40 cm, and 1 m. Soil moisture has been, monitored at one-week intervals from early March to November. Soil samples have been and will continue to be taken at three depths every three weeks and analyzed for nitrate, P, K, Ca, Mg, and pH at Kansas State University. Percentage organic matter will be determined for first and final samples. Later, we will correlate soil moisture and nutrient patterns with precipitation and periods of maximum growth, flowering, and fruiting.

Dr. Peter Kulakow is directing a major effort to evaluate the eastern gama grass germplasm and is in the process of describing the genetic variation in the germplasm collected from natural populations. We especially want to know if there is significant variability within natural populations, as well as between populations from different locations. This will help us to devise collecting strategies, and identify families and populations that have superior characteristics, including high seed yield. We will then select from this and other germplasm evaluations to develop base populations for our breeding program.

Kulakow is also directing a study to identify IBF germplasm collections that produce high seed yields over several years. It will be necessary to describe the genetic variability for adaptations to our region, as well as to describe the domestication characteristics. From this we will select plants to develop the base population for breeding IBF. We established 82 IBF accessions of wild populations from several states at The Land in 1988. Although it is only a portion of the natural distribution of the species, it is a good cross-section of the Kansas and Oklahoma populations. Data will be collected on several characteristics and examined for patterns of geographic diversity that would be used to plan future collections.

Kulakow is also directing a narrower study with IBF that involves the development of breeding lines resistant to seed shattering. Among the 100 germplasm collections of IBF assembled at The Land over the years, one collection from south of Knoxville, Tennessee, produces mature pods that are indehiscent. We grew this plant at The Land from 1984 to 1986, and in 1988 we planted seed from it in the germplasm study mentioned above. Among the 40 plants grown in 1988, 3 were apparently natural outcrosses to other accessions from the earlier planting, since plants were intermediate in appearance for legume dehiscence and in several other characteristics, including growth habit and legume shape. In addition, they were heterotic for large seed size. In 1989, therefore, we will raise the F_2 generation from hybrids between the indehiscent and dehiscent genotypes and develop F_3 seed.

In 1988, Kulakow directed the establishment of an experiment to assess the performance of germplasm accessions of IBF and EGG when they are grown in monocultures and when they are grown as an alternate row biculture subjected to interspecific competition. The degree to which genotypes vary in their performance, depending on the cropping system, will be used to assess when characteristics of these two species can be evaluated in monocultures, and which characteristics will have to be evaluated in multispecies stands, if our ultimate objective is to breed perennial seed crops adapted for use in polycultures.

Dr. Mary Handley has an experiment under way that should provide data for two of our four basic questions: the perennialism and high-yield question, and the problem of pest control in the polycultures of the future. The experiment has two components: seed yield estimate for the two sex-ratio variants of EGG, and disease assessments and correlations with growth and yield for 300 genotypes of this species.

The pistillate mutant appears to have the potential for high seed yield because of the greatly increased seed number. In 1988, however, the yield increase was relatively small in our plots at The Land. More information about agronomic performance of pistillate plants is needed, especially their seed yield, to determine if this variant will be useful in a perennial polyculture.

There are two major leaf diseases of EGG in our area, anthracnose and leaf rust. We have found that there is genetic variation for susceptibility to these diseases as well as for many agronomic traits, such as reproductive tiller number, seed yield, and growth rate. Anthracnose reduces growth and yield in Salina, Kansas, while leaf rust reduces growth but seems to have no effect on seed yield in our plots.

In 1989 we hope to identify high-yield plants that are not susceptible to disease for future polyculture experiments. In addition, a new planting of pistillate plants has been established in 1989. All the pistillate plants in a large planting from 1986 were measured and 100 of the largest plants were transplanted. Pistillate plants were dug and divided, yielding identical replicates. These should produce some seed in 1989 for an initial yield estimate. We intend to harvest them each year for several years to estimate long-term seed yields.

This will allow us to become more familiar with the pathology and the ecology of the plant since we will feature disease evaluation as well as the components of agronomic yield. We will then have a better basis of comparison for studying yield and disease differences between native plant communities and various cropping systems.

In 1989 we intend to describe the genetic variability in Leymus for characteristics needed to develop it as a perennial grain crop. Our evaluation of seed yield where the inputs are low reveals a dramatic decline. From 1986 to 1988 seed yield averaged in this order: 56.7 g/m^2, 18.7 g/m^2 and 6.5 g/m^2. We know this yield decline is particularly due to a decline in the numbers of reproductive culms per square meter each year. If Leymus is to be a viable perennial, we will have to maintain seed yields at a higher level over several years. We expect to achieve this through a combination of management and genetic selection. The object of this experiment is to assess the genetic variation for seed yield and other agronomic characteristics over several years.

Finally, Peter Kulakow will continue to direct our effort to develop a winter-hardy perennial sorghum over the next four years. His breeding strategy involves two stages. He will develop hybrid lines with increasing proportions of cultivated genes through backcrossing to the cultivated parent. These lines will be evaluated in 1991 to determine the best proportion of cultivated genes to incorporate in populations, in order to breed an agronomically useful perennial sorghum and to study the inheritance of rhizome production and its association with other traits. Second, he will develop a randomly mated base population to begin selecting a perennial cultivar that includes a diverse germplasm base, including sources of good adaptation to our region.

The Necessary Marriage of Agricultural Science and Ecology

Modern agriculture with its dependency on the greater extractive economy has increased the tendency for both researchers and farmers to understand and improve agricultural production in its own terms. The sustainable agriculture movement has the potential to help us acknowledge that, ultimately, agriculture comes out of nature.

In general, scientists—including ecologists—have given priority to the parts over the whole. But the discipline of ecology has a greater potential for broadening its outlook. It is not a simple repudiation of reductionism, the prime paradigm for an emphasis on cleverness, but rather an acknowledgment that we need to go beyond, to more fully explore the laws of nature at work at the ecosystem level. These laws must surround the emergent qualities that the ecosystem displays. It is not that the whole is greater than the sum of the parts so much as the whole is different than the sum of the parts. Even though ecology is perhaps too reductionistic, those who have studied bogs and alpine meadows, unplowed prairies and deciduous forests, rainforests and marshes are the best candidates for pondering and explaining what ecology does have to offer. How nutrients cycle to accommodate diversity or monoculture, and the pathways through which sunlight flows—these are ecosytem studies. Ecologists are evolutionary biologists, plant population biologists, quantitative geneticists, functional morphologists, physiological ecologists, and so on. There are many more subdisciplines, depending on how one wishes to identify the categories. What holds them all together is the fact that researchers in these fields want to understand how nature's life forms work in particular ways and how they respond to the larger systems of which they are a part. Again, most of their work is reductionistic, much of it necessarily so, although as we come to recognize the consequence of placing priority on part over whole, that ratio will change. Nevertheless, the knowledge these ecologists have placed on the shelves, mainly for its own sake, in the last 30 to 50 years is phenomenal. Unfortunately, much of it remains there.

What if more of this body of knowledge were applied to agriculture? And what if ecologists were to begin applying that knowledge to sustainable agriculture, using nature as the measure? The new field of agroecology, as it stands today, is not the hybrid I am talking about. Most of the workers in agroecology, whether they come out of agronomy or ecology, are still trying to understand agriculture against a standard that dates back 8,000 to 10,000 years at best and, even worse, to a standard made possible by the industrial revolution and the emergence of the chemical industry. By not using nature as the standard, agroecology is apt to be watered down to an emphasis on low inputs and soil erosion, and germplasm truncation will continue at alarming rates. I have no objection to this emphasis on low input in the short run, as long as it is recognized as an intermediate step only. We simply need to prepare now to be more attentive to that body of knowledge on the ecological shelves and to bring into sustainable agriculture the large numbers of workers responsible for putting much of it there. It is imperative that we begin raising a new generation of agroecologists who represent three-way hybrids: people willing to learn from farmers, who have much to teach in the way of tradition and experience, modern agricultural researchers, and ecologists.

We cannot expect this emphasis to come from agribusiness. Success in sustainable agriculture means that we would greatly reduce our reliance on nonrenewable resources, the very stock that agribusiness relies on. Furthermore, a success story would tell how we reduced our dependence on nature as a sink for pollutants. (Nature's sinks have been treated as givens by all of us interested in externalizing as many costs as possible.) This ecological emphasis will find the benefits running more to the farmer and the agricultural landscape. We cannot expect a push, therefore, to come from the business world, which supplies most of the inputs for agriculture in anything close to the way that biotechnology got its boost in the 1980s.

To provide the necessary emphasis will require a sustained effort for several years. The "nature as standard" idea needs high visibility in our country. The highly respected and widely known ecologists and agriculturalists can best articulate this need and thus encourage the young researchers who will pioneer this effort by researching and writing the classic papers in this exciting new discipline.

Acknowledgements

An effort of numerous Land Institute interns and staff members, spanning nearly a decade, is represented in this chapter. Nearly all of them are represented in the references to one or more of their papers published in The Land Institute Research Report. My greatest debt is to them. In the early stages it was Marty Bender, our first research associate at The Land, and the late Jim Peterson followed by Walter Pickett and Dr. Judy Soule who helped bring sense to lots of fuzzy thinking. Times are different now, and I thank our contemporary staff scientists Dr. Mary Handley, Dr. Peter Kulakow and Dr. Jon Piper who now carry most of the responsibility for our current research effort.

REFERENCES

Abrahamson, G. (1979). "Patterns of resource allocation in wild flower populations in fields and woods," *Amer. J. Bot.* **66(2)**: 72–79.

Adelman, P. (1984). "Comparison of sunflowers with different genomes," *Land Report Res. Suppl.* **1**:5–6.

Ahring, M. (1964). "The Management of Buffalograss for Seed Production in Oklahoma." Oklahoma State University *Agricultural Experiment Station Technical Bulletin* T-109.

Alessi, J., and J. F. Power. (1976). "Water use by dryland corn as affected by maturity class," *Agron, J.* **68**:547–50.

Altieri, M. A. (1987). *Agroecology*. Westview, Boulder, Colorado.

Baskin, J. M., and C. C. Baskin. (1977). "Predation of *Cassia marilandica* seeds by *Sennius abbreviatus* (Coleoptera: Bruchidae," *Bull. Torrey Bot. Club* **104**:61–64.

Bates, L. S., M. Bender, and W. Jackson. (1981). "Eastern gama grass, seed structure and protein quality," *Cereal Chem.* **58: 138**:138–41.

Berghoef, M., and J. Soule. (1984). "Third year yield in wild senna (*Senna marilandica*)," *Land Report Res. Suppl.* **1**:13.

Bodrov, M. S. (1960). "Hybridization between wheat and *Elymus*," *Wide Hybridization in plants*. Akademiya Nauk U.S.S.R. 238-241 (Translated from Russian, 1966). National Science Foundation, Washington, DC.

Bohlen, P. (1986a). "Effect of self-pollination on *Desmanthus Illinoensis* seed production," *Land Report Res. Suppl.* **3**:9–11.

Bohlen, P. (1986b). "Multiplication and evaluation of *Tripsacum dactyloides* accessions segregating for a recessive pistillate mutant," *Land Report Res. Suppl.* **3**:30–33.

Braun, L. (1985). "Overyielding in a perennial legume biculture," *Land Report Res. Suppl.* **2**:4–7.

Brown, H. R. (1943). "Growth and seed yields of native prairie plants in various habitats of the mixed prairie," *Trans. Kan. Acad. Sci.* **46**:87–99.

Bruns, M. (1985a). "Illinois bundle flower: A perennial food or feed crop?" *Land Report* **25**:6–9.

Bruns, M. (1985b). "Second year yields of Illinois bundle flower (*Desmanthus illinoensis*) accessions," *Land Report Res. Suppl.* **2**:27–28.

Burris, D. (1984). "Sexual reproductive effort of wild senna (*Senna marilandica*)," *Land Report Res. Suppl.* **1**:20–22.

Burritt, B. (1986). "Yield estimation and analysis of yield components in *Leymus racemosus*," *Land Report Res. Suppl.* **3**:25–30.

Calsbeek, J. (1984). "Seed yield comparison of *Helianthus grossesseratus* X *H. maximiliani* F_1 plants and *H. maximiliani*," *Land Report Res. Suppl.* **1**:19–20.

Carré, D. (1985). "Insect and pathogen observations," *Land Report Res. Suppl.* **2**: 15–16.

Carré, D. and M. Cavigelli. (1985). "Effects of density on yield of Illinois bundle flower (*Desmanthus illinoensis*) and wild senna (*Senna marilandica*)," *Land Report Res. Suppl.* **2**:7–9.

Cavigelli, M . (1985). "Effect of Maximilian sunflower (*Helianthus maximiliani*) on naturally occurring weeds," *Land Report Res. Suppl.* **2**:10–12.

Collins, M. (1986). "*Desmanthus illinoensis* density trial, 1986," *Land Report Res. Suppl.* **3**:18–19.

Cooper, H. W. (1957). "Producing and harvesting grass seed in the great plains," *USDA Farmers' Bull.* No. 2112.

Cornelius, D. R. (1950). "Seed production of native grasses under cultivation in eastern Kansas," *Ecol. Monogr.* **20**:1–29.

DeLoughery, R. L., and R. Crookston. (1979). "Harvest index of corn affected by population density, maturity rating and environment," *Agron. J.* **71**:577–80.

Dewald, C. L., and V. H. Louthan. (1979). "Sequential development of shoot system components in Eastern gama grass," *Range Manage.* **32**:147–151.

Dittman, D. (1987). "Soil moisture and nutrient patterns in agricultural plots and native prairie," *Land Report Res. Suppl.* **4**:14-16.

Duke, J. A. (1981). *Handbook of legumes of worldwide economic importance.* Plenum, New York.

Earle, F. R., and Q. Jones. (1962). "Analyses of seed samples from 113 plant families," *Econ. Bot.* **16(4)**:221–50.

Finley, K. (1988). "Patterns of soil moisture and nutrient depletion in plots of *Desmanthus illinoeinsis, Leymus racemosus*, and *Tripsacum dactyloides*," *Land Institute Res. Rep.* **5**:6–9.

Fisher, R. A., and C. Turner. (1978). "Plant production in arid and semiarid zones," *Ann. Rev. Plant Phys.* **29**:277–317.

Francis, C. A. (ed.) (1986). *Multiple Cropping Systems.* Macmillan, New York.

Gauthier, C., and J. Piper. (1988). "Vegetation patterns of native prairie," *Land Report Res. Rep.* **5**:1–5.

Gernes, M. (1986). "Relationship among cultural practices, stem density, and yield components in *Helianthus maximilianii* Schrad," *Land Report Res. Suppl.* **3**:20–22.

Gernes, M., and J. Piper. (1987). "Vegetation patterns in tallgrass prairie and their implications for sustainable agriculture," *Land Report Res. Rep.* **4**:1–8.

Gliessman, S. R. (1983). "Allelopathic interactions in crop-weed mixtures: Applications for weed management," *Ecol.* **9**:991–99.

Gliessman, S. R. (1984). "An agroecological approach to sustainable agriculture." In *Meeting the Expectations of the Land* (Eds. W. Jackson, W. Berry, B. Colman). pp. 160–71. North Point Press, Berkeley, California.

Great Plains Flora Association. (1986). *The Great Plains Flora.* Griffin, L. C., and R. M. Rowlett. (1981). "A lost Viking cereal grain," *Ethnobiol.* **1**:200–207.

Griffin, L. C., and R. M. Rowlett. (1981). "A lost Viking cereal grain," *Journal of Ethno. Botany.* **1**:200–207.

Gursky, M. (1984). "The performance and potential of *Desmanthus illinoensis* accessions," *Land Report Res. Suppl.* **1**:3–5.

Handley, M. (1986). "Survey of plant diseases in field plots at the Land Institute," Land Report Res. Suppl. **3**:36–38.

Handley, M. K., and P. Boehner. (1987). "Variability in resistance to anthracnose and leaf rust in *Tripsacum dactyloides* crosses segregation for sex ratio," *Land Report Res. Suppl.* **4**:30–35.

Heiser, C. B., and D. M. Smith. (1964). "Species crosses in *Helianthus*: II. Polyploid species," *Rhodora* **66**: 344–58.

Hickman, J. C., and L. Pitelka. (1975). "Dry weight indicates energy allocation in ecological strategy analysis of plants," *Oecologia* **21**:117–121.

Hitchcock, A. S. (1935). *Manual of the Grasses of the U. S.* USDA Misc. Pub. 200.

Jackson, W. (1985). *New Roots for Agriculture.* Univ. of Nebraska Press, Lincoln.

Jackson, W., W. Berry, and B. Colman. (1987). *Meeting the Expectations of the Land.* North Point Press, Berkeley, California.

Johnson, D. R., and D. J. Major. (1979). "Harvest index of soybean as affected by planting date and maturity rating," *Agron. J,* **71**:538–40.

Jones, D. G., and B. C. Clifford. (1978). *Cereal Diseases*, 2nd ed. John Wiley & Sons, Chichester, England.

Jones, Q., and E. R. Earle. (1966). "Chemical analyses of seed II: Oil and protein content of 759 species," *Econ. Bot.* **20(2)**:127–55.

Kendall, B. (1987). "Growth and seed yield within a polyculture of *Leymus racemosus, Tripsacum dactyloides,* and *Desmanthus illinoensis,*" *Land Report Res. Suppl.* **4**:36–40.

Klages, K.W.H., and R. H. Stark. (1949). "Grasses and grass seed production." University of Idaho *Agricultural Experiment Station Bulletin* No. 273.

Klebesadel, L. J. (1985). "Beach wildrye characteristics and uses of a native Alaskan grass of uniquely coastal distribution," *Agroborealis* **2**:31–38.

Kois, J. (1985). "Effect of density on seed yield and vegetative spread in first-year Maximilian sunflowers (*Helianthus maximiliani*)," *Land Report Res. Suppl.* **2**:1–4.

Komarov, V. L. (1934). "Flora of the USSR. Vol. II: Gramineae Botanical Inst. Academy of Sciences, USSR, Leningrad (Translated from Russian, 1963. " Office of Technical Services, US Dept. Commerce, Washington, DC.

Kulakow, P. A., L. L. Benson, and J. G. Vail. "Prospects for domesticating Illinois bundle flower," Proceedings of the First National Symposium for New Crops, Timber Press, Portland, Oregon. (in press).

LaLiberte, C. (1985). "Effect of Maximilian sunflower," *Land Report Res. Suppl.* **2**: 31–32.

Latting, J. (1961). "The biology of *Desmanthus illinoensis*," *Ecology* **42**:487–93.

Leonard, L. T. (1925). "Lack of nodule formation in a subfamily of the Leguminosae," *Soil Sci.* **20**:165–67.

Liebovitz, R. (1987). "Third year seed yield in plots of *Desmanthus illinoensis* and *Helianthus maximilianii*," *Land Report Res. Rep.* **4**:17–19.

Long, R. W. (1959). "Natural and artificial hybrids of *Helianthus maximiliani* X *H. grosseserratus*," *Amer. Bot.* **46**:687–92.

Lovett, J. V. (1982). "The effects of allelochemicals on crop growth and development." In *Chemical manipulation of crop growth and development* (Ed. J. S. McLoren). pp. 93–110. Butterworth Scientific, London.

McDaniel, M. E., and D. J. Dunphy. (1978). "Differential iron chlorosis of oat cultivars," *Crop Sci.* **18**:136–38.

Mecko-Ray, V. (1987). "Evaluation of *Desmanthus illinoensis* germplasm," *Land Report Res. Rep.* **4**:24–28.

Melderis, A., C. J. Humphries, T. G. Tutin, and S. A. Heathcote. (1980). "Tribe Triticeae Dumort." In: *Flora europaea, Vol. 5*, (Eds. T. G. Tutin, V. H. Heywood, N. A. Burges, D. M. Moore, D. H. Valentine, S. M. Walters, and D. A. Webb). pp. 190–200. Cambridge University Press, Cambridge, England.

Piper, J. (1986). "The prairie as a model for sustainable agriculture: A preliminary study," *Land Report Res. Suppl.* **3**:1–4.

Piper, J. (1987). "Population structure and seed yield within a three-year-old stand of *Cassia marilandica*," *Land Report Res. Suppl.* **4**:21–23.

Piper, J., and D. Towne. (1988). "Multiple year patterns of seed yield in five herbaceous perennials," *Land Report Res. Supp.* **5**:14–18.

Piper, J., J. Henson, M. Bruns, and M. Bender. (1988). "Seed yield and quality comparison of herbaceous perennials and annual crops." In *Global perspectives in agroecology and sustainable agricultural systems* (Ed. P. Allen and D. van Dusen). 715–719. Univ. of Calif., Santa Cruz.

Price, D. (1984). "The effect of *Helianthus maximiliani* on naturally occurring weeds," *Land Report Res. Suppl.* **1**:1–2.

Price, D., and M. Gursky. (1984). "Overyielding in a perennial legume biculture," *Land Report Res. Suppl.* **1**:10–12.

Putnam, A. T., and W. B. Duke. (1974). "Biological suppression of weeds: Evidence for allelopathy in accessions of cucumbers," *Science* **185**:370–72.

Rice, E. (1984). *Allelopathy*. Academic Press, New York.

Riley, K. (1984). "*Helianthus maximiliani* density trial," *Land Report Res. Suppl.* **1**: 6–8.

Rinehart, D. (1986). "Relationships among basal stem diameter, biomass, and yield," *Land Report Res. Suppl.* **3**:11–13.

Singh, I. D., and N. C. Stoskopf. (1971). "Harvest index in cereals," *Agron. J.* **63**:224–26.

Slater, M. (1986). "Survey of insects in native prairie and research plots," *Land Report Res. Suppl.* **3**:4–8.

Smith, M., and C. Francis. (1986). "Breeding for multiple cropping systems." In *Multiple Cropping Systems* (Ed. C. A. Francis). pp. 219–47. Macmillan, New York.

Steyermark, J. A. (1963). *Flora of Missouri*. Iowa State University Press, Ames.

Stiefel, V. (1985). "Yields of four and three year old wild senna (*Senna marilandica*)," *Land Report Res. Suppl.* **2**:29–32.

Taparia, A. L., T. W. Talmale, and V. V. Sharma. (1978). "Utilization of *Cassia tora* seeds in rations of buffalo calves," *Indian J. Ani. Sci.* **48**:804–10.

Thompson, K., and A.J.A. Stewart. (1981). "The measurement and meaning of reproductive allocation in plants." *Amer. Nat.* **117**:205–211.

Thornberg, A. (1971). "Grass and legume seed production in Montana," *Agricultural Experiment Station Bulletin* 333, Montana State University, Bozeman.

Tsitsin, N. V. (1978). *Remote hybridization in plants*. Nauca, Moscow, U.S.S.R.

USDA. (1948). *Grasses, Yearbook of Agriculture,* U.S. Government Printing Office, Washington, DC. 552 pp.

USDA. (1966). *Technical note 12/1/66*. Soil Conservation Service, Plant Materials Center, Bridger, Montana.

USDA. (1976). *Technical Report 1975-6*. Soil Conservation Service, Plant Materials Center, Coffeeville, Mississippi.

USDA. (1978). *Annual Report*. Soil Conservation Service, Plant Materials Center, Knox City, Texas.

USDA. (1980). *Agricultural Outlook*, Economics, Statistics, and Cooperative Service. June.

Wardlaw, I. F. (1980). "Translocation and source sink relationships," In Peter S. Carlson (ed.), *The biology of crop production*. Academic Press, New York. pp. 297–339.

Weaver, S. E., and P. B. Cavers. (1980). "Reproductive effort of two perennial weed species in different habitats," *J. Appl. Ecol.* **17**:505–13.

Weiner, M. A. (1980). *Earth medicine—Earth food*. Macmillan, New York.

Wenger, L. E., D. R. Cornelius, J. E. Smith, and A. D. Stoesz. (1943). "Methods of harvesting, storing, and processing native grass seeds." Subcommittee reports of the Southern Great Plains Revegetation Committee. March.

Wheeler, W. (1957). *Grassland Seeds*. Princeton University Press, Princeton. 434 pp.

Whyte, R. O. (1957). *Grasses in Agriculture*. Food and Agriculture Organization, Rome. 141 pp.

Williams, R. J. (1983). "Tropical legumes." In *Genetic resources of forage crops* (Eds. J. G. Melvan and R. A. Bray). CSIRO, East Melbourne, Australia.

15 MAJOR ISSUES CONFRONTING SUSTAINABLE AGRICULTURE*

WILLIAM LOCKERETZ
School of Nutrition,
Tufts University,
Medford, Massachusetts

EVOLUTION OF THE CONCEPT OF SUSTAINABLE AGRICULTURE

Although the term *sustainable agriculture* has come into common use in recent years, it still is only vaguely defined. In part, this may be because it derives from several sources, some recent, some going back several decades. Also, its connotations have been shifting as the term has been adopted by a wider range of people and applied to a wider range of agricultural systems. Its oldest precursor, to which the modern concept still owes a clear debt, is "organic farming". This term came into use sometime before World War II, although its originator is not known (Robert Rodale, private communication). Initially, organic farming emphasized recycling of farm-generated nutrient sources and discouraged bringing in nutrients in the form of livestock feeds and chemically processed fertilizers (Scofield, 1986). Today the term particularly emphasizes avoidance of synthetic pesticides.

This approach has evolved recently into a more general concept, variously called sustainable agriculture or any of several related terms (as discussed later). This evolution has come about as a broader range of people have become interested in the same goals for diverse reasons. The depressed farm economy has caused a shift in emphasis from "maximum

* This chapter, is reprinted in modified form from the article "Open Questions in Sustainable Agriculture" in *American Journal of Alternative Agriculture* **3(4)**: 174–181, 1988. Used with permission.

production" to "optimal production," meaning that a system should be evaluated not only by how much it produces, but also by the relative value of what comes out compared to what goes in. This has led to the concept of low-input agriculture, a reversal of a long-standing trend toward greater production through increased use of purchased inputs. In the United States, we tend to think of this concept as applying only to cultivated crops. However, it is equally applicable to pastures and forages, especially in Western Europe, for example, where conventional grassland management often involves a high level of inputs (Wagstaff, 1987).

Often, the purchased inputs that farmers can cut back on to save money also are environmentally damaging or hazardous to health. Thus, the trend toward low-input agriculture can help meet other important goals that have become more prominent in recent years, such as avoiding further contamination of groundwater by pesticides (Hallberg, 1986) and meeting consumers' growing demand for healthful, uncontaminated foods (Clancy, 1986). Farm families, who live closest to these problems, are especially concerned about nitrate in the water supply, pesticide residues in food, and other hazards to their health and safety.

These issues have led to much greater interest in sustainable agriculture, but even its basic principles remain to be worked out, not just specific questions of technique. Although the term is singular in form, *sustainable agriculture* really denotes a multidimensional concept, and it is not surprising that no single view of it has yet gained universal acceptance.

Therefore, it is important to keep in mind the many fundamental questions that still need to be discussed, analyzed, and debated before we can agree as to what *sustainable agriculture* really means. Some of these questions may never be answered, but sometimes they aren't even asked, or the answers are asserted rather than demonstrated. Until they are discussed explicitly, we cannot know whether we are using the same words to mean different things, whether a particular approach can simultaneously serve the several different goals we would like it to achieve, or even whether the various goals that have been thrown together under this single umbrella term might not have intrinsic contradictions. Nor will we know the best economic, institutional, and political environment in which the potential of sustainable agriculture can be fulfilled.

FUNDAMENTAL QUESTIONS

What Are the Differences between Sustainable, Alternative, Low-input, Ecological, Organic, and Regenerative Agriculture? In the past decade all these labels, plus several less common ones, have come into use for agricultural systems that share basic goals: reduced use of purchased inputs, especially toxic or nonrenewable ones; less damage to the environment;

and better protection of water, soil, and wildlife. But both the technical and popular literature are unclear about their exact connotations.

Sometimes they appear to be used synonymously, or almost so (Merrill, 1983). For example, a recent article published by the U.S. Department of Agriculture's Economic Research Service (1988) began by talking about " 'alternative agriculture', also known as 'sustainable, regenerative, organic', or 'low-input' agriculture." Other authors emphasize the differences among these terms. If defined according to everyday usage, these terms would have clearly differentiated meanings:

- *Sustainable* has a time dimension and implies the ability to endure indefinitely, perhaps with appropriate evolution.
- *Alternative* describes something that is different from the prevailing, or "conventional," situation. Another implication is that farmers should be able to choose among alternatives, or options, rather than having to follow a single prescription.
- *Low-input*, or *low external input*, to use its more precise form, means reduced use of materials from outside.
- *Ecological* refers to the principles and processes that govern the natural environment.
- *Regenerative* implies a continuing ability to re-create the resources that the system requires.
- *Organic* means like an organism, that is to say, a whole consisting of coherently functioning parts. (Historically, this meaning is older than the later application of the word *organic* to a particular class of chemicals. The older sense was intended in the term *organic farming* as well. Unfortunately, the multiple meanings of the word *organic* have resulted in pointless quibbling among organic farming's critics; it is irrelevant that the synthetic pesticides avoided in organic farming are, nevertheless, organic compounds.)

Regardless of their literal meanings, what about the current usage of these terms as actually applied to agriculture? All have acquired further connotations, and some have become almost like brand names. Thus *alternative agriculture* does not mean simply "a different kind of agriculture." As the term is now used, it must be different in a certain way. But in precisely what way? By being more sustainable? Ideally yes, but the way these terms are now used, it is not clear. For even though they carry specific implications when used in agriculture, terms like *sustainable* paradoxically have also come to be used in so many different ways by various groups that they seem in danger of becoming ambiguous. By now, *sustainable* may mean little more than *good*. If you like a particular practice, it's sustainable; you can criticize something you don't like simply by calling it nonsustainable.

What can we conclude from this semantic confusion, these ambiguities, this proliferation of labels that may all mean the same thing? Typically, authors writing about sustainable agriculture consider it necessary to start their articles by giving their definition of the word *sustainability*. Isn't something backwards here? It seems curious that a term is considered important even though we don't know what it means. Perhaps, as Frederick Buttel has suggested, sustainable agriculture "remains a solution in search of problems" (private communication).

Less pessimistically, there are three possibilities, of which two are mainly semantic problems. The first is that several fundamentally different concepts are involved, but authors are not always careful about choosing the most appropriate term. Thus, we really do need several labels, but imprecise usage has created the impression that these labels are more or less interchangeable.

Conversely, a second possibility is that, when applied to agriculture, all these words are referring to the same basic concept; they may have proliferated because each was put forth by someone who believed that the previous terms did not capture that concept accurately, or who for some other reason chose to shy away from an earlier term. For example, *organic* apparently is thought by some people to suffer from a negative image. That may be why an article that quoted verbatim the U.S. Department of Agriculture's oft-cited definition of organic farming never actually used the term, but instead claimed that "the U.S. Department of Agriculture defines *alternative farming* as follows" (Popkin, 1988 [emphasis added]).

A third possibility is more substantive: whether these terms are interchangeable might depend on the context, such as whether one is talking about broad principles or specific practices. Perhaps the basic concepts implied by these various terms are different. But when it comes to illustrating them by specific systems, everyone seems to end up using the same examples. For example, the first study of commercial organic farming in the United States (Lockeretz et al., 1981) only used the term "organic", and required only that the study farms not use synthetic pesticides and inorganic fertilizers. However, this same study has since been cited as dealing with "low-input," "sustainable," and "ecological" agriculture.

This is fine as long as a given production system indeed embodies several of these general concepts. But the confused semantic situation could reflect a more serious problem: people may assume, rather than demonstrate, that a given agricultural system can simultaneously fulfill several different goals and embody several different agricultural ideals. Actually, it is necessary to demonstrate that it even achieves its stated primary goal. Does calling a system "sustainable" guarantee that it is sustainable, that is, capable of enduring? And if it is "sustainable," is it necessarily "alternative"—meaning it is different from prevailing practices? Is an agricultural system that does not use so many inputs necessarily "ecological," that is, more like a natural ecosystem? In much of the literature, there is an implicit assumption that when one strives for any of these goals, the others somehow come along automatically. There is little recognition that these goals are substantially distinct and independent, so that each has to be achieved in its own right. And rarely is it acknowledged that in designing specific production systems, rather than talking about abstract concepts, there can be incompatibilities that actually make it impossible to achieve them all (Lockeretz, 1986).

Is Sustainable Agriculture Primarily a Matter of Reducing Use of Certain Inputs, Use of All Inputs, or Taking Positive Steps to Make Some Inputs Unnecessary? The common identification of "low-input" with "sustainable" agriculture raises some important questions. Organic farming avoids or greatly reduces the use of two important categories of purchased inputs, namely, synthetic pesticides and highly soluble inorganic fertilizers. The reasons for this involve soil productivity, the environment, and other biological and chemical considerations. An additional consequence is that organic farmers may have lower cash operating expenses (Lockeretz et al., 1981).

More recently, the idea of eliminating certain inputs regarded as particularly objectionable has been extended to reducing all kinds of purchased inputs, to the extent feasible. This extension has been motivated by the depressed economic conditions affecting much of American agriculture, which could be eased by reducing operating expenses. Still another school of thought, more related to the original idea of organic farming mentioned earlier, is that a farm should strive for self-sufficiency for its own sake, not just because a particular input is either damaging or expensive. In this view, the mere fact of being tied to another region or another sector of the economy can introduce risks or other problems. Interestingly, Rodale (1988) extends the distinction between a farm's internal and external resources—a key element in the concept of regenerative agriculture, of which he is the leading exponent—even to management decisions, although most authors confine themselves to physical and biological inputs.

Another view of the question of whether reducing inputs is intrinsically desirable is that sustainable agricultural systems should be based on positive steps that enhance soil fertility, control pests, and perform the many other functions that in conventional systems are largely performed by purchased inputs. Merely doing without such inputs is not the goal; the idea is to develop a system in which they would not be needed anyway. This concept emphasizes nutrient cycling, natural pest controls, diversity, continuous protection of the soil by living crops or residues, and wholesome housing and rations for livestock, as described in several previous chapters.

Interesting in this regard is the current use of the word "organic", once regarded as the narrowest and most proscriptive (in other words, "thou shalt not") of the terms under discussion here. The International Federation of Organic Agriculture Movements (1988), the only organization with sufficient status to presume to speak authoritatively on such matters internationally, begins its proposed technical standards for organic production with eight positive goals, ranging from high nutritional quality to maintenance of genetic diversity, to a decent return for farm workers. The standards go on to say that if techniques are chosen that will approach these goals, they will "make it possible to avoid . . . chemical fertilizers, pesticides, and other chemicals."

The differences among these attitudes toward reducing inputs cannot be dismissed as academic nit-picking. Organic farmers in the narrower sense of the term—those who do not use certain forbidden materials—do not automatically try to cut down on their use of all types of inputs (Wernick and Lockeretz, 1977). In fact, they may compensate by increasing their use of certain other inputs. For example, controlling weeds mechanically instead of with herbicides requires additional implements and fuel, as does applying manure rather than concentrated fertilizers. Some organic farmers buy organic fertilizer materials, which usually are more expensive than the same nutrients in inorganic form (Wernick and Lockeretz, 1977). If nonchemical pest control requires hiring professional scouts, then one expense (a service) is being substituted for another (a product or material). By environmental criteria, scouting is certainly preferable, but economically it could go either way.

In response to economic pressures, some farmers have been cutting back on pesticides and fertilizers, and perhaps other inputs as well, without doing anything else differently. According to one view, this qualifies as "sustainable." For example, the Wisconsin Rural Development Center (1988) puts the matter very simply: "Sustainable methods are those that use less commercial fertilizer, herbicide, and pesticide." In some systems, such as dryland wheat in the Great Plains before the introduction of herbicide-based chemical fallowing (important for erosion control), pesticides and fertilizers were never very important. Was this system sustainable? The Dust Bowl shows that it wasn't (Lockeretz, 1978). Was it organic? It would have come close to qualifying for certification as organic by several U.S. organizations (but not according to the proposed IFOAM standards), since certification programs generally emphasize what must not be done, and only secondarily mention what should be done. Was it the best possible system from the viewpoints of environment and resources? No. Other systems could have done more to enhance soil productivity, reduce erosion, preserve wildlife habitat, and increase economic returns, while continuing to avoid the materials considered undesirable. And this monocultural system certainly would not qualify as "ecological," since it hardly reflected the complex ecological structure of the shortgrass prairie it replaced.

Another interesting contradiction arises if a farm changes its fertilization system only because it happens to be near a source of organic wastes that can be substituted for purchased inorganic fertilizer. Such a farm fulfills one requirement of sustainability, in that it avoids using a material that is nonrenewable and potentially damaging to the environment. (Of course, the wastes must be applied in an environmentally suitable manner, or they will create just as much of a problem.) However, it is not a low-input system; only the form of the input is changed. Therefore, it is as vulnerable to external disruption as a system dependent on inputs made from nonrenewable resources. One source of vulnerability has merely been replaced by another; for example, the source of the waste may close. This point has been made by Vail and Rozyne (1982), who found that "organic" farmers near poultry plants in Maine, where the industry has been declining sharply, did not take positive steps to build soil fertility, but merely exploited the fortuitous availability of poultry manure as long as they could get it, to save some money on fertilizer expenses.

The relation between sustainability and some versions of low-input agriculture is also critical for understanding how important short-term economic factors have been in helping "sustainability" gain acceptance. I therefore take it up again in the next section.

Does Sustainable Agriculture Require Fundamental Changes in Either the Economic and Institutional Environment or Farmers' Motivations and Values? This question provokes widespread disagreement. Some authors confine their discussions of sustainable agriculture to agronomic, environmental, and biological factors, or to economic evaluations assuming prevailing conditions. They believe that the same farmers, operating the same farms, can switch production systems without a significant change in attitude or in the economic, political, and social setting in which farming occurs.

In other discussions, far-reaching socioeconomic transformations are emphasized even more than the technical differences between sustainable and conventional farming methods. These transformations may include any of the following: reduced linkage between farming and the industrial economy; more direct ties between producers and consumers; greater regional food self-sufficiency; a preference for family rather than corporate farms; higher employment in agriculture; equitable distribution of economic returns among different classes of farmers and between present and future generations; and the social and economic revitalization of rural communities (Douglass, 1984; Crosson, 1986).

The connection, if any, between such transformations and changes in specific practices can go in either direction. That is, a different socioeconomic environment could be a prerequisite for widespread adoption of sustainable methods, or it could be a consequence of this adoption. For example, intensive use of chemical pest control is sometimes said by advocates of alternative agriculture to result from the domination of farming by agrichemical interests, who are described as exerting a strong influence on farmers' decisions and on research priorities. Therefore, this domination must be reduced before farmers will be receptive to alternatives. On the other hand, if farmers decide to switch to reduced-chemical methods because they perceive problems with the particular agrichemicals now in use, such as high cost, fear of liability, or threats to their health, this switch will reduce the role of the agrichemical industry. Whatever the reason farmers made the change, some people would consider the reduced industry influence as desirable in its own right, apart from the undesirable properties of pesticides.

Another important structural issue that is still open concerns the most appropriate scale for sustainable agriculture. Its supporters often consider that it is most suited for small to moderate-sized family farms. However, the empirical support for this view is largely lacking, and the theoretical arguments are equivocal (Buttel et al., 1986). Certainly, the trend towards larger farms has been associated historically with specialization, whereas sustainable agriculture favors diversification over specialization. Also, sustainable agriculture may require greater attention to management, as discussed below. If so, the farmer can give more attention to each field or each animal if the farm is not too large. On the other hand, larger farms may be better able to afford to hire specialized expertise or to have better facilities and equipment.

The relation between the production system and the farmer's personal values is also unclear. Some authors see little connection. Today's farmers can, if they choose, adopt sustainable methods without any rethinking of motivations, values, or broader philosophical considerations. Moreover, they will do so, it is argued, if the alternative is more attractive from the viewpoints of economic return or health and safety.

But to others, the reason a farmer farms is an overriding consideration. In this view, farmers should be concerned not just with short-term profits, but also with the well-being of future generations, the rural communities in which they live and work, the natural environment, the aesthetic appeal of the landscape, and the resources consumed in farming (Fischer, 1978; Bidwell, 1986). Such considerations are sometimes regarded as the distinguishing characteristic of sustainable agriculture (Freudenberger, 1986); the specific choices of production methods, in turn, follow from these goals.

As with the previous discussion of the institutional and economic environment, change can be started from either direction. If farmers can be persuaded to be more concerned about environmental values, they will adopt environmentally sounder methods. But if they can be persuaded to adopt these methods, for economic or other reasons, the result will be a system that better protects the environment, even if that is not why they chose to make the change.

A possible objection to this last point is that changes made purely for economic reasons can be transient, given the variability of economic conditions. If farmers reduce pesticide use because crop prices have been too low to justify the cost, this environmentally beneficial change could be undone by the next sharp increase in crop prices, just as some good soil conservation work was undone by the exceptionally rapid price increases from 1972 to 1974 that led farmers to include even marginal land when planting "fencerow to fencerow" (Grant, 1975). The growing interest in low-input agriculture has been closely tied to the distress in the farm economy during the mid-1980s. This points to a key difference between low-input and sustainable approaches: even though both may have the short-term effect of reducing pesticide use, the latter does so for less ephemeral reasons than the temporary diseconomy of applying pesticides heavily when crop prices are low. The distinction between the two concepts appears to be eliminated when they are merged under the singular term "lower-input/sustainable agriculture." This term was proposed by Edwards (1987) for systems intended to solve problems of both overproduction and environmental degradation. (Unlike some authors, Edwards assumed that reduced use of inputs would lower total output.) But trying to solve two problems for the price of one raises another question: what happens when the problem becomes scarcity, not surpluses? Must we give up sustainability when we need to produce more food?

Similarly, the term "low-input/sustainable agriculture" has been adopted for a new program of research grants administered by the U.S. Department of Agriculture, described in an earlier chapter by Madden. It is not clear what is meant by joining the two components of this term. Are they to be regarded as equivalent, with one simply a different way of saying the other? Or is the composite term intended to cover any system that has at least one of the two characteristics? Or does it cover only those systems that have both characteristics (which may be only a subset of the previous category)?

It is difficult, but essential, to resolve the relationship between sustainability and the low-input approaches that have been adopted in direct response to current economic conditions. Although proponents of sustainable agriculture may prefer to emphasize longer-range considerations, short-range economic factors cannot be ignored: if a system doesn't return enough income to let the farmer remain in business, it isn't sustainable (Madden, 1987). The solution may lie in the earlier discussion of whether sustainable agriculture is primarily a matter of merely doing without certain materials, or of doing positive things that make these materials unnecessary. Ideally, with appropriate crop rotations, crop varieties, and tillage methods, the farmer who, for environmental reasons, doesn't want to use pesticides won't have to and won't be tempted even when crop prices rise again.

Does Understanding Sustainable Agriculture Involve Concepts That Are Fundamentally Different from Conventional Systems, or Do We Only Need to Extend the Application of Known Principles to the Conditions That Prevail under Sustainable Practices? The literature on sustainable agriculture offers various views regarding qualitative differences between the basic processes underlying sustainable and conventional production systems. One view is that both kinds of systems can be described and analyzed using the same concepts and that they differ only in the specific conditions created by differing practices. Another view is that the complex interactions among the components of a sustainable farming system cause new phenomena to emerge that are not observed in the more simplified structure of a conventional system, so that one cannot understand the former by merely extrapolating from the latter.

Each of these views has some validity, and the difference is often one of emphasis. An example is whether control of insect pests is fundamentally different in the two approaches. Sustainable control is generally described as using natural processes, such as predation on pests by other insects, or resistant varieties and hybrids. Conventional control, in contrast, is sometimes derided by sustainable agriculture advocates as a "magical bullet" approach (Hill, 1982) depending entirely on insecticides, which not only substitute for natural processes, but may even interfere with them. For example, an insecticide may wipe out a beneficial species that previously had controlled a pest.

Where this does happen, pest control differs in a basic way between the two systems: in one, it is a matter of toxicology, in the other, an ecological phenomenon. No matter how much one studied the pest's mortality at different pesticide application rates, this knowledge would not be enough to predict the survival rate at zero application, where a different mechanism (predation) comes into play.

But sometimes the pesticide adds another control mechanism to the existing natural controls, rather than replacing them. If so, then the two approaches partially overlap. Even under "conventional" practice, many potential pests are actually controlled naturally—without the farmer having to do anything—rather than by pesticides. Many elements of conventional pest control are the same as in sustainable practices: for example, resistant varieties, good residue management, and best choice of planting date. Therefore, it is not correct to describe sustainable and conventional practice as governed by fundamentally different principles and mechanisms—"natural" as opposed to "chemical"—despite the different relative importance of these mechanisms in the two approaches.

This question has important implications for agricultural research policy. Perhaps the established methods of agricultural research are appropriate for dealing with sustainable agriculture; all that might be needed is for more attention to be devoted to the particular techniques and conditions of sustainable agriculture. In contrast, some people believe that new approaches and methods are required, involving far-reaching differences in basic theories and concepts. The growing field of agroecology, as applied to sustainable agriculture, is an attempt to introduce new theoretical principles to analyze agricultural systems (new, that is, in the domain of agriculture, although well established regarding natural ecosystems).

This field is relatively immature, however, and basic agroecological research is not yet receiving very strong support at most colleges of agriculture. Thus, it has yet to fulfill the claims that its supporters have made. In the meantime, a considerable portion of the work labelled as sustainable agriculture research differs from earlier kinds of research that it is supposed to improve upon, mainly in that it generously uses terms like "systems approach."

To What Extent Do the Resource-conserving and Environmentally Sounder Techniques Being Developed at Mainstream Agricultural Institutions Already Represent Sustainable Agriculture? Two related points were discussed earlier: whether acceptance of sustainable agriculture requires far-reaching economic or attitudinal changes, and whether understanding it involves a fundamentally different scientific outlook. These points, in turn, raise the question of whether existing research and teaching institutions can deal with this area adequately. Long before the term *sustainable agriculture* came into common use, techniques with similar goals were already attracting attention. Examples include genetic resistance and integrated pest management to reduce pesticide use, improved methods for storing and applying livestock manures to maximize their fertilizer value and reduce water pollution, and reduced tillage systems to control soil erosion. More recently, many agricultural institutions have started programs specifically labelled sustainable agriculture. Such programs explicitly acknowledge the influence of the sustainable agriculture movement and show mainstream agriculture's interest in accommodating its ideas.

However, some people interested in sustainable agriculture—both within mainstream institutions and on the outside—do not view with undiluted optimism the changes that have already occurred in mainstream research, teaching, and extension. To some, the established research institutions are under very powerful constraints, especially the constraints imposed by disciplinary boundaries and by researchers' need to publish frequently. This, in turn, may discourage long-term projects, such as studies covering several cycles of a many-year rotation. These constraints may make it difficult or impossible for established institutions to organize agricultural research appropriately for dealing with sustainable systems (Bidwell, 1986). Also, to those who believe that sustainable agriculture involves fundamentally different principles, older ways of thinking are too firmly established among the current generation of researchers to permit newer ideas to flourish. Therefore, although such people may welcome the new interest in sustainable agriculture, they do not expect that the change will be able to go far enough in the current institutional environment. Buttel and Gillespie (1988) expect that the low-input work that mainstream research institutions are showing interest in will be done mainly by the same researchers—now "born again" agronomists—whose limited vision created the need for new approaches.

Some critics of current agricultural research believe that its limits are more fundamental and that they would not be eliminated even if appropriate institutional arrangements were provided and a new generation of researchers took over. This view considers that the problem is nothing less than the way scientists think about how the universe works. Our thinking, it is said, is too "mechanistic" or "reductionist" to encompass the complex relations of a sustainable agricultural system, a task for which a new scientific paradigm is needed (Cobb, 1984). However, this criticism may be making the mistake of attacking all science, rather than bad science. The most dramatic and exciting breakthroughs in the history of science include many that are profoundly "antireductionist": quantum mechanics, evolution, and general relativity, for example. To believe that one has a better way of understanding how the world works amounts to placing oneself above Bohr, Darwin, and Einstein.

Finally, certain critics—again, some of whom are at mainstream agricultural institutions—have argued that such institutions not only have failed to grasp the spirit of sustainable agriculture, but do not even want to. The flurry of recently instituted programs is said to be merely a way to gain more research funds, or to appear to be responding to outside pressures, and perhaps also to blunt the full thrust of the movement. In this view, advancing the cause of sustainable agriculture means challenging some far-reaching economic, social, or political constraints, a challenge that mainstream agricultural institutions are not likely to mount (Altieri, 1988). Similarly, Buttel and Gillespie (1988) argue that because a range of concepts have been advanced as "ecological agriculture," mainstream agriculture has been able to choose the least threatening version of it, sanitize it further to make it bureaucratically acceptable, and appropriate it as its own.

The relation between mainstream institutions and supporters of sustainable agriculture will undoubtedly clarify itself with time. The most discouraging outcome would be for the term "sustainable agriculture" to degenerate into just another bureaucratic buzz word used to show that something new and exciting is going on, even though nothing has really changed. But perhaps the mainstream will prove highly receptive to the ideas now labelled as alternative. If so, it might still be intrinsically advantageous to have someone whose role is to continue to challenge the established institutions and to keep prodding them to move further than institutional inertia would otherwise permit. On the other hand, if the receptiveness extends to the spirit of sustainable agriculture as a continuing quest, and not just to the specific details of systems being advocated today, then eventually it should be possible to drop the distinction between *alternative* and *conventional* as irrelevant and unnecessary.

Does Sustainable Agriculture Require a Higher Level of Management Ability among Farmers? Sustainable agriculture is commonly said to require more management than conventional practice, both in how much effort the farmer must expend and in the quality of management demanded. The explanation is that sustainable practices substitute knowledge and understanding for technological control of growing conditions (Stinner and House, 1987). For example, in conventional practice, a disease of livestock might be completely controllable by routine prophylactic administration of an antibiotic, whereas in sustainable practice the goal is to prevent the disease by reducing stresses that make the animals more susceptible and by housing them in conditions less conducive to communicating the disease. The farmer therefore must monitor the animals carefully to be able to start treatment promptly should it become necessary.

Another source of greater management requirements is said to be the need to make decisions on an integrated, whole-farm basis, in contrast to the more compartmentalized approach possible in conventional systems. Also, sustainable management is often depicted as the adaptation of general principles to the specific circumstances of the individual farm, whereas conventional practice is sometimes characterized as a "cookbook" approach—a set of prescriptions that can be applied anywhere, without much understanding of the agroecosystem (Friend, 1983; Ehrenfeld, 1987). The sustainable agriculture literature emphasizes the need for flexibility, and stresses that there is no one best method under all circumstances.

Although these arguments are persuasive, the question of comparative management requirements is still open. Certainly, sustainable agriculture will require farmers to acquire different kinds of knowledge and skills, but this does not mean it is necessarily more difficult. Some proponents of sustainable agriculture may exaggerate how much purchased inputs obviate the need for judgment in conventional practice, and may not take due account of the many decisions that still must be made even if one uses inorganic fertilizers, synthetic pesticides, and livestock antibiotics. For example, specialization, expansion, and heavy use of inputs in conventional approaches may require a *greater* management ability in handling financial and marketing affairs. Also, the stereotype of "cookbook" farming may indeed apply to poorer conventional managers, but not to the more discerning ones who bring both new knowledge and past experience to their operations to supplement the use of chemicals. Also, some of the expertise required for sustainable practices can be hired—for example, professional services for integrated pest management.

Finally, even if sustainable agriculture imposes greater management difficulties now, this problem may be reduced after farmers have gained more experience and after more effort has been focused on these systems by established research, extension, and teaching institutions. The lack of reliable information sources has been a common complaint among organic farmers (Wernick and Lockeretz, 1977; Baker and Smith, 1987), and undoubtedly applies to other sustainable approaches as well. Eventually, the mystique surrounding the complexities of sustainable management may disappear, and what now may seem bewildering very likely can be made much more fathomable (Coleman, 1985). Farmers have been called on many times in the past to take on new management challenges, and it seems plausible that with appropriate support, the challenges of sustainable management can be met as well.

CONCLUSION: DOING IT RIGHT

Sustainable agriculture is not so much a new idea as a synthesis of ideas originating from various sources, out of various motivations. It is continuously being modified and refined to reflect changing economic pressures on farmers and the increasing concern over agriculturally related environmental problems. Although its roots go back much further, sustainable agriculture as an explicitly formulated concept is young compared to the time it will take to explore its ramifications and to understand fully its basic principles.

What can be done to advance this understanding? Certainly, more research and development is needed to apply sustainable agriculture concepts to specific situations and specific problems. However, unless we move considerably beyond merely doing some additional detailed empirical studies, it will amount to little more than a faddish but empty slogan. To fulfill its true potential, sustainable agriculture needs greater intellectual rigor. Important conceptual questions are not being asked, let alone answered. Fundamental principles still need to be developed and refined. Too much that needs to be demonstrated is instead simply asserted, or unconsciously assumed, or removed from debate by being made a matter of definition. People with a particular view of what sustainable agriculture is all about sometimes are not willing to acknowledge that other versions may be equally legitimate.

Supporters of sustainable agriculture often claim—with considerable justification—that mainstream agricultural thinking is too reluctant to challenge basic assumptions, too dogmatic, and too quick to become immersed in technical minutiae even though fundamental questions remain unaddressed. Ironically, sustainable agriculture runs the risk of repeating these mistakes. Fortunately, the field still has time to take heed of previous experiences. If it does, it will not only generate new solutions to the particular problems now affecting agriculture, but also set a new and better standard for thinking about agriculture in the future.

REFERENCES

Altieri, M. A. (1988). "Agroecology: A new research and development paradigm for world agriculture." International Symposium on Agricultural Ecology and Environment, University of Padova, Italy, April.

Baker, B. P., and D. P. Smith. (1987). "Self-identified research needs of New York organic farmers," *Am. J. Alt. Agric.* **2(3)**:107–13.

Bidwell, O. W. (1986). "Where do we stand on sustainable agriculture?" *J. Soil Water Conserv.* **41(5)**:317–20.

Buttel, F. H., G. W. Gillespie Jr., R. Janke, B. Caldwell, and M. Sarrantonio. (1986). "Reduced-input agricultural systems: rationale and prospects," *Am. J. Alt. Agric.* **1(2)**:58–64.

Buttel, F. H., and G. W. Gillespie Jr. (1988). "Agricultural research and development and the appropriation of progressive symbols: Some observations on the politics of ecological agriculture." Bulletin No. 151, Dept. of Rural Sociology, Cornell University, Ithaca, New York.

Clancy, K. L. (1986). "The role of sustainable agriculture in improving the safety and quality of the food supply," *Am. J. Alt. Agric.* **1(1)**:11–17.

Cobb, J. B. (1984). "Theology, perception, and agriculture." In *Agricultural Sustainability in a Changing World Order* (Ed. G. K. Douglass). pp. 205–17. Westview Press, Boulder, Colorado.

Coleman, E. (1985). "Towards a new McDonald's farm." In *Sustainable Agriculture and Integrated Farming Systems* (Eds. T. C. Edens, C. Fridgen, and S. L. Battenfield). pp. 50–55. Michigan State University Press, East Lansing.

Crosson, P. (1986). "Sustainable food production: Interactions among natural resources, technology, and institutions," *Food Policy* **11(2)**:143–56.

Douglass, G. K. (1984). "The meanings of agricultural sustainability." In *Agricultural Sustainability in a Changing World Order* (Ed. G. K. Douglass). pp. 1–29. Westview, Boulder, Colorado.

Edwards, C. A. (1987). "The concept of integrated systems in lower input/sustainable agriculture," *Am. J. Alt. Agric.* **2(4)**:148–52.

Ehrenfeld, D. (1987). "Sustainable agriculture and the challenge of place," *Am. J. Alt. Agric.* **2(4)**:184–87.

Fischer, C. (1978). "Introduction to the conference theme 'Towards a Sustainable Agriculture.'" In *Towards a Sustainable Agriculture* (Eds. J. M. Besson and H. Vogtmann). pp. 11–17. Verlag Wirz AG, Aarau, Switzerland.

Francis, C. A., and J. W. King. (1988). "Cropping systems based on farm-derived, renewable resources," *Agric. Systems* **27**:67–75.

Francis, C. A., R. R. Harwood, and J. F. Parr. (1986). "The potential for regenerative agriculture in the developing world," *Am. J. Alt. Agric.* **1(2)**:65–73.

Freudenberger, C. D. (1986). "Value and ethical dimensions of alternative agricultural approaches: In quest of a just and regenerative agriculture." In *New Directions for Agriculture and Agricultural Research* (Ed. K. Dahlberg). pp. 349-364. Rowman and Allenheld, Totowa, New Jersey.

Friend, G. (1983). The potential for a sustainable agriculture. In *Sustainable Food Systems* (Ed. D. Knorr). pp. 28–47. AVI, Westport, Connecticut.

Grant, K. E. (1975). "Erosion in 1973–74: The record and the challenge," *J. Soil Water Conserv.* **30(1)**:29–32.

Hallberg, G. R. (1986). "From hoes to herbicides: Agriculture and groundwater quality," *J. Soil Water Conserv.* **41(6)**:357–64.

Harwood, R. R. (1985). "The integration efficiencies of cropping systems." In *Sustainable Agriculture and Integrated Farming Systems* (Eds. T. C. Edens, C. Fridgen, and S. L. Battenfield). pp. 64–75. Michigan State University Press, East Lansing.

Hill, S. B. (1982). "Steps to a holistic ecological food system." In *Basic Technics in Ecological Farming* (Ed. S. Hill). pp. 15-21. Birkhauser Verlag, Basel, Switzerland.

Hodges, R. D. (1982). "Agriculture and horticulture: The need for a more biological approach," *B.A.H.* **1(1)**:1–13.

International Federation of Organic Agriculture Movements. (1988). "Technical Committee proposal for new IFOAM basic standards for organic agricultural production," *Internal Letter* **31**(June):19–31.

Lockeretz, W. (1978). "The lessons of the Dust Bowl," *Am. Scient.* **66(5)**:560–69.

Lockeretz, W. (1986). "Alternative agriculture." In *New Directions for Agriculture and Agricultural Research* (Ed. K. Dahlberg). pp. 291–311. Rowman and Allenheld, Totowa, New Jersey.

Lockeretz, W. (1988). "Open questions in sustainable agriculture," *Am. J. Alt. Agric.* **3(4)**:174–81.

Lockeretz, W., G. Shearer, and D. Kohl. (1981). "Organic farming in the Corn Belt," *Science* **211**:540–47.

Madden, P. (1987). "Can sustainable agriculture be profitable?" *Environ.* **29(4)**:18–20,28–34.

Merrill, M. C. (1983). "Eco-agriculture: a review of its history and philosophy." *B.A.H.* **1(3)**:181–210.

Popkin, R. (1988). "Alternative farming: a report." *EPA J.* **14(3)**:28–30.

Rodale, R. (1988). "Agricultural systems: The importance of sustainability," National Forum, *Phi Kappa Phi J.* (Summer):2–6.

Scofield, A. M. (1986). "Organic farming: The origin of the name," *B.A.H.* **4(1)**: 1–5.

Stinner, B. R., and G. J. House. (1987). "Role of ecology in lower-input, sustainable agriculture: An introduction," *Am. J. Alt. Agric.* **2(4)**:146–47.

USDA, Economic Research Service. (1988). "Alternative agriculture gains attention," *Agric. Outl.*, April, pp. 26–28.

Vail, D., and M. Rozyne. (1982). "Contradictions in organic soil management practices: Evidence from thirty-one farms in Maine." In *Basic Technics in Ecological Farming* (Ed. S. Hill). pp. 32–40. Birkhauser Verlag, Basel, Switzerland.

Wagstaff, H. (1987). "Husbandry methods and farm systems in industrialized countries which use lower levels of external inputs: A review," *Agric., Ecosystems, and Environ.* **19**:1–27.

Wernick, S., and W. Lockeretz. (1977). "Motivations and practices of organic farmers," *Compost Sci./Land Util.* **18(6)**:20–24.

Wisconsin Rural Development Center. (1988). "WRDC leads study on sustainable payoff," *Newsletter* **5(3)**:1.

16 FUTURE DIMENSIONS OF SUSTAINABLE AGRICULTURE

CHARLES A. FRANCIS
Department of Agronomy,
University of Nebraska,
Lincoln, Nebraska

What will determine the direction of agricultural systems in the future? We have traditionally thought that food preferences, crop adaptation, economic incentives, and international trade would be the primary influences on crops, animal species, and production systems. Evaluation of historical trends has been used to project what will happen in the future. In light of new information and growing awareness of resource constraints and environmental impacts of conventional systems, we are adjusting not only our thinking about the future but also our potential influence on its direction. There is no doubt that we will continue to place high priority in the world community on security of food production. Yet new directions need to be determined in part by values rather than trends, by the types of systems that will help us meet broad human and society objectives. Unlike the past, the future can be under our control. The principal question in agriculture is how to design systems to increase their productivity consistent with the long-term need for marshaling nonrenewable resources for possible future demands. This must also be done within a context of preserving a livable environment if we are to assure the survival of our species.

Within this frame of reference, the details of production systems presented in previous chapters can be brought together into recommendations for future approaches in agriculture. There are potential genetic modifications in every system; environmentally sound methods of tillage and seeding are essential, and plants need adequate soil fertility. Pest protection is critical to success; new systems designed to make more efficient use of internal, renewable production resources will be one key to future solutions. Not all the needed information and recommendations are on the shelf to implement these systems. A dynamic and well-oriented research program is essential, along with an extension effort that will bring farmers and ranchers into the process of developing, testing, and evaluating new information.

Education and communication will become increasingly important in making it possible to implement tomorrow's efficient agricultural production systems. Agricultural, environmental, and resource policies in each country as well as in a world context will be important to future design and application of new systems. Specifically, the environmental and health dimensions of agriculture will grow increasingly important. To some degree our future production systems can be designed to emulate some of the diverse strategies found in natural ecosystems. These factors are discussed as they relate to future systems and how we can make them sustainable.

How do we achieve a sustainable agriculture? As a useful point of departure, the CAST (1988) report defines a long-term viable agriculture "as one providing safe, abundant, and nutritious food supplies at a reasonable cost while preserving the environment and the beauty and wholesomeness of our rural heritage." Written after careful study by a balanced group of experts from universities, private industry, and the nonprofit sector, the report contends that viability has economic and environmental dimensions. Among other conclusions, the group found that reliance on technology was not a substitute for careful stewardship of the resource base, and that ground water contamination from commercial fertilizers and pesticides is growing in many parts of the United States. These experts suggested that one of the best strategies for economic viability is flexibility, and that investment in human resources, science, technology, and wise use of soil, water, and other natural resources is critical. The CAST report does not address questions about the structure of agriculture or its impact on rural communities. These factors, as discussed by Flora (Chapters 12 and 13), must be a part of the frame of reference within which future systems are designed and implemented.

FUTURE MANAGEMENT DECISIONS AND PRACTICES

In design of a management strategy for future agricultural systems, it is useful to evaluate a wide range of options and consider the types of

TABLE 16.1 Agricultural Production Resources Derived from Internal and External Sources.

Internal Resources	External Resources
Sun—source of energy for plant photosynthesis	Artificial lights—used in greenhouse food production
Water—rain and/or small, local irrigation schemes	Water—large dams, centralized distribution, deep wells
Nitrogen—fixed from air, recycled in soil organic matter	Nitrogen—primarily from applied synthetic fertilizer
Other nutrients—from soil reserves recycled in cropping system	Other nutrients—mined, processed, and imported
Weed and pest control—biological, cultural, and mechanical	Weed and pest control—chemical herbicides and insecticides
Seed—varieties produced on-farm	Seed—hybrids or certified varieties purchased annually
Machinery—built and maintained on farm or in community	Machinery—purchased and replaced frequently
Labor—most work done by the family living on the farm	Labor—most work done by hired labor
Capital—source is family and community, reinvested locally	Capital—external indebtedness, benefits leave community
Management—information from farmers and local community	Management—from input suppliers, crop consultants

Source: Francis and King, 1988a.

inputs that are most likely to be available for crop and animal production. Farming and ranching systems use a mixture of production resources, some of which are renewable and derived from the local community and some of which are external or purchased from some distance (Rodale, 1985; Francis and King, 1988a). Table 16.1 describes a number of these contrasting sources of fertility, water, pest control, and other elements needed for growth and production of crops and animals. Internal or renewable resources are those available in the field or on the farm, or available from a source nearby. Examples are rainfall, nutrients from lower soil strata and recycled within the system, fixed nitrogen, family labor, and local management. External, fossil fuel derived resources include synthetic fertilizers, chemical pesticides, irrigation water from deep wells or a distant source, and crop or management consultants. Production systems that can maximize the efficient use of internal resources and thus minimize the need for purchase of external resources are viewed as potentially more sustainable.

Production systems dependent on external energy and material resources could be considered fragile and highly susceptible to many externalities. They can allow the farmer to achieve and maintain high yields only as long as (1) those outside resources are available in sufficient quantity and at the right time, (2) the farmer has cash or credit to purchase or otherwise access the inputs, and (3) the entire farm operation is profitable in the short run and continues to be profitable over a range of input and commodity prices, as well as a range of climatic conditions.

In contrast, systems that are more dependent on internal resources are less dependent on the vagaries of supply and price of inputs from outside the farm. They allow the farmer to better achieve and maintain adequate yields and profits, as long as careful attention is given to new information on biological structuring and to the efficient use of resources. It is important to design systems that employ a logical mix of needed inputs from both internal and external sources. However, any shift from reliance on a purchased input to supply the same growth factor with an internal source, while maintaining the same level of productivity, has potential to reduce production costs and increase profits. With the right balance and a continued quest for most efficient use of all inputs, we can enhance the long-term viability and thus the sustainability of profitable farming and ranching. One overriding principle that will contribute to viability of systems is greater genetic and landscape diversity in future cropping and crop/animal systems. What are some of the components of these diverse future technologies and systems? (See Francis et al., 1987.)

Potentials for Genetic Improvement

Improvement of crops through crossing and selection is a long-term process, which this author described in Chapter 2. When successful, the breeding strategy provides a cost-effective and sustainable solution to insect and disease problems and stress caused by unfavorable soil and climate. Breeding can lead to more efficient use of limited resources such as nutrients and water. Potentials may exist to incorporate genes that promote nitrogen fixation by grasses or other crops that generally do not have this capacity. There is ample variability in most crop and animal species for continued genetic improvement, and the potentials of molecular genetics are opening new doors for efficiently improving some heritable traits. Nutritional quality of food products is a growing concern, and one that can be addressed through crossing and selection. Whether a breeding program will be successful depends on identification and successful genetic manipulation of the characteristic in question, selection and testing in environments and systems where the new varieties or hybrids will be produced, and effective production and distribution of the new seed. Plant breeding already has made a major contribution to production sustainability and stability of food supply. More advances are on the horizon.

To further enhance the sustainability of future systems, researchers and farmers need to consider potentials in current and alternative crop species. Is it more cost-effective to improve a present crop through breeding or some modification in cropping system, or should a new species be used that has greater drought tolerance, higher nitrogen use efficiency, or better protein quality? Within a crop species, it is important to evaluate whether there is sufficient genetic variability for the characteristic desired, if the heritability is sufficiently high to guarantee success in a breeding program, and if consumers or processors will accept the new genetic types produced. Whether specific hybrids and varieties need to be developed for more sustainable production systems (for example, specific genetic adaptation to specific cropping systems) has yet to be established for most crops (Francis, 1986).

Another unresolved question is the desirability of broad or narrow adaptation in crop hybrids and varieties. To the extent that the search for broad adaptation brings a wide range of genetic potential into profitable public and private breeding programs, this is a useful objective. If the resulting products can be both relatively stable over a series of years and climatic conditions on each farm and responsive to favorable conditions in good years, broad adaptation can be a boon for the farmer. Yet each farmer is most concerned about the range of conditions that can occur on that one farm over a period of years, and not to the broad conditions across a wide geographical area. In this case, specific genetic adaptation to a narrow range of soil and climatic conditions could give greater profits to each farmer, depending, of course, on the price of the seed of these improved varieties. There are trade-offs between broad and narrow adaptation. In summary, genetic improvement provides a well-recognized and proven method of increasing productivity and reducing the effects of stress conditions, especially in well-managed crop production systems. This is a prime component in the development of a sustainable agriculture.

Land Preparation and Subsequent Tillage Practices

Primary objectives of tillage in agriculture have been seedbed preparation, weed control, and residue incorporation. More recently the objective of fertilizer and/or pesticide incorporation has been added to this list (Cruse, 1988). Uniform or blanket tillage over the entire field disturbs all soil in the surface layer, rather than only that soil where the seed is to be placed. According to Cruse, "This is important because soil disturbance by tillage, particularly when conducted at inappropriate soil water contents, has very negative impacts on soil structural or tilth conditions." He maintains that deteriorating soil structural condition leads to a dependence on tillage to loosen the soil and provide for soil aeration, a favorable seedbed, good water penetration, and reduced compaction. This is a short-term substitute for reduced tillage options that would not create the unfavorable structure in the first place! Land tillage almost always increases the potential for soil erosion, loss of soluble nutrients and organic matter, and soil crusting.

We estimate that over 80% of the row crop acres in the Midwest are now managed through some form of reduced tillage system. There are important consequences for soil conservation. A comparison of moldboard plow land preparation with five reduced tillage alternatives is shown in Table 16.2 from Dickey et al. (1988). Listed are typical field operations and

TABLE 16.2 Advantages, Disadvantages, and Typical Field Operations for Selected Tillage Systems.

System	Typical Field Operations	Major Advantages	Major Disadvantages
Moldboard plow (clean tillage)	Fall or spring plow; two spring diskings; plant; cultivate	Suited to most soil and management conditions. Fall plow excellent for poorly drained soils. Excellent incorporation. Well tilled seedbed.	Little erosion control. High soil moisture loss. Timeliness considerations. Highest fuel and labor costs.
Chisel plow	Fall or spring chisel; spring disk; plant; cultivate	Less erosion than from cleanly tilled systems. Less winter erosion potential than fall plow or fall disk. Fall chisel well adapted to poorly drained soils. Good to excellent incorporation.	Additional operations, often performed, result in excessive soil erosion and moisture loss. In heavy residues, stalk shredding may be necessary to avoid clogging.
Disk	Fall or spring disk; spring field cultivate; plant; cultivate	Less erosion than from cleanly tilled systems. Well adapted for lighter to medium textured, well drained soils. Good to excellent incorporation. Few residue clogging problems.	Additional operations, often performed, result in excessive soil erosion and moisture loss. Soil compaction associated with disking wet soils.

TABLE 16.2 (continued)

System	Typical Field Operations	Major Advantages	Major Disadvantages
Rotary-till	Rotary-till and plant; cultivate	Excellent erosion control up to planting time. Excellent incorporation when used full width. Well suited for furrow irrigated areas. Well tilled seedbed.	Depending on use: Low erosion control after planting. Possible soil crusting. Possible increased power requirement.
Ridge-plant (till-plant)	Stalk chopping; planting on ridges; cultivate to maintain ridges	Excellent erosion control if on contour. Well adapted to poorly drained soils. Excellent for furrow irrigated areas. Ridges warm up and dry out quickly. Low fuel and labor costs.	No incorporation. Creating and maintaining ridges. Keeping planter on top of ridge.
No-till	Spray; plant into undisturbed surface; postemergent spraying or cultivation as necessary	Maximum erosion control. Soil moisture conservation. Minimum fuel and labor costs.	No incorporation. Increased dependence on herbicides. Not suited for poorly drained soils or weed infested fields. Management is highly critical.

Source: Dickey et al., 1988.

TABLE 16.3 Typical Diesel Fuel and Labor Requirements for the Tillage Systems Discussed.

Tillage System	Fuel gal/ac	Labor hr/ac
Moldboard plow	5.28	1.22
Chisel	3.34	0.89
Disk	3.03	0.84
Rotary-till	3.00	0.88
Till-plant	2.69	0.91
No-till	1.43	0.49

Source: Dickey et al., 1988.

major advantages and disadvantages of each system. As tillage is reduced (from top of table to bottom), there is an elimination of trips across the field, an increase in amount of residue left on the soil surface, less potential for soil erosion, and reduction in fuel and labor costs. In some systems there is an increased reliance on chemical use for weed control. Much of the incentive for farmers to reduce tillage and increase residue on the soil surface has come from compliance regulations in government programs that specify certain levels of residue that must be maintained and soil erosion amounts that cannot be exceeded. In five-year dryland studies of six tillage systems for soybeans and grain sorghum and a four-year study of corn, there were no significant differences in grain yields (Dickey et al., 1988). The authors further showed fuel requirements for moldboard plow and no-till systems to be 5.28 and 1.43 gallons per acre, respectively; labor requirements for the two systems were 1.22 and 0.49 hours per acre (Table 16.3). Although there are clear environmental advantages to the reduced tillage systems, these require careful timing and management as noted in the table.

Future systems are highly likely to follow the trend in the Midwest toward reduced primary tillage. There is little reason to prepare 10,000 square meters of soil surface in each hectare for seed that will occupy less than 0.1% of that area, if the microenvironment for each seed can be made favorable for germination and establishment and if weeds can be controlled in some manner. Reducing soil disturbance actually reduces the potential for weed germination of many species, since they are not brought up to the surface or don't receive light to help them germinate and grow. Some details on weed growth and management are given by Liebman and Janke (Chapter 4) and by Thompson et al. (Chapter 9). Development of new generation herbicides that are applied in smaller quantities per hectare, that are more carefully targeted toward predominant weed species, and that are more readily degraded into ecologically harmless forms may provide environmentally acceptable and economically profitable alternatives for weed control in the future. Reduced tillage and increased residue on the soil surface help to reduce impact of raindrops, impede surface water runoff, and minimize erosion. Trapping and holding water until the crop can use this resource is a highly cost-effective tactic for reducing drought effects. These practices also reduce surface soil erosion and preserve nutrients in place in the field.

Promoting Adequate Soil Fertility

The largest major change anticipated in future thinking about soils and plant nutrition as related to crop growth will be a shift from dependence on applied fertilizers to management of the total soil environment for enhanced and sustained fertility. Due to inexpensive nitrogen and favorable crop prices for several decades, farmers began to equate fertility with applied fertilizers. Today and in the future, this simplified approach will not be economically feasible in the short term nor ecologically acceptable due to off-farm effects of excessive applications.

In recent years, much more has been learned about the potentials for legumes in cropping systems (see Power, 1987; also Chapter 6). Much empirical knowledge and on-farm experience has accumulated during the past century. Science is now providing more detailed understanding about how legumes function and contribute to fertility. There is much research under way on nutrient cycling in the soil solution/organic matter/plant/soil microorganism system and on how management can affect these complex interactions. Growing interest in the rhizosphere, the thin layer or zone that represents the interface between soil solution and plant root, has led to better understanding of how chemical, physical, and biological reactions in that zone influence availability and uptake of plant nutrients. The incredible complexity of population dynamics of soil microorganisms, their contributions to soil fertility, and how cropping sequences and practices including fertilization affect activity is a field that represents a new frontier in applied biology (see Doran and Werner, Chapter 7). Use of molecular genetic techniques to incorporate nitrogen fixation potential into grass species could also contribute substantially to system sustainability.

Future management decisions on soil fertility will depend on advances in research in these several areas. Greater reliance on crop rotations, especially of cereals with legumes, will undoubtedly be an important dimension of cropping systems. Use of cover crops, both between cycles of major cash crops and overlapping their cycles as overseeded species, will increase as a strategy to provide more fixed nitrogen and to capture and store more of the nutrients present in the relatively stable organic fraction. Details of these management strategies are given by King (Chapter 5) in the discussion of sustainable soil fertility.

Future Pest Management Strategies

Expansion of the size of farms has led to increased reliance on chemical controls of undesirable weeds, insects, and plant pathogens in agriculture.

Although this strategy has allowed management of more hectares with reduced labor cost, there have been a number of consequences that were not anticipated. One is an increasing genetic resistance, especially in weed and insect species, to applied chemicals. Another is the ineffectiveness of some applied products (for example, herbicides) when climatic conditions are not optimum for their activation. This results in expensive additional cultivation that might have replaced the chemical in the first place. A number of pesticides and their breakdown products are showing up in surface and ground water at some distance from the point of application. The growing awareness and concern about health and safety issues related to agricultural chemicals is causing many in the industry to search for alternatives that will be less costly, more effective, and more sustainable for the long term.

Reduced levels of herbicide are used by many farmers who apply a band over the row and then cultivate to control weeds between rows. Some farmers apply subrecommended herbicide rates or mixtures with fertilizer and report effective control. Minimum tillage or no-till planting in most cases reduces the number of germinating weeds and the cost of weed control. Some applications of the ridge-till system, using rotary hoe and cultivator to control weeds, allow complete elimination or drastic reduction of the need for herbicides. Crop rotations play an important role in weed management, since sequences of dissimilar species (cereal–legume, summer annual–winter annual, annual crop–perennial crop) can substantially reduce reproductive success of most weed species compared to continuous culture of the same crop and use of the same herbicide. Weed indexing and relating populations to economic thresholds of competition are providing more information about when and how herbicides should be used. Details of these practices are described by Liebman and Janke (Chapter 4), and future systems will be much more carefully designed to provide some level of weed management at a lower production and environmental cost than current conventional weed control approaches with herbicides.

Management of plant pathogens, insects, and nematodes also will benefit from current research on crop scouting, determination of economic threshold levels, and crop rotation. The same types of rotations described above to help control weeds generally are useful for reducing other pests in the crop environment. Future systems will also take advantage of another consequence of reduced chemical application—the buildup of populations of beneficial insects in the cropping ecosystem. Current pesticide-intensive approaches are usually not selective toward only the insect that is causing crop damage; they kill everything in the area where applied. In Chapter 3, by Bird et al., a number of strategies are described that employ more information and management to help the crop compete with undesirable pest species in the field. Genetic resistance in crops and animals, biological and cultural control measures are prominent among the alternatives to chemical application. Much is yet to be learned about the complex interactions among crops, insects, plant pathogens, nematodes, and beneficial soil microorganisms. Research in this area will provide new insight on how to design more cost-effective and sustainable cropping systems. Decision aids can be developed to help farmers carefully evaluate options and their economic and environmental implications.

Future Cropping Systems

Some of the complexities of interactions among components of a cropping system are illustrated in Figure 1.2. There is a growing appreciation of both the impacts of individual components of technology (new hybrid, insecticide formulation, tillage implement) and how they influence an incredible array of other parts of the production system. This has given rise to the study of farming systems and the complexities of factor interactions, as well as the management decisions that will help the farmer to design and implement a profitable strategy each season. Researchers have found that much of the excitement and potential for new advances is coming from the interfaces between disciplines and the interactions in systems. This is where much of our future research and management will need to focus, and interdisciplinary teams will become increasingly important for systems research.

Changes from one crop species to another (maize to grain sorghum), from chemical weed control to integrated weed management, from application of maximum rates of nitrogen to partial reliance on crop rotation and overseeded legumes—all have potential to reduce both production costs and undesirable impacts on the environment. Yet each of these changes has impacts within the farming system far beyond just the type of grain that will be harvested, the cost of applied chemical or fertilizer. Fertility level and source may affect the weed populations of different species and how they compete with crops, and changing the crop may change the competitiveness with weeds and the amount and timing of needs for nutrients. Thus, in this example of just three potential changes in the management of a crop production sequence, there are numerous interactions among those practices, even without exploring how these changes influence other parts of the crop ecosystem.

In the future there will continue to be valuable research and consequent management recommendations for specific components of technology—new soybean varieties, biological insect control products, more productive rotation sequences. In part, the development of more efficient and sustainable production systems depends on the availability of specific technologies to solve current constraints in the system. Yet increasingly, attention will focus on how these components are integrated and how they interact in the system. There is little chance of continuing to experiment with large numbers of potential changes in components in systems in an empirical manner using conventional approaches and designs in the field.

New research and management strategies that employ computer simulation and expert systems will give greater amounts of reliable information on the potential performance of new combinations of technologies and inputs. These simulations can be run under any specified set of conditions, for example, on a specific farm or over a 50-year set of climatic conditions, in a cost-effective way that would be impossible in field research. Progress in this new frontier will only be successful, of course, with a valid set of baseline information on individual crops and how they react to a range of conditions in the field. Potentials for integrating animal species in farming systems need to be evaluated and added to the data base. Building the necessary models for simulation will help us identify gaps in our understanding of present components of systems and their interactions. Success with these models will be impossible without practical validation procedures where the best systems are tested in the field and demonstrated for farmers to observe and evaluate. Their participation in implementing and interpreting data from these field experiences will be one part of tomorrow's research and extension agenda.

RESEARCH AGENDA FOR SUSTAINABLE AGRICULTURE

The agenda for research appropriate to building a basis for sustainable agriculture in the temperate zone can be built from information in previous chapters and the technologies described there. Although a number of component technologies are available that can help to reduce input costs, minimize environmental problems, and make systems more energy-efficient, there are still many details that need to be explored (Francis, 1988). As more is learned about the biological processes and interactions in natural ecosystems and cropping sequences, there will be new management strategies to be tested for their applicability in future crop and animal production schemes. Can we envision a research agenda for the short-term future that will enhance the sustainability of agriculture? A group was convened at North Carolina State University in 1987 that evaluated needs for alternative agriculture (Miller, 1988). This has been reprinted in full in the North Central Regional Conference Proceedings (Francis and King, 1988b). Five areas were identified for more intensive research.

Alternative Crop Nutrient Sources and Nutrient Cycling

An overall concern in the area of fertility is that much information exists in the literature and in specific research organization reports that is not easily accessible. There is need for a data base on nutrient values of legumes, manures, and other organic wastes and how these can be most efficiently utilized in reduced-chemical systems. Nutrient cycling will likely play a major role in the design of future systems. Thus, more information is needed on several topics:

- The role of deep-rooted crops in cycling nutrients from lower soil strata.
- The environmental impact of legume nitrogen sources, compared to fertilizer sources.
- Cycling processes of nitrogen and other nutrients in the organic and inorganic fractions as well as in the soil solution.
- The feasibility of alternative systems of tillage in legume based rotations.
- Reduced tillage and improved residue management.
- How animal enterprises can be integrated into systems to increase diversity and enhance nutrient cycling.

The role of the rhizosphere and the microorganisms active there is seen as a new and poorly understood frontier where much more work is needed. Specific research areas include the following:

- Dynamics of rhizosphere organisms in different cropping sequences and rotations.
- The basis of plant-microbe interactions to improve symbiotic fixation of nitrogen.
- Plant-microbe interactions of mycorrhizae and other associations that improve plant nutrient uptake under low-input conditions.
- Modification of the soil rhizosphere through introduction of different organisms that could improve crop growth, reduce toxins, and reduce populations of pathogens and other pests.
- The impact of perennial polycultures on microorganism and rhizosphere dynamics.

Strategies for Weed Management

The emphasis in cropping systems will shift from complete control of weeds to the management of unwanted species only when they are present in economically significant numbers. This will involve an integrated management strategy that includes cultural practices, competition from the crop, modifications in both crops and tillage, and possible biological control agents. This research agenda includes the following details:

- Studies of cultural control of weeds, including effects of primary tillage and cultivation on weed competition, populations, and weed dynamics. Also, crop planting dates, densities, row spacings, cover crops, intercropping, fertilization practices, and irrigation can all affect weed growth and reproduction. Such long-term factors as effect of

crop rotations, development and use of more competitive crop cultivars, and managing the weed/crop interactions have potential for manipulation through management.
- Basic research on weed biology and ecology, including weed competition with other weeds, bud dynamics and dormancy, allelopathic suppression of weeds, genetic diversity and size of the weed seed pool, weed interactions with insects and pathogens, relationships with micorrhizae, weed life cycles and hard seed survival, and influences of the environment on weeds and weed/crop water use.
- Biological control of weeds, including surveys of natural predators, effects of bioherbicides, and allelochemicals introduced into crops to help suppress weed germination and growth.

Strategies for Plant Pathogen and Nematode Management

Central to future strategies for cost-effective and environmentally sound management of plant diseases and nematodes are genetic resistance in crop cultivars, cultural management such as crop rotations, and promoting healthy plant growth with minimal stress. Research is needed on a number of components of this strategy:

- Understanding how the balance between beneficial and harmful microorganisms is affected by cropping patterns, rotations, irrigation and fertility, cultivation, and chemicals applied to the system. This will require better techniques to identify and quantify pathogen populations and antagonistic microflora with which they interact.
- Developing alternative methods to reduce or replace chemical pesticide use, including biocontrol agents and cultural practices that favor indigenous antagonists over plant pathogens.
- Continued emphasis on genetic resistance or tolerance to major plant pathogens and nematodes. There is ample genetic variation in disease and nematode reaction, and inheritance of resistance is often controlled by one or a small number of genes.

Strategies for Insect Management

With a reduction of chemical pesticide inputs, there will likely be a buildup of predators that will help control unwanted insects. With changes in tillage patterns, cultural management of crop residues, and crop rotations, there may be both positive and negative effects on insect survival and population dynamics. Research is needed in a number of specific areas:

- Improved understanding of the ecology of cropping and tillage systems as they influence insect pests, including analysis of the structure of cropping systems and pest response to modified and more sustainable management techniques. Importance of modified physiological reactions of plants to changes in growth, nutritional status, water reactions, and other insects in the system should be quantified. Role of soil arthropods may be affected by tillage, crop residues, organic matter, and other changes in the production system.

- Cultural control of insect pests through manipulation of the environment includes use of cover crops, modified fertilization and irrigation patterns, and use of manures and other organic wastes.
- Genetic resistance to insects has been found in many crop species, and considerable genetic variability exists that has not been utilized; in general, inheritance of resistance to insects is more complex than resistance to plant pathogens.
- Biological control of insects requires further study of natural enemies and introduction of nonindigenous new enemies, inundative and inoculative releases of natural enemies of the target insect.
- Better understanding of the regional dynamics of insect populations will help our understanding of how and where economically important infestations are likely to occur and aid in management of insect pests.

Management of Alternative Cropping Systems

One of the greatest immediate needs in the area of cropping systems is to develop an information base for available results from past cropping systems research. This allows valid comparison of alternative systems in the context of new available technology, current resource constraints, and concerns about the environment. Important to this data base is information on crop and livestock yields, labor requirements and costs, machinery and other input costs, expected prices and income from alternative crops, animal species, and production systems.

A more basic research area is determination of the biological/physical basis of rotation effects. Rotations increase genetic diversity on a given parcel of land over multiple years. Cover crops can increase genetic and species diversity at the same time. How they fill niches and promote nutrient cycling is not well understood. Most consequences of rotating crop species are positive, although there may be some allelopathic effects of crops and cover crop species on other desired plants. Some of the measured and speculated causes of rotation effects that need to be studied include nitrogen and other nutrient contributions, improved soil sturcture, decreased pest problems, promotion of growth promoting substances, and modification of soil microorganism populations that is favorable for succeeding crops. If these positive rotational effects are better elucidated, it will be possible to more fully exploit their potentials in new crop management systems.

Modeling should play an important role in the systematic evaluation of alternative cropping systems. This is the only efficient way to compare the potentials of a near infinite number of combinations of cropping sequences, cover crops, tillage and other cultural practices, fertilizer, and chemical input variables that could be combined in future systems. One dimension of current interest to farmers is the effect of a major change in the production system from one that is highly chemical- and fertilizer-dependent to one that is characterized by low chemical inputs, the so-called transition or conversion process (see Chapters 9 and 10). What are the major expected changes in pathogen, weed, and insect problems, in fertility and productivity of soils, and in the economics of production? By evaluating a number of farm operations that have been through this transition, it is possible to seek strategies and practices that minimize the potential negative effects of the change. Long-term economic effects of the transition or conversion, both on the individual farm and on the macro level, need to be evaluated. These are some of the research areas that will receive priority attention in the next decade.

FUTURE AGENDA FOR EXTENSION

The recent national study on future extension priorities has resulted in the formulation of eight areas of emphasis. These are currently being encouraged from the national level and through implementation by each state and local extension organization. The priorities reflect the changing nature of U.S. agriculture and the new types of challenges that face our farming industry and rural population. A number of these relate directly to the development of a sustainable agriculture throughout the temperate zone. The eight areas include

- Competitiveness and profitability.
- Alternative agricultural opportunities.
- Conserving and managing natural resources.
- Water quality.
- Improving nutrition, diet, and health.
- Building human capital.
- Family and economic well-being.
- Revitalizing rural America (communities).

We can strive to make agriculture both profitable and competitive by searching for ways to reduce input costs that do not reduce yields, or by cutting costs per unit of production through a range of strategies. Many of these approaches have been presented in detail in the preceding chapters: ways to reduce tillage, substitution of internal nutrient resources and fixed nitrogen for purchased fertilizers, integration of crop and animal enterprises, use of crop rotations and genetic resistance to manage pest populations. Alternative opportunities in agriculture include other crops and value-added products that can enhance farm income and the entire local economy. The discussion by Flora (Chapter 12) of community values and how these relate to the success of agriculture will find application in this extension initiative. Benefits of reduced chemical use and more judicious application of fertilizers will have obvious benefits in the areas of water quality and the conservation of natural resources.

In the implementation of the initiatives and the re-thinking of traditional extension activities, there are several strategies that have been proposed to keep this endeavor on the cutting edge. It is certain that extension will need to maintain a diversity of programs, with emphasis on activities in which this public sector program has a comparative advantage. Since the farmer and rancher today are confronted with a multiplicity of information sources, one of the key roles of extension in the future will be to provide the methods and guidelines on how to sort out this information, to bring together those components of available technology into systems and production packages that are specific to each farm. Here are some of the possible strategies (Lucas et al., 1988; Francis et al., 1988).

Focus on Systems Rather than Components

Although progress will continue to be achieved through introduction of new hybrids and varieties, fertilizer rates and placement options, and new herbicide formulations, for example, many of the quantum changes in the future will depend on combining these components into workable systems for specific situations and environments. One of the ways that individual producers can best tap into these new combinations is to run a simulation of optional and available technologies under the management and resource constraints of each farm or for each field. One of extension's roles could be to develop and provide access to expert systems as an aid to management and long-term planning.

Focus on Efficient Use of Resources

Finite reserves of fossil fuels and their increasing cost in the future will make it imperative for those in agriculture to use them carefully and efficiently. Other needs of society for fossil fuels—from heating houses to personal transportation—often prevail over those of agriculture; people are willing to pay a high price to fill these desires, and it is difficult for agriculture to compete. One solution proposed in the research section above is greater reliance on internal, renewable resources, and the substitution wherever possible of an internal resource for a purchased input. A number of examples were presented. With no specific product to sell, extension is uniquely placed to help farmers review alternative management practices based on a balance of internal and external production inputs, while providing methods for a smooth transition to greater reliance on those resources that are renewable in nature.

Focus on Information as a Key Production Input

Information about efficient management comes from a wide range of sources, including the prior experience of the manager, industry recommendations, farm press, neighbors, and extension. Although a new piece of information may be external to the farm, the moment it is received and processed within the context of a specific set of resource constraints and economic goals, it becomes an internal resource on that farm. There are a number of unique properties or characteristics of information: it is a creatable resource, it is expandable, diffusive, transportable, substitutable, and shareable. Once applied in helping make a management decision, information is not used up! These dimensions of information as a unique renewable resource are discussed in a later section of this chapter.

Focus on Participatory Systems for Developing Information

On-farm demonstrations and joint activities for validation of recommendations have long been an important part of university/farmer and industry/farmer cooperation. Potentials for future development of this type of participatory information development and validation are being explored by key groups such as the Practical Farmers of Iowa, the Nebraska Sustainable Agriculture Association, and the Northern Plains Sustainable Agriculture Society (Kirschenmann, 1988).

Using long strips that are one width or multiple widths of standard commercial equipment, with randomized and replicated treatments in the field, farmers are able to collect credible data that can be analyzed using standard statistical techniques. Recent analyses of a large number of these trials have shown that a high level of confidence can be attained from their results, with coefficients of variation most frequently below 5% (Rzewnicki et al., 1988). This is a credible level for plots of any size. When farmers are involved with the elaboration of project objectives, choice of experimental treatments for the field, management of a trial and collection of data, and interpretation of the results, there is a high degree of identity with the work. These participating farmers become extenders of the results within the local community.

Focus on Process Rather than Product

We currently have a generation of researchers, extension specialists, farmers, and ranchers who too often seek to develop and use a "formula" or "product in a package" to solve most production constraints. This pertains to high-input chemical farmers as well as many "organic farmers" who also insist on buying something in a package as a substitute for good management. In the future, we need to stress the importance of evaluating a series of alternative solutions to each problem as well as seeking an understanding of the entire production system and environment in which it operates. The *process* of solving constraints leads us through a series of steps that includes consideration of the costs and impacts of all possible alternatives, as compared to an automatic "pulling a package off the shelf" to solve each challenge.

Focus on Diversity of Enterprises and Products

We will be moving from an era of extreme specialization to one of diversity and flexibility on the farm and ranch. Biological diversity in crop rotations and crop/animal integrated enterprises lead to greater integration efficiencies in the total farming system (Harwood, 1985). This biological diversity helps in the control of severe outbreaks of insects and plant pathogens, and gives some competitive advantage for crops over unwanted weed species. To some degree, this mimics the survival strategies and patterns for plant and animal species found in natural ecosystems. Greater diversity in enterprises and in the landscape can lead to greater biological and economic sustainability. Producers need to be able and willing to react to new economic realities, changes in market windows and opportunities, prices of inputs and income received for products. Value-added products from farm or local enterprises can bring greater income and income stability to the farming operation and community.

Focus on Community as Well as Farming and Ranching

Viable communities and services are critical to the long-term viability of farm families. When value-added industries are located near the source of agricultural raw materials, the local economy is enhanced with more jobs and the promotion of greater commercial vitality in the community. More locally produced fresh foods could enhance their nutritional quality for the community. Extension efforts by public agencies and by industry contribute to the development of rural communities through educational and practical training opportunities. These include not only agricultural practices but also small business development and community action. This catalyzing role is one that public sector extension specialists have played since the system began. The Cooperative Extension Service is pursuing a series of new national priorities, and must seek new ways to provide relevant services at the community level in the future. Neighbor-to-neighbor information sharing, farmer-researcher dialog, sister communities, and extension support groups all help to build rural community and increase rural-urban communication. These alternative communication and education techniques promote sharing of information as well as enhancing the spirit of community. Many aspects of community importance are discussed by Flora (Chapter 12).

FUTURE EDUCATIONAL CHALLENGES

Education and training will be central building blocks in the foundation for a long-term, sustainable agriculture. To this end, it will be possible to work with existing programs and institutions as well as seek new ways to efficiently prepare people for responding to evolving challenges. There are a number of component courses and nontraditional activities within the classical university community that provide some direction toward a more sustainable agriculture. There are other noteworthy efforts outside the mainstream. New organizational patterns are being explored for putting together interdisciplinary courses across traditional department lines. Finally, there is a critical need to reach through the entire spectrum of the educational process and provide an integrated and future-oriented perspective from preschool through postgraduate years. How can this be achieved?

First, there are relatively few clearly identified programs in today's university community that deal with sustainable agriculture. Some noteworthy exceptions include

- A professorship position and course curriculum in sustainable agriculture at University of Maine, with key activities in the Plant and Soils Department, but close collaboration across several relevant departments; there is also a large organic farm to be used as a national research and demonstration center.
- Endowed chair positions in sustainable agriculture at the University of Minnesota and Michigan State University, including support for fellowships in several departments and some operating funds.
- Campus committees addressing sustainable agriculture questions at many state universities, including Nebraska, Ohio, Missouri, and Kansas.
- Recognized centers for sustainable agriculture, such as the Leopold Center at Iowa State University.
- Research farm activities related to the teaching program at state universities in California, Pennsylvania, New York, and other states.
- Comprehensive alternative agriculture curriculum at University of Vermont that centers on obtaining sustained yields through technologies that are ecologically sound, including nutrient and organic matter cycling, closed energy flows, decreased soil erosion, integrated pest management, and reduced production costs.

Where there are no identified programs, some courses are offered in agronomy, soil science, entomology, plant pathology, and resource economics that address many of the questions of sustainability. There are well-established activities such as the agroecology program at U.C. Santa Cruz that maintains a research farm, has supervised graduate thesis research, and hosts interns for long-term practical training in the principles of low-input and nonchemical farming and gardening. This is a fast-growing activity in our university community, and one that will undoubtedly emerge as a major focus area in the next few years in most states. A recent set of guidelines was developed by Lamm (1989) to help universities develop these programs.

Outside the traditional university, there are private and nonprofit centers for research and training that broaden the opportunities for people to continue education in sustainable agriculture. Among these are the Land Institute (Salina, Kansas; see Chapter 14), the Rodale Research Center (Kutztown, Pennsylvania; see Chapter 10), the New Alchemy Institute (East Falmouth, Massachusetts), the Center for Rural Affairs and Small Farm Resources Project (Walthill and Hartington, Nebraska), and the Meadowcreek Project (Fox, Arkansas). There are many others. A good source of information and announcements about activities of these centers is the "Alternative Agriculture News", published by the Institute for Alternative Agriculture (9200 Edmunston Road, Suite 117, Greenbelt, MD 20770). A number of nongovernmental organizations have recently formed a consortium, The Sustainable Agriculture Working Group, to work actively toward common goals. This working group will meet occasionally and communicate about common interests and strategies to coordinate their work and make it more effective; for example, current emphasis is on the 1990 farm legislation at the national level (reference: Center for Rural Affairs, Box 405, Walthill, NE 68067).

The organization of centers or areas of emphasis across departmental lines is a relatively new approach to bringing people and resources to focus on a priority area. Education in the area of sustainable agriculture could benefit from an approach such as that used at University of California, Davis, in the graduate teaching program. In this innovative approach, a graduate teaching faculty, representing a specific interest area, is recruited from traditional deparments and organizes a curriculum that takes advantage of people and facilities from a number of sites. For example, the genetics program includes input from Agronomy and Range Science, Horticulture, Pomology and Viticulture, and Animal Science as well as the Department of Genetics. Team teaching of courses, such as the International Agriculture classes at Cornell University, provides another option to bring people together in complex and interdisciplinary fields. In some respects, sustainable agriculture is such a pervasive concept that it would be ill-advised to create one new course in any single department and declare that the issue had been addressed. This is a philosophy that in the future will pervade our total teaching programs in the universities, and the concepts of environmental soundness, efficient resource conservation and use, and human health and safety will be central to the entire curriculum.

Finally, there is need for creative and long-term planning in our total educational system with regard to natural resources and how we use them to human advantage. Many habits are developed in the early years, and a healthy concern about food safety, caution in resource use, recycling of wastes, and creating fewer wastes can all begin as early as preschool years. At the very least, we need to encourage people of all ages to learn about where food comes from and how it is produced. Now that the large majority of people are distant from the production and processing of food, there is little appreciation of what this industry means to our very survival. Demonstrations and experiments with plants and how different environmental factors affect their growth are useful through the elementary and middle school years. There are many good teaching materials that illustrate the hydrologic cycle and the ways that nutrients are acquired and used by plants. These all need to be related to environmental issues at each step in the process. In high school, there are many more opportunities to teach conservation and efficient resource use, and agriculture can play a key role in science courses such as biology, chemistry, physics, geology, and climatology. Food and agricultural issues can become part of the study of geography, history, languages, and other parts of the total curriculum.

University and postgraduate studies are obvious places to emphasize the interrelated nature of agriculture and use of technology with broader issues of conservation, resource use, and the environment. Again, we too often separate these issues into specific discipline-related treatment of variables and specific production systems. There is potential for greater integration across department lines in the existing curriculum. Education and training should not stop with a high school, university, or postgraduate degree.

We need to recognize the emerging importance of nontraditional educational routes. There is substantial growth of community colleges, trade schools, and adult education programs. Industry is playing an increasingly important role in preparing people for specific jobs related to new technologies. Delivery systems such as off-site classes, correspondence, and satellite delivery may revolutionize our thinking about education in the future. Home schooling is an option used by some parents to provide a holistic approach to learning.

Given the critical role of information and the dynamic changes that are occurring in this industry, everyone will have to be a part of the lifelong learning process. This is the only way to keep up to date, much less make any progress toward the goals outlined throughout this book. In the overview, the more consistent and organized these programs are throughout the entire formal educational curriculum and into later years, the better people will be able to pull information together and make it understandable and useful. Just as we organize a progressive set of concepts and courses in mathematics, our future programs related to food, agriculture, and the environment should build from simple factors and relationships to the more complex interactions and interconnectedness of systems. This will be a part of tomorrow's educational environment, and plans to promote a sustainable agriculture must pursue a broad strategy within this framework.

ROLE OF INFORMATION AS A RESOURCE

How can a communication strategy be designed for a sustainable agriculture? What would have to be unique about such a system, as compared to existing channels and methods of moving information? We operate in the information age. Both the types of data and interpretations are rapidly changing, along with the methods of communicating that information. People today are exposed to more well-packaged news and commercial messages each day than our grandparents received in a month. Part of this is our obsession with news, from local to international, and a well-organized industry to provide that news. In addition, a large part of what we receive is advertising for products. It is said that we are innundated with information, but starved for knowledge, interpretation, and wisdom. One of the greatest challenges facing the farmer and rancher today is how to sort out information and recommendations, and how to use this resource effectively rather than be drowned by it.

Information has certain key characteristics that make it especially valuable as a production input or resource (King and Francis, 1988a, 1988b). We could call information anything that reduces uncertainty. Internal information on the farm or ranch is that experience of the farmer and family that has come from both formal education and long years of trial and error in management of the operation. External information comes from off the farm, from commercial, university, government, or other sources. It normally is not locally based. Farmers use a combination of these two types of information to make management decisions.

Several unique features of information make it a particularly valuable resource for developing a sustainable agriculture (King and Francis, 1988a, 1988b). Information can be transmitted or communicated from one place or person to another. It is created and expanded on the farm and in the community through application and experience with the consequences. At times information about the cropping system can substitute for other purchased, external production inputs. It can be shared with others because information is mobile and can be distributed. It can be expanded, renewed, and re-created—and information is not expendable. Information is integrated into the existing base of experience and is carried forward to the next year and growing season. It grows with use. This is why education and the ability to use information wisely are such critical components of sustainable agriculture.

Communication is the movement of information from one person or place to another. Today's communication environment is different from the past in a number of ways. One is the sheer quantity of information and messages that reach the farmer and rancher each day. The big challenge is how to screen these messages and to sort out those that are relevant in a local situation and profitable within the operation. Messages often are in conflict. How do we sort out which is correct, and under what circumstances?

One example is the major role of government regulations, particularly for those producers who participate in various support payment programs. Within the public sector, there may be substantial disagreement between program requirements and extension recommendations. To maintain base acres for feed grain subsidies may require a farmer to plant virtual monoculture, while researchers demonstrate conclusively that rotations boost productivity. In order to promote more resource-efficient and sustainable systems, there must be a "vertical integration" and agreement between the results of research and the formulation of policy.

Localization and market segmentation are two approaches now used by industry to bring information to local markets for products. In the area of alternative agriculture, there are new products—nonconventional soil fertility amendments, biological pest control agents, water treatment products, fertility enhancing agents—that are not well known and often not tested in a community. It is difficult to sort out the often conflicting claims for these products.

Another major change in communication is the refinement of message design and psychology of advertising. What is seen on television and in glossy advertisements is slick and convincing. Use of products is tied to often unrelated values such as community or family, national traditions, or what the neighbors think. Farmers and others in the agricultural community need to be equally sophisticated in their reception and interpretation of these messages. One critical role for universities and for extension specialists in the future will be to help people sort out information, providing tools and methods that will help each manager to evaluate information and recommendations as to how they relate to the specific operation. Will they make that operation more profitable and more environmentally sound? This is why information and communication are central to farming and ranching today, and why it is important to focus on this area in education and extension for a more sustainable agriculture.

POLICY DIMENSIONS IN AGRICULTURE

During the formulation of farm legislation in Congress, there is an ongoing debate about which agricultural policies and support will best achieve the goals of long-term food production and a competitive posture in international markets. There is general agreement that the erosion provisions of the 1985 farm bill have helped to put some highly erodible areas into a reserve, and that compliance with reduced-erosion requirements has helped to conserve the valuable soil resource. But there are specific aspects of this policy that appear to be counterproductive for long-term sustainability of farming and the resource base on which it depends. Patrick Madden has discussed these aspects in Chapter 11, and Cornelia Flora expanded the policy dimensions in Chapter 13. Similar challenges are facing other governments in the temperate zone.

There is general accord in the United States that more flexibility is needed in the determination and preservation of base areas of feed grains, so that farmers can use rotations with greater frequency. Government program regulations need to be in agreement with best management recommendations for resource-efficient production. The continued culture of the same species with the same or similar pest control chemical application is seen as detrimental to the long-term sustainability of productivity. There is also a need to promote genetic diversity and more variety of crops within fields. Rotations of cereals with legumes, greater diversity of the cropping landscape within fields, and alternatives to large, monoculture fields can be promoted to some degree by validating and publicizing alternatives to current conventional practices. Although it is more desirable to achieve these goals through encouragement and education, there is a role for regulation if the current economic incentives do not solve critical problems.

Regulations and controls of pesticide and fertilizer applications are strongly resisted by farmers and by industry. Those active in agriculture generally feel that they understand best the need for these products and the most cost-effective and safest way to apply them. It is important that policy be designed to set up general goals to be achieved in terms of minimal contamination of surface and ground water, as well as the atmosphere, and that details on how to achieve these goals be left in the hands of those who best understand the technology. Yet these rules have to be sufficiently stringent, and the enforcement carried out in a realistic way, so that the goals really can be achieved. It will require creative planning and management to find ways to eliminate current and future pesticide and nitrate problems from our drinking water supplies, for example, and farmers need to work together with industry and university specialists to fully understand the broad impacts of products used in agriculture and how we can successfully use alternatives.

This will require a rethinking of how we develop and apply technology, and often the best "product" or solution to a problem will be a management strategy that eliminates or minimizes the problem before it develops. There is general agreement among farmers that government should stay out of agriculture, yet there is broad participation in programs that enhance incomes or maintain base prices for major commodities. We need to design and implement policies that take into account the total health and well-being of our population, rural and urban alike. Often this can only be achieved by regulation, but educational approaches provide a more positive and lasting approach whenever possible.

FUTURE OF SUSTAINABLE AGRICULTURE AND THE ENVIRONMENT

The overview definition of sustainable agriculture given by Harwood (1988) expresses well the intent of the chapters that describe specific practices and their consequences as they relate to the human future and the environment. He advocates "an agriculture that can evolve indefinitely toward greater human utility, greater productivity, and a balance with the environment that is favorable both to humans and to most other species." This definition encompasses the biological, the economic, and the social objectives of the challenge. We cannot focus on any single dimension of this challenge without taking into account the interactions with other potential impacts. Increased productivity is not useful if it is not economical in both the short and the long term. A practice that is economically useful for a year or two is not sustainable if it creates a drastic resource shortage or environmental problem in the long term. Use of specific technologies that bring down the cost of food production are most useful if there is some equity involved in the final results of the use of those technologies. In our discipline-organized and narrowly-oriented activities in the past, it has been difficult to deal with these broad issues.

New instruments and measurement techniques are helping us understand the role of specific substances in our environment. We are increasingly aware of the connection between applications of certain technologies and human health. As we become more informed about these interconnections and interactions with the natural and the human-altered environment, it is important that we act with caution in making decisions about what levels of chemicals to accept and how to deal with immediate problems. It is not a time to wait until more data can be collected, to see with certainty what the long-term effects of a particular chemical will be, or to trust in technology to solve all current and future problems. It is a time for prevention, for anticipation of impacts of synthetic products, and for awareness of the multitude of ways in which our current industrialization paradigm affects both humans and other species.

Human quality of life is directly enhanced by such practices as crop rotations, reduced pesticide use, and diversified tree plantings in filter strips and set-aside lands. In addition to obvious health benefits, there is an increase in wildlife and locally available recreation areas. Our concern for the welfare of other species—both plant and animal—is one indication of human awareness of the interconnectedness of large biological systems.

There is little debate today about whether the environment is important. We have learned through difficult experience the effects of massive oil spills, inadvertant releases of chemicals from manufacturing plants, and even cumulative effects from the application of some products that followed well-tested recommendations from the best scientists and laboratories. In the future, we will be well advised to anticipate these effects, through the application of past experience and cautious speculation drawn from other or similar technologies. Current reaction to global warming trends illustrates the difficulty of dealing with critical global issues. The environment is ever important to the very survival of our species, and learning to work with natural processes in designing new food production strategies is an essential dimension of our future agenda in agriculture. This is the central theme of sustainable agriculture. Whatever the definition and interpretation of terms, as described by Lockeretz (Chapter 15), we need to seek systems that will provide adequate food and other products that will sustain the human species while preserving and improving the viability of systems that allow other species to survive. We need to maintain and enhance productivity, but this cannot be done at the long-term expense of the environment. To the best of our abilities, we need to preserve our future options and the capacity to evolve indefinitely. This is a challenge for those in research, extension, industry, and farming. It's also one of the most compelling future challenges for everyone.

Acknowledgments

The reviews and comments of several colleagues are sincerely appreciated: Drs. Dan Walters, Max Clegg, and James King (University of Nebraska), Dr. Cornelia Flora (Kansas State University), and Ms. Sarah Ebenreck (Institute for Alternative Agriculture). The word-processing skills and diligent revisions of several drafts of this chapter and other sections of the book by Ms. JoAnn Collins are gratefully acknowledged.

REFERENCES

CAST. (1988). *Long-Term Viability of U.S. Agriculture.* Council for Agricultural Science and Technology, Report No. 114, June. 48 pp.

Cruse, R. M. (1988). "Tillage and soil erosion." In *Sustainable Agriculture in the Midwest,* (Eds. C. A. Francis and J. W. King). pp. 31–34., North Central Regional Conference, Coop. State Res. Service, USDA and Inst. of Agr. and Natural Resources, Univ. Nebraska, Lincoln.

Dickey, E. C., P. J. Jasa, A. J. Jones, and D. P. Shelton. (1988). "Conservation tillage systems for row crop production," *Conservation Tillage Proceedings,* 7:1–6 Coop. Extension Service, Univ. Nebraska, Lincoln. (in NebGuide G80-535).

Francis, C. A. (1986). "Variety development for multiple cropping systems," *CRC Critical Rev. Plant Sci.* **3**:133–68.

Francis, C. A. (1988). "Research agenda for sustainable agriculture," *Sustainable Agriculture in the Midwest: North Central Regional Conference Proceedings* (Eds.

C.A. Francis and J. W. King). pp. 77–84. Coop. Ext. Service & Agr. Res. Div., Univ. Nebraska, Lincoln. March 22.

Francis, C. A., and J. W. King. (1988a). "Cropping systems based on farm-derived, renewable resources," *Agric. Systems* (U.K.) **27**:67–77.

Francis, C. A., and J. W. King (eds). (1988b). *Sustainable Agriculture in the Midwest: North Central Regional Conference Proceedings.* Coop. Ext. Service & Agr. Res. Div., Univ. Nebraska, Lincoln. March 22. 102 pp.

Francis, C. A., D. Sander, and A. Martin. (1987). "Search for a sustainable agriculture: Reduced inputs and increased profits," *Crops & Soils*, August–September, pp. 12–14.

Francis C. A., J. W. King, D. W. Nelson, and L. E. Lucas. (1988). "Research and extension agenda for sustainable agriculture," *Amer. J. Alt. Agric.* Spring/Summer, **3**:123–26.

Harwood, R. R. (1985). "The integration efficiencies of cropping systems." In *Sustainable Agriculture and Integrated Farming Systems* (Eds. T. C. Edens, C. Fridgen, and S. L. Battenfield). pp. 64–75. Michigan State Univ. Press, East Lansing.

Harwood, R. R. (1988). "History of sustainable agriculture: U.S. and international perspective." In *Sustainable Agricultural Systems* (Ed. C. A. Edwards, R. Lal, P. Madden, R. H. Miller, and G. House). Soil & Water Cons. Soc., Ankeny, IA (in press).

King, J. W., and C. A. Francis. (1988a). "Back to the future: the power of communication and information." In *Farming Systems Research/Extension Symposium Proceedings*, Univ. Arkansas, Fayetteville. October 11.

King, J. W., and C. A. Francis. (1988b). "Information and technology: Unlearning the old, integrating the new, and moving toward sustainable agricultural systems." In *Farming Systems Research/Extension Symposium* Proceedings, Univ. Arkansas, Fayetteville. October 11.

Kirschenmann, F. (1988). *Switching to a Sustainable System.* Northern Plains Sustainable Agriculture Society, Windsor, ND.

Lamm, T. (1989). *Guidelines for Developing University Sustainable Agriculture Programs.* Wisconsin Rural Devel. Center, Black Earth.

Lucas, L. E., C. A. Francis, J. W. King, and D. W. Nelson. (1988). "Extension agenda for sustainable agriculture." *Sustainable Agriculture in the Midwest: North Central Regional Conference Proceedings* (Eds. C. A. Francis and J. W. King, pp. 85–88. Coop. Ext. Service & Agr. Res. Div., Univ. Nebraska, Lincoln.

Miller, R. (1988). "Planning conference on research and extension needs for alternative agriculture." Draft report, August 24–26, 1987, workshop, Raleigh, NC.

Power, J. F. (ed.). (1987). *The Role of Legumes in Conservation Tillage Systems.* Soil & Water Cons. Society, Ankeny, IA. 153 pp.

Rodale, R. (1985). "Internal resources and external inputs: The two sources of all production needs," *Workshop Rept. on Regenerative Farming Systems*, USAID, Washington DC December 10–11. Rodale Institute, Emmaus, PA.

Rzewnicki, P. E., R. Thompson, G. W. Lesoing, R. W. Elmore, C. A. Francis, A. M. Parkhurst, and R. S. Moomaw. (1988). "On-farm experiment designs and implications for locating research sites," *Am. J. Alt. Agric.* **3**:168–173.

AUTHOR INDEX

Note: Bold page numbers indicate author of chapter.

SUBJECT INDEX